Structure and Properties of Multiphase Polymeric Materials

PLASTICS ENGINEERING

Founding Editor

Donald E. Hudgin

Professor
Clemson University
Clemson, South Carolina

1. Plastics Waste: Recovery of Economic Value, *Jacob Leidner*
2. Polyester Molding Compounds, *Robert Burns*
3. Carbon Black-Polymer Composites: The Physics of Electrically Conducting Composites, *edited by Enid Keil Sichel*
4. The Strength and Stiffness of Polymers, *edited by Anagnostis E. Zachariades and Roger S. Porter*
5. Selecting Thermoplastics for Engineering Applications, *Charles P. MacDermott*
6. Engineering with Rigid PVC: Processability and Applications, *edited by I. Luis Gomez*
7. Computer-Aided Design of Polymers and Composites, *D. H. Kaelble*
8. Engineering Thermoplastics: Properties and Applications, *edited by James M. Margolis*
9. Structural Foam: A Purchasing and Design Guide, *Bruce C. Wendle*
10. Plastics in Architecture: A Guide to Acrylic and Polycarbonate, *Ralph Montella*
11. Metal-Filled Polymers: Properties and Applications, *edited by Swapan K. Bhattacharya*
12. Plastics Technology Handbook, *Manas Chanda and Salil K. Roy*
13. Reaction Injection Molding Machinery and Processes, *F. Melvin Sweeney*
14. Practical Thermoforming: Principles and Applications, *John Florian*
15. Injection and Compression Molding Fundamentals, *edited by Avraam I. Isayev*
16. Polymer Mixing and Extrusion Technology, *Nicholas P. Cheremisinoff*
17. High Modulus Polymers: Approaches to Design and Development, *edited by Anagnostis E. Zachariades and Roger S. Porter*
18. Corrosion-Resistant Plastic Composites in Chemical Plant Design, *John H. Mallinson*
19. Handbook of Elastomers: New Developments and Technology, *edited by Anil K. Bhowmick and Howard L. Stephens*

20. Rubber Compounding: Principles, Materials, and Techniques, *Fred W. Barlow*
21. Thermoplastic Polymer Additives: Theory and Practice, *edited by John T. Lutz, Jr.*
22. Emulsion Polymer Technology, *Robert D. Athey, Jr.*
23. Mixing in Polymer Processing, *edited by Chris Rauwendaal*
24. Handbook of Polymer Synthesis, Parts A and B, *edited by Hans R. Kricheldorf*
25. Computational Modeling of Polymers, *edited by Jozef Bicerano*
26. Plastics Technology Handbook: Second Edition, Revised and Expanded, *Manas Chanda and Salil K. Roy*
27. Prediction of Polymer Properties, *Jozef Bicerano*
28. Ferroelectric Polymers: Chemistry, Physics, and Applications, *edited by Hari Singh Nalwa*
29. Degradable Polymers, Recycling, and Plastics Waste Management, *edited by Ann-Christine Albertsson and Samuel J. Huang*
30. Polymer Toughening, *edited by Charles B. Arends*
31. Handbook of Applied Polymer Processing Technology, *edited by Nicholas P. Cheremisinoff and Paul N. Cheremisinoff*
32. Diffusion in Polymers, *edited by P. Neogi*
33. Polymer Devolatilization, *edited by Ramon J. Albalak*
34. Anionic Polymerization: Principles and Practical Applications, *Henry L. Hsieh and Roderic P. Quirk*
35. Cationic Polymerizations: Mechanisms, Synthesis, and Applications, *edited by Krzysztof Matyjaszewski*
36. Polyimides: Fundamentals and Applications, *edited by Malay K. Ghosh and K. L. Mittal*
37. Thermoplastic Melt Rheology and Processing, *A. V. Shenoy and D. R. Saini*
38. Prediction of Polymer Properties: Second Edition, Revised and Expanded, *Jozef Bicerano*
39. Practical Thermoforming: Principles and Applications, Second Edition, Revised and Expanded, *John Florian*
40. Macromolecular Design of Polymeric Materials, *edited by Koichi Hatada, Tatsuki Kitayama, and Otto Vogl*
41. Handbook of Thermoplastics, *edited by Olagoke Olabisi*
42. Selecting Thermoplastics for Engineering Applications: Second Edition, Revised and Expanded, *Charles P. MacDermott and Aroon V. Shenoy*
43. Metallized Plastics: Fundamentals and Applications, *edited by K. L. Mittal*
44. Oligomer Technology and Applications, *Constantin V. Uglea*
45. Electrical and Optical Polymer Systems: Fundamentals, Methods, and Applications, *edited by Donald L. Wise, Gary E. Wnek, Debra J. Trantolo, Thomas M. Cooper, and Joseph D. Gresser*
46. Structure and Properties of Multiphase Polymeric Materials, *edited by Takeo Araki, Qui Tran-Cong, and Mitsuhiro Shibayama*

Additional Volumes in Preparation

Plastics Technology Handbook: Third Edition, Revised and Expanded, *Manas Chanda and Salil K. Roy*

Handbook of Radical Vinyl Polymerization, *edited by Munmaya K. Mishra and Yusuf Yagci*

Photonic Polymer Systems: Fundamentals, Methods, and Applications, *edited by Donald L. Wise, Gary E. Wnek, Debra J. Trantolo, Thomas M. Cooper, and Joseph D. Gresser*

Handbook of Polymer Testing: Physical Methods, *edited by Roger Brown*

Structure and Properties of Multiphase Polymeric Materials

Edited by

Takeo Araki
Qui Tran-Cong
Mitsuhiro Shibayama
Kyoto Institute of Technology
Kyoto, Japan

MARCEL DEKKER, INC. NEW YORK • BASEL • HONG KONG

Library of Congress Cataloging-in-Publication Data

Structure and properties of multiphase polymeric materials / edited by Takeo Araki, Qui Tran-Cong, Mitsuhiro Shibayama.
p. cm.
Includes bibliographical references and index.
ISBN 0-8247-0142-9
1. Plastics 2. Polymers—Structure. I. Araki, Takeo. II. Tran-Cong, Qui. III. Shibayama, Mitsuhiro.
TP1120. S796 1998
668.9—dc21 97-52811
CIP

The publisher offers discounts on this book when ordered in bulk quantities. For more information, write to Special Sales/Professional Marketing at the address below.

This book is printed on acid-free paper.

MARCEL DEKKER, INC.
270 Madison Avenue, New York, New York 10016
http://www.dekker.com

Current printing (last digit):
10 9 8 7 6 5 4 3 2 1

PRINTED IN THE UNITED STATES OF AMERICA

Preface

Studies on multiphase and multicomponent polymeric materials have been growing intensively in the past three decades. Such studies are activated not only by the ceaseless industrial need for developing new materials but also by the long-standing academic interest in the complex fluids to which polymeric systems belong.

In the early stages of these studies, tremendous efforts were made to obtain homogeneous polymer blends by seeking suitable polymer pairs with mutual specific interactions. In modern technology, the great advantages offered by heterogeneous blends with varying phase-separated structures have been recognized in achieving superior physical properties. Hence, the stucture–property relationship in multiphase polymeric materials is of crucial importance in developing new polymer materials satisfying the diverse requirements of industry. For this purpose, morphology control has frequently been performed by adding specific polymeric emulsifiers such as block or graft copolymers to adjust the interfacial tension between different phases.

The chapters in this book have been carefully selected to emphasize the physical aspects of the structures and properties of multiphase polymers. This book explores these topics in scientific depth as in other books dealing with polymer blends and/or polymer alloys. However, the unique feature of this book is that it emphasizes the correlation between molecular parameters (degree of polymerization, the Flory–Huggins interaction parameter, etc.) and macroscopic phenomena (e.g., phase separation). Furthermore, modern methodologies are also addressed such as small-angle X-ray and neutron scattering, which are necessary to investigate heterogeneous structures in multiphase polymers in the submicron to nanometer scales.

The main purpose of this book is to offer an updated overview of recent advances in multiphase polymeric materials to scientists and engineers currently working in the field of polymer materials science. The contents of each chapter are also accessible to graduate students majoring in polymer science. In addition, this book is designed to serve as a quick guide to polymer materials science for nonpolymer specialists with different backgrounds.

The book is divided into three parts: I, Theory; II, Structures; and III, Properties. The thermodynamics as well as the kinetic aspects of phase separation of polymer blends, block copolymers, and homo/block copolymer blends are described in Part I. Chemical synthesis, morphology control, and structure analysis are presented in Part II. Finally, Part III reviews the bulk, surface, and interfacial properties of these multiphase polymer materials, together with their corresponding thermal, rheological, and mechanical properties.

Chapters 1 and 3 summarize the theoretical achievements in microphase and macrophase separations in multicomponent polymers. These theories allow one to synthesize tailor-made multiphase polymer materials with known χ parameters and/or phase diagrams. The analysis of thermodynamic instabilities of multicomponent polymer by using small-angle X-ray and neutron scattering is provided in great detail in Chapter 3. Chapter 4 demonstrates the chemical techniques for the precise synthesis of block and graft copolymers with specific architecture. Since block and graft copolymers can be classified as amphiphilic polymers because of the presence of unlike chains connected to each other, applications of these polymers to the modification of polymer–polymer interfaces are widely spread to micelles, surfactants, vesicles, etc. The most important feature of Chapter 5 is to demonstrate the roles of thermodynamically open conditions in the phase separation of polymer mixtures. With the exception of Chapter 6, all of the chapters deal with bulk polymers in which van der Waals interaction plays a dominant role. On the other hand, as shown in Chapter 6, in gel systems there exist competing molecular interactions such as hydrophobic, electrostatic, and hydrogen bonding interactions. Since these interactions can be driven antagonistically, drastic changes can emerge in morphology, stability, and physical properties of the gels. Chapters 7 and 9 show that the morphology and the viscoelastic behavior of multiphase polymers can be controlled by a shear field. Here, the couplings between the thermodynamic instabilities and hydrodynamics are described in detail. The information provided in these chapters is related strongly to industrial problems because morphology control is one of the essential problems of polymer processing. Chapters 8 and 10, on the other hand, demonstrate the remarkable achievements in the surface–interface characterization. The techniques introduced here, such as ellipsometry, neutron reflectometry, and atomic force microscopy, ensure a promising future for multiphase polymer materials in controlling surface/interface properties. Thermal properties of multiphase polymers have not been extensively studied thus far compared with other physical properties. The examples given in Chapter 11 show that the thermal diffusivity and thermal conductivity are sensitive to changes in the phase behavior of multicomponent polymeric systems. This promises a new methodology to investigate nondestructively multiphase polymers. In Chapter 12, two principal mechanisms of plastic deformation are described: crazing and shear yielding. These mechanisms and the materials toughness are predicted from the morphology and molecular parameters of the blends. Chapter 13 discusses future directions in multiphase polymeric materials research. Since all the chapters deal with molecular properties of polymers, multiphase materials in the macroscopic sense, composites such as fiber-reinforced plastics (FRPs) and others were omitted from this book.

The contributors and chapters were selected to cover a broad area ranging from chemical synthesis to the structure–property relationships of multiphase polymer materials. Also included are the recent developments in polymer materials (e.g., characterization of multiphase polymer surfaces and interfaces) and phase separation phenomena under external constraints such as light irradiation, shear flow, and the presence of networks. It is hoped that this volume will provide the reader with a valuable tool for carrying out related studies, and that it will serve as a comprehensive textbook or guidebook. It is anticipated that further advances in this highly important field of new polymer materials will be realized through the readership. The editors would like to thank all the contributors and the publisher for their kind cooperation and patience in the preparation of this volume.

Takeo Araki
Qui Tran-Cong
Mitsuhiro Shibayama

Contents

III. Properties

Contributors

D. Dompas Department of Chemistry, Laboratory of Macromolecular Structure Chemistry, Catholic University of Leuven (KU Leuven), Heverlee, Belgium

Hiroshi Furukawa Institute of Physical Sciences, Faculty of Education, Yamaguchi University, Yamaguchi, Japan

G. Groeninckx Department of Chemistry, Laboratory of Macromolecular Structure Chemistry, Catholic University of Leuven (KU Leuven), Heverlee, Belgium

Charles C. Han Polymers Division, National Institute of Standards and Technology, Gaithersburg, Maryland

Ian Hopkinson Department of Physics, University of Cambridge, Cambridge, England

Masami Kawaguchi Department of Chemistry for Materials, Faculty of Engineering, Mie University, Mie, Japan

Yushu Matsushita Neutron Scattering Laboratory, Institute for Solid State Physics, University of Tokyo, Tokyo, Japan

Alan I. Nakatani Polymers Division, National Institute of Standards and Technology, Gaithersburg, Maryland

Monica Olvera de la Cruz Department of Materials Science and Engineering, Northwestern University, Evanston, Illinois, and Service de Chimie Moléculaire, CE-SACLAY, Gif-sur-Yvette, France

Randal W. Richards Interdisciplinary Research Centre in Polymer Science and Technology, University of Durham, Durham, England

Shinichi Sakurai Department of Polymer Science and Engineering, Kyoto Institute of Technology, Kyoto, Japan

Mitsuhiro Shibayama Department of Polymer Science and Engineering, Kyoto Institute of Technology, Kyoto, Japan

Qui Tran-Cong Department of Polymer Science and Engineering, Kyoto Institute of Technology, Kyoto, Japan

Naoto Tsutsumi Department of Polymer Science and Engineering, Kyoto Institute of Technology, Kyoto, Japan

Hiroshi Watanabe Institute for Chemical Research, Kyoto University, Kyoto, Japan

1

Phase Segregation in Copolymer and Homopolymer Multicomponent Mixtures

Monica Olvera de la Cruz
Northwestern University, Evanston, Illinois, and CE-SACLAY, Gif-sur-Yvette, France

I. INTRODUCTION

New polymeric materials are created by blending polymers. Polymers, however, are strongly incompatible and nonhomogeneous microstructures result upon mixing. Macroscopic domains generally lead to poor mechanical properties. The mechanical properties can be improved by adding compatibilizer agents, such as copolymers or another third polymeric component, which modify the interfaces.

Even simple ternary blends, which only contain chemically homogeneous polymers, have complicated phase-separation kinetics with complicated interfaces. The microstructure during the decomposition process will be discussed by Furukawa in Chapter 2 [1]. The equilibrium interface will be briefly discussed below.

The addition of a minority C component to A–B immiscible binary blends, reduces the interfacial tension and increases the interfacial thickness [2–5]. A minority C component added to an A–B immiscible polymer blend is adsorbed at the interface. Incompressible binary blends are described by the local free energy of mixing [6],

$$f(\phi(r)) = f_0(\phi(r)) + c(\nabla\phi(r))^2/2$$

where f_0 is the local free energy of an homogeneous blend of composition $\phi(r) = \phi_A(r)$ and $\phi_B(r) = 1 - \phi_A(r)$. The coefficient c determines the contribution to the free energy due to local concentration fluctuations, and it is computed using the *random phase approximation (RPA)* [7]. In the RPA, the free energy is obtained as an expansion in the Fourier components of the local concentration fluctuations around the mean composition of the blend ϕ_k,

$$\Delta(F(\phi_k)/kT = (1/2!)\sum \phi_k\phi_{-k}/S_0(k)$$
$$+ (b/V3!)\sum \phi_k\phi_{k'}\phi_{k''}\delta(k+k'+k'')$$
$$+ (d/V4!)\sum \phi_k\phi_{k'}\phi_{k'''}\delta(k+k'+k''+k''')$$

where $S_0(k)$ is the scattering intensity and $d > 0$. The scattering function has a maximum at $k = 0$, as in binary liquids or alloys, $S_0(k) = 1/(a + ck^2)$. The constant a is equal to the second derivative of $f_0(\phi(r))$ with respect to $\phi(r)$ evaluated at the mean composition ϕ_0, and the RPA gives $a \sim 1/(\phi_0 N_A) + 1/((1-\phi_0)N_B) - 2\chi$, where χ is the Flory interaction parameter and N_I is the degree of polymerization of the component $I =$ A,B. The above expansion is extremely useful to describe continuous ($b = 0$) transitions from a one-phase region to a two-phase region occurring when $a < 0$. That is, the above free energy has instabilities to infinite long wavelength fluctuations when $a < 0$, suggesting the decomposition of the homogeneously mixed one-phase state into two compositions given by the minimum value of $\Delta(F(\phi_0)$ with respect to ϕ_0. The constant c in polymer blends is temperature (T) independent, $c \sim 1/(\phi(1-\phi)18)$. Meanwhile, in small molecule mixtures undergoing phase separation it is T dependent and of the order of χ.

In the presence of a third component, the incompressible ternary blend free energy is a function of two concentration variables, say the local concentrations of the A and B components, $\phi_A(r)$ and $\phi_B(r)$, respectively ($\phi_C(r) = 1 - \phi_A(r) - \phi_B(r)$). When there is phase separation, the minority component is adsorbed in the interface between the A-rich and the B-rich phases. The degree of adsorption is strongly dependent on the coefficient of the gradient in compositions, c_{AA}, c_{BB}, and c_{AB} [5], which can be obtained straightforwardly by the RPA. The adsorption of the minority component is higher in small molecules than in blends because these coefficients are generally smaller in small molecules allowing less increase in the free energy from the gradient in the C composition at the interface.

When the minority component is a copolymer, the microstructure during phase separation is richer, and the interface is more complex [8]. It is important to understand the physical properties of the copolymers in the melt state before analyzing their behavior in more complex mixtures. Two types of copolymers will be discussed here, *random copolymers* and *block copolymers.*

Random A–B copolymers are very difficult to analyze due to the random chemical heterogeneity along the chains. They have traditionally been treated in terms of the *global average* monomer composition, expressed here as the overall fraction f of monomer A in the system. It is usually assumed that each chain has the same fraction f of A repeat units, i.e., the copolymer system is compositionally uniform. *Random copolymers* (RCs) are, however, mixtures of chemically distinct chains, many of which have individual compositions $f_i \neq f$. Since RC are naturally multicomponent systems, their behavior in the molten state, as well as in polymer A–B mixtures and in solutions, is highly complex. In dilute solutions, for example, if the solvent is selective the collapse of the chains can be very difficult to analyze [9,10]. Furthermore, even in nonselective solvents when the concentration of copolymer increases, the physical properties of the system are highly nontrivial, as discussed in the last section of this chapter.

In block copolymer melts, since two chains are chemically linked to form a single molecule, all the chains have equal (or nearly equal) compositions. The net repulsive interaction between the A and B monomers ($\chi > 0$) of the chemically linked chains, however, drives the system to segregate locally. The degree of polymerization N, the block copolymer composition f and the interaction parameter $\chi(\sim 1/T)$ determine the physical properties of the melt. When χN is small the block copolymer melt is isotropic. At $\chi N = (\chi N)_t$ a transition to a periodic structure occurs, known as the *microphase separation transition* (MST). At the MST the copolymer blocks segregate into A-rich and B-rich domains that form periodic arrays termed microphases. The free energy expansion in the local composition fluctuations in block copolyers is very different than in binary blends. Block copolymers are unstable to finite wavelength fluctuations. That is, $S_0(k)$ has its maximum value at $k_0^* \sim 1/R_0$, where R_0 is the unperturbed chain radius of gyration R_0, implying local segregation or microphase segregation.

The possibility of nearly continuous transitions from an istropic state to a periodic structure in block polymer melts has led to numerous studies of the equilibrium structure at the transition point [11–18]. If at the transition point the periodic structure can be described by a single density wave with an infinitesimally small amplitude (the first harmonic amplitude of the periodic structure) continuously changing from zero in the istropic state to a nonzero value at the transition, we say the transition is continuous. In a nearly continuous transition to a periodic structure, on the other hand, the amplitude of the wave at the transition, though small, changes from zero in the disordered state to a discrete nonzero value at the transition (in such a case, the density in the periodic state may not be able to be described by a single wave, and the higher-harmonic amplitudes of the periodic structure may be required to describe the density distribution in the period state). The

Landau mean-field approach (LMFA) [11,12] of Leibler [15] predicts transitions to a lamellar structure in symmetric diblocks and to a body-centered cubic (bcc) structure in asymmetric diblocks. When the density fluctuations are self-consistently included in the analysis [14] (the *Hartree Approximation*) a lamellar structure is obtained even in asymmetric diblocks of finite molecular weight, followed by a two-dimensional hexagonal structure as the degree of asymmetry increases [16]. The common experimentally observed morphologies in diblocks are alternating lamellar domains (lam), hexagonally packed cylinders (hpc), bcc lattice of spheres, and bicontinous nets [19], previously described as a bicontinuous double diamond structure and recently shown to form a gyroid structure (gyr) [20,21]. Also, a lamellar-catenoid (lc) structure has been proposed and observed experimentally [18]. The approaches used to describe nearly continuous transitions to periodic structures in block copolymer melts is described in Sec. II. This section is not a review of the large existing literature on block copolymer microphases. We do not discuss, for example, the strong segregation regime [22–24]. A review on this subject was recently written [25].

II. NEARLY CONTINUOUS TRANSITIONS TO PERIODIC STRUCTURES IN BLOCK COPOLYMER MELTS

A. The Landau Mean Field Approach (LMFA)

The microphase transition if it is continuous or nearly continuous is analyzed by expanding the free energy of the isotropic state in a power series of the density fluctuations $\delta\rho(r)$ around the mean density ρ_0 [1],

$$\begin{aligned}\Delta(F(\rho_k)/kT &= (1/2!)\sum \rho_k\rho_{-k}/S_0(k)\\ &+ (1/V3!)\sum \Gamma_3(k, k'k'')\rho_k\rho_{k'}\rho_{k''}\delta(k+k'+k'')\\ &+ (1/V4!)\sum \Gamma_4(k, k', k'', k''')\rho_k\rho_{k'}\rho_{k''}\rho_{k'''}\delta(k+k'+k''+k''')\end{aligned} \tag{1}$$

where ρ_k are the Fourier components of $\delta\rho(r)$. In incompressible copolymer melts, $\delta\rho(r)$ is the local deviation of the concentration of A monomers from its mean value f [15]. The $S_0^{-1}(k)$, $\Gamma_3(\{k\})$ and $\Gamma_4(\{k\})$ in Eq. (1) are functions of the bare two, three, and four-monomer correlation functions G_{ij} G_{ijk}, and G_{ijkl}, $(i, j, k, l = \text{A, B})$ calculated by the RPA [15,26,27]. In the RPA, $S_0(k)$ is the scattering intensity in the disordered state [5]: $1/S_0(k) = Q(x)/N - 2\chi$, where $x = k^2Nl^2/6$ and N is the number of segments of length l (for simplicity we assume here $l_A = l_B$) per chain. Since $Q(x)$ has a minimum at $x = x_0^*$, $S_0(k)$ has a χ-independent peak at k_0^*. The peak position scales as $k_0^* \sim 1/R_0$, where R_0 is the unperturbed radius of gyration $R_0 = (N/6)^{1/2}l$.

For diblock copolymers with $f = f_c = 0.5$ (when $\Gamma_3 = 0$), $S_0(|k| = k_0^*)$ diverges at $(\chi N)_c = 10.495$, leading to an instability of the isotropic state to density fluctuations of wavevector magnitude equal to k_0^*. The microphase transition in the LMFA is a function of the molecular architecture (diblock, triblock, etc.) and composition of the copolymer. We discuss here diblock and triblock copolymer melts only.

At the transition, $\delta\rho$ has the symmetry of the periodic structure. It can therefore be expanded in plane waves whose vectors are the reciprocal lattice vectors of the periodic structure, the compositional harmonics [11]. The largest contribution to the free energy in nearly continuous transitions ($\Gamma_3 \sim 0$) is from the first *harmonic vectors* [12] $|K_r^{(1)}| = k^* r = 1, \ldots, n_1$, where n_1 is the number of nearest-neighbors (n.n.) in the reciprocal lattice. Since higher harmonic vectors have larger magnitudes $(k_r^{(i)}| = \alpha_i k^*$, $r = 1, \ldots, n_i$, with $\alpha_i > 1$ for $i > 1$), their contribution is often neglected.

There are structures whose first and second harmonic vectors have equal magnitude. In a trigonal or hexagonal structure ($\mathrm{tri}_{c/a}$ or $\mathrm{hex}_{c/a}$) with $c/a = \sqrt{3}$, for example, one finds $|K_r^{(1)}| = |K_1^{(2)}| = k^*$. Therefore, to analyze these structures, two harmonics have to be considered in $\delta\rho$. Notice that these structures have actually two first harmonics and not one first and one second harmonic. It is misleading to call the oscillation perpendicular to the hexagonal plane a second harmonic when its magnitude is k^*. The reasoning for this notation is so that structures other than the commonly analyzed structures can be described. For example, one could easily take a square lattice with a superposed lamellar-type oscillation orthogonal to the plane of the square fluctuations describing a tetragonal structure. It is clear that in a tetragonal or an orthorhombic structure the resulting minima will be given by all the amplitudes of the harmonics being equal and the lattice spacing along the different directions being equal, which means they will give a simple cubic structure. The $\mathrm{tri}_{c/a}$ and the $\mathrm{hex}_{c/a}$ are identical structures when $c/a = \sqrt{3}$ if only the first two harmonics are included. We denote them here simply as the 1c structure.

Consider an expansion of $\delta\rho$ up to the m-th harmonic,

$$\delta\rho(r) = \sum_{i=1}^{m} \left(\frac{a_i}{\sqrt{n_i}} \right) \sum_{r=1}^{n_i} \left(\exp\left[i\left(rK_r^{(i)} + \phi_r^{(i)} \right) \right\} + c.c. \right) \tag{2}$$

The harmonic vectors of various three-dimensional structures and their values of n_i and α_i are listed in Table 1 (the values for the 1am and hpc are given in Ref. 11).

Replacing the wavevector summations in Eq. (1) by sums over the allowed vectors in Eq. (2) one obtains

Table 1 Harmonic Vectors for Three-Dimensional Periodic Structures

$lc_{c/a=\sqrt{3}}$			bcc = bcc2			fcc = nfcc			bdd			gyr			sc		
$(hkil)$	n_i	α_i	(hkl)	n_i	α_i	(hkl)	n_i	α_i	(hkl)	n_i	α_i	(hkl)	n_i	α_i	(hkl)	n_i	α_i
(1010)	3	1	(110)	6	1	(111)	4	1	(110)	6	1	(211)	12	1	(100)	3	1
(0002)	1	1	(200)	3	$\sqrt{2}$	(200)	3	$\sqrt{4/3}$	(111)	4	$\sqrt{3/2}$	(220)	6	$\sqrt{(4/3)}$	(110)	6	$\sqrt{2}$
(1012)	3	$\sqrt{2}$	(211)	12	$\sqrt{3}$	(220)	6	$\sqrt{8/3}$	(200)	3	$\sqrt{2}$	(310)	12	$\sqrt{(5/3)}$	(111)	4	$\sqrt{3}$
(1120)	3	$\sqrt{3}$				(211)		$12\sqrt{3}$				(312)	24	$\sqrt{(7/3)}$			

$$\Delta F/KT = \sum_{i=1}^{m} S_0^{-1}(\alpha_i k^*)a_i^2 + \sum_{ijk}^{m}(\Gamma_3)_{ijk}a_i a_j a_k + \sum_{ijkl}^{m}(\Gamma_4)_{ijkl}a_i a_j a_k a_l \tag{3}$$

Contributions to $(\Gamma_3)_{ijk}$ in (3) are possible only when three-vector circuits of the form $\pm K_r^{(i)} \pm K_s^{(j)} \pm K_t^{(k)} = 0$ are allowed in the structure. Similarly, to have a contribution $(\Gamma_4)_{ijkl} \neq 0$, four-vector circuits of the form $\pm K_r^{(i)} \pm K_s^{(j)} \pm K_t^{(k)} \pm K_u^{(l)} = 0$ are required. The phases dictate the sign of the various contributions in these terms. The LMFA free energy of a given morphology where only two harmonics are included, for example, is given by

$$F/KT = F_0/KT + F_1(a_1) + F_2(a_2) + F_{1,2}(a_1, a_2) \tag{4a}$$

where

$$F_i(a_i) = S_0^{-1}(\alpha_i k^*)a_i^2 + (\Gamma_3)_i a_i^3 + (\Gamma_4)_i a_i^4 \tag{4b}$$

and

$$F_{1,2}(a_1, a_2) = (\Gamma_3)_{112}a_1^2 a_2 + (\Gamma_4)_{1112}a_1^3 a_2 + (\Gamma_4)_{1122}a_1^2 a_2^2 \tag{4c}$$

The $(\Gamma_3)_i$ and $(\Gamma_4)_i$ in Eq. (4b) result from evaluating Eq. (1) for the i-th harmonic. In the quartic term there are $6n_i$ linear arrangements $K_m^{(i)} + K_m^{(i)} - K_m^{(i)} - K_m^{(i)} = 0$ denoted by $(\Gamma_4)_i^{\mathrm{L}} = 6n_i\Gamma_4(0, 0; \alpha_i k^*)$, $12n_i(n_i - 1)$ planar contributions of the form $\Gamma_4(0, h_{jm}; \alpha_i k^*)$ from the arrangements $K_j^{(i)} - K_j^{(i)} - K_m^{(i)} + K_m^{(i)} = 0$ with $j \neq m$ and $|K_j^{(i)} + K_m^{(i)}|^2 = h_{jm}(\alpha_i k^*)^2$ denoted by $(\Gamma_4)_i^{\mathrm{P}}$, and the nonplanar contributions $\Gamma_4(h_{ml}, h_{mq}; \alpha_i k^*)$ from the arrangements $K_m^{(i)} + K_l^{(i)} + K_j^{(i)} + K_q^{(i)} = 0$ with $q \neq j \neq l \neq m$ denoted by $(\Gamma_4)_i^{\mathrm{NP}}$. Therefore $(\Gamma_4)_i = \{(\Gamma_4)_i^{\mathrm{L}} + (\Gamma_4)_i^{\mathrm{P}} + (\Gamma_4)_i^{\mathrm{NP}}\}/n_i^2 4!)$. The terms in $F_{1,2}(a_1, a_2)$ are evaluated in a similar way. The sign of $(\Gamma_3)_i$ and the mean value of a_i are determined minimizing the free energy with respect to the phases $\{\phi_m^{(i)}\}$ and a_i. The maximum value of the third-order term, $(\Gamma_3)_i$ is given by $n_t\Gamma_3(k^*, k^*, k^*)/n_i\sqrt{n_i}3!)$, where n_t is the number of triangles formed by the vectors $K_j^{(i)} - K_k^{(i)} - K_m^{(i)} = 0$, because $\Gamma_3(k, k' + k'')\delta(k + k' + k'')$ is only a function of the magnitude of the vectors and not the angles between them. Leibler considered only the first harmonic of standard structures in diblock copolymer melts [15]. Therefore, various subscripts are not necessary in his approach. He computed the free energy only for $\alpha_1 = 1$, and in his paper $\Gamma_4(h_{ml}, h_{mq}, k^*)$ is denoted by $\Gamma_4(h_1, h_2)$ (the dependence on k^* is not written explicitly in the function because he evaluated the functions at k_0^*).

The scaling properties of $1/S_0(k)$, $(\Gamma_3)_{ijk}$ and $(\Gamma_4)_{ijkl}$ in block copolymers dictate that $N\Delta F/KT$ in Eq. (3) for copolymers is a function of $x^* = k^{*2}R_0^2$. Consider the case of $\Gamma_3 = 0$ for diblock copolymers ($f = 0.5$). Using Leibler's notation, $n_i = n$, where $n = 1, 2, 3, 4, 5$ and 6 correspond to the lam, square, hcp, nfcc (here we denote by fcc a structure with a different phase minimization than the fcc or hexagonal of Leibler, as discussed below), and bcc, respectively, and $\psi_n = a_1$, the free energy is simply

$$(N\Delta F/K\Gamma)_n = a\psi_n^2 + \beta_n\psi_n^4 \tag{5}$$

where $\beta_n = N(\Gamma_4)_{1111}$ for the various morphologies are given by

$$\beta_1 = \frac{N\Gamma_4(0,0)}{4} \tag{6a}$$

$$\beta_2 = \frac{N[\Gamma_4(0,0) + 2\Gamma_4(0,2)]}{8} \tag{6b}$$

$$\beta_3 = \frac{N[\Gamma_4(0,0) + 4\Gamma_4(0,1)]}{12} \tag{6c}$$

$$\beta_4 = \frac{N[\Gamma_4(0,0) + 6\Gamma_4(0,4/3) - 2\Gamma_4(4/3,4/3)]}{16} \tag{6d}$$

(in this paper if the off-planar term, the last term on the left-hand side of Eq. (6d), is negative the structure is termed nfcc and if it is positive it is termed fcc)

$$\beta_6 = \frac{N[\Gamma_4(0,0) + 8\Gamma_4(0,1) + 2\Gamma_4(0,2) + 4\Gamma_4(1,2)]}{24} \tag{6e}$$

(if the off-planar contribution, the last term in Eq. (6e), is negative the structure is termed bcc2; Leibler did not analyze the bcc2 structure). For the simple cubic structure $n = 3$ also and

$$\beta_{sc} = \frac{N[\Gamma_4(0,0) + 4\Gamma_4(0,2)]}{12} \tag{6f}$$

Another important structure is the gyr. The gyr has $n = 12$ and for the phase minimization with the largest third-order term, $(\Gamma_3)_{111} = \Gamma_3(k^*, k^*, k^*)/(3\sqrt{3})$,

$$\begin{aligned}\beta_{gyr} = {} & N[\Gamma_4(0,0) + 2\Gamma_4(0,8) + 4\Gamma_4(0,9) + 8\Gamma_4(0,5) + 4\Gamma_4(0,2) \\ & + 4\Gamma_4(0,1) + 2\Gamma_4(2,8) + 4\Gamma_4(9,1) + 4\Gamma_4(2,5)]/48\end{aligned} \tag{6g}$$

(other possible phase minimizations for the gyroid lead to smaller $(\Gamma_3)_{111}$ and β_{gyr} values). In Eq. (5) a is morphology independent, and it is given by,

$$a = N/(2S_0(k^*)) = Q(x^*)/2 - \chi N = (\chi N)_s - \chi N \tag{7}$$

After minimizing the free energy with respect to the amplitude of the single wave,

$$F(\phi_n) = \frac{-a^2}{4\beta_n} \tag{8}$$

So the smallest β_n minimizes the free energy. Since in the periodic structures the chains are perturbed [28], a further minimization of Eq. (8) with respect to x^*, where $x^* = (R_0 k^*)^{12}$, for each morphology is required, giving the periodicity $d = 2\pi/k^*$, as a function of χ and N [29].

The free energy of the lc at this level of approximation for $\Gamma_3 = 0$ is given by

$$(N\Delta F/KT)_{lc} = a\psi_1^2 + \beta_1\psi_1^4 + a\psi_3^2 + \beta\psi_3^4 + N\Gamma_4(0, 2)\psi_1^2\psi_3^2 \tag{9}$$

where a, β_1 and β_3 are given in Eqs. (7), (6a), and (6c), respectively. One can clearly see that the free energy of the lc cannot be minimized unless the wavevector dependence of the $\Gamma_4(0, h_2)$ is such that it has a strong minimum at $h_2 = 2$. When $\Gamma_3 = 0$ one would need at least $\Gamma_4(0, 2) = \Gamma_4(0, 0)/3$, in which case by symmetry $\Gamma_4(0, 1) = \Gamma_4(0, 0)/2$, to be able to minimize the free energy of the lc (otherwise the lowest free energy is for the lam, giving $\psi_3 = 0$); i.e., only if $\Gamma_4(0, 2) < \Gamma_4(0, 0)/3$ does the lc have lower energy than the lam. Notice, however, that even if this were the case for certain copolymer architectures (it is not for diblock copolymer melts), the sc would have lower energy than the lc. Actually, the sc would be more preferable than the lam even if a less strong angular dependence of $\Gamma_4(0, h_2)$ was observed: one would need only $\Gamma_4(0, 2) = \Gamma_4(0, 0)/2$, in which case $\Gamma_4(0, 1) = (3/4)\Gamma_4(0, 0)$, for the sc to have equal energy to the lam. Furthermore, other three-dimensional structures would probably have even lower energy than the sc and/or lam structures if the $\Gamma_4(h_1, h_2)$ is strongly wavevector dependent when h_1 and h_2 are nonzero, such as the nfcc and/or bcc2 structures, as discussed by Brazovskii et al. [30]. They found that indeed structures other than lam, hpc and bcc can be obtained when $\Gamma_4(K_r^{(1)}, K_s^{(1)}, K_t^{(1)}, K_u^{(1)})$ is dependent on the angles between the vectors $\pm K_r^{(1)} \pm K_s^{(1)} \pm K_t^{(1)} \pm K_u^{(1)} = 0$. They analyzed all possible structures described by vectors with $|K_r^{(m)}| = k^*$ when $\Gamma_3 = 0$, including the lc structure.

In diblock copolymer melts $\Gamma_4(0, h_2)$ has a minimum at $h_2 = 0$ for all values of k^*, and $\Gamma_4(h_1, h_2)$ with nonzero h_1 and h_2 values is not large enough to give the lowest free energy for the nfcc and/or the bcc2 structure. Therefore the lam has lowest energy when $\Gamma_3 = 0$ if only harmonics with magnitude k^* are included in the analysis. The values of $N\Gamma_4(h_1, h_2)$ at x_0^* required to evaluate the free energies of various morphologies in diblock copolymer melts are given in Ref. 5. In Table 2 we give the values of

Table 2 Numerical Factors A_{ijk} and B_{ijkl} for Various Structures[a]

	bcc	fcc(nfcc)	bdd	sc(nsc)
A_{111}	48	0	48	0
A_{222}	0	0	0	48
A_{333}	−48	−48	0	0
A_{112}	−72	72 (0)	0	72
A_{113}	−144	−72	−72	0
A_{133}	144	0	0	0
A_{223}	0	−72 (0)	−72	0
A_{233}	−144	0	0	0
A_{123}	144	0	0	−144 (0)
B_{1111}	540	216 (5/9)	540	90
B_{2222}	90	90	216	540
B_{3333}	2376	540	90	216
B_{1122}	576	576 (0)	1296	1152 (1/4)
B_{1133}	4752	1152	576	432
B_{2233}	1296	1152 (1/4)	576	1296 (1/9)
B_{1123}	2304	−1440 (0)	0	0
B_{1223}	−1152	0	0	−1728 (0)
B_{1233}	−2880	0	0	0
B_{1112}	−576	0	0	0
B_{1113}	−1728	0	−576	−192
B_{1333}	−3456	0	0	0
B_{2333}	1152	0	0	0

[a]($A_{122} = 0$, $B_{1222} = 0$ and $B_{2223} = 0$ for all structures)

$N\Gamma_4(h_1, h_2)$ for various f values at x_0^* for triblock copolymers with $\tau = 2.8$ (where we are using the notation of Ref. 16; i.e., the composition of the A component of the A_1–B–A_2 triblock is $f = f_1 + f_2$ and $\tau = f_1/f$) in the melt state. While $\Gamma_4(0, h_2)$ has a minimum at $h_2 = 0$ in diblocks, it has a minimum at $h_2 = 2$ in triblocks with $\tau = 2.8$ regardless of the value of f. If only harmonics of magnitude k^* are included in the LMFA, however, the microphases obtained in diblocks and triblocks with all τ values are lam, hcp, and bcc only, as reported in Refs. 5 and 16, respectively. This is true even if we minimize the resulting free energy with respect to x^*, which generates a χN-dependent inverse periodicity [29]: the inverse periodicity decreases, suggesting the chains stretch in the microphase state. When higher harmonics are included, however, the phase diagrams change considerably.

It is important to notice that the function $N\Gamma_4(h_1, h_2; x^*)$ for diblock copolymer melts is very badly behaved as x^* decreases (this is evident in Ref. 19: when x^* is approximately less than 2, the free energy cannot be minimized because the fourth-order term goes to minus infinity). For example, $N\Gamma_4(0, h_2)$ becomes negative for x^* small and it tends to minus infinity when x^* tends to 0. Meanwhile, $\Gamma_4(h_1, h_2)$ with nonzero h_1 and h_2 tends to plus infinity in this limit. In Ref. 11 the function $\Gamma_4(h_1, h_2)$ was parameterized using the relation $\Gamma_4(0, h_2) = a(0, h_2) + b(0, h_2)x^*$. We set, however, $a(0, h_2) = a(0, 0) = a$, ignoring its strongly h_2 dependence (it has as a minimum at $h_2 = 0$) while $b(0, h_2)$ was left to be a function of h_2. Therefore, we artificially generated a minimum of $N\Gamma_4(0, h_2)$ at $h_2 = 2$, favoring three-dimensional structures at $f = 0.5$ (there were also typos in the computer program used in Ref. 11). The LMFA breaks down when the fourth-order term is negative, suggesting that the RPA Gaussian chain approximation breaks down in the strong segregation regime, $x^* \ll x_0^*$, where the chains are stretched.

We discuss next the effect of including harmonics of magnitude greater than k^*. First of all we describe the phase minimization in the presence of harmonics of magnitude higher than k^* in the LMFA analysis. In order to simplify the analysis let us assume Γ_3 and Γ_4 in Eq. (1) are wavevector-independent constants. In this case, $(\Gamma_3)_{ijk} = A_{ijk}\,\Gamma_3/3!$ and $(\Gamma_4)_{ijkl} = B_{ijkl}\,\Gamma_4/4!$, where A_{ijk} and B_{ijkl} are numerical factors evaluated by counting the possible combinations of harmonic vectors that add to zero, weighted by the phase factors. For example, the terms A_{112} and B_{1122} for the square structure are evaluated counting the closed circuits formed with the first harmonic vectors $\{K_r^{(1)}\}_{r=1,2} = \{(1, 0), (0, 1)\}$ and second harmonic vectors $\{K_r^{(2)}\}_{r=1,2} = \{(-1, 1), (1, 1)\}$ with phase $\{\phi_1^{(1)}, \phi_2^{(1)}\}$ and $\{\phi_1^{(2)}, \phi_2^{(2)}\}$ respectively:

$$A_{112} = 12\left\{\cos\left(\phi_1^{(1)} - \phi_1^{(2)} - \phi_2^{(1)}\right) + \cos\left(-\phi_1^{(1)} - \phi_1^{(2)} + \phi_2^{(2)}\right)\right\}$$

and

$$B_{1122} = 96 + 24\left[\cos\left(\phi_2^{(2)} + \phi_1^{(2)} - 2\phi_2^{(1)}\right) + \cos\left(-2\phi_1^{(1)} + \phi_1^{(2)} + \phi_2^{(2)}\right)\right]$$

The resulting free energy (3) is minimized with respect to the phases $\{\phi_r^{(i)}\}$. In centrosymmetric structures the phases within a harmonic are all equal, $\{\phi_r^{(i)}\} = \phi_i$ for $r = 1, \ldots, n_i$. Centrosymmetric structures with ϕ_i equal to 0 or π generally minimize the free energy, since they lead to the largest or smallest possible A_{ijk} and B_{ijkl} values. The coefficients A_{ijk} and B_{ijkl} for the first three harmonics for various morphologies when $\Gamma_3 < 0$ (all phases shift by π when $\Gamma_3 > 0$) are given in Table 3. The free energy of morphologies with phases $\phi_1 = \phi_2 = 0$ and $\phi_3 = \pi$ are given in the first column of Table 3. The energy of the second minima of the fcc and sc, denoted by nfcc and nsc respectively in Table 3, is obtained by multiplying by the factor in parenthesis in the left-hand sides of the fcc and sc columns respectively. For example, if

Table 3 Values of x^*, $(\chi N)_s$, $N\Gamma_4(h_1, h_2)$, and $N\Gamma_3(x^*)$ for Triblock Copolymer Melts with $\tau = 28$ for Various Values of f. f Goes from 0.46 to 0.62. In Each Set of Data we Give f, x^* $(\chi N)_s$, $N\Gamma_4(0, 0; x^*)$, $N\Gamma_4(0, 1; x^*)$, $N\Gamma_4(0, 2; x^*)$, $N\Gamma_4(1, 2; x^*)$, $N\Gamma_4(0, 4/3; x^*)$, $N\Gamma_4(4/3, 4/3; x^*)$, $N\Gamma_3(x^*, x^*, x^*)$

0.46	5.81	16.76	364.93
363.12	334.94	361.77	328.13
284.22	−28.14		
0.48	5.81	16.53	341.99
341.39	311.94	340.93	304.94
260.11	−22.27		
0.50	5.82	16.39	327.00
327.48	296.53	327.74	289.32
243.15	−16.74		
0.52	5.84	16.35	319.24
320.70	287.99	321.57	280.54
232.54	−11.41		
0.54	5.87	16.41	318.59
320.95	286.21	322.33	278.49
228.24	−6.12		
0.56	5.90	16.57	325.16
328.38	291.26	330.19	283.23
230.22	−0.74		
0.58	5.95	16.83	339.60
343.67	303.79	345.86	295.41
239.07	4.90		
0.60	6.00	17.20	363.10
368.04	324.93	370.57	316.15
255.88	10.98		
0.62	6.06	17.70	397.38
403.25	356.31	406.10	347.06

the first and second harmonic phases of the square are $\{0, 0\}$ and $\{\pi, 0\}$ (or all shift by a factor π), then $A_{112} = 0$ and $B_{1122} = 48$. Notice that the nfcc has lower energy than the fcc when only the first harmonic is included in the analysis (since $A_{111} = 0$, the smallest B_{1111} minimizes the free energy), as pointed out by Leibler and others. When Γ_3 is large the fcc with $A_{112} = 72$ minimizes the free energy, as discussed below.

The phase minimization for the bcc structure that has a negative off-planar contribution, denoted by bcc2, gives $A_{111} = 0$. Therefore, it may be the lowest energy structure when $\Gamma_3 = 0$ if the Γ_4 is strongly wavevector dependent, as pointed out by Brazovskii et al. [30]. We found that the bcc2 is never the lowest energy phase in block copolymers near the weak segregation regime.

Consider Γ_3 and Γ_4 constants and $S_0^{(-1)}(|k|) = \tau^* + (|k| - k^*)^2$. The phase diagrams τ^*/Γ_4 versus $|\Gamma_3|/\Gamma_4$ for all $k^{*2}/\Gamma_4 > 0.05$ are very similar to the one-harmonic phase diagram if τ^* is small: direct transitions to the lam are observed when $|\Gamma_3|/\Gamma_4 = 0$, and as $|\Gamma_3|/\Gamma_4$ increases transitions to bcc structures are obtained. Direct transitions to fcc structures occur when k^{*2}/Γ_4 decreases. In this region, however, the LMFA is not valid. Also, when $\Gamma_3 = 0$ transition to a lam occurs at $\tau^* = 0$ and as τ^* decreases, a transition to an nfcc structure is obtained. Clearly this is not possible because transitions from lower to higher symmetry cannot occur as τ^* decreases.

Therefore, when $\Gamma_3 = 0$ only harmonics with $|K_r^{(m)}| = k^*$ can be included in the analysis. Otherwise, transitions from lam to nfcc structures will be predicted. In diblock copolymer melts with $f = 0.5$ at $\chi N \sim 11$, the nfcc has a lower energy than the lam because in the nfcc there are no couplings between the first and second harmonics (see Table 3) and when $S_0^{-1}(\alpha_2 k^*) < 0$ the second harmonic grows, decreasing the free energy of the nfcc considerably. Actually, even when $S_0^{-1}(\alpha_i k^*) > 0$ they are couplings between the harmonics that can make $a_i \neq 0$. The lowest-order coupling which leads to growth of a harmonic with $|K_r^{(m)}| > k^*$ is proportional to $\Gamma_4 a_1^3 a_i$ when $\Gamma_3 = 0$, generating after minimization with respect a_i terms of order higher than a_1^4 in the free energy, which are inconsistent with the truncation of the free energy expansion up to fourth order. When $\Gamma_3 \neq 0$, however, the i-th harmonic must be included if $A_{11i} \neq 0$, since in this case the minimization with respect to a_i leads to $-(\Gamma_3)_{11i} a_1^4 S_0^{-1}(\alpha_i k^*)/2$, which is of the same order as $(\Gamma_4)_{1111}$. Therefore, the effective $(\Gamma_4)_{1111}$ is

$$(\Gamma_4)_{1111} - (\Gamma_3)_{11i} a_1^4 S_0^{-1}(\alpha_i k^*)/2 \tag{10}$$

All the couplings that generate terms of order a_1^5 and greater should be set to zero upon minimizing the free energy. Therefore, only the terms $S_0^{-1}(\alpha_i k^*)$ and A_{11i} when $\alpha_i >$ can be added to the free energy of the harmonics with magnitude k^*. When one studies structures in regions below the transition where $S_0^{-1}(\alpha_i k^*) < 0$ one can include the contributions of these harmonics even when $A_{11i} = 0$. In this region, however, the LMFA breaks down.

The above conclusion is also true when the Γ_3 and Γ_4 are wavevector-dependent functions. Therefore, the LMFA of Leibler must be modified to include the i-th harmonic with $(\Gamma_3)_{11i} \neq 0$. The generalized functions $\Gamma_3(K_r^{(i)}, K_s^{(j)}, K_t^{(k)})$ and $\Gamma_4(K_r^{(i)}, K_s^{(j)}, K_t^{(k)}, K_u^{(l)})$ required to evaluate $(\Gamma_3)_{ijk}$ and $(\Gamma_4)_{ijkl}$ in the multiharmonic approach were calculated by Mayes using the RPA [31]. Since $\Gamma_3(K_r^{(1)}, K_s^{(1)}, K_t^{(i)})$ is nonzero only if $K_r^{(1)} + K_s^{(1)} + K_t^{(i)} = 0$, then harmonics with magnitude up to $2k^*$ must be included. After minimizing the free energy with respect to a_i, the analysis is identical to the one-harmonic approximation but with an effective β_n,

$$\beta_n^{\text{eff}} = N(\Gamma_4)_{1111}^{\text{eff}} \tag{11a}$$

with

$$(\Gamma_4)_{1111}^{\text{eff}} = (\Gamma_4)_{1111} - \sum(\Gamma_3)_{11i} a_1^4 S_0^{-1}(\alpha_i k^*)/2 \tag{11b}$$

We computed the $(\Gamma_4)_{1111}^{\text{eff}}$ for various morphologies. In diblock copolymer melts we find direct transitions to an fcc structure as soon as $f > 0.6$. When the free energy is minimized with respect to x^*, however, $(\Gamma_4)_{1111}^{\text{eff}} < 0$ so the LMFA is not valid in this region. Also, fcc structures are found even at $f \sim 0.53$ when $\chi N \sim 10.69$. When the free energy is minimized with respect to x^*, however, the $(\Gamma_4)_{1111}^{\text{eff}} < 0$ in all the regions where the one-harmonic phase diagram is altered. Therefore, the LMFA is not valid in this region. The gyroid and the lamellar catenoid are never the lowest energy structure with this approach. Though in some regions the gyroid has lower energy than the bcc, the fcc in these regions is always the lowest energy structure. In triblock copolymer melts, the situation is very different because there $\Gamma_3(K, K', K'')$ is never equal to zero for all K, K', and K'' vectors. Therefore, the higher harmonics always alter $(\Gamma_4)_{1111}^{\text{eff}}$. For the triblocks described in Table 2 we find that the lamellar catenoid has the lowest free energy when $f \sim 0.6$.

B. The Self-Consistent Mean-Field Approach: The Hartree Approximation

In the isotropic state ($\chi \to 0$) the chains are random coils. In the ordered phases the chains are expected to be perturbed. In the weak segregation limit where the interface is large, Leibler theory [15] assumed that d scales with N^ν, $\nu = 1/2$; i.e., the chains remain unperturbed Gaussian coils near the transition. Minimizing the free energy of Leibler we found that in this limit the chains are actually stretched [29], $\nu \sim 1$, in agreement with the experiments [19,28]. It was also found experimentally that in the isotropic state [32] the chains are also stretched, $\nu = 0.8$, larger than in the strong segregation limit (sharp interface), where $\nu = 2/3$ [22–24].

The LMFA neglects the effects of fluctuations. When the concentration fluctuations are self-consistently included in the analysis (the *Hartree approximation*) [14], a first-order transition is induced at f_c. This approach has been applied to block copolymers [16], considering $\Gamma_3(\{k\})$ and $\Gamma_4(\{k\})$ in

Eq. (1) as $\{k\}$ independent (and setting $k^* = k_0^*$). When the wavevector dependence of $\Gamma_3(\{k\})$ and $\Gamma_4(\{k\})$ is included in the analysis, a shift of k^* to lower values is observed as one approaches the transition [17]. The shift in k^* to smaller values suggests that the chains stretch in the disordered state. We calculated k^* as a function of χ and N, and the periodicity of the ordered morphology in the weak segregation regime: while in the lam and hpc the chains stretch, in the bcc they contract.

In the Hartree approximation, transitions with $k^* \neq 0$ are never continuous. Therefore, a single set of plane waves cannot describe the density distribution in the periodic structure near the transition. In Ref. 7 two harmonics were included in the Hartree analysis, neglecting the fluctuations of the second harmonic. The fluctuations of the second harmonic were included in systems described by simple Hamiltonians [33]. Here we review the Hartree analysis including only harmonics of magnitude equal to k^*.

The Hartree equations can be obtained adding an external field $U(r)$ coupled to the order parameter in the free energy functional Eq. (1) as $\sum U_k \rho_{-k}$, and replacing ρ_k by $g_k + \xi_k$, $F(\rho_k) = F(g_k) + G(g_k, \xi_k)$ where g_k is the mean value of the order parameter, and ξ_k are the fluctuations about it, $\langle \xi_k \rangle = 0$ [4]. Setting $\langle \xi_k \xi_{k'} \rangle = S(k)\delta(k + k')$ in the relation $\langle \delta G(g_k, \xi_k)/\delta \xi_k \rangle = 0$, one obtains the Hartree equation for U_{-k}. The scattering function in the disordered state, $S(k)$ is given by

$$S^{-1}(k) = S_0^{-1}(k) + H(k)/2 \tag{12a}$$

where

$$H(k) = (2\pi)^{-3} \int \Gamma_4(0, h_{kk'}; |k + k'|)S(k')\, d^3k' \tag{12b}$$

Since the angular dependence of $N\Gamma_4(k, k', -k, -k') = \Gamma_4(0, h_{kk'}; |k + k'|)$ is very small in copolymers [34], and the integration over the angle between k and k' results in a linear function of $(k^2 + k'^2)Nl^2/6 = x + x'$, $N\Gamma_4(k, k', -k, -k')$ can be approximated to $N\Gamma_4(0, 0; (k^2 + k'^2)/2)$ (where $N\Gamma_4(0, 0; k)$ is defined in Ref. 5 as $N\Gamma_4(0, 0)$, a function of $x = k^2 R_0$). Expanding $N\Gamma_4(0, 0; (k^2 + k'^2)/2)$ around $k'^2 = k^{*2}$ we find that only the first term $N\Gamma_4(0, 0; (k^2 + k^{*2})/2))$, contributes to $H(k)$. The linear term $(k^2 - k^{*2})$ vanishes because $S(k')$ is symmetric around k^{*2} and the higher-order terms are zero since $N\Gamma_4(0, 0; (k^2 + k'^2)/2))$ is nearly linear in k'^2. When $k' \ll k_0^*$, however, $N\Gamma_4$ is no longer linear and higher-order terms in the expansion need to be considered. Since the smaller the N the smaller the value of k^*, the approximation fails for N small. With this approximation,

$$H(k) \sim \Gamma_4(0, 0; (k^2 + k^{*2})/2)2\alpha/\sqrt{r} \tag{12c}$$

where $\alpha = k^{*2}/4\pi c$,

$$c^2 = x^*[\partial^2(Q(x) + NH(k)/2)/\partial x^2]_{x^*}/3 \sim [\partial^2 Q/\partial x^2]_{x^*}/12 \tag{12d}$$

and $r = S^{-1}(k = k^*)$ is the inverse susceptibility. The approximation works very well for large N (we tested it by solving numerically the integral in Eq. (12b). (Note a typo in Ref. 7: c should be divided by four.) We assume that near $k = k^*$, $S^{-1}(k)$ can be expanded as

$$NS^{-1}(k) = Nr_d + 6c^2(x - x^*)^2/x^* \tag{12e}$$

(there is no linear term $x - x^* = (k^2 - k^{*2})Nl^2/6$ in the expansion of $S^{-1}(k)$, because $NS^{-1}(k = k^*)$ has a minimum at $k^2 = k^{*2}$). The inverse susceptibility in the disordered state, $r_\mathrm{d} = S^{-1}(k = k^*)$,

$$r_\mathrm{d} = S_0^{-1}(k^*) + \Gamma_4(0, 0; k^*)\alpha/\sqrt{r_\mathrm{d}}$$

is obtained by self-consistently finding the k^* for which $S^{-1}(k)$ in Eq. (12e) is a minimum using Eq. (12d) and $\partial NS^{-1}/\partial x = 0$. The value of $x^* = k^{*2}Nl^2/6$ versus χN is plotted in Fig. 1. With this approach one finds that the

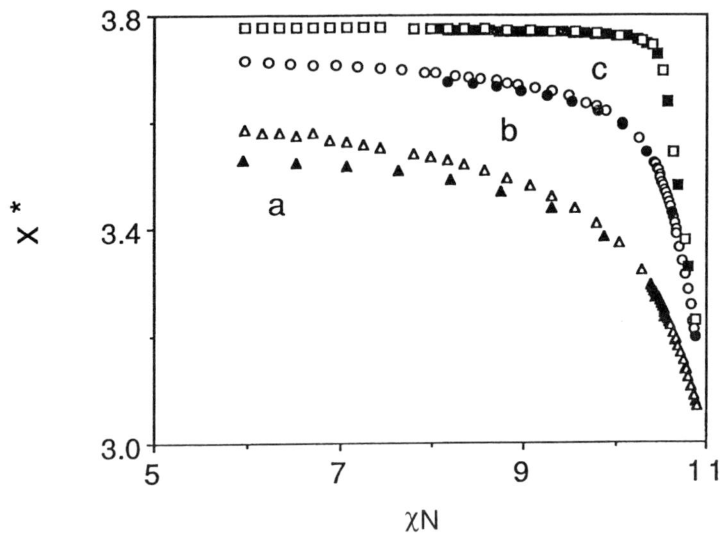

Figure 1 $x^* = k^{*2}Na^2/6$ versus χN in the disorder state for $f = f_c = 0.5$ (in the LMFA $x^* = 3.7849$). The open symbols correspond to varying χ keeping N fixed (in a, $N = 10^5$; in b, $N = 10^6$; in c, $N = 10^8$), and the solid symbols correspond to varying N keeping χ fixed (in a, $\chi = 11.215 \times 10^{-5}$; in b, $\chi = 10.885 \times 10^{-6}$; in c, $\chi = 11.49 \times 10^{-8}$.)

corrections in x^* from the LMFA value x_0 decrease as N increases. Near the transition $k^* \sim k_0^*(1 - O(1/N)$, in agreement with the experiments.

In the ordered phase the order parameter takes the form given in Eq. (2) with $m = 1$. Evaluating U_{-k} at $k = k_m^{(1)}$,

$$U_1 = r_1 a_1 + 3(\Gamma_3)_1 a_1^2/2 + (\Gamma_4)_1 a_1^3/12 - n_1 \Gamma_4(0, 0; k^*) a_1^3/2 \tag{13}$$

where r_1 is the inverse susceptibility in the *ordered* state

$$r_1 = S_0^{-1}(k^*) + \Gamma_4(0, 0; k^*)\alpha/\sqrt{r_1} + B_1 a_1^2 \tag{14}$$

where $B_1 = (\Gamma_4)_1^{\rm P}/12 + n_1 \Gamma_4(0, 0; k^*)$ and $(\Gamma_4)_1^{\rm P}$ are $12n_1(1 - n_1)$ terms that result from the arrangements $K_j^{(1)} - K_j^{(1)} - K_m^{(1)} + K_m^{(1)} = 0$ in $\Gamma_4(\{k\})$ and $(\Gamma_4)_1^{\rm NP}$ are the terms that result from the arrangements $K_i^{(1)} + K_j^{(1)} + K_m^{(1)} + K_h^{(1)} = 0$ with all different vectors in $\Gamma_4(\{k\})$.

The free energy of the ordered phase $A(k^*)$ is obtained by integrating Eq. (13) from $a_1 = 0$ to $a_1 = \underline{a}_1$ the above equations. In Fig. 2 we plotted the value of x^* that minimizes $A(k^*)$ versus χN. It is common to fit the experimental data to a power law with N and χ. Though a single power law cannot describe our results, in order to compare with the experiments we found the best fit near the transition: for the hpc and lam $d \sim N^\nu \nu \in [0.83,1]$, and $d \sim \chi^\beta(\beta = \nu - 1/2)\beta \in [0.335,0.5]$ close to the experimental values [19].

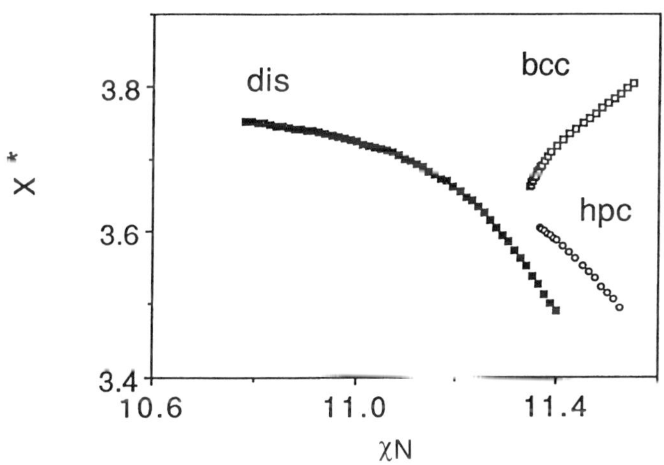

Figure 2 $x^* = k^{*2}Na^2/6$ versus χN keeping $\chi = 12.195 \times 10^{-7}$ in the disordered, hpc and bcc states for $f = 0.6$. In the LMΓA, as in previous Hartree approximations, $x^* = 3.8429$, which is the value for unperturbed Gaussian chains. In our approach in the hpc the chains are stretched, while in the bcc they are contracted.

In the bcc structure the chains contract, $\nu < 1/2$ (Fig. 2). We have obtained k^* for other metastable structures (higher energy). In the fcc and simple cubic structures, the chains stretch as in the hpc. In the quasi-crystal with $n_1 = 15$ and $(\Gamma_3)_1 = -20\Gamma_3(k^*)/(15\sqrt{15})$ the chains contract as in the bcc structure. When α is set to zero in the Hartree equations (the LMFA) the d scaling with N for the bcc is the same as for the hpc and lam structures. Therefore, the increase in x^* with increasing N for the bcc is due to Hartree effects which appear as soon as $\alpha \neq 0$. The Hartree approximation suggests that in 3D structures whose reciprocal lattice vectors form large numbers of triangles the chains contract (Semenov [23], using a completely different self-consistent approach, predicted a power-law dependence $\nu = 1/3$, in the weak segregation regime for spherical domains regardless of the structure.)

The equilibrium structure for $f \in [0.5, 0.5 + \Delta]$ is the lam [16], the structure with smallest n_1 (and the smaller the N the larger the Δ). As f increases the transition is to hpc and then to bcc (the nfcc is obtained for larger (χN), when the approach breaks down). The minimization in k^* yields lower values of $(\chi N)_t$ than the ones obtained by setting $k^* = k_0^*$; for the lam, $(\chi_t - \chi_c) \sim (\Gamma_4\alpha)^{2/3} \sim N^{-4/3}(1 - O(1/N))$. When N decreases, however, higher-order corrections need to be included in the analysis [33].

We conclude that the concentration fluctuations induce a shift in k^* in the disordered state near the transition, and that in the ordered state the chain deformation depends on the structure.

C. Semidilute Block Copolymer Solutions

The *microphase transition* and the equilibrium properties of the microphases in nonselective good solvents was studied following the approach of Broseta et al. [35] for semidilute solutions of A and B homopolymers using scaling analysis [36].

Let ϕ be the concentration of block copolymer, $\phi = \phi_A + \phi_B$, and $(1 - \phi)$ the solvent concentration in a block copolymer solution of a nonselective good solvent. The block copolymer composition is $f = \phi_A/\phi$. Three regimes with different physical properties are distinguishable in block copolymer solutions: dilute, semidilute, and concentrated [35]. In dilute solutions ($\phi < \phi^*$, where ϕ^* is the concentration at which the chains begin to overlap) the interpenetration of unlike monomers is prevented by the excluded volume [37], so that A and B monomers cannot tell each other apart and their effective interaction in the dilute solution is zero (there are only corrections to scaling which vanish in the limit of large molecular weight. [37] However, in the semidilute regime, $\phi^* \ll \phi \ll 1$, the interactions are only screened at distances $r < \xi$, where

$$\xi \sim \phi^{-\nu(3\nu-1)}$$
$$\nu \sim 0.588 \tag{15}$$

For distances larger than ξ the chains overlap, while at shorter distances the chains are swollen [36]. In the asymptotic semidilute region, the system is described as a molten state of "blobs" of size ξ, and the number of blobs per polymer is

$$Z = N\phi^{1/(3\nu-1)} \tag{16}$$

In semidilute solutions of block copolymer in a nonselective good solvent, the solvent is homogeneously distributed within the A and B blocks. The system can be regarded as a molten state of block copolymers made out of blobs of size ξ. The free energy of a semidilute block copolymer solution is calculated using scaling arguments. The scaling hypothesis, extensively used in polymer semidilute solutions [36], assumes that the free energy density expressed in terms of the properly scaled parameters (these scaled parameters take into account the swelling of the chains) has the same functional form as the mean-field free energy.

The effective interaction between blobs of A and B monomer, denoted by u, can be obtained from the osmotic pressure of the block copolymer in a semidilute solution [37], and is given by

$$u = \chi'\phi^{\alpha(3\nu-1)} = \chi'\phi^{0.29} \tag{17}$$

where ξ' is proportional to χ and $\alpha = 0.22$ [35].

The MST in the asymptotic semidilute limit can be analyzed using the above scaling arguments. Using the analog of Leibler's mean-field analysis one finds for a symmetric diblock copolymer, $f = 0.5$, a line of critical points occurring at [38–39]

$$(uZ)_c = 10.495 \tag{18}$$

where Z and u are given in Eqs. (16) and (17), respectively. As for $f \neq 0.5$ the MST is a first-order transition, a coexistence curve is generated by the MST (one phase microphase segregated and the other homogeneously mixed). In the semidilute regime, the width of the coexistence curve is negligible, and the phase diagram (the equilibrium microphase versus f) for a block copolymer solution is given by the mean-field phase diagrams replacing χ by u and N by Z given in Eqs. (17) and (16), respectively. The analysis can also be extended to A–B–A triblock copolymers in semidilute solutions. The microphase transition in block copolymer semidilute solutions is found at

$$(\chi'\phi^{(\alpha+1)/(3\nu-1)}N)_t = (\chi'\phi^{1.61}N)_t = \text{constant} \tag{19}$$

where the constant is determined by the composition f and the copolymer architecture.

When the Hartree approximation is used [16,17], rather than the LMFA, there are corrections to the microphase transition value at $f_{c'}(uZ)_t$, of order $Z^{-1/3}$. The Hartree phase diagrams for diblock and triblock copolymers in nonselective good solvents can be obtained from the melt state phase diagrams if χ is replaced by u and N by Z in the vertical axis. In the Landau and Hartree approaches, R_0 in x^* should be replaced by

$$R_0 \sim \{N\phi^{(1-2\nu)/(3\nu-1)}\}^{1/2} \tag{20}$$

In order to obtain the correct scaling for the domain periodicity $D \sim 1/k^*$, we must again replace χ by u and N by Z in the same LMFA relation for the periodicity

$$x^* \sim 1/N\chi \tag{21}$$

and use $x^* = (R_0 k^*)^2$, where R_0 is given by Eq. (20).

1. Semidilute Solution of Block Copolymers in Nonselective θ Solvents

The analysis to determine the MST described in the previous section can also be applied to θ solvents, provided that the solvent is nonselective. The statistics of θ solvents are properly described by setting $\nu = 1/2$ in Eq. (16) [36]. The value of the effective interaction between A and B blobs u in θ solvents was calculated by Onuki and Hashiomoto and it was used in Ref. 28.

$$u(\theta) \sim \chi/(\phi + \chi/u^*) \tag{22}$$

where u^* is a constant.

Following the steps outlined in the previous section for solvents (setting $u = u(\theta)$ given by Eq. (22) and $\nu = 0.5$) one finds that the MST in the Landau approach is given by,

$$(N\phi^2\chi/(\phi + \chi/u^*))_t = \text{constant} \tag{23}$$

Therefore, the concentration for *microphase transitions* in strongly incompatible blocks, ϕ_t, scales as

$$\phi_t \sim N^{-1/2} \tag{24}$$

The width of the coexistence curve, however, will not be reduced.

Expression (23) in the limit of concentrated solutions, $\phi \to 1$, where the chains obey Gaussian statistics, gives

$$(\phi\chi N) = \text{constant} \tag{25}$$

as expected by the mean field approximation [39]. This result, however, can only be obtained if the solvent is assumed to be a θ solvent. That is, it cannot be recovered if the solvent is a good solvent. Recent experimental results [40] indicate that the semidilute good solvent results also hold in the concentrated regime, suggesting that the mean-field result is only valid in concentrated solution if the solvent is a θ solvent.

In this chapter we neglect the contributions to the free energy density from the solvent concentration fluctuations. We assume that the solvent is homogeneously mixed throughout the blocks, an assumption which is only valid below and at the MST for continuous or slightly first-order transitions, for $(f \sim f_c)$. In the strong segregation limit the solvent concentration is expected to be larger at the interphase [39].

III. RANDOM COPOLYMERS

A. Statistics

The thermodynamics of multicomponent mixtures of random A–B copolymer melts was addressed by Scott [41] and by Erukhimovich [42] who evaluated the instability condition (the *spinodal*) for a single phase. The instability limit of the homogeneously mixed melt state (spinodal) is found as χ increases to

$$\chi_s = (2f(1-f))^{-1} \tag{26}$$

Bauer [43] extended the calculation to equilibrium between multiple phases. The case of ideal random copolymer mixtures, discussed below, was studied in detail by Nesarikar et al. [44]. Multiphase equilibria were determined for various composition distributions of the copolymer chains. More detailed statistical arguments were employed by Shakhnovich and Gutin [45] to evaluate the free energy of mixing for ideal random copolymers in the limit of $N = \infty$. They found that the system undergoes a continuous transition from a single phase to a microscopically segregated state, an MST, at $\chi = 2$. A similar approach was used by Fredrickson and Milner [46] who considered chains of finite N. They found a continuous transition when $f = 0.5$ from one phase to two phases when N is finite, in agreement with Bauer, and a continuous transition to a mesophase when $N = \infty$, in agreement with Shakhnovich and Gutin.

We consider here a monodisperse system of ideal A–B random copolymers in which the overall fraction of A monomers is f. A and B monomers have equal volumes. Thus the mole fractions and the volume fractions are equivalent. While all chains in the system have the same degree of polymerization N, ideally random copolymerization yields $(N + 1)$ different *compositions* $f_i (f_i = i/N,\ i = 0, 1, \ldots, N)$. For example, when $N = 30$ there are 31 components or chains with compositions $f_i = 0/30, 1/30, 2/30, \ldots, 30/30$. The *concentration* of each chain type is expressed as the mole (or volume) fraction ϕ_i. Global concentration ϕ_i of any component is equal to the probability of forming a copolymer of N units with composition f_i, subject to the overall composition being f. For the case of equal reactivity ratios for monomers A and B this results in the binomial distribution.

In Ref. 4 we investigated the equilibrium states of statistical or ideal random copolymer systems by equating the chemical potential of each component in the various phases. This multicomponent thermodynamic approach, originally used by Bauer, determines the number and volume fractions of equilibrium phases as well as the distribution of different chain types in each phase. As the interaction parameter χ is increased, transitions occur sequentially from one phase to two phases, from two to three phases, from three to four phases, etc.; we defined χ at the transition from $y - 1$ phases to y phases as χ_{ty}. We determined the order of the transition at χ_{ty} and the dependence of χ_{ty} on N and f. The transition from one phase to two phases was also studied using a Landau approach following Fredrickson and Milner. When $f = 0.5$ and N is large, the Landau values for χ_{t2} (χ for the transition from one phase to two phases) and equilibrium phase concentrations for $\chi > \chi_{t2}$ are in good agreement with those from the multicomponent thermodynamic approach. When $f \neq 0.5$, however, the Landau analysis and the thermodynamic model disagree. The mean-field approaches are not expected to be valid, however, because of the complexity of these systems.

If the sequence of monomers types along the chains is random, a Gaussian probability for f_i, $P(f_i)$, with width $(f(1 - f)/N)^{1/2}$, results. The possible random sequences generate mixtures of $2^{N-1} + 2$ chemically different molecules, making the continuous transition when $f = 0.5$ to two phases, α and β, at χ_{t2}, highly unusual. Indeed, the resulting nonGaussian distributions in the α and β phases, obtained by equating the chemical potential of the $N + 1$ components in the two phases, strongly overlap even when their mean compositions f^{α} and f^{β} are not infinitesimally close (i.e., not necessarily close to $\chi_{t2} = 2$). Furthermore, since transitions to three, four, and more macroscopic phases are predicted as χ increases, and the transitions come closer together as N increases, inducing overlap of the probabilities of f_i, the probability of finding a given f_i value in the n-phase-separated system is flat. This analysis suggests that when N goes to infinity an infinite number of phases with infinitesimally close mean compositions (a pure interface morphology) results when the system is quenched into the unstable region.

The LMFA in these systems predicts local segregation for infinite length chains and $f = 0.5$ to a lamellar *microphase* structure at $\chi_m = 2$ with periodicity $D \sim 1/k^*$ [45]. The amplitude of the largest composition fluctuations $m(k^*)$ scales as $\chi - \chi_m$ and k^* as $(\chi = \chi_m)^{1/2}$. The model extension to finite chains [46] predicts a two-phase region of width $\chi_m - \chi_T \sim 1/N$ and a *microphase transition* to a disordered microphase at $\chi_m = 2 + \epsilon/N$ was postulated because the Hamiltonian for these systems has a fourth-order contribution equal for all the morphologies (that is, β_n in Eq. (6) is equal for all values of n, as well as for the sc, fcc, bcc2, etc. given equal free energy for all possible periodic structures when $f = 0.5$).

Dobrynin and Erukhimovich [47] also found tha the structure is a disorder microphase. They, however, by including the fluctuations in the analysis (Hartree approximation) self-consistently, found a first-ordered transition to a disordered microphase at $\chi > 2$, and a shift of k^* to larger values even in the isotropic phase, $k^* \neq 0$ at $\chi = 2$ (opposite to the fluctuation effect in block copolymers, where k^* shifts to smaller values and the degree of shifting is N dependent [17]). It was later shown that the mean-field microphase transition does not occur [48]: a system quenched to $\chi > 2$ stays in a disordered phase; its morphology resembles a mixture of unconnected A and B units undergoing spinodal decomposition, [49]. When an additional term neglected in Refs. 34–37 is included in the Hamiltonian, however, a first-order transition to a lamellar microphase is found with the Hartree approximation [50]. In the next section we analyze RC solutions by computer simulations.

It is important to point out before describing the statistics that the dynamics of these systems are unclear due to the disorder along the chains. Bouchaud and Cates [51] calculated the effect of the potential created by a frozen RC melt on the dynamics of an ideal *reptating* chain with $N_{\text{eff}} = N/N_e$ tube segments ("kinks") of length $N_e^{1/2}a$ where $N_e a^2$ is the square of the distance between entanglements and a is the persistence length, hereafter we set a equal to one. They found that when $\chi N^{1/2}/N_e^{1/4} > 1$, a kink diffusing along the tube path a distance s experiences a random potential due to the RC chemical disorder ($\langle \Delta V(s)^2 \rangle_{\text{RCP}} \sim s\sigma^2, \sigma \sim \chi KT$). Therefore, s scales logarithmic with time t [52], leading to a center of mass square displacement,

$$R^2(t) = s^* \ln^2 t/t^* \qquad t^* \ll t \ll t^{**} \tag{27a}$$

$$R^2(t) \sim D_{\text{CM}} t \qquad t^{**} \ll t \tag{27b}$$

where $t^* \sim s^{*2}/D$, $D \sim 1/N$, $s^* \sim (KT/\sigma)^2$ and the relaxation time t^{**} is given by the time it takes a kink to move a distance s equal to the tube length [41] $(N_{\text{eff}}N_e^{1/2})$. Bouchaud and Cates approximate $t^{**} = t(s = N_{\text{eff}}N_e^{1/2})$ assuming that $t(s)$ is given by the time it takes a kink to surmount a typical potential barrier,

$$t(s) \sim (s^2/2D)\exp(\langle \Delta V(s)^2 \rangle_{\text{RCP}}^{1/2)}/KT \tag{28a}$$

The resulting diffusion of the center of mass (given by setting $R^2(t^{**}) = N$ in Eq. (27b)) is strongly reduced,

$$D_{\text{CM}} \sim D_{\text{rep}} \exp(-\chi N^{1/2}/N_e^{1/4}) \tag{28b}$$

They state that even though in an RC melt the medium is not frozen, the motion of the RC chains will not destroy the quasi-frozen state, Eq. (27).

Next we examine the statistics of short chains (up to $N = 64$) which obey Rouse-type diffusion. In the last section we test the conjecture of Bouchaud and Cates for reptating long chains ($N_{\text{eff}} = N/N_e = 64$; in melts [11] $N_e \sim 200$).

We study concentrated and semidilute solutions of symmetric RC ($f = 0.5$) in nonselective solvents as a function of the interaction energy $w_{\text{AB}} > 0$ (we set $w_{\text{AA}} = w_{\text{BB}} = 0$). In order to compare our results to theoretical predictions, we compute the effective net interaction energy χ_{eff}. We assume the mean-field result is valid; that is, $\chi_{\text{eff}} = \chi\phi$, where ϕ is the polymer volume fraction and $\chi = z\Delta w_{\text{AB}}/KT$, where z is the coordination number of the lattice and the next-nearest-neighbor interaction between unlike monomers $\Delta w_{\text{AB}}/KT = w_{\text{AB}}/KT - (w_{\text{AA}} + w_{\text{BB}})/(2KT)$, and is set equal to w_{AB} in the simulation. As mentioned in Sec. II.C, the mean-field result for the scaling of χ_{eff}, equivalent to u, is not expected to hold in nonselective good solvents. Therefore, strong corrections to this scaling are expected, especially in semidilute solutions.

The statistics are studied using the *bond fluctuation model* (BFM) [53] in three dimensions [54]. We extend the model to treat the case of RCs with short-range repulsive interactions between unlike monomers. We weight simulation steps according to Metropolis Importance Sampling [55]. After athermal initialization of the system physical properties are measured at regular intervals determined by the relaxation time of the system τ_r [56].

The radius of gyration R_g is plotted in Figs. 3a and 3b for concentrated ($\phi = 0.5$) and semidilute solutions ($\phi = 0.25$), respectively. Though in both regimes the chains have random walk statistics, in semidilute solutions the coils are swollen by the repulstive interaction between the A and B units, and they are more swollen when the number of AB contacts increases.

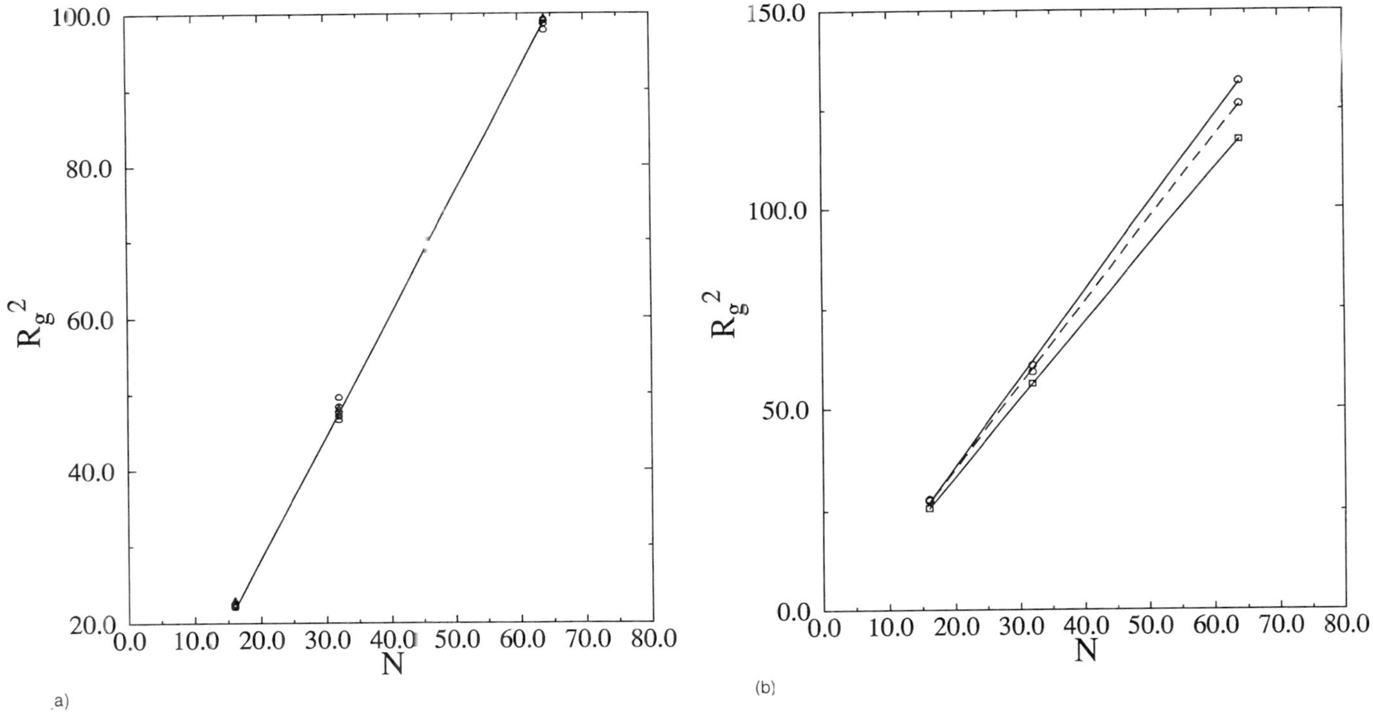

Figure 3 Plot of R_g versus N for $f = 0.25$ and $f = 0.5$ and $w_{AB} = 0, 1.5$. In a, $\phi = 0.5$ and the solid line is a linear fit of systems with no energetics applied and $f = 0.5$. In $b\phi = 0.25$ and the solid lines are linear fits for systems with $f = 0.5$ and the dashed line represents linear fits of systems with $f = 0.25$; the squares are data points for systems with no energetics applied and the circles for systems with $w_{AB} = 1.5$.

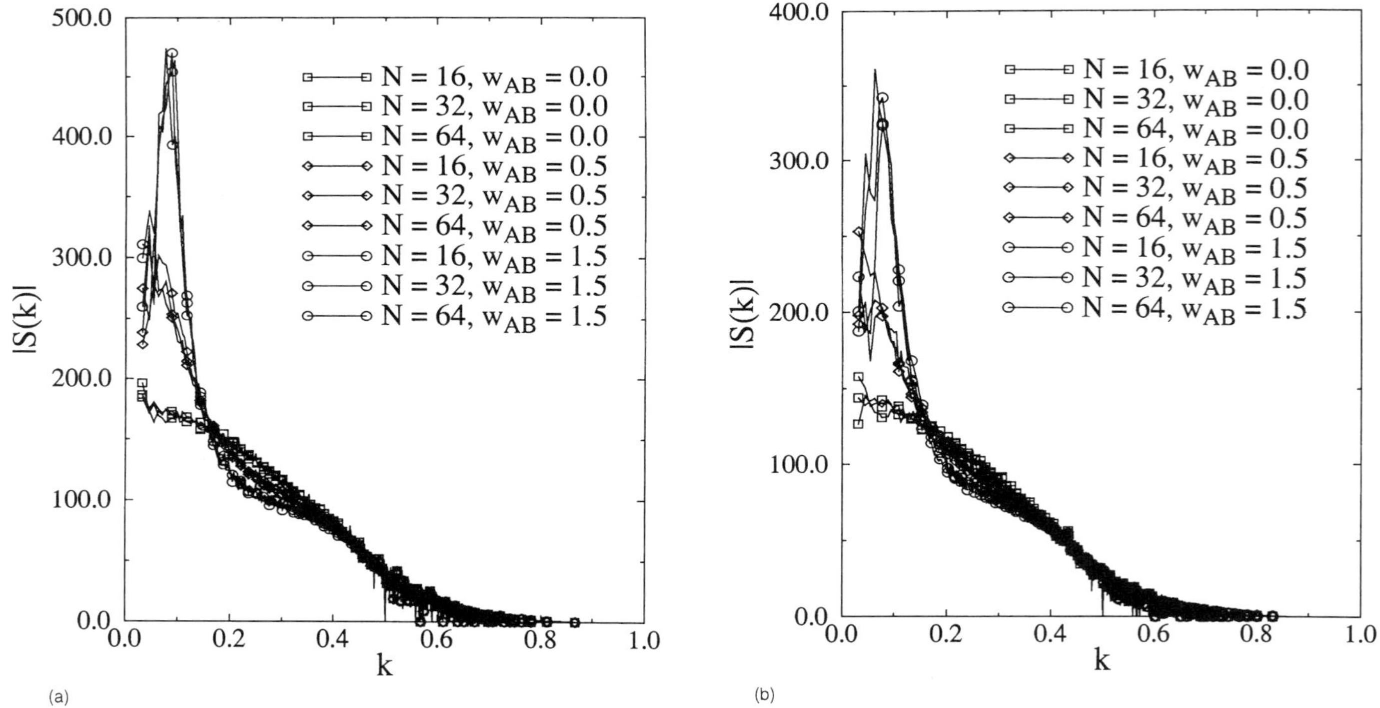

Figure 4 The A monomer scattering function for $\phi = 0.5$ at various values of N and w_{AB}. In a, $f = 0.5$ and in b, $f = 0.25$. The morphology is N-independent.

The A monomer scattering function $S(k)$, and the local A monomer density, ρ_A, are computed as a function of w_{AB} for $\phi = 0.5$ and $\phi = 0.25$. Notice that $w_{AB} = 0, 0.25, 0.5, 1$, and 1.5 correspond to $\chi_{eff} = 0, 1, 2, 4$, and 6 for $\phi = 0.5$ and to $\chi_{eff} = 0, 0.5, 1, 2$, and 3 for $\phi = 0.25$, respectively. In Figs. 4a and 4b we plot $S(k)$ for $f = 0.5$ and $f = 0.25$, respectively, for $\phi = 0.5$. Our results are N independent. When $\chi_{eff} \sim 2$, the maximum scattering intensity $S(k^*)$ shifts to $k^* \neq 0$, giving a maximum wavelength fluctuation of the order of 11 for $f = 0.5$ and of 14 for $f = 0.25$. The peak intensity increases with increasing χ_{eff}, and the intensity is larger for the system with a larger number of AB contacts ($f = 0.5$). In Fig. 5 we plot $S(k)$ for $\phi = 0.25$. In the semidilute regime we observe a peak at $k^* = 0$ even when the AB repulsive interactions are strong, suggesting macroscopic phase separation.

We plot the local monomer A density distribution function $P^l(\rho_A)$ for $l = 5$ for $f = 0.5$ and $\phi = 0.5$ in Fig. 6 and for $\phi = 0.25$ in Fig. 7. We

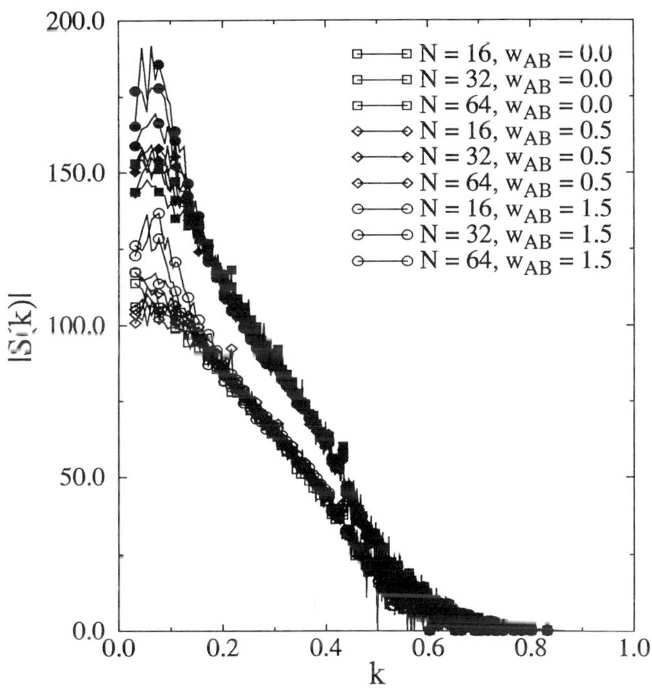

Figure 5 The A monomer scattering function for $\phi = 0.25$ at various values of N and w_{AB}. Filled symbols are for $f = 0.5$ and unfilled symbols for $f = 0.25$. The morphology is N-dependent.

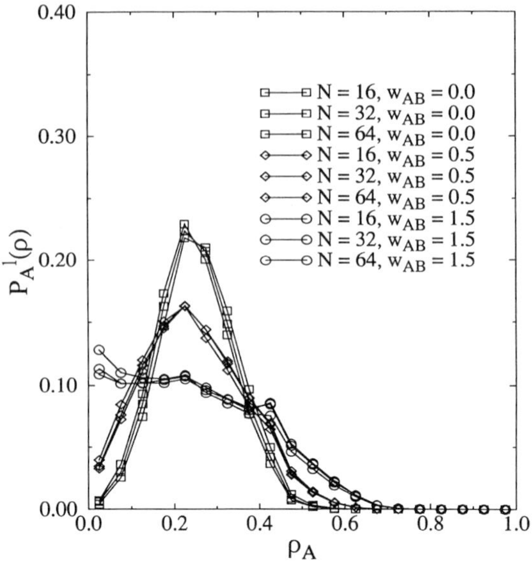

Figure 6 The local A monomer density distribution function for cells of size $l = 5$ at various repulsive interactions for $\phi = 0.5$ and $f = 0.5$.

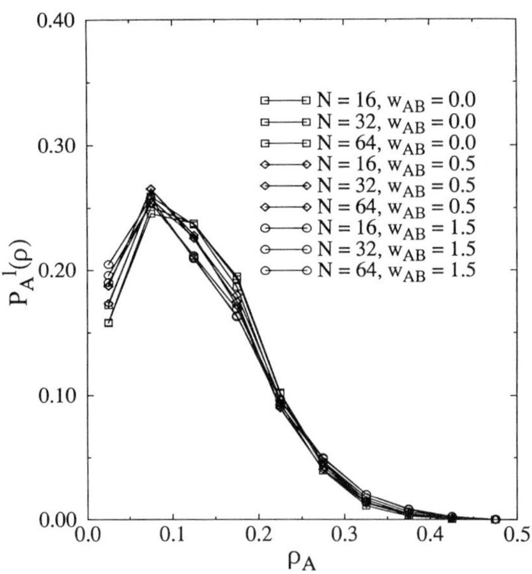

Figure 7 The local A monomer density distribution function for cells of size $l = 5$ at various repulsive interactions for $\phi = 0.25$ and $f = 0.5$.

generate $P^l(\rho_A)$ by selecting cubic cells with l lattice spacings per edge and calculating ρ_A inside them. For systems with no A–B interactions, $P^l(\rho_A)$ is Gaussian, with the peak centered at $\rho_A = f\phi$ (say at 0.25 for $\phi = 0.5$), as expected. We discuss only the case $\phi = 0.5$, which seems to be undergoing an MST as the net repulsive interaction increases above 2. For small χ_{eff} the peak remains centered at $\rho_A = 0.25$, however, the maximum decreases and a broadening of the peak width is observed. For $\chi_{\mathrm{eff}} = 6$ the peak vanishes, and over the range $\rho_A \in (0, 0.5)$ the curve is almost flat, indicating that the local density assumes all possible values with equal probability when $l < 1/k^*$. This is in contrast to the classic microphase separation behavior observed in diblock copolymers where $P^l(\rho_A)$ has two distinct peaks at the mean densitities of the A- and B-rich microphase domains [57]. When $\phi = 0.25$, the strength of the repulsive interactions has a less pronounced effect on the distribution of A monomers in the system. In Ref. 48 we showed the local solvent density ρ_S as a function of ρ_A in cubic cells with $l = 5$. For $\chi_{\mathrm{eff}} = 0$, ρ_S decreases linearly with increasing ρ_A as expected. For $\chi_{\mathrm{eff}} > 0$ there is a decrease in ρ_S in regions where $\rho_A \gg \rho_B$ or $\rho_A \ll \rho_B$ and solvent enrichment for $\rho_A \sim \rho_B$, because the number of AB contacts within a cell, roughly proportional to $n_{AB} \sim \rho_A \rho_B$, decreases as ρ_S increases.

The morphology and $S(k)$ of RC are very similar to the morphology and $S(k)$ of unconnected A–B units with the same $\chi_{\mathrm{eff}} = \chi\phi, f = 0.5$ and ϕ values frozen at the early stages of spinodal decomposition, except when $\chi_{\mathrm{eff}} \approx 2$. The scaling of χ_{eff} with ϕ may not be the same in both systems (in block copolymers solutions, for example, it is more reduced, as discussed in Sec. II.C, and that near $\chi_{\mathrm{eff}} = 2$ the systems cannot be compared [49]. When $\chi_{\mathrm{eff}} = 2$ the unconnected monomers are already segregated (a peak at $k^* \neq 0$ is clearly observed in the nonequilibrium $S(k)$), suggesting that χ_{eff} is underestimated, while in RCs the scattering intensity is not strongly peaked at a given value k^* (it is rather flat around $k = 0$). We do not find a periodic microphase forming in RC. Even if a lamellar microstructure is induced by placing a wall of A monomers at one edge, when the wall is removed, the system relaxes to the disordered equilibrium morphology described above (in diblock copolymers, on the other hand, the lamellar structure does not disappear when the wall is removed) [58].

B. Dynamics

The BFM for the above studies gives Rouse-type dynamics regardless of the value of χ_{eff}. In Fig. 8 we plot the squared monomer displacement as a function of time for $\phi = 0.5$, $f = 0.5$ and $w_{AB} = 1.5$ to show that Brownian diffusion is obtained in the systems studied in the previous section. Since to observe reptation dynamics we must have at least $N > 600$ at

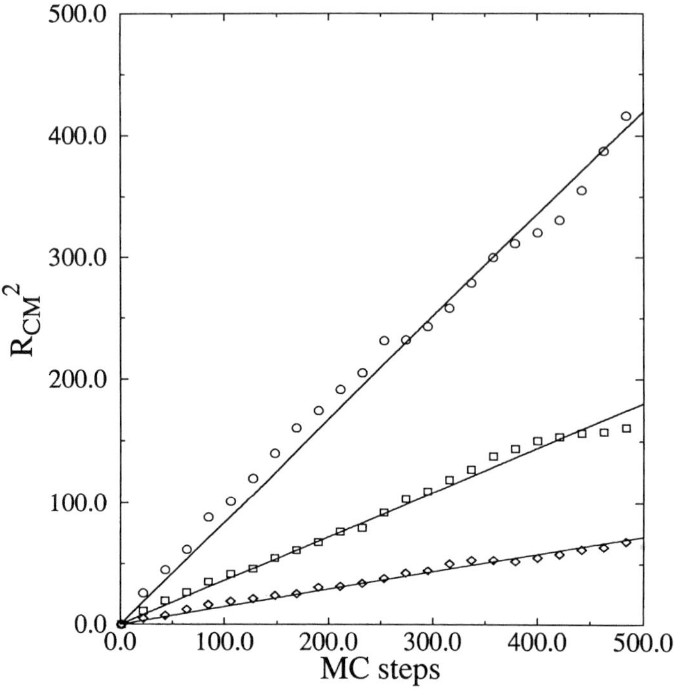

Figure 8 Log plot of the mean squared monomer displacement as a function of log time for $w_{AB} = 1.5, f = 0.5$, and $\phi = 0.5$ Circles are for $N = 16$, squares for $N = 32$ and diamonds for $N = 64$.

the concentrations studied by the BFM, the BFM is not efficient to simulate reptating chains. We therefore use the Cage [59] model, which artificially induces reptation. For simplicity we set $N_e = 1$, the Cage model with unit spacing. Interactions between adjacent monomers and soft core repulsions between monomers occupying the same site are applied to prevent the collapse of the chains and the build-up of like monomers onto the same site. In Fig. 9 we plot the squared monomer displacement as a function of time for $\phi = 0.5$ and $f = 0.5$ for various w_{AB} values. Two sets of data are plotted, one for a single chain in a frozen RC medium (unfilled symbols) and one for many-chain statistics, all undergoing motion (filled symbols). The two sets of data have the same shape, supporting the conjecture of Bouchaud and Cates [51]. For the control case of no repulsive interactions between unlike monomers, we observe the expected results (a region with slope 1/2, which moves to slope 1). In this case the curve for a single chain in a frozen potential is higher (indicating more diffusion) than the curve for the mul-

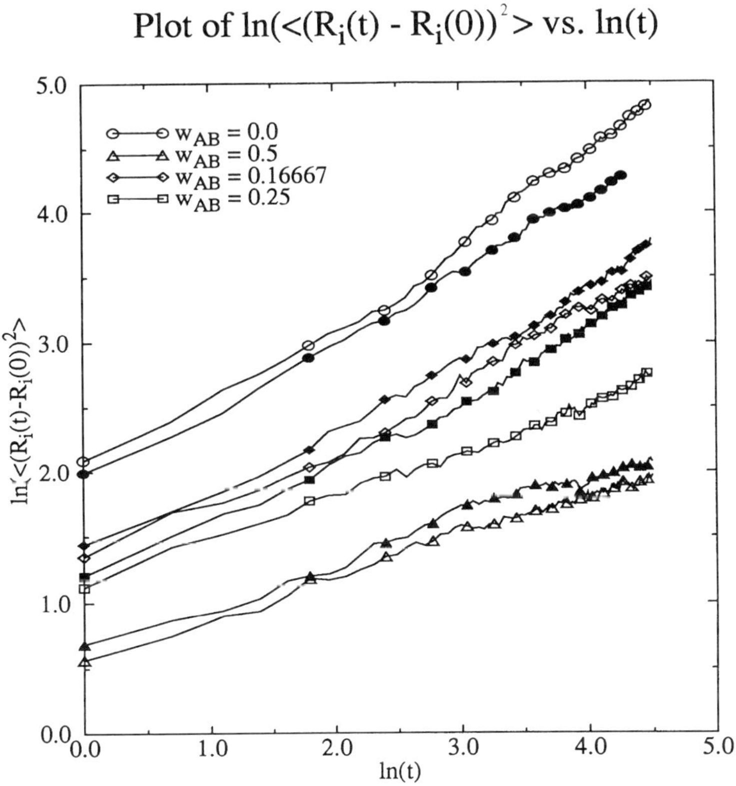

Figure 9 Log plot of the mean squared monomer displacement as a function of log time for chains of $N_{\mathrm{eff}} = 64, f = 0.5$, and $\phi = 0.5$, and various repulsive interactions. Unfilled symbols represent data for a single chain diffusing through a frozen RC medium. Solid symbols represent data for systems with all the chains undergoing motion.

tiple-chain system, due to the soft core repulsions applied. In systems with A–B repulsive interactions, however, multiple-chain systems diffuse faster. We could not check the validity of Eq. (28) due to the enormous computer time required to reach t^{**}. It is important to point out, however, that it is dangerous to use scaling arguments to compute t^{**} and D_{CM} in this problem because the mapping into a one-dimensional diffusion in a random potential leads to $\ln t$ and not power law diffusion. For example, the average value of $t(s)$ in Eq. (28a) scales as [60] exp $(s(\sigma/KT)^2)$, as if it is dominated by the time it takes a monomer to surmount an atypical potential barrier, [61] which grows as $\langle \Delta V(s)^2 \rangle_{\mathrm{RCP}} \sim s$, and not a typical barrier.

In conclusion, unlike the classic microphase of diblock copolymer systems where locally A- or B-rich regions of the microphase have distinct values of ρ_A, in RCs a broad range of values of ρ_A are obtained. Our results indicate the absence of well-defined domains with narrow interfaces. This is unique to RC systems. We observe the same trends in $S(k)$ when $f \neq 0.5$ although the peak position and intensity change. In addition, more drastic solvent enrichment/depletion is observed [58]. Finally we find that for systems with A–B repulsions, a dramatic reduction in the diffusion is observed when N is very large, suggesting that very long chains are quasi-frozen [63].

ACKNOWLEDGMENTS

Part of the work in this chapter has been done in collaboration with A.M. Mayes and B.W. Swift. This research was supported by grants from the National Institute of Health, the National Science Foundation (PYI Grant No. DMR 9057764), and the David and Lucile Packard Foundation.

REFERENCES

1. H Furukawa. Chapter 2 of this book.
2. HR Brown. J Macromolecules 26:1666, 1993.
3. D Gersappe, D Irvine, AC Balasz, L Guo, M Rafailovich, J Sokolov, S Schwarz, D Peiffer. Science 265:1073, 1994.
4. CH Huang, M Olvera de la Cruz. Phys Rev E 53:812, 1996.
5. CH Huang, M Olvera de la Cruz. Europhysics Lett 34:171, 1996.
6. J Cahn, JE Hilliard. J Chem Phys 28:258, 1958.
7. PG de Gennes. J Phys (Paris) Lett 40:69, 1979. K Binder. J Chem Phys 79:6387, (1983).
8. D Gersappe et al. Phys Rev E (in press).
9. SP Obukhov. J Phys A Gen 19:3655, 1986.
10. HS Chan, KA Dill. Physics Today 46:24, 1993. SW Swift, M Olvera de la Cruz. J Chem Phys 100:774, 1994.
11. LD Landau. In: D Ter Haar, ed. The Collected Papers of LD Landau. Oxford: Pergamon Press, 1965, pp 212–214.
12. S Alexander, J McTague. Phys Rev Lett 41:702, 1978.
13. CM Marques, ME Cates. Europhys Lett 13:267, 1990.
14. SA Brazovskii. Sov Phys JETP 41:85, 1975.
15. L Leibler. Macromolecules 13:1602, 1980.
16. GH Fredrickson, E Helfand. J Chem Phys 87:697, 1987.
17. M Olvera de la Cruz. Phys Rev Lett 67:85, 1991.
18. EL Thomas, DM Anderson, CS Henkee, D Hoffman. Nature 334:598–601, 1988, and references therein.

19. T Hashimoto, M Shibayama, H Kawai. Macromolecules 13:1237, 1980.
    T Hashimoto, M Shibayama, H Kawai. Macromolecules 16:1093, 1983.
    G Hadziioannou, A Skoulios. Macromolecules 15:258, 1982.
    R-J Roe, M Fishkis, JC Chang. Macromolecules 14:1091, 1981.
20. DA Hajduk, PE Harper, SM Gruner, C Honeker, EL Thomas, LJ Fetters. Macromolecules 27:4063, 1994.
    MF Schultz, FS Bates, K Almdal, K Mortensen. Phys Rev Lett 73:1, 1994.
21. M Olvera de la Cruz, AM Mayes, BW Swift. Macromolecules 25:944, 1992.
22. E Helfand. Macromolecules 8:552, 1975.
23. AN Semenov. Zh Eksp Teor Fiz 88:1242 1985 [Sov Phys JETP 61:733, 1985].
24. MW Matsen, M Schick. Phys Rev Lett 72:2660, 1994.
25. FS Bates, GH Frederickson. Ann Rev Phys Chem 41:525, 1990 (HL Strauss, ed.).
26. AM Mayes, M Olvera de la Cruz. J Chem Phys 91:7228, 1989.
    Star copolymers were studied with the same approach: AV Dobrynin, IYa Erukhimovich. Macromolecules 26:276, 1993.
27. M Olvera de la Cruz, IC Sanchez. Macromolecules 19:2501, 1986.
28. JN Owens, IS Gancarz, JT Koberstein, TP Russell. Macromolecules 22:3380, 1989.
29. AM Mayes, M Olvera de la Cruz. Macromolecules 24:3975, 1991.
30. SA Brazovskii, et al. Sov Phys JETP 66:625, 1988.
31. AM Mayes. PhD thesis, Northwestern University, 1991.
32. K Almdal, JH Rosedale, FS Bates, GD Wignall, GH Frederickson. Phys Rev Lett 65:1112, 1990.
33. JL Jones, M Olvera de la Cruz. J Chem Phys 100:5272, 1994.
34. AM Mayes, M Olvera de la Cruz. In: EA Buckley, G Gallagher-Daggitt, FE Karasz, DA Ulrich, ed. Proceedings of the Materials Research Society Symposium 175. 1990, p 345.
    AM Mayes, M Olvera de la Cruz. J Chem Phys 95:4670, 1991.
35. D Broseta, L Leibler, JF Joanny. Macromolecules 20:1935, 1987.
36. PG deGennes. Scaling Concepts in Polymer Physics. Ithaca, NY: Cornell University Press, 1979.
37. JF Joanny, L. Leibler, R. Ball. J Chem Phys 81:4640, 1984.
38. M Olvera de la Cruz. J Chem Phys 90:1995, 1989.
39. Ching-I Huang (private communication).
40. TP Lodge, C Pan, X Jim, Z Liu, J Shiao, WW Maurer, FS Bates. J Polym Science Polym Phys Ed 33:2289, 1995.
41. RL Scott, J Chem Sci 9:423, 1952.
42. IYa Yerukhimovich. Polym Sci USSR 24:2223, 1982.
43. BJ Bauer. Polym Engg Sci 25:1081, 1985.
44. A Nesarikar, M Olvera de la Cruz, B Crist. J Chem Phys 98:7385, 1993.
45. EI Shakhnovich, AM Gutin. J Phys France 50:1843, 1989.
46. GH Fredrickson, ST Milner. Phys Rev Lett 67:835, 1991.
47. AV Dobrynin, IYa Erukhimovich. Sov Phys JETP Lett 53:570, 1991.

48. IYa Erukhimovich, AV Dobrynin. Sov Phys JETP Lett 57:125, 1993.
AV Dobrynin, IYa Erukhimovich. Sov Phys JETP 77:307, 1993.
CD Sfatos, A Guntin, EI Shakhnovich. J Phys A 27:L411, 1994.
49. AV Dobrynin, IYa Erukhimovich. J Phys I (France) 5:265, 1995.
50. AM Gutin, CD Sfatos, EI Shakhnovich. J Phys A 27:7957, 1994.
51. JP Bouchaud, ME Cates. J Phys II (France) 3:1171, 1993.
52. YaG Sinai. Theor Prob Appl 27:256, 1982.
53. I Carmesin, K Kremer. Macromolecules 21:2819, 1988.
54. HP Deutsch, K Binder. J Chem Phys 94:2294, 1991.
55. N Metropolis, AW Rosenbluth, MN Rosenbluth, AH Teller. J Chem Phys 21:1087, 1953.
56. N Madras, AD Sokal. J Stat Phys 50:109, 1988.
57. A Weyersberg, TA Vilgis. Phys Rev E 48:377, 1993.
58. BW Swift. Ph D thesis, Northwestern University, 1995.
59. KE Evans, SF Edwards. J Chem Soc Faraday Trans 77:1891, 1981.
60. PG de Gennes. J Stat Phys 12:463, 1975.
61. M Olvera de la Cruz, JM Deutsch, SF Edwards. Phys Rev A 33:2047, 1986.
62. PG de Gennes. Scaling Concepts in Polymer Physics. Ithaca, NY: Cornell University Press, 1979.
M Doi, SF Edwards. The Theory of Polymer Dynamics. Oxford: Clarendon Press, 1986.
63. BW Swift, M Olvera de la Cruz. J Europhysics Lett 35:487, 1996.

2

Dynamics of Phase Separation and Its Application to Polymer Mixtures

Hiroshi Furukawa
Yamaguchi University, Yamaguchi, Japan

I. INTRODUCTION

There are two main motivations to study phase separation. The first is from the viewpoint of materials science. By mixing we can obtain materials with various kinds of properties, and a local phase separation sometimes gives rise to a change in the properties of a mixture. For instance, duralumin, which is an alloy mainly consisting of aluminum, hardens with microscopic phase separation. The second motivation is from the viewpoint of basic science. There is no analytic continuation between the outside and the inside of the coexistence curve. We can approach the coexistence curve from the outside, keeping the time-independent thermal equilibrium state, but if we cross the coexistence curve, the system can no longer be time-independent. The inside and the outside of the coexistence curve are qualitatively different from each other. Two phenomena, the critical phenomena of the second-order phase transition near the critical point and the dynamics of the phase separation inside the coexistence curve, are closely related to each other but cannot be treated as the same problem.

In Fig. 1 we show a typical example of a phase diagram in the temperature–composition plane, where the coexistence line and the spinodal line are shown with the critical temperature T_c. The spinodal line is the limit of the metastability. The square indicates an initial stable homogeneous state. When the temperature is suddenly changed by bringing the system into an unstable state inside the coexistence curve, phase separation takes place. The

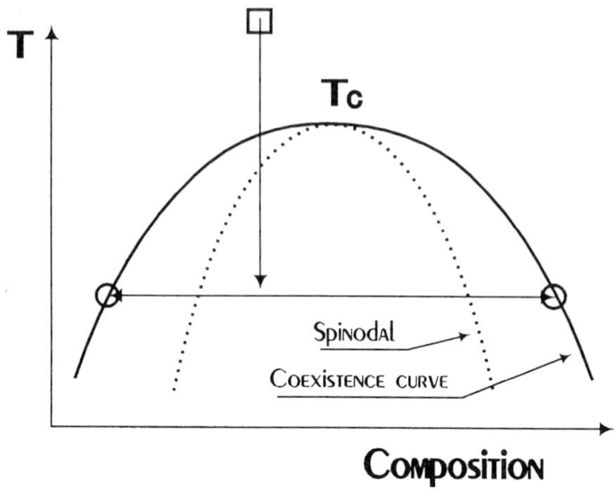

Figure 1 The phase diagram in the temperature-composition plane.

final state is a coexistence state indicated by circles. The dynamics of the phase separation is to examine this unmixing process. The phase separation is the unlimited temporal change in the state governed by the minimum principle of free energy. No ergodicity is satisfied in this problem and this temporarily unlimited process is what we are going to discuss.

The main purpose of this chapter is to review the separation of polymer mixtures. However, reviewing the phase separation of polymer mixtures is almost equivalent to reviewing the phase separation of a general system. Therefore, we focus our attention not only on the fluid system but also on the solid system. However, we are not interested in individual properties of materials. We are concerned only with the generic behavior of the phase separation with symmetric spinodal and binodal curves with an upper critical temperature. Characteristics of polymers are represented through the dependences of a few physical quantities on molecular size N.

There are many review articles about phase separation in general systems or small molecule systems [1–6] as well as in polymer systems [7–12]. We focus our attention mainly on generic phase separations. There are many other interesting aspects of phase separations in polymer systems. For instance, the shear-induced phase separation [13–20] or phase separation with chemical reactions [21–24] is interesting because the shear or the chemical reaction is a phenomenon competing with or accelerating the phase separation. However, to discuss such problems is beyond the scope of the present review. The chapter is organized as follows. In the next section we present basic equations in binary fluid systems. In Sec. III we present a phenomenological description of phase separation in the late stage. Several kinds of growth laws are presented. In Sec. IV the dynamical scaling assumption for phase separation is discussed. In Sec. V energy dissipation in a fluid system under phase separation is discussed. In Sec. VI the application of phase separation to the polymer systems will be discussed. A short summary is presented in Sec. VII.

II. DESCRIPTION OF UNSTABLE BINARY MIXTURES

In this section we present a suitable dynamical model of a binary mixture undergoing phase separation. The model must include the following two properties. First, the free energy must represent a thermodynamically unstable state. Second, the kinetic equation must describe hydrodynamic motions as well as particle diffusions. We consider the case where the phase separation proceeds isothermally. Such an isothermal process is appropriate in the slowly varying phase separation found in polymers, in alloys or in a late stage of any phase separation. The relevant free energy in such an isothermal process is the Helmholtz free energy.

A. Free Energy

Consider a system of a binary mixture consisting of A and B species. The system we are going to describe is unstable, inhomogeneous, and time-dependent. Therefore, the free energy F of the system is a functional of local variables such as two particle densities $n_A(\mathbf{r})$ and $n_B(\mathbf{r})$ and local fluid velocity $\mathbf{v}$. These variables are assumed to be slowly varying in space and time. The total free energy is the sum of the local free energy density f:

$$F = \int f \, d\mathbf{r} \tag{1}$$

We introduce convenient variables ψ and n defined by

$$\psi \equiv n_A - n_B \qquad n \equiv n_A + n_B \tag{2}$$

The quantity ψ is called the order parameter. The order parameter represents the degree of mixing of two species. n is the total particle density. The free energy density f is a functional of $\mathbf{v}$, n and ψ. In Fig. 2 schematic aspects of the local free energy density are shown. In this figure the vertical axis represents the free energy density f and the horizontal axis represents the order parameter ψ. The minimum of the upper curve corresponds to the homogeneous stable state where two species are mixed up stably. When the system is quenched, this state becomes unstable. Therefore, $\psi = 0$ corre-

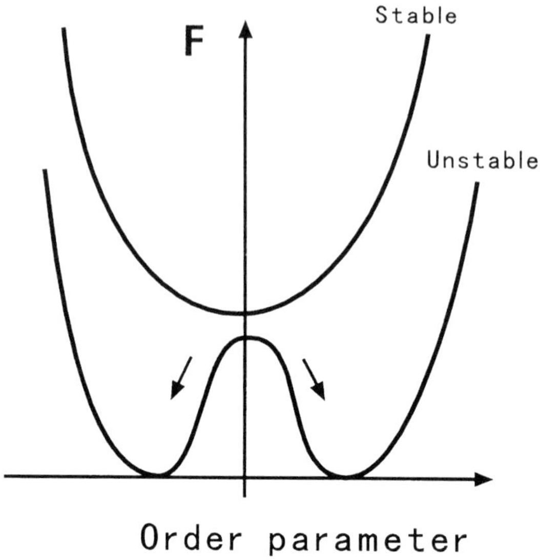

Figure 2 Schematics of the free energy of a polymer mixture.

sponds to the local maximum of the free energy represented by the lower curve. This is the most important property of the free energy describing the phase separation. In addition, the free energy contains the kinetic energy of drift motion in fluid as well as the energy associated with the compression. We shall employ the following phenomenological free energy density

$$f = \frac{\rho}{2}v^2 + \frac{1}{2}\chi_0^{-1}(n - n_0)^2 + \frac{g}{2}|\vec{\nabla}\psi(\mathbf{r})|^2 - \frac{\alpha}{2}\psi^2 + \frac{\beta}{4}\psi^4 \tag{3}$$

where χ_0 is the compressibility, and g, α, and β are constants. Constants χ_0, g, and β should be positive. Eq. (3) is the *Ginzburg–Landau free energy* amended by adding the translational kinetic energy and the energy from the compression. Here ρ is the mass density of the fluid and is given as $\rho = nm$, where m is the mass of molecule. The first term is the kinetic energy density of drift motion. The velocity field $\mathbf{v}$ is defined as the average of all particles in a small region of space. Therefore, the original kinetic energy density is written as

$$\frac{m}{2}\sum_i(\mathbf{v} + \delta\mathbf{v}_i)^2 = \frac{1}{2}\rho v^2 + \frac{m}{2}\sum_i \delta v_i^2$$

where $\delta\mathbf{v}$ is the deviation from an average, and the sum is taken over a small unit volume. No linear term in $\mathbf{v}$ appears. The second term of the right-hand side is the thermal energy and is not written explicitly in the free energy description. The second term of Eq. (3) is the energy coming from the compression. The remaining terms are due to the inhomogeneity of the composition. The sum of the last two terms corresponds to the free energy in Fig. 2. The term containing $\vec{\nabla}$ originates from the interaction among constituent particles. We assume that $\chi_0^{-1} \gg 1$, which guarantees the constantness of the total particle density n. Setting $\alpha > 0$, the free energy takes minima at $\psi = \pm\sqrt{\alpha/\beta}$. In the case the two-phase state is stable.

B. Kinetic Equations

The first important kinetic equation describing the dynamics of an unstable mixture was given by Cahn and Hilliard many years ago [see Ref. 25]. This takes the form

$$\frac{d\psi}{dt} = L_\psi \nabla^2 \frac{\delta F}{\delta\psi} = L_\psi \nabla^2(-g\nabla^2\psi - \alpha\psi + \beta\psi^3) \tag{4}$$

where δ indicates a functional derivative. The quantity L_ψ is a so-called kinetic coefficient, which is positive. Strictly speaking, the free energy in Eq. (4) should not contain the velocity field $\mathbf{v}$. Although this equation does not contain $\mathbf{v}$ nor the total density n, this has played an important role in understanding and examining the nature of the phase separation. Here we extend the kinetic equation (4) to include the velocity field $\mathbf{v}$ and the total density n. Phenomenological kinetic equations of fluid mixtures have been developed by many authors (see for instance [26–28]). We here derive kinetic equations based on the thermodynamics of irreversible process. Under constant temperature, constant volume, and constant total particle number, the free energy does not increase. Therefore,

$$\begin{aligned} \frac{dF}{dt} &= \int \left(\dot{\mathbf{v}}\frac{\delta F}{\delta \mathbf{v}} + \dot{n}\frac{\delta F}{\delta n} + \dot{\psi}\frac{\delta F}{\delta \psi} \right) d\mathbf{r} \\ &= \int \left(\dot{\mathbf{v}}\vec{\mu}_v + \dot{n}\left(m\frac{v^2}{2} + \mu_n\right) + \dot{\psi}\mu_\psi \right) d\mathbf{r} \leq 0 \end{aligned} \tag{5}$$

where the dot denotes the time derivative, μ_n and μ_ψ are static chemical potentials, and $\vec{\mu}_v$ is a kind of a chemical potential for the velocity field:

$$\vec{\mu}_v \equiv \frac{\delta F}{\delta \mathbf{v}} = \rho\mathbf{v} \qquad \mu_n \equiv \frac{\delta F^{(0)}}{\delta n} = \chi_0^{-1}(n - n_0)$$
$$\mu_\psi \equiv \frac{\delta F}{\delta \psi} = -g\nabla^2\psi - \alpha\psi + \beta\psi^3 \tag{6}$$

and $F^{(0)}$ is the static part of the free energy, i.e., $F^{(0)} \equiv F(\mathbf{v}=0)$. Since $n_A = (n+\psi)/2, n_B = (n-\psi)/2$, we find that $\mu_n = (\mu_A + \mu_B)/2$, and $\mu_\psi = (\mu_A - \mu_B)/2$, where $\mu_A \equiv \delta F^{(0)}/\delta n_A$ and $\mu_B \equiv \delta F^{(0)}/\delta n_B$ are the static chemical potentials of species A and B, respectively. Because n and ψ should satisfy conservation laws, we obtain

$$\frac{dn}{dt} = -\vec{\nabla}\cdot\mathbf{j}_n \qquad \frac{d\psi}{dt} = -\vec{\nabla}\cdot\mathbf{j}_\psi \tag{7}$$

The above equations are the continuity equations and **j**s are local currents. A local current consists of time-reversible current $n\mathbf{v}$ or $\psi\mathbf{v}$, which arises due to the fact that a system is moving, and the diffusion current, which is proportional to the second derivative of the free energy, μs. Namely,

$$\mathbf{j}_n = n\mathbf{v} - L_n\vec{\nabla}\mu_n^* \qquad \mu_n^* \equiv \frac{\delta F}{\delta n} = m\frac{v^2}{2} + \mu_n \tag{8}$$

$$\mathbf{j}_\psi = \psi\mathbf{v} - L_\psi\vec{\nabla}\mu_\psi \tag{9}$$

where the second term in each equation represents the diffusion. The kinetic coefficients Ls, which are assumed to be constant, should be positive. μ_n^* is the chemical potential in a dynamical sense.

For the velocity field **v** we may write the equation of motion as

$$\rho\frac{d\mathbf{v}}{dt} = \mathbf{F}^* + \eta\nabla_{\mathbf{v}}^2 \tag{10}$$

Here the first term on the right-hand side represents the time-reversible part and the second term represents the dissipative term, which may be of more complicated form but here we use this simplest one. In order to determine $\mathbf{F}^*$ we use an identity:

$$\int [\mathbf{v}\cdot(\mathbf{F}^* + n\vec{\nabla}\mu_n^* + \psi\vec{\nabla}\mu_\psi)]\,d\mathbf{r} = 0 \tag{11}$$

To prove this identity let us remember that the time derivative of the free energy Eq. (5) is identically zero for $L_n = L_\psi = \eta = 0$, because all equations are now time-reversal symmetric. By substituting Eqs. (7) and (10) for $L_n = L_\psi = \eta = 0$ into Eq. (5), and performing a partial integration once for each term, we arrive at Eq. (11). Eq. (11) is transformed into

$$\int [\mathbf{v}\cdot(\mathbf{F}^* + \rho(\mathbf{v}\cdot\vec{\nabla})\mathbf{v} + n\vec{\nabla}\mu_n + \psi\vec{\nabla}\mu_\psi)]\,d\mathbf{r} = 0 \tag{12}$$

Here we have used $\frac{1}{2}\mathbf{v}\cdot\vec{\nabla}v^2 = \mathbf{v}\cdot[(\mathbf{v}\cdot\vec{\nabla})\mathbf{v}]$ in the second term on the right-hand side of Eq. (11). In order for Eq. (12) to hold identically we choose

$$\mathbf{F}^* = -\rho(\mathbf{v}\cdot\vec{\nabla})\mathbf{v} - n\vec{\nabla}\mu_n - \psi\vec{\nabla}\mu_\psi \tag{13}$$

Thus a desired equation of motion for $\mathbf{v}$ is given by

$$\rho \frac{d\mathbf{v}}{dt} = -\rho(\mathbf{v} \cdot \vec{\nabla})\mathbf{v} + \eta \nabla^2 \mathbf{v} - n\vec{\nabla}\mu_n - \psi \vec{\nabla}\mu_\psi \tag{14}$$

Some comments are given on the above results.

1. The above equation of motion (14) is equivalent to the *Navier–Stokes equation* at constant temperature. To show this one may use the Gibbs–Duheim relation, $s\,dT - dP + \sum n_i\,d\mu_i = 0 (i = \mathrm{A, B})$, where s is the entropy density, T is the temperature and P is the pressure. This can be rewritten as $s\,dT - dP + n\mu_n + \psi\,d\mu_\psi = 0$ or as $s\vec{\nabla}T - \vec{\nabla}P + n\vec{\nabla}\mu_n + \psi\vec{\nabla}\mu_\psi = 0$. We set $\vec{\nabla}T = 0$. Then we find that the force term of Eq. (14), the sum of the last two terms, is given as $-\vec{\nabla}P$ giving the Navier–Stokes equation.
2. If the correlation among physical quantities in a fluid is regarded as local, then the pressure is a well-defined thermodynamic quantity, i.e., $dP = n\,d\mu_n + \psi\,d\mu_\psi$ is a total derivative. However, if the correlation is nonlocal, then the derivative of the local pressure $\delta P(\mathbf{r})$ is not a total derivative, in general. Let us explain this for a single variable $\psi(\mathbf{r})$. Let us consider a simplified case where $\mu_\psi(\mathbf{r}) = \int h(|\mathbf{r} - \mathbf{r}'|)\psi(\mathbf{r}')\,d\mathbf{r}'$. Then $\delta P(\mathbf{r}) = \psi(\mathbf{r})\delta\mu_\psi(\mathbf{r})$. Thus $(\delta/\delta\psi(\mathbf{r}''))(\delta/\delta\psi(\mathbf{r}'))P(\mathbf{r}) = \delta(\mathbf{r} - \mathbf{r}'')h(|\mathbf{r}'' - \mathbf{r}'|)$. This is not equivalent to $(\delta/\delta\psi(\mathbf{r}'))(\delta/\delta\psi(\mathbf{r}''))P(\mathbf{r})$, if the function $h(r)$ is not of a local form $\propto \delta(\mathbf{r})$. Therefore, the local pressure $P(\mathbf{r})$ is not a conserved quantity if the correlation is nonlocal.
3. By substituting Eqs. (7)–(9) and (14) into (5) and using some manipulation we find that the inequality in Eq. (5) is satisfied, i.e.,

$$\frac{dF}{dt} = -\int (-\eta\mathbf{v} \cdot \nabla^2\mathbf{v} + L_\psi|\vec{\nabla}\mu_\psi|^2 + L_n|\vec{\nabla}(m\frac{v^2}{2} + \mu_n)|^2)\,d\mathbf{r} \leq 0 \tag{15}$$

   Therefore, the present dynamical equations are consistent with the second law of thermodynamics.

The dynamical equations take simpler forms in specific cases. In a solid no drift motion exists. This corresponds to setting $\nu \to \infty$ and $\mathbf{v} = \mathbf{0}$. As a result the equation of motion for ψ is of a closed form:

$$\frac{d\psi}{dt} = L_\psi \nabla^2(-g\nabla^2\psi - \alpha\psi + \beta\psi^3) \tag{16}$$

which is the *Cahn–Hilliard equation (4)*. Even in a fluid polymer mixture this equation is useful, because the nonlinear term can be neglected in the early stage of the phase separation in the polymer system. This will be discussed later.

In a fluid the assumption of incompressibility eliminates the velocity field $\mathbf{v}$. Then we obtain a closed equation for ψ, if inertial terms, the left-hand side, and the first term on the right-hand side of Eq. (14), are neglected. This equation is written as [29].

$$\frac{d\psi}{dt} = L_\psi \nabla^2 \mu_\psi - \vec{\nabla}\psi \int : \mathbf{T}(\mathbf{r}-\mathbf{r}') : (\vec{\nabla}\psi(\mathbf{r}'))\mu_\psi(\mathbf{r}')\, d\mathbf{r}' \tag{17}$$

where $\mathbf{T}$ is the Oseen tensor:

$$\begin{aligned}\mathbf{T}(\mathbf{r}-\mathbf{r}') &\equiv \eta^{-1}\nabla^{-2}(\mathrm{I}-\nabla^{-2}\vec{\nabla} : \vec{\nabla})\delta(\mathbf{r}-\mathbf{r}') \\ &= (8\pi^3\eta)^{-1}\int k^{-2}(\mathrm{I}-\hat{\mathbf{k}} : \hat{\mathbf{k}})e^{i\mathbf{k}\cdot\mathbf{r}-\mathbf{r}'}\, d\mathbf{k}\end{aligned} \tag{18}$$

where $\mathbf{k}$ and $\hat{\mathbf{k}} \equiv \mathbf{k}/k$ are the wavenumber vector and unit wavenumber vector, respectively.

C. Instability of a Mixture

Let us consider a stable binary mixture. The free energy of this system has one minimum which corresponds to the stable homogeneous state. When the system is quenched, the stable state corresponding to $\psi = 0$ of the upper curve of Fig. 2 becomes unstable, because the free energy changes to have a maximum at $\psi = 0$ as shown by the lower curve. Then the mixture moves to new stable states. The arrows indicate the possible direction of this change. One of two directions is statistically chosen so that small domains of two species are randomly distributed in space. Such initial inhomogeneities are amplified by diffusion of particles. This diffusion is different from the ordinary diffusion in the stable state. In a stable state, particles diffuse from dense parts to thin parts. On the other hand in an unstable state particles diffuse from thin parts to dense parts. This is schematically shown in Fig. 3. In this figure the arrows indicate the direction of diffusion, the solid curves indicate the particle density profile in space, and the broken curves indicate the previous density profile. In the unstable state initial density fluctuations grow, whereas they decrease in stable states. Such behavior is consistent with the free energies in Fig. 2. When the initial fluctuations are small, the time evolution of the fluctuation can be treated by the linearized kinetic equations. When the higher-order fluctuation can be neglected, the first term on the right-hand side of Eq. (9) can be neglected. As a result even

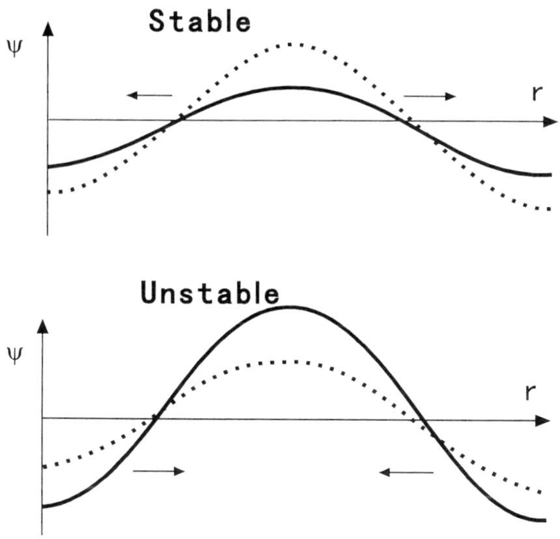

Figure 3 Two kinds of diffusion of particles. "Stable" and "Unstable" indicate stable and unstable states.

for a fluid the time evolution of ψ is described by the *linearized Cahn–Hilliard equation*:

$$\frac{d\psi}{dt} = L_\psi \nabla^2(-g\nabla^2\psi - \alpha\psi) \tag{19}$$

Here we have assumed that $\langle\psi\rangle = 0$, but this assumption is not essential. The above equation can be easily solved by means of Fourier analysis.

$$\psi_{\mathbf{k}}(t) \equiv \int \psi(\mathbf{r}, t)e^{i\mathbf{r}\cdot\mathbf{k}}\, d\mathbf{r} = e^{-k^2(gk^2-1)t}\psi_{\mathbf{k}}(0) \tag{20}$$

The mean average of fluctuation, which is called the *structure function*, is

$$S_k(t) = \langle|\psi_{\mathbf{k}}|^2\rangle = e^{-2k^2(gk^2-1)t}\langle|\psi_{\mathbf{k}}(0)|^2\rangle \tag{21}$$

If ψ is initially random, then $\langle|\psi_{\mathbf{k}}(0)|^2\rangle = \langle|\psi_0(0)|^2\rangle$. Let k_m be the wave-number at which the structure function is largest. Then

$$k_m = \sqrt{\frac{\alpha}{2g}} \qquad \left(\frac{d}{dk}k^2(gk^2 - \alpha) = 0 \text{ for } k = k_m\right) \tag{22}$$

$S_k(t)$ increases exponentially for $k < \sqrt{2}k_m$ and decreases exponentially for $\sqrt{2}k > k_m$. Such an exponential change in the structure function $S_k(t)$ has been sought for alloys, but was not observed. Later, the exponential growth was observed in polymer mixtures. This will be discussed later.

In a short initial period, growths of local density fluctuations saturate. However, this is not a state corresponding to the absolute minimum of the free energy. Small domains or precipitates gradually coarsen until the system reaches a completely phase-separated state. This process is reflected in the peak position k_m of the structure function i.e., k_m gradually decreases. In these stages the linearized approximation of the kinetic equation cannot be applied. A conventional treatment of nonlinear terms of the kinetic equations (16) is a mean-field-like approximation such as

$$\psi^3 \approx A\psi$$

where A is to be determined self-consistently as in usual mean field approximations. For the phase separation, however, transient local phase separations should be suitably taken into consideration [30,31]. Then the coefficient A is time-dependent, resulting in a gradual decrease of k_m as a function of time.

In the late stage of phase separation a sharp domain interface is formed. The width of the interface is of the order of the thermal correlation length $\xi \sim \sqrt{\alpha/g}$. Then for a large domain size the interfacial width may be set equal to zero. This singularity in the domain interface makes it difficult for us to develop a further analytic method, though an idea to hide the nonlinearity in the free energy was used in Ref. 32.

III. GROWTH LAWS OF LENGTH SCALE IN LATE STAGES

In the late stages several coarsening mechanisms of domains are known. In Fig. 4 three typical coarsening mechanisms are illustrated. Fig. 4(a) represents a so-called evaporation–condensation mechanism. Due to the irregularities in the interfaces, the local curvatures of clusters or domains are irregular. Near domain interfaces with smaller curvatures the pressure and the chemical potential are larger than the averages of such quantities. Particles therefore evaporate in such parts and condense at other interfaces with larger curvatures where the pressure and the chemical potential are smaller. This process continues until the system becomes completely phase-separated. Fig. 4(b) represents the coarsening mechanism in fluid. The pressure gradient produced by local curvatures of domain interfaces induces convective motion of the fluid. This motion coarsens domains, too. If two kinds of droplets are interconnected, then such a coarsening continues until the system becomes completely phase-separated. This greatly accelerates the domain growth. In the case where the cluster is not inter-

(a)

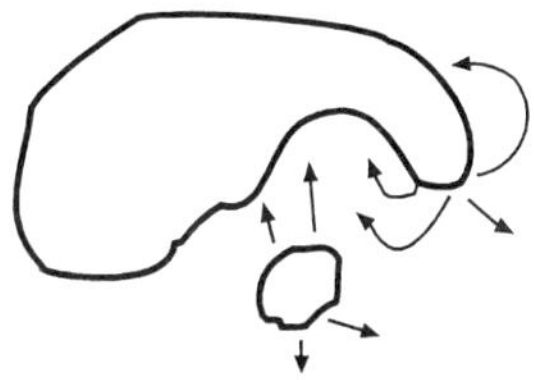

(b)

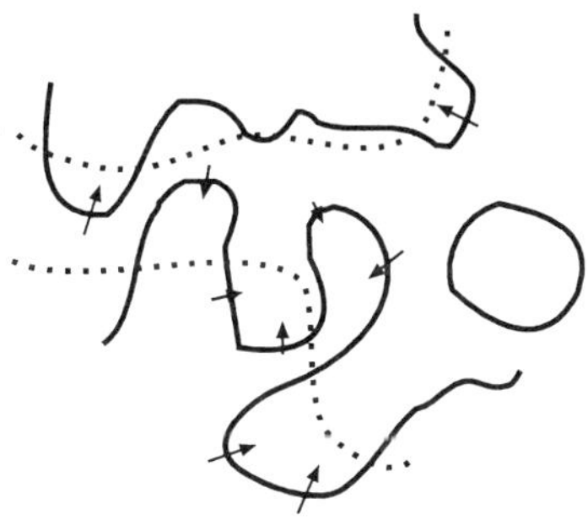

(c)

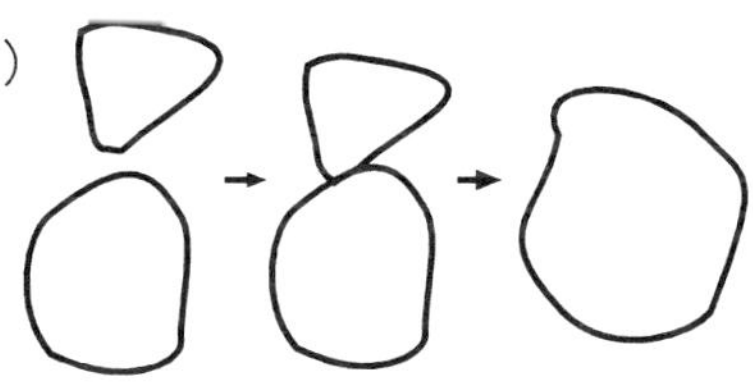

Figure 4 Three typical coarsening mechanisms of domains.

connected, the convective motion induced by the surface tension is not effective in the domain growth. In such a case the domain coarsening depends on the frequency of collisions among isolated domains as shown in Fig. 4(c). Isolated domains, i.e., clusters or droplets, randomly move as Brownian-like particles. If two such clusters collide then they coalesce into a larger one. The growth law of clusters does not depend on the surface tension in this case. For each coarsening mechanism there is a corresponding characteristic growth law. Here we shall review several growth laws. Discussions are only qualitative.

A. Growth by the Evaporation–Condensation Process

To examine the growth of domains at phase separation, it is convenient to introduce an average length scale R. The length scale is defined as the mean radius of the cluster or the mean inverse curvature of interface. Then the average inverse curvature of the interface and the average distance between neighboring clusters are of the order R. Any quantities are then assumed to be scaled using R. For instance, the deviation of the local chemical potential is of the order $1/R$ The local particle current density is then

$$\mathbf{j} \sim M\vec{\nabla}\mu \sim \sigma M/R^2 \tag{23}$$

where M is the particle mobility and $\mu \approx \sigma/R$ is the chemical potential with σ being the surface tension. The growth of the volume of a cluster is proportional to the volume of net gain of particles which condense at interfaces. Therefore, we find that

$$\frac{dR^d}{dt} \sim jR^{d-1} \sim M\sigma R^{d-3} \tag{24}$$

where d is the spatial dimensionality. This gives

$$R^3/M \sim \sigma t \tag{25}$$

If M is a constant independent of R, i.e.,

$$M = M_0 \tag{26}$$

then this gives the well-known Lifshitz–Slyozov $t^{1/3}$ growth law [33]. If the mobility M depends on R this growth law is modified. If the particle can move only on cluster interfaces the mobility is smaller by a factor of $1/R$ than the bulk mobility:

$$M = \frac{S_0}{R} \tag{27}$$

where S_0 is a constant. In this case we observe the $t^{1/4}$ growth law. This growth law cannot be applied to isolated clusters. Generally, the surface diffusion gives rise to the Brownian movement of isolated clusters. Therefore, the surface diffusion contributes to the cluster coagulation through Brownian motion, as illustrated in Fig. 4(c).

B. Growth by Drift Motion of Fluid

In fluid the change in local curvatures of domain interfaces gives rise to the pressure gradient, and as a result gives rise to drift motion. This motion readily prompts growths of clusters. In fluid the diffusion constant D is not that of an individual particle but that of collective motion. This gives the same R-dependence as the Stokes' law, i.e.,

$$D \sim \frac{k_B T}{\eta R^{d-2}} \tag{28}$$

The corresponding mobility is given by

$$M \sim \frac{R^d}{k_B T} D \sim \frac{R^2}{\eta} \tag{29}$$

By substituting Eq. (29) into Eq. (25) we obtain a growth law

$$R \sim \frac{\sigma}{\eta} t \tag{30}$$

which is the surface-tension-driven and dissipation-controlled growth law [34] and was well established first in simple liquid mixtures near the critical point [35–39]. This growth law is also observed for polymer mixtures.

C. Growth of Isolated Clusters

Evaporation and condensation of particles within the same cluster only shifts its center of gravity, producing random motion. Such a random motion also occurs in other cases. For instance, isolated clusters floating in a fluid also move randomly. Due to these random motions clusters coalesce into larger ones. Such mechanisms are important when the surface tension is not the relevant driving force. In such cases the growth of clusters occurs once in each collision. Let D be the diffusion constant of a cluster. Then the time τ from collision to collision is about

$$\tau \sim \frac{R^2}{D} \tag{31}$$

and during this time interval the cluster grows by a factor of the order R due to the coalescence. Therefore,

$$\frac{dR}{dt} \sim \frac{\nabla R}{\tau} \sim \frac{D}{R} \tag{32}$$

Namely,

$$\frac{D}{R^2} \sim t \tag{33}$$

If we put $D \propto 1/R^d$, which corresponds to the bulk diffusivity of particles in a solid, we obtain $R \propto t^{1/(d+2)}$. In a fluid we may use Eq. (28). Then we obtain

$$R \sim \left(\frac{k_B T}{\eta}\right)^{1/d} t^{1/d} \tag{34}$$

which is the cluster coagulation by Brownian motion [40]. There are other possibilities for growth laws. For instance the surface diffusivity in a solid gives $R \propto t^{1/(d+3)}$.

D. Growth Controlled by Inertia in Fluid

When R becomes larger or if the viscosity is small, the effect of inertia on the fluid motion and therefore on the growth of droplets becomes dominant. In such a case the diffusional current Eq. (23) is not effective. Neglecting the dissipation term in Eq. (10) and approximating $\vec{\nabla} P \sim \sigma R^{-2}$, we find that

$$j = \rho v \sim \sigma t / R^2 \tag{35}$$

This gives

$$R \sim \left(\frac{\sigma}{\rho}\right)^{1/3} t^{2/3} \tag{36}$$

which is the surface-tension-driven and inertia-controlled growth law [2,41].

For isolated droplets the effect of the inertia is also important. If clusters are floating in vapor, then the cluster diffusivity does not obey Stokes' law. The average cluster velocity is determined by the equipartition law

$$\rho R^d \left(\frac{R}{t}\right)^2 \sim k_B T \tag{37}$$

Here R/t is regarded as the average velocity of clusters of size R. From this we obtain

$$R \sim \left(\frac{k_B T}{\rho}\right)^{1/(d+2)} t^{2/(d+2)} \tag{38}$$

which is applied to clusters floating in vapor [2,41].

Table 1 Several Growth Laws in Binary Solid and Fluid Mixtures. Thermal and Surface Driving Forces Respectively Indicate Thermal Fluctuation and Surface Tension. "Bulk" and "Surface" Mobility Indicate Solid Mobilities. Fluid Mobility Is Based on Stokes' Law. "Inertial" Indicates Growth in the Case Where the Inertia Is Effective

	Dissipative			
	Mobility			
Driving force	Fluid	Bulk	Surface	Inertial
Thermal	$1/d$	$1/(d+2)$	$1/(d+3)$	$2/(d+2)$
Surface	1	1/3	1/4	2/3

The crossover between the dissipative and the inertial growth mechanisms occurs where two growth laws give the same cluster size R. In Table 1 we summarize several growth exponents in solids and liquids. Some of them are discussed in Ref. 2.

IV. DYNAMICAL SCALINGS

The dimensional-like analyses in Sec. III are based on a more fundamental concept. If R is much larger than any microscopic quantities, then spatial patterns at any time may be self-similar, i.e., patterns scaled by the length scale $R(t)$ cannot be statistically distinguishable. It is also expected that time-dependent processes themselves may be scale invariant. These properties in the late stage of the phase separation are called *dynamical scalings*.

A. Scaling of the Same Time Correlation Function

The dynamical scaling is often represented using the *structure function*:

$$S_k(t) = \int G(r, t)e^{-i\mathbf{k}\cdot\mathbf{r}}\, d\mathbf{r} \tag{39}$$

where

$$G(r, t) \equiv \langle \psi(\mathbf{r}, t)\psi(0, t)\rangle \tag{40}$$

is the same time correlation function. The dynamical scaling for the structure function is represented as

$$S_k = R(t)^d \bar{S}(kR(t)) \tag{41}$$

together with the power growth law of the length scale $R \propto t^{\alpha}$. An implicit idea of such a scaling as Eq. (41) is found in the evaporation–condensation model [32,33] and the cluster diffusion-reaction model [40], which gave power growth laws. In order to analyze the kinetic equation [32] and in order to analyze numerical simulation [42] this scaling assumption was used. This scaling assumption has been examined in a large number of articles and it is considered to be an appropriate assumption. Similar scalings hold for other physical quantities. The evaluation of the kinetic equation with this scaling leads to the same growth laws as those obtained by the dimensional analysis. The structure function itself contains much information about the phase separations. Although the analytic calculation of the structure function is not tractable, there are a few universal properties of the structure functions in the late stage of the phase separation. Since the order parameter ψ is the conserved quantity and the contribution from the thermal fluctuation can be neglected the structure function at a small wavenumber k should vanish

$$\lim_{x\to 0} \bar{S}(x) = 0 \tag{42}$$

At large wavenumbers the structure function should also vanish because of the sum rule $\sum_k S_k = \langle \psi(\mathbf{r})^2 \rangle =$ constant:

$$\lim_{x\to\infty} \bar{S}(x) = 0 \tag{43}$$

Expanding $\bar{S}(x)$ by x we have

$$\bar{S}(x) = Ax^{\delta} + \ldots \qquad (x < 1) \tag{44}$$

For a certain kind of kinetic equation, e.g., for the Cahn–Hilliard equation, it was shown that $\delta \geq 4$ [43]. The exponent δ should be equal to 4 [44–46]. However, this does not hold for fluids. In Sec. II we have shown that the velocity field $\mathbf{v}$ obeys a continuity equation. Therefore, $\psi(t)\mathbf{v}(t)$ is of the form $\psi(t)\vec{\nabla}\int^t G(t')\,dt'$, and cannot be written in the form $\vec{\nabla}\phi$. For this reason the expansion of the structure function may have an x^2 term.

The expansion of the structure function at large wavenumbers is known to be of the form [47]

$$\bar{S}(x) = x^{-d}(B_1 x^{-1} + B_2 x^{-3} + B_3 x^{-5} + \ldots) \qquad (x > 1) \tag{45}$$

The expansion starts with x^{-d-1}, and $\bar{S}x^{d+1}$ has no odd power of x. This expansion is due to the fact that the width of the domain interface can be regarded as being sharp in the large limit of the length scale R. The first term of the expansion can be understood as follows. The correlation in real space $\langle \psi(\mathbf{r})\psi(0) \rangle$ has nontrivial values only near the sharp interfaces. The volume of this region is equal to (interfacial area $\times$ width r). The contribution to the correlation function from this region is proportional to r (= ratio of this volume to the total volume). The Fourier component of r gives k^{-d-1}. This is called the *Porod law*. A simple form of $\bar{S}(x)$ which satisfies the lowest orders of expansions at small x and large x is [44–50]

$$\bar{S}(x) = \frac{(\alpha + \beta)x^{\alpha}}{\beta + \alpha x^{\alpha+\beta}} \tag{46}$$

We have set $\bar{S}(1) = 1$ and $d\bar{S}/dx|_{x=1} = 0$. Constants α and β represent suitable values, e.g., $\alpha = 4$ and $\beta = d + 1$. In Fig. 5 (46) $\bar{S}(x)$ for several values of α and β are shown. It is difficult for us to calculate a general form of the scaling structure function, which reflects composition dependence and other properties of the system, such as fluid or solid (see for instance Refs. 51 and 52, and references therein). In particular, for large compositions only an

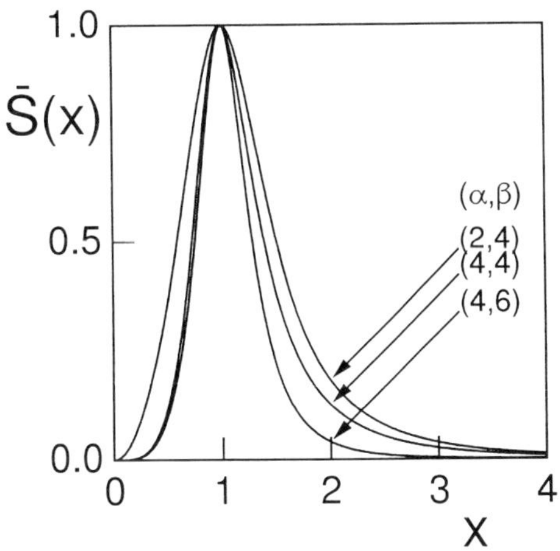

Figure 5 Plots of scaling structure function (46).

approximate theory [53] is known, which can be compared to the experimental data rather satisfactorily [54].

Although the scaling assumption Eq. (41) is known to be appropriate for several systems, this assumption relies on the assumption that the interfacial thickness is very narrow. Therefore, if this assumption is not satisfied, then Eq. (41) may not hold. In fact there is an example of the breakdown of such a *single length* scaling as (41) [55].

B. Scaling of Two-Time Correlation Function

Another interesting concept, which can also be examined by experiment, is the *autocorrelation function* defined by

$$G(r, t, t') = \langle \psi(\mathbf{r}, t)\psi(0, t') \rangle \tag{47}$$

or its Fourier component:

$$S_k(t, t') = \int G(r, t, t')e^{-i\mathbf{k}\cdot\mathbf{r}}\, d\mathbf{r} \tag{48}$$

These quantities obey a scaling similar to that of the structure function (41) [44,56]. Let us explain this. In Fig. 6 different patterns of the same system at different times are shown. For simplicity we show the case of an assembly of isolated clusters. A simple dynamical assumption states that these patterns are statistically similar. Therefore, the structure function in each time satisfies the scaling (41) with the same function $\bar{S}(x)$. We now explain another type of scaling assumption. This is the assumption of similarity between

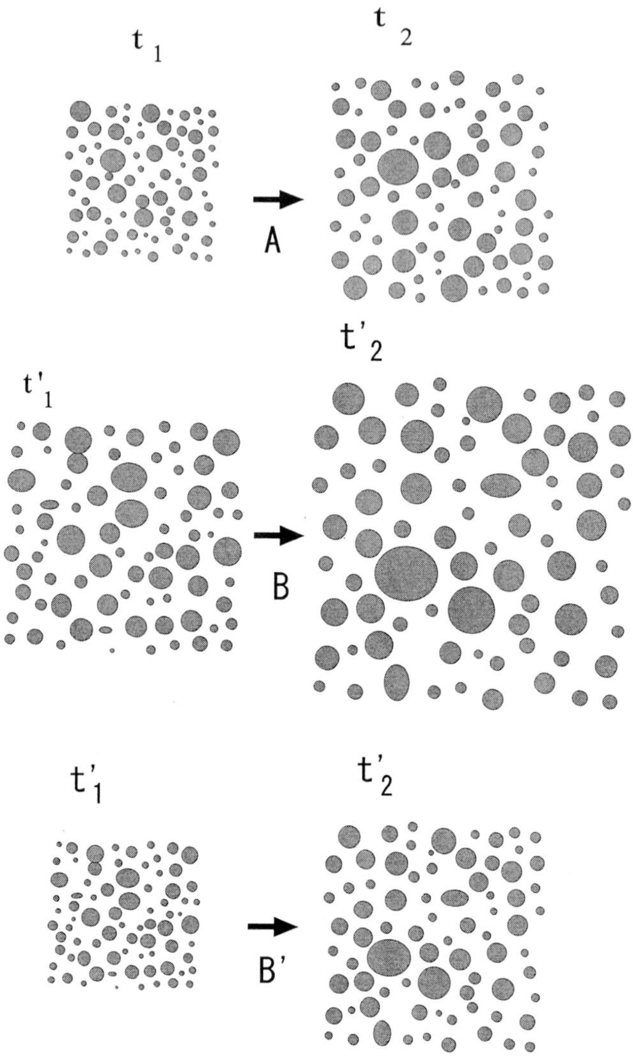

Figure 6 Explanation of two-time scaling.

transitions. Here the transition means the dynamical transition between states at different times t_1 and time $t_2(> t_1)$. The processes labeled by A and labeled by B represent transitions between different sets of times. With a suitable scaling the pattern at time t_1' of the process B can be set statistically identical to the pattern at time t_1. This is represented by the pattern at t_1' of the process labeled by B$'$. Using the same scaling factor we scale the pattern at t_2 of the process B. This is the pattern at time t_2' of the process B$'$. If the pattern at time t_2' of the process B$'$ also has the same length scale as the pattern at time t_2 of the process A, then we assume that the transition of the process A and the transition of the process B$'$ are not statistically distinguishable from each other. Then the auto-correlation function (47) or (48) should be the same both for the process A and the process B ($= B'$). This holds only if the auto-correlation function $G(\mathbf{r}, t, t')$ is a function of only $r/R(t)$ and $R(t)/R(t')$. Thus

$$G(r, t, t') = \bar{G}'\left(\frac{r}{R(t)}, \frac{R(t)}{R(t')}\right) = \bar{G}\left(\frac{r}{R(t)}, \frac{r}{R(t')}\right) \tag{49}$$

Here, the set of independent variables $r/R(t')$ and $R(t)/R(t')$ is changed into the set of independent variables $r/R(t)$ and $(r/R(t)) \times (R(t)/R(t')) = r/R(t')$ to transform the second side into the last side in (49). Since the strength of the order parameter ψ is independent of time t, the function $\bar{G}$ has no extra coefficient depending on time nor the length scale R. The correlation function for $r = 0$ is a function of only $R(t)/R(t')$. If the two times are greatly different from each other, e.g., $t \gg t'$, the correlation vanishes between two time points even at the same spatial position. Then we find that

$$\lim_{t/t' \to \infty} G(0, t, t') = \left(\frac{R(t')}{R(t)}\right)^{\lambda} = \left(\frac{t'}{t}\right)^{a\lambda} \to 0 \tag{50}$$

Here the exponent λ cannot be obtained by a simple dimensional analysis [57]. The Fourier coefficient (48) is written as

$$S_k(t, t') = R^d \bar{S}_3\left(kR(t), \frac{R(t)}{R(t')}\right) = R(t)^d \bar{S}(kR(t), kR(t')) \tag{51}$$

The same limiting procedure as (50) may be possible for $S_k(t, t')$. It should be noted that the scaling assumption for the auto-correlation function (51) is more general than the scaling assumption for the single time correlation function (41). There is an example for which the scaling for the single time correlation holds but it does not hold for the auto-correlation function [58]. This states that the scaling for the pattern and scaling for the process are not the same.

V. ENERGY DISSIPATION IN A FLUID

When the inertial term is not effective, the surface energy released by the coarsening is dissipated simultaneously. On the other hand, if the inertial term cannot be neglected then the released surface energy by coarsening is dissipated at larger wavenumbers. Therefore, there exist two different characteristic length scales. We here examine roles of these two length scales in the coarsening process in fluid. For simplicity we write the hydrodynamic equation in the following form

$$\rho \frac{d\mathbf{v}_\mathbf{k}}{dt} + \eta k^2 \mathbf{v}_\mathbf{k} = \mathbf{F}_\mathbf{k} \tag{52}$$

where $\mathbf{F}$ is the force term. If the friction term is small, then it can be treated as a perturbation. Therefore, we set

$$\rho \frac{d\mathbf{v}_\mathbf{k}}{dt} = \mathbf{F}_\mathbf{k} \qquad \text{for } \nu k^2 t \quad (= \nu k^2 u/\dot{u}) < 1 \tag{53}$$

The energy dissipation is then evaluated as

$$\dot{\varepsilon} = \sum_{k<k_l} \eta k^2 |\mathbf{v}_\mathbf{k}|^2 \sim \sum_{k<k_l} \eta \rho^{-2} k^2 t^2 |\mathbf{F}_0|^2 \tag{54}$$

where we have set $\mathbf{F}_\mathbf{k} = \mathbf{F}_0$. Here k_l is the upper limit of the integration and it is assumed to be smaller than k_c where k_c is the critical wavenumber determined by $\nu k_c^2 t = 1$. The fact that $\mathbf{F}_0$ does not vanish can be shown as follows. We may consider only the nonlocal gradient term of the chemical potential. Then we can obtain $\mathbf{F}_0 = \sum_\mathbf{q} \mathbf{q}^3 |\psi_\mathbf{q}|^2$. We find that $\langle \mathbf{F}_0 \rangle = 0$, and that $\langle |\mathbf{F}_0|^2 \rangle = \sum_{\mathbf{q},\mathbf{q}'} \mathbf{q}^2 \cdot \mathbf{q}'^3 \langle \Delta|\psi_\mathbf{q}|^2 \Delta|\psi_{\mathbf{q}'}|^2 \rangle$, where $\Delta X \equiv X - \langle X \rangle$. If we may assume that $\langle \Delta|\psi_\mathbf{q}|^2 \Delta|\psi_{\mathbf{q}'}|^2 \rangle \sim \langle (\Delta|\psi_\mathbf{q}|^2)^2 \rangle \delta_{\mathbf{q},\mathbf{q}'}$, then we find $\langle |\mathbf{F}_0|^2 \rangle \neq 0$.

We find that

$$\dot{\varepsilon} \sim |\mathbf{F}_0|^2 \eta \rho^{-2} t^2 k_l^{d+2} \sim |\mathbf{F}_0|^2 \eta^{-1} k_l^{d+2} / k_c^4 \qquad \text{for } k_l < k_c \tag{55}$$

The upper limit of the integration k_l is determined as follows. Let the average length scale be R. When the surface tension is effective, the total energy density is given by $\varepsilon \sim \sigma/R$, and the force F in a real space is of the order $\varepsilon/R \sim \sigma R^{-2}$. Therefore, in Fourier space we have $|\mathbf{F}_0|^2 \sim \sigma^2 R^{d-4}$. The energy dissipation is also given by $\varepsilon/t \sim \sigma/(Rt)$. Substituting these quantities into (55) we have

$$(k_l R)^{d+2} \sim \frac{\eta}{\sigma t} (R k_c)^4 R \qquad \text{for } k_l < k_c \tag{56}$$

The wavenumber k_c is not observable, but k_l is the cut-off of the velocity field and is observable and gives another length scale. That is, the released surface energy at the scale R is dissipated at a smaller scale $1/k_l$: The kinetic energy remains in the fluid for this period. Due to this remaining kinetic energy, it is possible that clusters may separate into pieces again. If such a re-separation really occurs, then the limit of the wavenumber k_l of the remaining kinetic energy should be the characteristic length scale R. Then we observe that $R \sim 1/k_l$ and $\varepsilon \sim \sigma k_l$. By setting $k_l R = 1$ in the relation (56) we obtain

$$R \sim \left(\frac{\sigma \nu}{\rho}\right)^{1/5} t^{3/5} \tag{57}$$

This growth law was not discussed in Sec. III. This type of growth law has not been observed and seems unlikely to be observed. Generally, by setting $R \propto t^a$, the Reynolds number $R_e \equiv \nu R^2/t$ increases if $a > 1/2$. The inertial term becomes effective for Reynolds number larger than 1. Three growth laws $R \propto t$, $t^{2/3}$, and $t^{3/5}$, give the same value of the length scale

$$R_c = \frac{\rho \nu^2}{\sigma} \tag{58}$$

for $R_e = 1$ (crossover point).

Before closing this subsection we notice the importance of the inertia in two dimensions. In certain relaxational systems the energy dissipation is crucial to the coarsening mechanisms [59–61]. In a fluid the dissipation of kinetic energy may also be crucial to the coarsening mechanisms in a particular case. By rewriting Eq. (15) with the use of the equations of motions (7)–(9) and (14) we obtain

$$\begin{aligned}\frac{dF}{dt} = &-\eta^{-1}\int\left(\rho\frac{d\mathbf{v}}{dt} + n\vec{\nabla}\mu_n + \psi\vec{\nabla}\mu_\psi\right)\nabla^{-2}\left(\rho\frac{d\mathbf{v}}{dt} + n\vec{\nabla}\mu_n + \psi\vec{\nabla}\mu_\psi\right)d\mathbf{r}\\ &- L_n^{-1}\int\left(\frac{dn}{dt} + \vec{\nabla}\cdot n\mathbf{v}\right)\nabla^{-2}\left(\frac{dn}{dt} + \vec{\nabla}\cdot n\mathbf{v}\right)d\mathbf{r}\\ &- L_\psi^{-1}\int\left(\frac{d\psi}{dt} + \vec{\nabla}\cdot \psi\mathbf{v}\right)\nabla^{-2}\left(\frac{d\psi}{dt} + \vec{\nabla}\cdot \psi\mathbf{v}\right)d\mathbf{r}\end{aligned} \tag{59}$$

Here we examine the first term on the right-hand side. This is rewritten by means of Fourier coefficients as

$$\int \eta^{-1}k^{-2}\left|\rho\frac{d\mathbf{v}}{dt} + n\vec{\nabla}\mu_n + \psi\vec{\nabla}\mu_\psi\right|^2 k^{d-1}\,dk = \int \eta|\mathbf{v_k}|^2 k^{d+1}\,dk. \tag{60}$$

The integration on the right-hand side of the above equation does not diverge. In order for the integration at small wavenumbers on the left-hand side to converge, one demands, in the limit $k \to 0$, that

$$\rho\frac{d\mathbf{v}}{dt} + n\vec{\nabla}\mu_n + \psi\vec{\nabla}\mu_\psi \to 0 \qquad \text{for } d < 2 \tag{61}$$

and

$$\left| \rho \frac{d\mathbf{v}}{dt} + n\vec{\nabla}\mu_n + \psi\vec{\nabla}\mu_\psi \right| < \infty \qquad \text{for } d > 2 \tag{62}$$

Generally, $n\vec{\nabla}\mu_n + \psi\vec{\nabla}\mu_\psi$ may be a nonconserved quantity. If $\lim_{k\to 0} |n\vec{\nabla}\mu_n + \psi\vec{\nabla}\mu_\psi|^2$ remains finite, then this means that the inertial term cannot be neglected in two dimensions.

VI. PHASE SEPARATION OF POLYMER MIXTURES

Most of the discussion presented above can be applied to the case of polymers. However, several physical quantities appearing in the discussion, e.g., the free energy and the viscosity, have not yet been specified for polymer. We here examine these quantities for polymers.

A. Free Energy

The free energy of the polymer system is described by the Flory–Huggins–de Gennes free energy [62–64]. But this is not the free energy in the sense of coarse-graining. The coarse-grained free energy generally has different system-size-dependent coefficients than the original Flory–Huggins–de Gennes. Our primary aim is not to describe the atomic level behavior, but molecular level behavior. Therefore, we use Eq. (3), choosing a suitable set of coefficients g, α, and β. The minimum length scale of the coarse-graining is the radius of the gyration R_g which is regarded as the effective molecular size. The explicit form of the free energy is crucial to describing the *initial stage of the phase separation*. We therefore focus our attention on the free energy suitable for the initial stage of the phase separation. Under such a consideration we may assume that the coefficients of coarse-grained free energy for polymer mixtures are such that

$$g = O(1) \qquad \alpha \sim O\left(\frac{\epsilon}{N}\right) \qquad \beta \sim O\left(\frac{1}{N}\right) \tag{63}$$

where ϵ is not the same as ϵ in Eqs. (54) and (55) but is given by

$$\epsilon \equiv \frac{|T - T_c|}{T_c} \tag{64}$$

and where T_c is the critical temperature. Here $O(1)$ means a quantity of atomic size. The first condition of Eq. (63) is a direct consequence of the Flory–Huggins–de Gennes free energy. The second and the third are due to the demand that the length scale of the characteristic fluctuation is of the order of an effective molecular size, which is given by the gyration radius of a polymer molecule in a dense solution or in a melt [63,65]:

$$R_g = O(\sqrt{N}) \tag{65}$$

We mean therefore that short scale fluctuations are renormalized. Let us show that the above assumption gives a suitable fluctuation of the order parameter. Outside the coexistence curve, both α and β are positive. The characteristic lenth scale of the concentration fluctuations outside the coexistence curve is then given by equating the third term and the fourth term of the free energy (3) as

$$\xi \sim (N/\epsilon)^{1/2} \sim R_g \epsilon^{-1/2} \tag{66}$$

The fluctuation amplitude of the order parameter for the wavenumber $k = 1/\xi$ is, with the *Ornstein–Zernike* approximation,

$$\langle |\psi_{\mathbf{k}}|^2 \rangle |_{k=1/\xi} \sim \frac{1}{g\xi^{-2} + \alpha} \sim \left(\frac{\xi}{r_0} \right)^2 \tag{67}$$

where

$$\psi_{\mathbf{k}} \equiv \frac{1}{\sqrt{V}} \int \psi(\mathbf{r}) e^{i\mathbf{k}\cdot\mathbf{r}}\, d\mathbf{r} \tag{68}$$

and r_0 represents a length of atomic size. (See [66–68] for more details of static properties of a general system near the critical point.) Dividing the system into cells with volume ξ^d we define the average of the order parameter in a cell:

$$\bar{\psi} \equiv \xi^{-d} \int_{\xi^d} \psi(\mathbf{r})\, d\mathbf{r} \tag{69}$$

Even if the correlation between ψ is larger than $1/k$, the correlation length between $\psi e^{i\mathbf{k}\cdot\mathbf{r}}$ is not larger than $O(1/k)$ due to the factor $e^{i\mathbf{k}\cdot\mathbf{r}}$. Therefore,

$$\begin{aligned} \langle |\psi_{\mathbf{k}}|^2 \rangle |_{k=1/\xi} &\sim \frac{1}{V} \xi^{2d} \sum_{j,j'} \langle \Delta\bar{\psi}_j \Delta\bar{\psi}_{j'} \rangle \delta_{j,j'} \\ &\sim \frac{1}{V} \xi^{2d} (V/\xi^d) \langle \Delta\bar{\psi}^2 \rangle \sim \xi^d \langle \Delta\bar{\psi}^2 \rangle \end{aligned} \tag{70}$$

where $\Delta\bar{\psi} \equiv \bar{\psi} - \langle \bar{\psi} \rangle$ and therefore

$$\langle \Delta\bar{\psi}^2 \rangle \sim \xi^{-d} \langle |\psi_{\mathbf{k}}|^2 \rangle |_{k=1/\xi} \sim \xi^{2-d} \sim \left(\frac{N}{\epsilon} \right)^{(2-d)/2} \tag{71}$$

Inside the coexistence curve we also assume that $\alpha < 0$ and $\alpha \sim \epsilon/N$. Then we are dealing with the fluctuation with a length scale larger than $R_g \sim \sqrt{N}$. Smaller-scale motion is assumed to be averaged out.

Far from the critical temperature the above evaluations of coefficients are not appropriate to describing the late stage of the phase separation. In the *late stage of the phase separation* sharp interfaces are created. The shape of a polymer molecule near the domain interface is strongly deformed. The width of the interface is not semi-macroscopic of order R_g but is microscopic of the order of the atomic distance r_0 due to a strong repulsion between monomers [69–71]. The evaluation of interfacial energy cannot be performed by the present simplified free energy of polymer. The energy required to create a surface is given by [72–75]

$$F_s = g \int \left(\frac{d\psi}{dx} \right)^2 d\mathbf{r} \tag{72}$$

Let the width of the domain interface be λ. Then the above interfacial free energy has a $1/\lambda$ dependence. If we assume that $g = O(1)$, $-\alpha \sim \beta = O(1/N)$, then we find that $\lambda \sim R_g \sim N^{1/2}$. Therefore, the surface energy and the surface tension σ depend on N as $\sigma \sim 1/\lambda \sim N^{-1/2}$. This obviously contradicts the physical consideration that the surface tension should be a quantity of $O(1)$ far from the critical temperature. The shape of a polymer molecule near the domain interface is strongly deformed by strong segregation. Such a situation is not taken into consideration in our free energy. The formation of a sharp interface is suitably represented by choosing an appropriate surface tension $\sigma = O(1)$.

B. Viscosity and Diffusivity

The dynamical aspect of polymers is characterized by the kinetic viscosity ν or the diffusion coefficient D. There is a standard theory for the dependence of such quantities on the size of a polymer molecule, i.e., the polymerization index N. Let us here summarize the necessary quantities. By the so-called *reptation mechanism* [76,77] a polymer chain is assumed to move only along its own curvilinear shape. The friction for this motion is then $\gamma_{\text{rep}} \propto N$. Then the diffusion coefficient of the curvilinear motion, D_{rep}, is proportional to $1/\gamma_{\text{rep}} \propto 1/N$. Then the time needed for a molecule to move a distance of the same order as the molecular length L is

$$\tau_N \sim L^2/D_{\text{rep}} \propto N^3 \qquad (L \propto N) \tag{73}$$

In this time interval the center of gravity of the molecule moves for a distance of the order of its own radius $R_G \propto N^{1/2}$ and the molecule loses the memory of its previous position. Therefore, the diffusion constant D_N of the center of gravity is estimated as

$$D_N \sim \frac{R_g^2}{\tau_N} \sim D_0 N^{-2} \tag{74}$$

where D_0 is the quantity corresponding to the diffusion constant of a small molecule. This can also be written in the same way as Eq. (28) by using the corresponding kinetic viscosity ν_N as

$$D_N \sim \frac{k_B T}{\rho \nu_N R_G^{d-2}} \propto \frac{k_B T}{\rho \nu_N N^{(d-2)/2}} \tag{75}$$

Therefore, the kinetic viscosity ν_N is given by:

$$\nu_N \sim \nu_0 N^{(6-d)/2} \tag{76}$$

where ν_0 is the corresponding kinetic viscosity for small molecule. It is well known that the experimental observation for the N-dependence on τ_N is slightly different from (73). However, we disregard such a difference because qualitatively this effect is insignificant.

C. Initial Fluctuation and Linear Approximation

Let ℓ_0 be the correlation length associated with ψ before quenching and ℓ be the length scale at which the initial instability occurs. We consider the case where $\ell \approx \ell_0 \approx \xi$. In the critical phenomena, fluctuations should be smaller than the average value so that the mean field approximation can be used. By remembering that $\langle\bar{\psi}\rangle \approx \sqrt{\epsilon}$ and using (71) we find that the mean-field approximation can be used in the case

$$\langle\bar{\psi}\rangle^2 \sim \epsilon = R_G^2 \xi^{-2} \gg \langle\Delta\bar{\psi}^2\rangle \sim \xi^{2-d} \tag{77}$$

This is transformed into

$$\epsilon^{2-d/2} \gg R_G^{2-d} \sim N^{1-d/2} \tag{78}$$

This is the *Ginzburg criterion* [78–80], and was applied to the dynamics of phase separation [81]. For a small particle system Eq. (78) simply gives $d > 4$ as $\epsilon \to 0$, i.e., the mean-field approximation is valid for dimensions higher than four. In polymer systems, however, (78) is satisfied for $d > 2$ if ϵ is not very small.

Consider the validity of the linearization of the kinetic equation to describe the dynamics. In order for the nonlinear fluctuation to be neglected we demand that:

$$\langle \Delta \bar{\psi}^2 \rangle \gg \langle \Delta \bar{\psi}^4 \rangle \approx \langle \Delta \bar{\psi}^2 \rangle^2 \tag{79}$$

Let us assume that the initial state of the phase separation is the thermal equilibrium state and Eq. (71) is valid. Then Eq. (79) is equivalent to

$$d > 2 \tag{80}$$

For $d > 2$ and for a large N the nonlinear fluctuation of ψ can be neglected in the initial stage of the phase separation. The two-dimensional case is marginal.

The estimation of the fluctuation of the initial velocity field is easier. Using the equipartition law, we have

$$\langle \bar{v}^2 \rangle \approx \frac{k_B T}{\rho \xi^d} \sim \frac{k_B T}{m(\xi/r_0)^d} = v_0^2 (r_0/\xi)^d \tag{81}$$

where $\bar{v}$ is the average velocity in the volume ξ^d. Therefore initial velocity fluctuations can be neglected if ξ is large in any dimensions.

According to these evaluations, fluctuations of ψ of any order cannot be neglected in two-dimensional systems, $d = 2$. On the other hand, in three-dimensional systems the lowest-order fluctuation is largest for $\xi \gg r_0$. This condition is generally satisfied for polymers, but is not satisfied for small molecular systems far from the critical point. This is the reason why the exponential growth of the structure function (21) is observed not in alloys [82–85] but in polymer mixtures [86–95].

D. Late Stage of Phase Separation in Polymer Systems

In order to compare a small molecule system and a polymer system we now scale the hydrodynamic equation by both the correlation length ξ and the molecular size N. Let us introduce the following reduced time and space:

$$\tilde{\mathbf{r}} \equiv \frac{\mathbf{r}}{\xi} \qquad \tilde{t} \equiv \frac{t}{\tau} \tag{82}$$

Here τ means τ_N. Using these dimensionless variables we rewrite the hydrodynamic equation as

$$\frac{d\tilde{\mathbf{v}}}{d\tilde{t}} = -(\tilde{\mathbf{v}} \cdot \tilde{\nabla})\tilde{\mathbf{v}} + \tau \nu \xi^{-2} \tilde{\nabla}^2 \tilde{\mathbf{v}} + \frac{\tau^2}{\xi} \frac{\psi_0 \mu_0}{\rho \xi} \tilde{\mathbf{F}} \tag{83}$$

where $\tilde{\mathbf{v}} \equiv d\tilde{\mathbf{r}}/d\tilde{t}$, $\tilde{\nabla} \equiv \xi \nabla$ and $\tilde{\mathbf{F}}$ is the scaled force term represented by scaled variables. This equation is further transformed into a reduced form:

$$\frac{d\tilde{\mathbf{v}}}{d\tilde{t}} = -(\tilde{\mathbf{v}} \cdot \tilde{\nabla})\tilde{\mathbf{v}} + \hat{\nu}^2 \hat{\xi}^{d-2} [\tilde{\nabla}^2 \tilde{\mathbf{v}} + \tilde{\mathbf{F}}] \tag{84}$$

where $\hat{\xi} \equiv r_0^{-1}\xi, \hat{\nu} \equiv \nu_0^{-1}\nu_N$. Here coefficients of Eq. (83) have been evaluated as $\tau\nu\xi^{-2} = \nu/D_T = \rho\nu^2\xi^{d-2}/k_B T \sim \hat{\nu}^2 m r_0^{-d} r_0^4/t_0^2 \xi^{d-2}(k_B T)^{-1} \sim \hat{\nu}^2(\xi/r_0)^{d-2}\ [m(r_0/t_0)^2/k_B T] = \hat{\nu}(\xi/r_0)^{d-2}$, $(\tau^2/\xi)(\psi_0\mu_0/\rho\xi) = (\tau^2/\rho\xi^{d+2})k_B T = (\tau^2/m r_0^{-d}\xi^{d+2})m(\frac{r_0}{t_0})^2 = \frac{\tau^2}{t_0^2}(\xi/r_0)^{-d-2} = \frac{\nu_0^2}{D_T^2}(\xi/r_0)^{-d+2} >= \hat{\nu}^{-2}(\frac{\nu}{D_T})^2(\xi/r_0)^{-d+2} \sim \hat{\nu}^2(\xi/r_0)^{d-2}$. These evaluations are only qualitative. The reduced hydrodynamic equation (84) clearly shows how the polymerization index N and the thermal correlation length ξ affect the hydrodynamic behavior. Notice that the reduced kinetic viscosity $\hat{\nu}$ appears in the square form and therefore the hydrodynamic equation is purely dissipative for large N:

$$-\nu\nabla^2\mathbf{v} = \mathbf{F}$$

The inertial term becomes effective only when it becomes larger than the dissipative term:

$$\left|\frac{d\tilde{\mathbf{v}}}{d\tilde{t}}\right| \gg C\left|\tilde{\nabla}^2\tilde{\mathbf{v}}\right| \tag{85}$$

where $C \equiv \hat{\nu}^2\hat{\xi}^{d-2}$. The critical length corresponding to (58) is given by $\tilde{R}_c \sim C$. Therefore, if $C \gg 1$, then the growth law is dissipative when the surface tension is a relevant driving force. But this analysis neglecting the inertia does not hold in two-dimensional systems if the driving force has a long wavelength component (see Eq. (61)).

There are many studies of phase separations in two-dimensional fluids both numerical [28,96–106], and experimental (in polymer system) [107–116]. Eq. (14) is supported by many numerical studies [28,96,97,103–106]. One experiment for polymer mixtures [115] exhibits the growth law $R \propto t^{0.5}$ for off-critical concentration in two dimensions, and another experiment [116] exhibits growth law (36). These two experiments are interesting because their growth laws are those for ideal two-dimensional fluids. Generally, in a real system it is difficult to achieve an ideal two-dimensional system because of the interaction between fluid and wall. If the friction between fluid and wall is effective, then the growth laws are the same as those in a simple diffusional system (solid).

E. Short Summary

We here summarize differences in phase separations between small molecule systems and polymer systems.

1. *Initial fluctuations*: For $\hat{\xi}^{2-d} \gg 1$ the initial fluctuation is described by the exponential growth (21). Otherwise, the initial fluctuation is nonexponential. For a polymer $\hat{\xi} \gg 1$, whereas for a simple liquid $\hat{\xi} \sim 1$ except near the critical point.

2. *Growth laws*: For $\hat{\nu}^2\hat{\xi}^{2-d} \gg 1$ the growth laws are dissipative. Otherwise, growth laws are inertial. For very large length scales, however, any growth laws are inertial. For a polymer $\hat{\nu} \gg 1$, whereas $\hat{\nu} \sim O(1)$ for a simple liquid. This criterion cannot be applied to the two-dimensional fluid if the driving force has non-vanishing components with small wavenumbers, i.e., $\lim_{\tilde{k}\to 0} \tilde{\mathbf{F}}_{\tilde{\mathbf{k}}} \neq 0$. In this case the growth laws are expected to be inertial in two dimensions.

VII. SUMMARY

We have reviewed the dynamics of phase separation of binary mixtures. Several growth laws for binary systems are shown. A brief discussion on the dynamical scaling was presented. Two different types of dynamical scaling, for pattern and for process, are discussed. These two types of dynamical scalings can be experimentally examined. Especially for polymer mixtures, the length scale are generally large. Then the patterns can be observed in real space. This makes it possible to obtain the auto-correlation function experimentally.

Applications of these discussions to polymer mixtures are examined. The differences between the simple fluid mixture and polymer mixtures are twofold. One difference is the static aspect. A polymer molecule consists of a very large number of monomer units. By this reason a relative fluctuation in the initial stage is very small compared with that of simple fluid mixtures. Then the linearized approximation of the kinetic equation is available in the initial stage of the phase separation of polymer mixtures. However, the linearization of the kinetic equation is not valid in two-dimensional systems, even for a polymer system. This is because higher-order fluctuations cannot be neglected compared to the lowest-order fluctuations in two dimensions.

The second difference is the dynamical aspect. Due to the long shape nature of a polymer molecule, the diffusion of a polymer molecule is very slow. This effect is important for spatial dimensions less than 6. Then the difference between polymer mixtures and simple fluid mixtures is that for polymers the inertial term may not play an important role. The two-dimensional case is marginal, however.

In conclusion, polymer mixtures can be used to study the dynamics of phase separation. This system exhibits a clear initial stage, where the linearized kinetic equation, which is essentially applied to nonhydrodynamic phenomena, is valid. In the late stage of the phase separation the dynamics obeys that of a viscous fluid up to enormously large length scales. Then spatial patterns and even the auto-correlation function can be observed in real space.

REFERENCES

1. JD Gunton, M San Miguel, PS Sani. In: C Domb, JL Lebowitz, ed. Phase Transition and Critical Phenomena. London: Academic, Vol 8, 1983.
2. H Furukawa. Adv Phys 34:703, 1985.
3. K Binder. In: RW Cohen, P Haasen, EJ Kramer, ed. Material Sciences and Technology. Weinheim: VCH, Vol 5, 1991, p 405.
4. GF Mazenko. Phys Rev B 43:8204, 1991.
5. A Shinozaki and Y Oono. Phys Rev E 48:2622, 1993.
6. AJ Bray. Adv Phys 43:357, 1994.
7. K Binder. Colloid Polymer Sci 265:273, 1987.
8. T Hashimoto. In: RM Ottenbrite, LM Utracki, S Inoue, ed. Current Topics in Polymer Science. Munich: Hasner, Vol 2, 1987, p 199.
9. T Nose. Phase Transition 8:245, 1987.
10. T Hashimoto. Phase Transition 12:47, 1988.
11. T Hashimoto. In: RW Cohen, P Haasen, EJ Kramer, ed. Materials Science and Technology. Weinheim: VCH, Vol 12, 1993, p 252.
12. K Binder. Advances in Polymer Sci 112:181, 1994.
13. A Onuki. Phys Rev A 34:3528, 1986.
14. A Onuki, K Kawasaki, ed. Dynamics and Patterns in Complex Fluids, Berlin: Springer, 1990.
15. T Ohta, H Nozaki, M Doi. Phys Lett A 145:304, 1990.
16. T Ohta, H Nozaki, M Doi. J Chem Phys 93:2664, 1990.
17. T Hashimoto, T Takebe, K Asakawa. Physica A 194:338, 1993.
18. J Casas-Vazquez, M Criado-Sancho, D Jou. Europhys Lett 23:469, 1993.
19. T Hashimoto, K Matsuzaka, E Moses, A Onuki. Phys Rev Lett 74:126, 1995.
20. J Läuger, C Laubner, W Gronski. Phys Rev Lett 75:3576, 1995.
21. SC Glotzer, DS Stauffer, N Jan. Phys Rev Lett 72:4109, 1994.
SC Glotzer, A Coniglio. Phys Rev E 50:4241, 1994.
SG Glotzer, EAD Marzio, M Muthukumar. Phys Rev Lett 74:2034, 1995.
22. Q Tran-Cong, A Harada. Phys Rev Lett 76:1162, 1996.
23. T Tamai, A Imagawa, Q Tran-Cong. Macromolecules 27:7486, 1994.
24. T Kyu, J Lee. Phys Rev Lett 76:3746, 1996.
25. JW Cahn. Trans. AIME 242:166, 1967.
26. K Kawasaki. Ann Phys 85:3556, 1986.
27. PC Hohenberg, BI Halperin 49:435, 1977.
28. JE Farrel, OT Valls. Phys Rev B 40:7027, 1989; 42:2353, 1990.
29. K Kawasaki. In: H Haken, ed. Synergetics. Stuttgart: Teubner, 1973, p 35.
K Kawasaki, T Ohta. Physica A 118:175, 1983.
30. JS Langer, M Bar-on, HD Miller. Phys Rev A 11:1417, 1975.
31. K Kawasaki, T Ohta. Prog Theor Phys 59:362, 1978.

32. H Furukawa. Prog Theor Phys 59:1072, 1978; Phys Rev Lett 43:136, 1979; Phys Rev A 23:1535, 1981.
33. IM Lifshitz, VV Slyozov. J Phys Chem Solids 19:35, 1961.
M Toluyama, K Kawasaki. Physica A 123:386, 1984.
34. ED Siggia. Phys Rev A 20:595, 1979.
35. NC Wang, CM Knobler. J Chem Phys 69:725, 1978; Phys Rev A 24:3205, 1981.
36. YC Chou, WI Goldburg. Phys Rev A 20:2105, 1979; A 23:858, 1981.
37. N Kuwahara, M Tachikawa, K Hamano, Y Kenmochi. Phys Rev A 25:3449, 1982.
38. D Beysens, P Guenoun, F Perrot. In: S Komura, H Furukawa, ed. Dynamics of Ordering Processes in Condensed Matter. New York: Plenum Press, 1988, p 373.
39. P Guenoun, R Gastaud, F Perrot, D Beysen. Phys Rev A 36:4876, 1987.
40. K Binder, D Stauffer. Phys Rev Lett 33:1006, 1974.
41. H Furkawa. Phys Rev A 31:1103, 1985; Physica A 204:237, 1994.
42. J Marro, JL Lebowitz, MH Kalos. Phys Rev Lett 43:282, 1979.
JL Lebowitz, J Marro, MH Kalos. Acta Metall 30:297, 1982.
43. C Yueng. Phys Rev Lett, 61:1135, 1988.
44. H Furukawa. J Phys Soc Jpn 58:216, 1989).
45. H Tomita. In: K Kawasaki, M Suzuki, A Onuki, ed. Formation, Dynamics and Statistics of Patterns. Singapore: World Scientific, 1989, p 113.
46. P Fratzl, JL Lebowitz, O Penrose, J Amar. Phys Rev B 44:4794, 1991.
47. G Porod. Kolloid Z 124:83, 1951; 125:51, 1952.
H Tomita Prog Theor Phys 72:656, 1984; 75:482, 1986.
48. H Furukawa. Physica A 123:497, 1984.
49. S Komura, K Osamura, H Fujii, T Takeda. Phys Rev B 31:1278, 1985.
50. M Takenaka, T Izumitani, T Hashimoto. J Chem Phys 92:4566, 1990.
51. M Tokuyama, Y Enomoto. Physica A 204:693, 1984.
52. GF Mazenko, RA Wickham. Phys Rev E 51:2886, 1995.
53. T Ohta, H Nozaki. In: F Tanaka, M Doi, T Ohta, ed. Space-Time Organization in Macromolecular Fluids. Berlin: Springer, 1989, pp 51–77.
H Tomita. Prog Theor Phys 90:521, 1993.
54. T Hashimoto, M Takenaka, H Jinnai. J Appl Cryst 24:454, 1991.
55. A Coniglio, M Zannetti. Europhys Lett 10:575 1989.
56. H Furukawa. Phys Rev B 40:2341, 1989. See also Ref. 6.
57. DS Fisher, DA Huse. Phys Rev B 38:373, 1988.
N Mason, AN Pargellis, B Yurke. Phys Rev Lett 70:190, 1993. See also Ref. 6.
58. TJ Newman, AJ Bray, MA Moore. Phys Rev B 42:4514, 1990. See also Ref. 6.
59. A Onuki. Prog Theor Phys 74:1155, 1985.
60. H Hayakawa, T Koga. J Phys Soc Jpn 59:3542 1990.
61. AJ Bray, AD Rutenberg. Phys Rev E 49:R27, 1994.
62. PJ Flory. Principles of Polymer Chemistry. Ithaca: Cornell University Press, 1953.

63. PG de Gennes. Scaling Concepts in Polymer Physics. Ithaca: Cornell University Press, 1979.
64. PG de Gennes. J Chem Phys 72:4756, 1980.
65. P Flory. J Chem Phys 17:303, 1949.
66. LD Landau, EM Lifshitz. Statistical Physics. Oxford: Pergamon Press, 1980.
67. HE Stanley. An Introduction to Phase Transitions and Critical Phenomena. Oxford: Oxford University Press, 1971.
68. ME Fischer. Rev Mod Phys 46:597, 1974.
69. E Helfand, AM Sapse. J Chem Phys 62:1327, 1975.
70. T Hashimoto, M Shibayama, H Kawai. Macromolecules 13:1237, 1980.
71. H Furukawa. Phys Rev A 40:6403, 1989.
72. JW Cahn, JE Hilliard. J Chem Phys 28:258, 1958.
73. E Helfand, Y Tagami. J Chem Phys 56:3592, 1971.
74. JE Joanny, L Leibler. J Physique 39:951, 1978.
75. K Binder, HL Frisch. Macromolecules 17:2927, 1984.
76. M Doi, SF Edwards. The Theory of Polymer Dynamics. Oxford: Clarendon, 1986.
77. PG de Gennes. J Chem Phys 55:572, 1971.
78. VL Ginzburg. Sov Phys Solid State 1:1824, 1960.
79. PG de Gennes. J Phys Lett (Paris) 38:L44, 1977.
80. JF Joanny. J Phys A 11:L117, 1987.
81. K Binder. J Chem Phys 79:6387, 1983; Phys Rev A 29:341, 1984.
82. AB Bortz, MH Kalos, JL Lebowitz, MA Zendejas. Phys Rev B 10:535, 1974.
83. J Marro, AB Bortz, MH Kalos, JL Lebowitz. Phys Rev B 12:2000, 1975.
84. M Rao, MH Kalos, JL Lebowitz, J Marro. Phys Rev B 13:4328, 1976.
85. A Sur, JL Lebowitz, J Marro, MH Kalos. Phys Rev B 15:3014, 1977.
86. HL Snyder, P Meakin, S Reich. Macromolecules 16:757, 1983.
87. T Hashimoto, T Izumitani. J Chem Phys 83:3694, 1985.
88. M Okada, CC Han. J Chem Phys 85:5317, 1986.
89. H Meier, GR Strobl. Macromolecules 20:649, 1987.
90. JS Higgins, H Fruitwala, PE Tomlins. Macromolecules 22:3674, 1989.
91. D Schwahn, H Yee-Madeira. Colloid Polym Sci 265:867, 1987.
D Schwahn, T Springer, K Mortensen, H Yee-Madeira. In: S Komura, H Furukawa, ed. Dynamics of Ordering Processes in Condensed Matter, New York: Plenum, 1988, p 445.
92. D Schwahn, T Springer, H Yee-Madeira, K Mortensen. In: D Richter, T Springer, ed. Polymer Motion in Dense Systems (Springer Proceedings in Physics, Vol 29). Berlin: Springer, 1988, p 296.
93. T Sato, CC Han. J Chem Phys 88:2057, 1988.
94. P Wiltzius, FS Bates, WR Heffner. Phys Rev Lett 60:1538, 1988.
95. FS Bates, P Wiltzius. J Chem Phys 91:3258, 1989.
96. OT Valls, GF Mazenko. Phys Rev B 38:11643, 1988.
97. Y Wu, FJ Alexander, T Lookman, S Chen. Phys Rev Lett 74:3852 1995.
98. WJ Ma, A Maritan, J Banavar, J Koplik. Phys Rev A 45:5347, 1992.
99. E Valasco, S Toxvaerd. Phys Rev Lett 71:388, 1993

100. P Ossadnik, MF Gryre, HC Stanley, S Glotzer. Phys Rev Lett 72:2498, 1994.
101. G Leptoukh, B Strickland, C Roland. Phys Rev Lett 74:3636, 1995.
102. FF Abraham, SW Koch, RC Desai. Phys Rev Lett 49:923, 1982.
103. FI Alexander, S Chen, DW Grunau. Phys Rev B 48:634, 1993.
104. S Bastea, JL Lebowitz. Phys Rev E 52:3821 1995.
105. WR Osborn, E Orlandini, MF Swift, JM Yeomans, JR Banavar. Phys Rev Lett 75:4031, 1995.
106. PBS Kumar, M Rao. Phys Rev Lett 77:1067, 1996.
107. RAL Jones, LJ Norton, EJ Kramer, FS Bates, P Biltzius. Phys Rev Lett 66:1326, 1991.
108. F. Bruder, R Brenn. Phys Rev Lett 69:624, 1992.
109. G Reiter. Phys Rev Lett 68:75 1992.
110. M Motowoka, H Jinnai, T Hashimoto, Y Qiu, CC Han. J Chem Phys 99:2095, 1993.
111. H Tanaka. Europhys Lett 24:665, 1993; Phys Rev Lett 72:1702, 1994.
112. U Steiner, J Klein, LJ Fetters. Phys Rev Lett 72:1498, 1994.
113. W Straub, F Bruder, R Brenn, G Krausch, H Bielefeldt, A Kirsch, O Marti, J Mlynek, JF Marko. Europhys Lett 29:353, 1995.
114. M Geoghegan, RAL Jones, AS Clough. J Chem Phys 103:2719, 1995.
115. L Sung, A Karim, JF Douglas, CC Han. Phys Rev Lett 79:4368, 1996.
L Sung, CC Han. J Polym Sci Polym Phys Ed 37:2405, 1995.
116. CK Haas, JM Tolkelson. Phys Rev Lett 75:3134, 1995.

3

Miscibility and Critical Phenomena in Binary Polymer Blends

Shinichi Sakurai
Kyoto Institute of Technology, Kyoto, Japan

I. INTRODUCTION

Controlling the structure and properties of multicomponent polymers has been one of the most significant issues of the past several decades in polymer science [1,2]. This significance has been mentioned previously in the Preface of this book. Since the properties of the multicomponent polymers depend intensively on the underlying structures, control and design of these structures is one of the most effective ways to produce novel functions or properties. For this purpose, it is important to find some general rules for structure–property relationships. Polymer blends provide simple models to study these relationships. This is one reason why polymer blends attract not only academic but also technological interest [3]. Phase separation and spinodal decomposition phenomena [4] in binary polymer mixtures are also utilized for designing multiphase structures [5]. To understand the fundamentals of these phenomena, we need to know thermodynamics [6], phase transition, and critical phenomena [7,8] in polymer blends. From this viewpoint, thermodynamics and critical phenomena are reviewed in Secs. II and III, respectively. Moreover, with regard to phase transition in polymer blend, it is important to evaluate quantitatively the degree of miscibility between the two component polymers. Therefore in Sec. IV, the methodology of determining the degree of miscibility particularly using scattering techniques [7] is described.

II. OVERVIEW

A. Phase Diagram

A brief overview on studies of miscibility assessment in the past ten years is presented in this section. The statistical thermodynamics to evaluate the free energy of mixing of two different kinds of polymers has been established by the pioneering work of Flory [6]. Based on the Flory–Huggins (FH) lattice theory, the free energy of mixing ΔF_{mix} can be given by:

$$\frac{\Delta F_{\mathrm{mix}}}{k_{\mathrm{B}}T} = \frac{\phi_{\mathrm{A}}}{N_{\mathrm{A}}} \ln \phi_{\mathrm{A}} + \frac{1-\phi_{\mathrm{A}}}{N_{\mathrm{B}}} \ln (1-\phi_{\mathrm{A}}) + \phi_{\mathrm{A}}(1-\phi_{\mathrm{A}})\chi_{\mathrm{AB}} \tag{1}$$

where ϕ_{A} denotes the volume fraction of component A, and incompressibility is assumed. k_{B} is the Boltzmann constant and T is temperature in kelvin. N_{A} and N_{B} are the degrees of polymerization of the A and B components, respectively. Furthermore, the parameter, χ_{AB}, denotes the degree of interaction between two types of segments, A and B, and is a quantitative measure of the miscibility. According to the original FH theory, χ_{AB} is defined by:

$$\chi_{\mathrm{AB}} = \frac{\Delta H_{\mathrm{mix}}}{k_{\mathrm{B}}Tn_{\mathrm{A}}N_{\mathrm{A}}(1-\phi_{\mathrm{A}})} \tag{2}$$

which means that the χ_{AB} parameter is ascribed to the enthalpic interaction upon mixing. ΔH_{mix} is the enthalpy of mixing and n_{A} is the number of A molecules. Note here that the degree of polymerization, N_{A}, in the denominator reduces the interaction parameter per segment. χ_{AB} is also shown to be related to the second virial coefficient, A_2, as [9]:

$$\chi_{\mathrm{AB}} = \frac{1}{N_{\mathrm{A}}}\left(\frac{1}{2} - \frac{V_{\mathrm{A}}M_{\mathrm{B}}^2}{V_{\mathrm{B}}^2}A_2\right) \tag{3}$$

where V_{A} and V_{B} denote the molar volume of the A and B components, respectively, and M_{B} is the molecular weight of component B. The weight-average molecular weight, M_{w}, and A_2 can be simultaneously evaluated with the z-averaged radius of gyration by scattering techniques through the so-called Zimm plot (see any textbook) for dilute polymer solutions. Eq. (3) indicates that a conventional Zimm plot is very useful to evaluate χ_{AB} even for polymer mixtures [9]. Technically, χ_{AB} can be roughly estimated from the solubility parameter, $\delta_i (i = \mathrm{A}$ or B), by using the Hildebrand equation [10], as:

$$\chi_{\mathrm{AB}} \cong \frac{v_0}{k_{\mathrm{B}}T}(\delta_{\mathrm{A}} - \delta_{\mathrm{B}})^2 \tag{4}$$

where v_0 is the reference cell volume. Since the value of δ_i can be extracted from the literature [11], the Hildebrand equation is useful for a rough estimation of the value of χ_{AB} for the polymer pair where the value of χ_{AB} is unknown.

Binary polymer mixtures undergo phase separation when the total free energy in the system decreases upon the phase separation. In this situation, two separated phases coexist in the system. The thermodynamic condition for the coexistence of two phases is given by equilibration of the chemical potential between two phases. Using Eq. (1), one can calculate the coexistence (binodal) condition by equating the chemical potential in the two separated phases, or similarly by using the following relationship,

$$\left.\frac{\partial(\Delta F_{\text{mix}}/k_{\text{B}}T)}{\delta\phi_A}\right|_{\phi_A=\phi_A^{\text{I}}} = \left.\frac{\partial(\Delta F_{\text{mix}}/k_{\text{B}}T)}{\delta\phi_A}\right|_{\phi_A=\phi_A^{\text{II}}} \tag{5}$$

and thus,

$$\frac{1}{N_{\text{A}}} ln\frac{\phi_{\text{A}}^{\text{I}}}{\phi_{\text{A}}^{\text{II}}} - \frac{1}{N_{\text{B}}} ln\frac{1-\phi_{\text{A}}^{\text{I}}}{1-\phi_{\text{A}}^{\text{II}}} - 2\chi_{\text{b}}(\phi_{\text{A}}^{\text{I}} - \phi_{\text{A}}^{\text{II}}) = 0 \tag{6}$$

is obtained, where $\phi_{\text{A}}^{\text{I}}$ and $\phi_{\text{A}}^{\text{II}}$ denote the volume fractions of the component A in the separated phases I and II, respectively ($\phi_{\text{A}}^{\text{I}} < \phi_{\text{A}}^{\text{II}}$), and χ_{b} designates the χ_{AB} value at the coexistence point. For the special case of $N_{\text{A}} = N_{\text{B}} = N$ (symmetric mixture), the coexistence condition can be obtained by:

$$\frac{1}{N} ln\frac{\phi_{\text{A}}}{1-\phi_{\text{A}}} - (2\phi_{\text{A}} - 1)\chi_{\text{b}} = 0$$

$$(\text{for } \phi_{\text{A}} \leq 0.5, \text{ symmetric with respect to } \phi_{\text{A}} = 0.5) \tag{7}$$

For a given value of χ_{b}, a set of values of $\phi_{\text{A}}^{\text{I}}$ and $\phi_{\text{A}}^{\text{II}}$ (or ϕ_{A} and $1 - \phi_{\text{A}}$) may be evaluated (coexistence points) from Eqs. (6) or (7). As a consequence of changing the value of χ_{b}, the coexistence points form a parabolic upward curve ($\partial^2\chi_{\text{b}}/\delta\phi_{\text{A}}^2 > 0$) in the plot of χ_{b} versus ϕ_{A}. This is called a coexistence curve. Experimentally, the phase separation is accessible by changing temperatures. Therefore, phase-separation temperatures determined by cloud-point measurements are obtained as a function of ϕ_{A} (the so-called phase diagram or cloud point curve). Correlation between χ_{AB} and temperature, i.e., the temperature dependence of χ_{AB}, can be determined from a comparison of the coexistence curve (χ_{b} versus ϕ_{A}) with the cloud-point curve (temperature versus ϕ_{A}). Namely, from the cloud-point measurements, it is possible to evaluate χ_{AB} values. However, one has to be aware that the thus obtained χ_{AB} values are generally different from the values evaluated from measurements of scattering functions of homogeneous mixtures (see Sec. II. B). Using this procedure, it has been shown for a solution of a polydisperse polymer that the χ_{AB} parameter has an entropic contribution (i.e., temperature-independent term) and a concentration-dependent term [12,13].

Homogeneous mixtures of binary polymer blends spontaneously separate into two phases when the condition $\partial^2 \Delta F_{mix}/\partial \phi_A^2 < 0$ is satisfied. Therefore, the condition $\partial^2 \Delta F_{mix}/\partial \phi_A^2 = 0$ gives the stability limit for the homogeneous mixtures. This is referred to as the spinodal condition and can be calculated from:

$$\frac{\partial^2(\Delta F_{mix}/k_B T)}{\partial \phi_A^2} = 0 = \frac{1}{N_A \phi_A} + \frac{1}{N_B(1-\phi_A)} - 2\chi_s \tag{8}$$

where χ_s is the χ_{AB} value at the spinodal point and is given by:

$$\chi_s = \frac{1}{2}\left[\frac{1}{N_A \phi_A} + \frac{1}{N_B(1-\phi_A)}\right] \tag{9}$$

Since the critical point corresponds to the minimum of the spinodal curve (χ_c versus ϕ_A), the mathematical expression for the critical point is as follows:

$$\frac{\partial \chi_s}{\partial \phi_A} = 0 = -\frac{1}{N_A \phi_{A,c}^2} + \frac{1}{N_B(1-\phi_{A,c})^2} \tag{10}$$

where $\phi_{A,c}$ denotes the critical composition. Hence, $\phi_{A,c}$ and the χ_{AB} value at the critical point (χ_c) can be given respectively by:

$$\phi_{A,c} = \frac{\sqrt{N_B}}{\sqrt{N_A} + \sqrt{N_B}} \tag{11}$$

and

$$\chi_c = \frac{1}{2}\left(\frac{1}{\sqrt{N_A}} + \frac{1}{\sqrt{N_B}}\right)^2 \tag{12}$$

Finally, note here that the spinodal and cloud-point curves contact each other only at the critical point, i.e., $\chi_c = \chi_b(\phi_{A,c})$.

From Eq. (1), it is clear that a two-component system spontaneously mixes when $\Delta H_{mix} \leq 0$. On the other hand, the system shows partial miscibility for $\Delta H_{mix} > 0$. While the thermodynamic relationships outlining the phase boundaries are well-defined, two distinct types of temperature dependence are observed. In some cases, at low temperatures, the sample is phase-separated and becomes a single phase on heating (the upper critical solution temperature (UCST) type). In other cases, the opposite behavior is observed and the sample phase separates on heating (the lower critical solution temperature (LCST) type. These two different types of behaviors are ascribed to the temperature dependence of χ_{AB}. According to Eq. (2), the χ_{AB} value decreases monotonically with T, and therefore, the contribution of the enthalpic term to the free energy of mixing becomes negligible at high temperatures. This is a possible explanation for a UCST-type phase diagram. Although χ_{AB} is expected to be inversely proportional to T from Eq. (2), from many experimental results it has been shown that χ_{AB} has additionally a temperature-independent term. Namely, the temperature dependence of χ_{AB} is of the form:

$$\chi_{AB} = A + \frac{B}{T} \qquad (T \text{ in kelvin}) \tag{13}$$

This clearly indicates that the χ_{AB} parameter has an entropic contribution (A) as well as an enthalpic contribution (B/T), where the entropic contribution may be attributed to the anisotropic shapes of the segments [14].

In addition to the deviation from the originally introduced idea of the χ_{AB} parameter in terms of the temperature dependence, as defined by Eq. (2), more anomalous phenomena are the LCST phase behaviors which are observed for many polymer blends including polystyrene (PS)/poly(vinyl methyl ether) (PVME) blends [15]. For such a system, we completely fail to explain the LCST behaviors with only the energetic contribution to the χ_{AB} parameter. Moreover, a few systems were found to have both the UCST and the LCST [16]. For this kind of behavior, the temperature dependence of χ_{AB} should not be of the simple form $\chi_{AB} = A + B/T$. An example is a blend of deuterated polystyrene (DPS) with poly(butyl methacrylate) (PBMA) and for this system nonsimple temperature dependence of χ_{AB} has been confirmed by a small-angle neutron scattering (SANS) study [17]. To explain the LCST behavior, Flory et al. [18] introduced equation-of-state (EOS) theories [19] which treat the polymeric liquid as a low molecular weight liquid [20]. The EOS approaches have had some success in explaining the pressure dependence of the miscibility [21] as well as the LCST behavior. Here, the key factor in explaining the phase separation at high temperatures is the effect of free volume.

B. Scattering Function

In the early stages of scattering studies on multicomponent polymers, various theories were developed to describe the scattering function for homogeneous mixtures such as polymer blends or block copolymers [22]. Among them, the random phase approximation (RPA) theory is now widely and generally used and is introduced in this chapter. The elastic scattering from homogeneous polymer mixtures can be expressed by the RPA theory which was first developed by deGennes [7] as:

$$\frac{1}{S(q)} = \frac{1}{N_A \phi_A g_D(x_A)} + \frac{1}{N_B(1-\phi_A) g_D(x_B)} - 2\chi_{AB} \tag{14}$$

$S(q)$ denotes the structure factor which is proportional to the scattering intensity, $I(q)$ [$= k \times S(q)$]. q is the magnitude of the scattering vector, defined by:

$$q = |\mathbf{q}| = (4\pi/\lambda)\sin(\theta/2) \tag{15}$$

where λ and θ are the wavelength of the radiation and the scattering angle, respectively, and $g_D(x_K)$ is the Debye scattering function [23] for single Gaussian chain, which is given by

$$g_D(x_K) = \frac{2}{x_K^2}[x_K - 1 + exp(-x_K)] \tag{16}$$

with

$$x_K = \frac{q^2 N_k b_k^2}{6} = (qR_{g,K})^2 \tag{17}$$

Eq. (14) clearly shows the dependence of the scattering intensity, $I(q)$, on the value of χ_{AB}. Hence, by analyzing the q-dependent scattering intensities (the so-called scattering profile) one can quantitatively evaluate χ_{AB}. Development of the RPA theory to extract χ_{AB} has been instrumental in utilizing SANS for miscibility assessment. More recently, many modifications to RPA theories have been included (some of these modifications are described in Sec. IV. B), and have had some success in accounting for the effects of molecular weight distribution, asymmetry in the segment size, and compressibility on scattering functions and evaluation of χ_{AB}. Moreover, the RPA theory has been extended to multicomponent systems (three or more components). Some of these modifications are discussed in more detail in Sec. IV. B. In the limit of $q \to 0$, since the Debye function $g_D(x) \to 1$, Eq. (14) converges on

$$\frac{1}{S(q \to 0)} = \frac{1}{N_A \phi_A} + \frac{1}{N_B(1-\phi_A)} - 2\chi_{AB} \tag{18}$$

Then, from Eq. (8), it follows that the inverse of the zero-angle scattering, $S^{-1}(q=0)\,[\propto I^{-1}(q=0)]$, is equal to the second derivative of the free energy of mixing with respect to concentration:

$$S^{-1}(q=0) = \frac{\partial^2(\Delta F_{\text{mix}}/k_B T)}{\partial \phi_A^2} \tag{19}$$

In the past ten years, it has been shown that the χ_{AB} parameter obtained from measurements of the elastic scattering of a homogeneous mixture based on Eq. (14) has a composition dependence, i.e., $\chi_{AB} = \chi_{AB}(\phi_A)$. For this situation, Sanchez [24] has considered that the

χ_{AB} values obtained by the scattering measurements differ from the original Flory–Huggins χ_{AB} parameter. To avoid confusion, the χ_{AB} parameters from the scattering measurements are designated by χ_{SC} and the original definition by χ_{FH} in this paragraph. Since the composition dependence of χ_{FH} has been observed experimentally, χ_{AB} in Eq. (1) for the free energy of mixing should be replaced by the composition dependent $\chi_{FH} = \chi_{FH}(\phi_A)$. Then, from the following relationship:

$$\chi_{sc} = -\frac{1}{2}\frac{\partial^2(\Delta F_{mix}/k_B T)}{\partial\phi_A^2} \tag{20}$$

we obtain a complicated expression for χ_{sc}

$$\chi_{sc} = \chi_{FH}(\phi_A) + (2\phi_A - 1)\frac{d\chi_{FH}(\phi_A)}{d\phi_A} - \frac{1}{2}\phi_A(1-\phi_A)\frac{d^2\chi_{FH}(\phi_A)}{d\phi_A^2} \tag{21}$$

Thus, χ_{sc} is not identical to χ_{FH} because of the composition dependence of χ_{FH}. To demonstrate the relationship of the χ_{AB} parameters obtained from the scattering measurements and the cloud points (or more rigorously the coexistence temperatures) with respect to χ_{FH}, Sanchez [24] has evaluated the composition dependences of χ_{sc}, $\chi_{\mu 1}$, $\chi_{\mu 2}$, and χ_{FH}, for the model case of a DPS/PVME blend (1:PVME; 2:DPS) [25,26]. Here, $\chi_{\mu 1}$ and $\chi_{\mu 2}$ are the χ_{AB} parameters appearing in the expressions for the chemical potentials of the A and B components, respectively. As depicted in Fig. 1, the composition dependence of χ_{sc} is much stronger than that of χ_{FH}. The composition dependence of χ_{sc} has been confirmed by small-angle neutron and X-ray scattering experiments in many blend systems including isotopic (deuterated and protonated) polymer mixtures and is discussed extensively in Sec. IV. C.2.

Here, we demonstrate a typical procedure for determining χ_{AB} and a phase diagram (cloud-point temperature and the spinodal temperature). Fig. 2(a) shows the SANS intensity, $I(q)$, as a function of q for the LCST-type deuterated polybutadiene (DPB; H-9)/polyisoprene (HPI; GL-4) blend [27]. Initially, the scattering intensity at a given q clearly increases with an increase in temperature. This behavior is consistent with the predictions of Eqs. (13) and (14) which indicate that as T increases, χ_{AB} increases, hence $I(q)$ also increases. The SANS data can be well described by the theory, as depicted in Fig. 3 where the theoretical curve is obtained from the least squares fit to the data (23°C) by Eq. (14) and the χ_{AB} parameter is one of the fit parameters. As temperature is increased further, the intensities in the small-angle region suddenly decrease above a certain temperature (88.3°C). The decrease indicates the onset of the phase separation. Fig. 2(b) shows the scattering intensity as a function of temperature

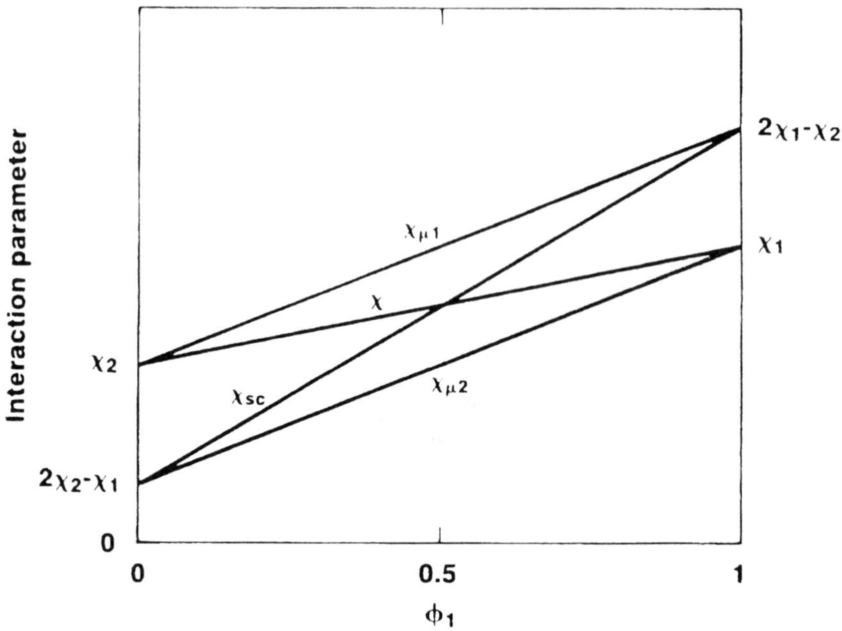

Figure 1 Schematic demonstration of the relationship of the composition-dependent interaction parameters obtained from the scattering measurements and the cloud points (or more rigorously coexistence temperatures), χ_{SC}, $\chi_{\mu 1}$, $\chi_{\mu 2}$, and χ (identical to χ_{FH} in the text), for the model case of the deuterated polystyrene/poly(viny methyl ether) blend (1:PVME; 2:DPS). Here, χ_{SC} denotes the interaction parameter obtained by the scattering technique. $\chi_{\mu 1}$ and $\chi_{\mu 2}$ are the interaction parameters appearing in the expressions for the chemical potentials for the components 1 and 2. From Ref. 24.

to demonstrate the intensity depletion. From the inflection points, the so-called "neutron cloud-point temperature" can be estimated.

Finally, we discuss the evaluation of the spinodal temperature, T_s. The spinodal point is the stability limit where the homogeneous mixture becomes unstable and can be defined by several criteria. The susceptibility (zero-angle scattering intensity, $I(q = 0)$) and the correlation length of the concentration fluctuations both diverge at the spinodal point [7]. The correlation length, ξ, of the concentration fluctuations in the binary polymer mixture denotes the length within which the concentration fluctuation correlates. From Eqs. (9) and (18), one obtains:

$$S^{-1}(q = 0) = 2(\chi_s - \chi_{AB}) \tag{22}$$

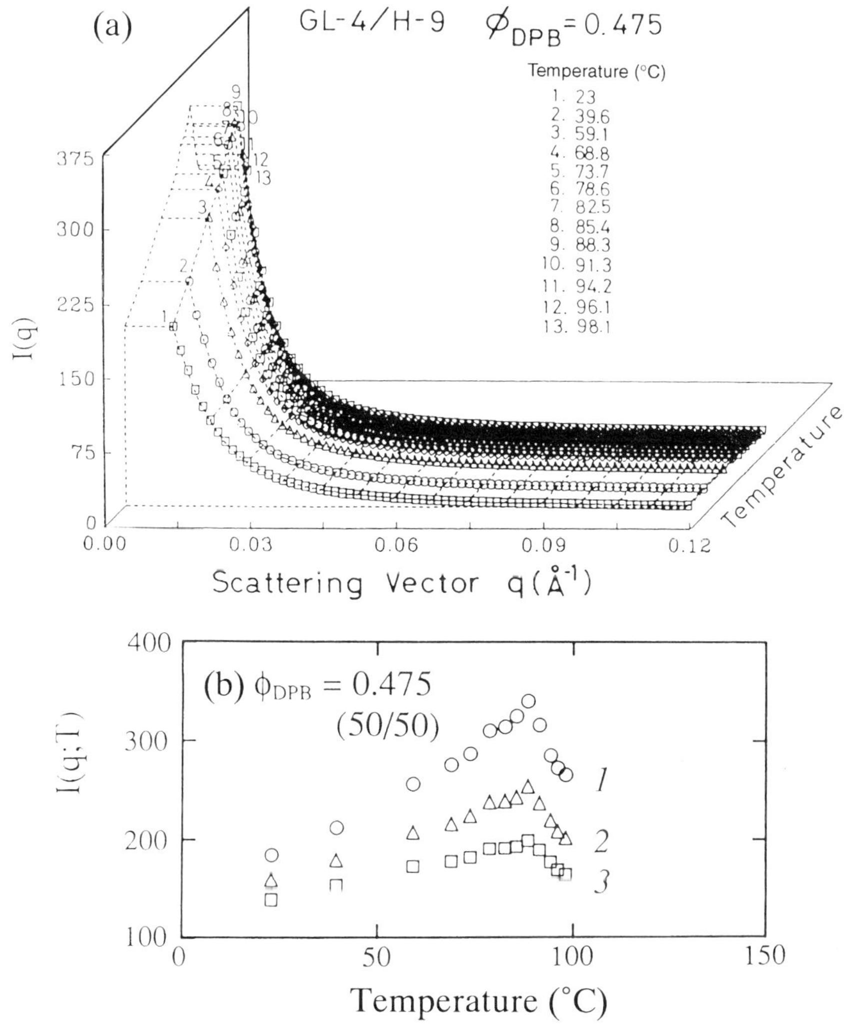

Figure 2 (a) SANS profiles as a function of temperature for the LCST-type DPB/HPI (H-9/GL-4) blend with $\phi_{DPB} = 0.475$. (b) Plot of the scattering intensity $I(q; T)$ versus temperature for various q values (1—0.093 nm^{-1}; 2—0.108 nm^{-1}; 3—0.122 nm^{-1}). From Ref. 27.

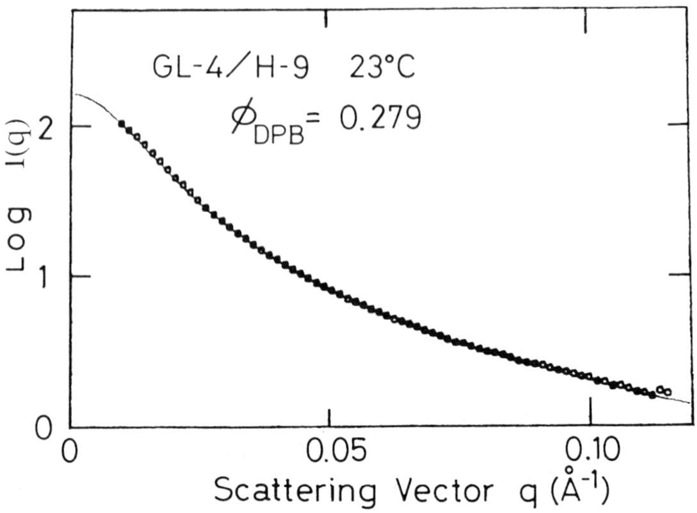

Figure 3 An example of the result of the nonlinear regression curve fitting for the SANS profile with the RPA formula. The interaction parameter χ_{AB}, is a fit parameter and is numerically determined as a result of the best fitting. The profile is for the DPB/HPI (H-9/GL-4) blend with $\phi_{DPB} = 0.279$ at 23°C. From Ref. 27.

If the scattering function of Eq. (14) is approximated in the small-angle region by the Lorentzian-type function as written by Eqs. (27) or (28), then the correlation length, ξ, is obtained. Note that ξ is related to χ_{AB} by the following equation [7]:

$$\xi^{-2} = 36(\chi_s - \chi_{AB})\left(\frac{b_A^2}{\phi_A} + \frac{b_B^2}{1-\phi_A}\right)^{-1} \tag{23}$$

In the mean-field region, it is well known that the temperature dependence of the interaction parameter, χ_{AB}, is given empirically by Eq. (13). Thus the following behavior [7,25,28] is expected for a mean-field system:

$$S^{-1}(q=0) \sim |T^{-1} - T_s^{-1}| \tag{24}$$

and

$$\xi^{-2} \sim |T^{-1} - T_s^{-1}| \tag{25}$$

Alternatively, if the temperature dependence of the interaction parameter, χ_{AB}, can be empirically expressed by Eq. (13), substitution of the value of χ_S into Eq. (13) yields the spinodal temperature, T_S. Here, the value of χ_{AB} is

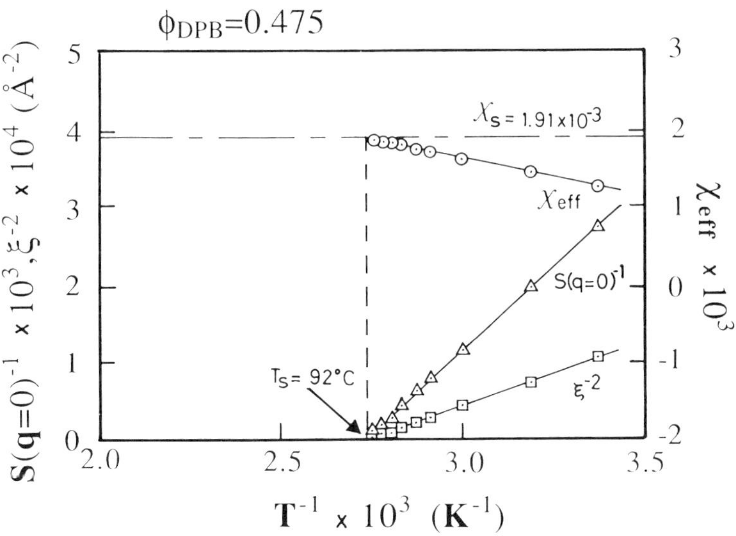

Figure 4 Plots of the inverse of the susceptibility, $S^{-1}(q=0)$, inverse of the square of the correlation length ξ^{-2}, and the effective interaction parameter, χ_{eff} (identical to χ_{AB} in the text), as a function of the inverse absolute temperature, T^{-1}, for the DPB/HPI (H-9/GL-4) blend with $\phi_{\text{DPB}} = 0.475$. Extrapolations from those three sets of data give a single value of the spinodal temperature, $T_S = 92°\text{C}$. From Ref. 27.

evaluated from the RPA fitting, and χ_S is calculated from the molecular weight information through Eq. (9). Therefore, extrapolations of $S^{-1}(q=0)$ and ξ^{-2} to zero in their plots against T^{-1} yield T_s^{-1}. Fig. 4 shows an example of the evaluation of T_S based on the above three criteria. As expected, the three different extrapolation techniques yield an identical T_S. In the vicinity of a critical point, the mean-field behavior is no longer observed. Therefore, the expression of the diverging functions and the methods of extrapolation are affected by critical phenomena, which are discussed in detail in Sec. III. Thus obtained T_S and SANS cloud points (see for example Fig. 2(b)) are plotted as a function of ϕ_{DPB} in Fig. 5.

C. Recent Advances in Instruments and Facilities for Neutron and X-Ray Scattering

The SANS technique is now widely used for the study of miscibility and critical phenomena in binary polymer mixtures. One of the main reasons for its preference is that the contrast for neutrons can be obtained easily by deuterium substitution of the hydrogen atoms with only trivial effects on the physiochemical properties of the polymer. This is very different from X-ray scattering where the contrast is due to differences in the density of free electrons. A similar technique to improve the contrast for X-rays is referred to as the "tagged polymer" method [29], where heavy atoms such as iodine are chemically attached to the polymer to increase the contrast. (In this regard, the deuterium labeling technique is a kind of tagged-polymer method.) However, as one would expect, the labeling with heavy atoms changes the thermodynamic properties of the polymer substantially. On the other hand, the deuterium substitution was initially believed to have trivial effects on the fundamental properties. However, it has been demonstrated that the phase diagram and miscibility are affected. In some cases, phase separation happens to occur in high molecular weight mixtures of deuterated and undeuterated polymers of the sample species. The effects of deuteration on miscibility are discussed later, as well as the effects of partial deuteration. Another important reason why the SANS technique is widely used is ascribed to the recent improvements in instrumentation and the increase in the number of assessible facilities. At present, more than ten facilities are available in the world, such as the National Institute of Standards and Technology (Maryland, USA) [25,26], Oak Ridge National

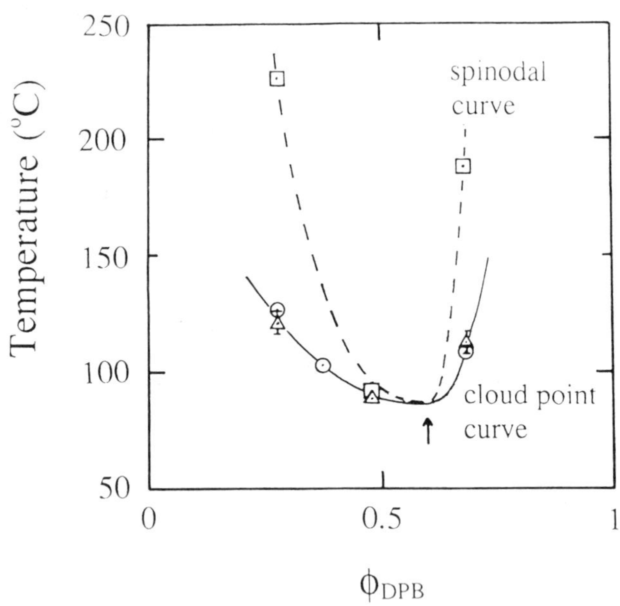

Figure 5 LCST-type phase diagram obtained experimentally for the DPB/HPI (H-9/GL-4) blend based on the SANS (including SANS cloud points; △) and LS (cloud points; ⊙) results. From Ref. 27.

Laboratory (Tennessee, USA) [30], Argonne National Laboratories (USA) [31], Los Alamos National Laboratory (New Mexico, USA) [32,33], Riso National Laboratory (Roskilde, Denmark) [34], the Japan Atomic Energy Research Institute (Ibaraki, Japan) [35], Institut Laue-Langevin (Grenoble, France) [9], KFA (Jülich, Germany) [36], Rutherford–Appleton Laboratory (Chilton, Oxfordshire, UK) [37].

The SANS technique is not the only method used to perform experimental studies on the miscibility and critical behavior of binary polymer mixtures. High-intensity X-rays improve the feasibility of small-angle X-ray scattering (SAXS) experiments on homogeneous polymer mixtures without using "tagged polymers." Recently, a number of synchrotrons, which are high-intensity sources of X-rays, have begun operating worldwide and are providing an alternative method to neutron scattering. Brookhaven National Laboratory (New York, USA) [38], Photon Factory (Tsukuba, Japan) [39], European Synchrotron Radiation Facility (ESRF) (France) [40], Cornell High Energy Synchrotron Source (CHESS) (Cornell, USA) [41], Synchrotron Radiation Source (SRS) at the SERC Daresbury Laboratory (Warrington, UK) [42] are some examples of facilities now being used for the studies of multicomponent polymers. While the deuterium substitution always promises high contrast for neutrons, utilizing the synchrotron facility does not necessarily ensure good statistics in the scattering intensity, since the intensity depends on the difference in the free electron density. Therefore, there are natural limitations in the choice of polymer pairs for SAXS studies. Additionally, absorption of the X-rays by samples and window materials for a vacuum flight path are crucial so that one should try to minimize thicknesses of the samples and windows. This also restricts some experimental possibilities, for instance, measurements under shear flow, which are actively performed by SANS (see Chapter 7 of this book), are totally impossible. However, deuterium labeling requires specialized synthetic techniques, and moreover, deuterated monomers are expensive and certain deuterated monomers are not available. For these cases, the SAXS technique is an attractive alternative and, moreover, the results are completely free from labeling effects. In this chapter, the results from SANS measurements are mainly discussed, but occasionally SR (synchrotron)-SAXS, and ordinary SAXS results are used for discussion.

III. CRITICAL PHENOMENA IN BINARY POLYMER MIXTURES

The critical behavior of binary polymer mixtures [7] is itself a very interesting research subject. Analyzing the critical behavior enables us to discuss the universality class of the polymer mixture, and hence, one can investigate to what extent the mean-field approximation, which was first introduced to polymer systems by Flory [6], is correct. Actually, the crossover from mean-field to three-dimensional (3D) Ising (fluctuation-dominated) behavior has been confirmed by numerous studies in the vicinity of the critical point [34,38,42–47]. This means that long extrapolation to the critical temperature with the assumption of mean-field behavior (as seen in Sec. II. B) would lead to a misjudgment of the location of the critical temperature. Therefore, knowledge of the critical behavior characteristic of polymer mixtures is important from a technical point of view when a precise determination of the critical temperature is required. In this section, the critical behavior is discussed exclusively based on the results of the critical exponents for the susceptibility and correlation length.

A. Critical Exponents for Susceptibility and Correlation Length

Studies of the critical behavior and the classification of the universality class are usually discussed and performed based on the critical exponents for the susceptibility and correlation length [7,25,43]. The susceptibility is proportional to the scattering intensity at $q = 0$, i.e., $I(q = 0)$. Generally, the correlation function, $g(r)$, of the concentration fluctuations can be expressed by the following Ornstein–Zernike (OZ)-type relation [7,48]:

$$g(r) = c\frac{\xi}{r}\exp\left(-\frac{r}{\xi}\right) \tag{26}$$

where r denotes the spatial coordinate. The scattering function obtained by the Fourier transformation of Eq. (26) is of Lorentzian shape:

$$I(q) = \frac{I(q=0)}{1+q^2\xi^2} \qquad \text{for } q\xi \ll 1 \tag{27}$$

For evaluation of both the susceptibility, $I(q = 0)$, and the correlation length, ξ, it is convenient to rewrite the scattering function as:

$$I^{-1}(q) = I^{-1}(q=0)(1+q^2\xi^2) \qquad \text{for } q\xi \ll 1 \tag{28}$$

The plot of $I^{-1}(q)$ versus q^2 (the so-called OZ plot) may be approximated by a straight line for $q\xi \ll 1$, and the susceptibility and ξ are obtained from the intercept and slope according to Eq. (28). Typical examples are shown in the inset of Fig. 6 for a polyisoprene (PI)/deuterated poly(ethylene propylene) (DPEP) blend with a nearly critical composition of 0.61 (= volume fraction of PI) [32]. The degree of polymerization for PI is 29, while that for DPEP is 73.

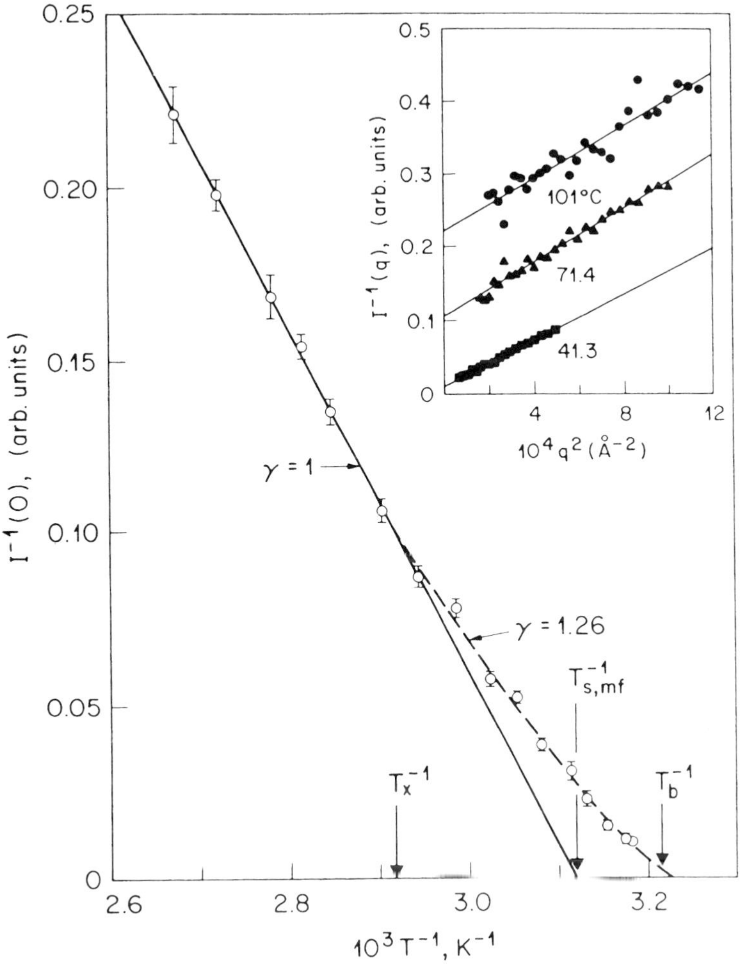

Figure 6 Plot of the inverse susceptibility, $I^{-1}(q=0)$, as a function of the inverse absolute temperature, T^{-1}, showing the clear crossover from the mean-field behavior with critical exponent, $\gamma = 1$, to 3D Ising behavior with $\gamma = 1.26$. $T_{s,mf}$ denotes the spinodal temperature based on the assumption of the mean-field behavior. T_b stands for the independently determined binodal temperature. T_x is a deviation temperature which is the onset of deviation from a linear dependence of $I^{-1}(q=0)$ with respect to T^{-1} (mean-field behavior). The inset shows the plots of the inverse of the scattering intensity, $I^{-1}(q)$, as a function of q^2. The linearity of the data points qualitatively indicates that the Lorentzian-type scattering function holds for these binary mixtures of DPS/HPS. From Ref. 32.

Fig. 6 shows the plots of the inverse of the susceptibility, $I^{-1}(q = 0)$, as a function of the inverse of the absolute temperature, T^{-1}. The plot is normally used in the estimation of the spinodal (T_S) and the critical (T_C) temperatures, by assuming the mean-field behavior, as demonstrated in Sec. II. B. One can see that only the data points far from the spinodal temperature fall on a straight line, indicating that the mean-field behavior of $I^{-1}(q = 0)$ (i.e., Eq. (24)) is observed for temperatures far from the spinodal temperature. Although the spinodal temperature would be evaluated from the extrapolation, $I^{-1}(q = 0) \to 0$, in the vicinity of the critical temperature, a deviation from linearity of the data points is observed, indicating a break down of the mean-field behavior at the deviation temperature, T_X, $(= 70°C)$, which is 32°C above the binodal temperature, Tb, $(= 38°C)$. The crossover is discussed in terms of the critical exponent for the susceptibility, $I(q = 0)$, which is evaluated with respect to the reduced temperature, ε, as:

$$I(q = 0) = I_0 \varepsilon^{-\gamma} \tag{29}$$

and

$$\xi = \xi_0 \varepsilon^{-\nu} \tag{30}$$

where

$$\varepsilon = \left| \frac{T - T_C}{Tc} \right| \tag{31}$$

The mean-field behavior is accompanied by $\gamma = 1$ and $\nu = 0.5$. On the other hand, it has been reported that in the vicinity of the critical temperature, T_C, the exponents deviate from the mean-field values to $\gamma = 1.26$ (see also Fig. 6) and $\nu = 0.63$ [34,38,42–47]. These exponents are equivalent to those expected for the 3D Ising behavior, where critical fluctuations are important. There are numerous experiments showing similar crossover from the mean-field to 3D Ising behavior [34,38,42–47].

B. Crossover from Mean-Field to Three-Dimensional Ising Behavior

The crossover from the mean-field to 3D Ising behavior is usually discussed in context with the Ginzburg criterion, which was introduced by deGennes [7] and further discussed by Binder [49]. The Ginzburg number, *Gi*, quantitatively defines where mean-field behavior is expected ($\varepsilon > Gi$). One of our interests regarding critical behavior is to what extent the Flory mean-field picture is valid with respect to the degree of polymerization. Bates et al. [32] were the first to define the *Gi* criterion as:

$$Gi = Cv_{\mathrm{m}}^{2}\frac{[N_{\mathrm{A}}^{-1}\phi_{\mathrm{A}}^{-3} + N_{\mathrm{B}}^{-1}(1-\phi_{\mathrm{A}})^{-3}]^{2}}{[N_{\mathrm{A}}^{-1}\phi_{\mathrm{A}}^{-1} + N_{\mathrm{B}}^{-1}(1-\phi_{\mathrm{A}})^{-1}][R_{\mathrm{A}}^{2}N_{\mathrm{A}}^{-1}\phi_{\mathrm{A}}^{-1} + R_{\mathrm{B}}^{2}N_{\mathrm{B}}^{-1}(1-\phi_{\mathrm{A}})^{-1}]^{3}} \tag{32}$$

where C is a system-independent constant, v_{m} is the segmental volume, and $R_{\mathrm{g,K}} = b_{\mathrm{K}}(N_{\mathrm{K}}/6)^{1/2}$ is the radius of gyration of the unperturbed chain for the K(= A or B) component with b_{K} being the statistical segment length. Gi is approximately proportional to N^{-1}, indicating that the crossover from mean-field to 3D Ising behavior would be simply governed by the molecular weight. The temperature window for the 3D Ising behavior is expected to be very narrow for a large molecular weight polymeric system and therefore one can assume entirely mean-field behavior. However, even for large molecular weight polymers, Schwahn et al. [34,44] have observed 3D Ising behavior in a blend of DPS/PVME (DPS; $M_{\mathrm{w}} = 232{,}000$, PVME; $M_{\mathrm{w}} = 89{,}000$) in the vicinity of the critical temperature (within 1 K) since they could control the temperature with an accuracy better than 10 mK. The situation is quite different for the oligomeric polymer blends because the molecular weights are small enough to expand the temperature window for 3D Ising behavior accessible experimentally with the ordinary accuracy of most temperature controls. Bates et al. reported that the temperature window is 32 K for PI/DPEP blends with molecular weights 2000 and 5800 (see Fig. 6, where the binodal temperature, T_{b}, is 38°C, and the deviation temperature, T_{x}, is 70°C, and therefore, the temperature range for the 3D Ising behavior is 32°C) [32].

At first glance, the above argument supports the idea that the Ginzburg number is a universal quantity. However, very recently it has been shown experimentally [35,50] and theoretically [51,52] by different groups that Gi depends on the segment structures [35] as well as pressure (compressibility) [50]. Thus, the crossover temperature is not simply scaled by the molecular weight. Hair et al. [46] have alternatively defined Gi by including the system-dependent parameters, ξ_0 and A as,

$$Gi = Cv_{\mathrm{m}}^{2}\frac{[N_{\mathrm{A}}^{-1}\phi_{\mathrm{A}}^{-3} + N_{\mathrm{B}}^{-1}(1-\phi_{\mathrm{A}})^{-3}]^{2}}{[N_{\mathrm{A}}^{-1}\phi_{\mathrm{A}}^{-1} + N_{\mathrm{B}}^{-1}(1-\phi_{\mathrm{A}})^{-1} - 2A]^{4}\xi_0^{6}} \tag{33}$$

where ξ_0 is the experimentally observed bare correlation length in Eq. (30) and $A(\equiv \partial(T\chi_{\mathrm{AB}})/\partial T)$ is the entropic term of $\chi_{\mathrm{AB}}(= A + B/T)$. Moreover, the deviation from the mean-field behavior is observed experimentally further from the critical point than predicted by the Ginzburg criterion. As pointed out by Hair et al., the Ginzburg criterion should only be used for an order of magnitude estimation of the temperature window of the 3D

Ising behavior [46]. It is important in practice to introduce a realistic crossover function to describe the variation of $I(0)$ or ξ with T [35,47].

C. Other Related Topics

Very recently it has been reported that a lamellar diblock copolymer (A–B) blended with two homopolymers (A and B) shows isotropic Lifshitz behavior at the Lifshitz point where the order–disorder transition line coincides with the macrophase transition line on the phase diagram (temperature versus ϕ, the net volume fraction of A and B homopolymers, where the net volume fraction of A in the mixture, $\overline{\phi_A}$, is fixed at about 0.5) [53]. The Ginzburg–Landau free-energy density can be given by [53]

$$F(\psi) = a_2\psi^2 + a_4\psi^4 + a_6\psi^6 + \ldots + c_1(\nabla\psi)^2 + c_2(\nabla^2\psi)^2 + \ldots \tag{34}$$

At an ordinary critical point, $a_2 = 0$, and at the Lifshitz point, $a_2 = c_1 = 0$. Since the scattering function, $S(q)$ is in general given by [54]:

$$S^{-1}(q) = 2a_2 + 2c_1q^2 + 2c_2q^4 + \ldots \tag{35}$$

then at the Lifshitz point composition where $c_1 = 0$, the scattering function becomes

$$S^{-1}(q \to 0) = 2a_2 + 2c_2q^4 \tag{36}$$

Recalling that the scattering function of an ordinary binary mixture has a q^2 term (the OZ function), it is interesting to see that the scattering function for the A–B/A/B mixture ($\overline{\phi_A} \cong 0.5$) at the Lifshitz point has a q^4 term instead of the q^2 term. Even in the vicinity of the Lifshitz point, not exactly at the critical composition of the Lifshitz point, the scattering function is found to be of the form of Eq. (36) [55]. Therefore, distinction of the true Lifshitz point by scattering may be difficult. More experimental studies are required to check the scattering function. For this system, the correlation length, ξ, of the concentration fluctuations is defined by [54];

$$\xi = (4c_2/a_2)^{1/4} \tag{37}$$

Bates et al. [53] have recently reported the mean-field behavior with the Lifshitz critical exponents $\gamma = 1$ and $\nu = 0.25$ [56] for the polyethylene (PE)/poly(ethylenepropylene) (PEP)/DPE–PEP blend ($N_{PE} = 392$, $N_{PEP} = 409$, $N_{DPE-PEP} = 1925$) with the overall homopolymer volume fraction of 0.916 (at the Lifshitz point). Note that the temperature range examined is 4 to 100 K above the Lifshitz point temperature (T_{LP}) [$0.01 < (T - T_{LP})/T_{LP} \le 0.25$] with $T_{LP} = 399.2$ K.

Schwahn et al. [57] have examined the critical exponents both above and below the critical point. Namely, they have discussed the exponents

ν, ν', γ, and γ' in the following critical scaling relationships for the UCST-type poly(phenylmethyl siloxane) (PPMS)/deuterated polystyrene (DPS) blend:

$$S(q=0) = C^{+}\varepsilon^{-\gamma}, \xi = \xi^{+}\varepsilon^{-\nu} \qquad \text{for } T > T_c \tag{38}$$

$$S(q=0) = C^{-}\varepsilon^{-\gamma'}, \xi = \xi^{-}\varepsilon^{-\nu'} \qquad \text{for } T < T_c \tag{39}$$

The exponents for $T > T_c$ and for $T < T_c$ are found to agree with each other and with the theoretical values within 3% (typically $\gamma = 1.23$, $\gamma' = 1.25$, $\nu = 0.62$, and $\nu' = 0.60$ for their experimental results, and $\gamma = 1.24$ and $\nu = 0.63$ for the theoretical results).

Schwahn et al. [35] have also reported that the value $\gamma = 1.39$ is obtained for a blend of PS with high 1,2-content deuterated polybutadiene (DPB) (91% of 1,2). The data seem to show crossover from $\gamma = 1.24$ (3D Ising) to 1.39 by approaching to the critical point. They mentioned that this behavior with $\gamma = 1.39$ might result from the occurrence of "Fisher renormalized" (FR) critical behavior [58]. The FR behavior is predicted in ternary systems such as a polymer blend in solvent near T_c, where a solvent plays an important role as a third component which is chemically inert for polymer components. The analogy between a polymer blend solution and a blend of DPS/DPB (high 1,2) implies that a large number of vacancies or free volume in the DPS/DPB mixture could behave like a solvent. To check this curious crossover behavior, further work is required.

IV. EVALUATION AND CONTROL OF MISCIBILITY

A. Classical Methods for Evaluation of χ_{AB}

The SAXS and SANS techniques are widely and effectively utilized to evaluate the interaction parameter, χ_{AB}, in order to quantitatively discuss the miscibility. The random phase approximation (RPA) theory [7] is generally employed to express the scattering function (Eq. (14)), as described in Sec. II. B While the details for the recent advances in the analyses based on the RPA theory will appear in Sec. IV. B, this section is mainly concerned with the classical method for the determination of the χ_{AB} parameter. As already mentioned, χ_{AB} can be evaluated from ΔH_{mix} [59] and A_2 [9]. To obtain ΔH_{mix} experimentally, the most popular method utilizes an oligomeric mixture, on which thermodynamic measurements can be made. However, it is much more desirable to obtain χ_{AB} values which describe the segmental interaction in the polymeric mixture. As described in Sec. IV. C. 2, χ_{AB} has phenomenological molecular weight dependence. Therefore, the methodology utilizing oligomeric mixtures is rather limited. One of the most commonly employed methods to obtain χ_{AB} is a parameter fitting to experimentally determined phase diagrams [12,13]. The binodal curve calculated using Eqs. (5) or (6) is fitted to the phase diagram with χ_{AB} as a fit parameter. In early 1970, it was already found that the evaluated χ_{AB} parameter depends on the blend composition, as well as temperature. In addition to the composition dependence of χ_{AB}, the validity of the FH lattice theory is the key issue in this analysis. Therefore, reasonable modifications to the FH lattice theory are required for more accurate evaluation of χ_{AB}. Freed and coworkers [60,61] have generalized the FH theory to include compressibility by introducing the contribution of voids, and more recently they have developed a new lattice model (Huggins–Guggenheim–Miller-type theory) to take into account a branched lattice [62], and a segment occupying several lattice sites [63]. The latter is relevant to extensions of the lattice model to more realistic segment shapes. Hobbie and Han [64] have presented a lattice model of a hydrogen-bonded polymer blend. Thus, actual polymer blend systems, which are usually accompanied by various types of specific interactions, require an appropriate model. It is important to minimize practical inconvenience and hence it is important for future models to be generally applicable to a variety of shapes or interactions.

One of the other notable methods to evaluate χ_{AB} is the Zimm method [65] which is utilized for analysis of scattering experiments. This method is generally useful for polymer solutions to determine simultanteously the weight-average molecular weight, the radius of gyration, and the second virial coefficient, A_2 [66–68]. The scattering function from a blend containing a small amount of deuterated polymer A with a major component of a protonated polymer B can also be analyzed by the Zimm equation in the small-angle region [65]. Since the second virial coefficient, A_2, is related to the interaction parameter, χ_{AB}, through Eq. (3), the interaction parameter, χ_{AB}, between A and B segments is evaluated for low concentrations of the A polymer in the A/B blends. The two methods mentioned above, however, are restricted to narrow temperature and composition ranges, respectively. Namely, the former is relevant to the value of χ_{AB} at only one temperature (binodal temperature) for a given composition, and the latter is only applicable to low concentrations of A (or alternatively B). On the other hand, the composition and temperature dependence of the interaction parameter, χ_{AB}, can be evaluated independently using the RPA method when the scattering profile is measured at a given temperature for any given composition [26,69].

B. Scattering Technique for Miscibility Evaluation

1. General Expression for Structure Factor

The scattering function is calculated according to the RPA theory as follows:

$$I(q) = K\left(\frac{b_A}{v_A} - \frac{b_B}{v_B}\right)^2 \left[\frac{\overline{S(q)}}{\overline{W(q)}} - 2\chi_{AB}\right]^{-1} \tag{40}$$

where K is a proportional constant and the factor $(b_A/v_A - b_B/v_B)^2$ is the contrast between A and B segments for the radiation, and the quantities b_A and b_B are, respectively, the scattering lengths [70] of A and B segments for neutrons or the electron densities for X-rays. Here, v_K and v_0 denote the molar volumes of K (=A or B) segments and of the reference cell, respectively. v_K is calculated as $v_K \equiv Mu_K/\rho_K$, where Mu_K is the molecular weight of the K segment, and ρ_K is the mass density of the K polymer. We assume $v_0 = \sqrt{v_A v_B}$. The functions $\overline{S(q)}$ and $\overline{W(q)}$ are given by the following equations using the volume-averaged correlation functions, $\langle S_{AA}(q)\rangle_v$, $\langle S_{AB}(q)\rangle_v$, and $\langle S_{BB}(q)\rangle_v$ for polydisperse samples (see Eqs. (47)–(51) below)

$$\overline{S(q)} = \langle S_{AA}(q)\rangle_v + 2\langle S_{AB}(q)\rangle_v + \langle S_{BB}(q)\rangle_v \tag{41}$$

$$\overline{W(q)} = \langle S_{AA}(q)\rangle_v \langle S_{BB}(q)\rangle_v - \langle S_{AA}(q)\rangle_v^2 \tag{42}$$

with

$$S_{AA}(q) = \left(\frac{v_A}{v_0}\right)^2 \frac{1}{r_K} \sum_{i,j=1}^{N_K} \delta_i \delta_j P_{ij}(q) \tag{43}$$

$$S_{BB}(q) = \left(\frac{v_B}{v_0}\right)^2 \frac{1}{r_K} \sum_{i,j=1}^{N_K} (1-\delta_i)(1-\delta_j) P_{ij}(q) \tag{44}$$

$$S_{AB}(q) = \left(\frac{v_A}{v_0}\right)\left(\frac{v_B}{v_0}\right) \frac{1}{r_K} \sum_{i,j=1}^{N_K} \delta_i (1-\delta_j) P_{ij}(q) \tag{45}$$

where

$$P_{ij}(q) = \exp[-\langle R^2\rangle_{ij} q^2] \tag{46}$$

and q is the magnitude of the scattering vector, defined by Eq. (15). In Eqs. (43)–(45), δ_i (or δ_j) is equal to unity when the A segment occupies the i-th (or j-th) segmental position on the polymer chain from one end and δ_i (or δ_j) is equal to zero when the B segment occupies the i-th (or j-th) position. Here N_K is the polymerization index of a component polymer. For polymer blends, N_K should be taken as N_A or N_B and for A–B diblock copolymers, N_K should be equal to $N_A + N_B$. Moreover, r_K is the reduced degree of polymerization of the component polymer, and is defined as $r_K = (v_K/v_0)N_K$ (K = A or B) for polymer blends and $r_K = r_{bcp}(\equiv r_A + r_B$, Eq. (59)) for A–B diblock copolymers. It is therefore easy to see that the function $S_{AB}(q)$ necessarily is equal to zero when there is no connectivity between the A and B segments, as in A/B polymer blends. On the other hand, for the block copolymers, $S_{AB}(q)$ has nonzero values. The quantity $\langle R^2 \rangle_{ij}$ in Eq. (46), designates the mean-squared distance between the i-th and j-th positions. Thus, the scattering function $I(q)$ can be obtained analytically when Eq. (40) is formulated. Structure factors (corresponding to the correlation functions $S_{AA}(q)$, $S_{BB}(q)$, and optionally $S_{AB}(q)$) are available in the literature for various types of polymeric chain architectures, for instance, randomly crosslinked polymer networks as formed by, e.g., transesterification [71], star-shaped molecules [72], starburst-dendrimers [73], multiblock copolymers [74], regular polymer gels and networks [75], and flexible/stiff polymer blends [76], as well as various other types of chain architecture [77–79]. Note that the Debye scattering function given by Eq. (16) is the resultant formulation of Eq. (46) for a single Gaussian chain. Consequently, Eq. (14) is obtained for a monodisperse A/B polymer blend with $v_A = v_B = v_0$.

2. Effects of Polydispersity and Asymmetry in Segment Size on the RPA

Next, we consider the effects of polydispersity and asymmetry in the segment size on the RPA formula. For polydisperse samples, the correlation functions $S_{AA}(q)$, $S_{AB}(q)$, and $S_{BB}(q)$ should be volume-averaged using the Schultz–Zimm distribution function [80] for the molecular weight as:

$$\begin{pmatrix} \langle S_{AA}(q) \rangle_v \\ \langle S_{BB}(q) \rangle_v \\ \langle S_{AB}(q) \rangle_v \end{pmatrix} = \int_0^\infty \int_0^\infty dN_A \, dN_B \, \phi(N_A) \, \phi(N_B) \begin{pmatrix} S_{AA}(q) \\ S_{BB}(q) \\ S_{AB}(q) \end{pmatrix} \tag{47}$$

where $\phi(N_K)$ denotes the normalized volume distribution function of the polymer having the degree of polymerization N_K, which is given by:

$$\phi(N_K) = \frac{v_K N_K \Psi(N_K)}{\int_0^\infty v_K N_K \Psi(N_K) \, dN_K} \cong \frac{N_K \Psi(N_K)}{\int_0^\infty N_K \Psi(N_K) \, dN_K} \tag{48}$$

with

$$\Psi(N_K) = \frac{\nu_K^{h_K+1} N_K^{h_K}}{\Gamma(k_K + 1)} \exp(-\nu_K N_K) \tag{49}$$

$$\nu_K = \frac{h_K}{N_{K,n}} = \frac{h_K + 1}{N_{K,w}} \tag{50}$$

$$\frac{N_{K,w}}{N_{K,n}} = \frac{h_K + 1}{K_K} \equiv \lambda_K \tag{51}$$

In the following sections (IV. B.3–IV. B.6), the scattering functions are formulated for various systems by considering the effects of polydispersity and asymmetry in the segment size. The procedure is given by Eq. (47). The effects on the scattering function are not straightforward in the various systems. In Ref. 81, the effects are shown for an A–B–A triblock copolymer. It is demonstrated that the evaluated values of χ_{AB} without considering these effects are about twice as large as those evaluated including these effects [81]. It is also noteworthy that the effect of polydispersity on the evaluation of χ_{AB} is significant even for a small molecular weight distribution ($M_w/M_n = 1.2 \pm 0.15$) [81].

3. Polymer Blends

According to Eqs. (40)–(46), the scattering function for a polydisperse A/B polymer blend is formulated as:

$$\frac{1}{I(q)} = \frac{1}{v_A N_A \phi_A \overline{g_D(x_A)}} + \frac{1}{v_B N_B (1 - \phi_A) \overline{g_D(x_B)}} - \frac{2\chi_{AB}}{v_0} \tag{52}$$

which is a generalized version of Eq. (14) by taking into account the effects of polydispersity and the asymmetry in the segment size [69]. In Eq. (52), the function $\overline{g_D(x_K)}$ is given by

$$\overline{g_D(x_K)} = \frac{2}{x_{K,n}^2} \left[x_{K,n} - 1 + \left(\frac{h_K}{h_K + x_{K,n}} \right)^{h_K} \right] \tag{53}$$

with

$$x_{K,n} \equiv \frac{q^2 N_{K,n} b_K^2}{6} = q^2 \langle Rg_K^2 \rangle_n \tag{54}$$

and

$$h_K \equiv \left(\frac{N_{K,w}}{N_{K,n}} - 1 \right)^{-1} = \frac{1}{\lambda_K - 1} \tag{55}$$

where $N_{K,n}$ and $N_{K,w}$ designate number- and weight-average degrees of polymerization of the K polymer. Note that the third term in the square brackets of Eq. (53) converges on $\exp(-x_K)$ for the monodisperse limit. Thus, in the monodisperse limit, the function $\overline{g_D(x_K)}$ is identical to the Debye function (Eq. (16)).

4. Diblock Copolymers

For polydisperse A–B diblock copolymers, the scattering function in the disordered state is formulated as [81,82]:

$$S_{AA}(q) = r_{bcp}\phi_A^2\,\overline{g_D(x_A)} \tag{56}$$

$$S_{BB}(q) = r_{bcp}(1-\phi_A)^2\,\overline{g_D(x_B)} \tag{57}$$

$$S_{AB}(q) = r_{bcp}\phi_A(1-\phi_A)\,\overline{g_1(x_A)}\;\overline{g_1(x_B)} \tag{58}$$

r_{bcp} is the reduced degree of polymerization for the entire diblock copolymer, and is defined as:

$$r_{bcp} = \frac{v_A N_A + v_B N_B}{v_0} \tag{59}$$

ϕ_A should read as the volume fraction of A-block chains in the A–B diblock copolymer, and is given by:

$$\phi_A = \frac{r_A}{r_{bpc}} = \frac{v_A N_A}{v_A N_A + v_B N_B} \tag{60}$$

The volume-averaged single chain correlation function $\overline{g_1(x_K)}$, which appears in Eq. (58), is given by:

$$\overline{g_1(x_K)} = \frac{1}{x_{K,n}}\left[1 - \left(\frac{h_K}{h_K + x_{K,n}}\right)^{h_K}\right] \tag{61}$$

for a Gaussian chain using the Schultz–Zimm molecular weight distribution function [80]. Since the second term in square brackets in Eq. (61) converges on $\exp(-x_K)$ for the monodisperse limit, $\overline{g_1(x_K)}$ converges on

$$\lim_{\lambda_K \to 1}\left[\overline{g_1(x_K)}\right] = \frac{1}{x_K}[1 - \exp(-x_K)] \tag{62}$$

In Eqs. (55) and (62), λ_K characterizes the polydispersity of the K-block chain in the diblock copolymer, i.e., $\lambda_K \equiv (M_w/M_n)_K$. Since the values of λ_K are not always individually available, we calculated λ_K from the polydispersity index for the entire diblock copolymer, M_w/M_n ($\equiv \lambda_0$), using the relationship expressed by Eq. (63) assuming $\lambda_A = \lambda_B$ [81]:

$$\lambda_0 - 1 = (\lambda_A - 1)w_A^2 + (\lambda_B - 1)w_B^2 \tag{63}$$

where w_K denotes the weight fraction of the K component (K = A or B).

As depicted in Fig. 7, the scattering function for a block copolymer as a function of temperature is demonstrated. The data were obtained by SAXS measurements at the labeled temperatures for a polystyrene–polybutadiene–polystyrene (SBS) triblock copolymer ($M_n = 6.31 \times 10^4$ and $M_w/M_n = 1.15$) in dioctyl phthalate (DOP) with a polymer volume fraction $\phi_p = 0.68$ [81]. The solid curves are the best fit theoretical scattering profiles using the RPA formula developed for polydisperse A–B–A triblock copolymers (see Ref. 81 for the detailed formulae which are not identical to Eq. (58) for $S_{AB}(q)$). The upturns in the low-angle region are not describable by the RPA theory and may reflect segregated structures. However, further discussion on the upturn scattering is beyond the scope of this chapter. The scattering profiles exhibit a single peak, in contrast with the polymer

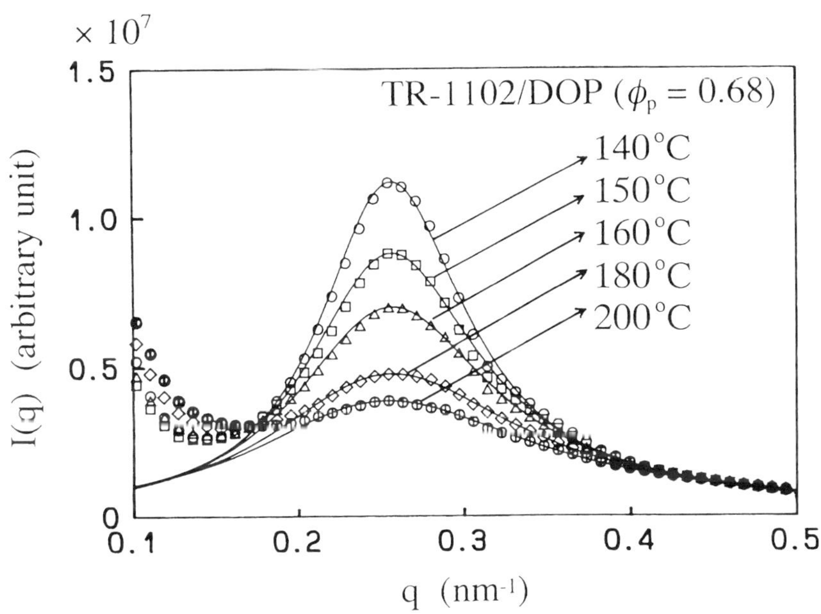

Figure 7 SAXS profiles for the polystyrene–polybutadiene–polystyrene (SBS) triblock copolymer (code name TR-1102; $M_n = 6.31 \times 10^4$ and $M_w/M_n = 1.15$) in dioctyl phthalate (DOP) with the polymer volume fraction, $\phi_p = 0.68$, measured at various temperatures. The solid curves are the best fit theoretical scattering profiles using the RPA formula developed for polydisperse A–B–A triblock copolymers (see Ref. 81 for the detailed formulae). In the region of $0.2 \leq q < 0.5$ nm^{-1}, the agreement is quite well, and thus the evaluated χ_{AB} values show the temperature dependence of $\chi_{AB} = A + B/T$. From Ref. 81.

blend shown in Fig. 3. Note that the SAXS peak from the block copolymers in the disordered state does not indicate the formation of any segregated structure. The peak results from concentration fluctuations, as clearly demonstrated by the RPA procedure [83]. The appearance of the peak can be understood as a consequence of the suppression of the large size concentration fluctuations as compared to the chain dimension of the entire block copolymer. The scattering intensity is proportional to the square of the amplitude of the concentration fluctuations, and hence the scattering is suppressed in the very small-angle region, which causes the appearance of a single peak at a finite q value roughly corresponding to the inverse of the chain dimension. In this respect, such an effect is called a "correlation hole" effect [7]. As can be seen in the region of $0.2 \leq q \leq 0.5\text{nm}^{-1}$, the agreement is quite good, and thus evaluated χ_{AB} values show the temperature dependence of $\chi_{AB} = A + B/T$ [81].

5. Binary Mixture of A–B Diblock Copolymers

To calculate the scattering function for binary mixtures of diblock copolymers $[\alpha : (A - B)_1$ and $\beta : (A - B)_2]$ in the disordered state, we only have to modify the correlation functions S_{AA}, S_{BB}, and S_{AB} by volume-averaging as follows [77,84]:

$$S_{AA}(q) = \phi_\alpha S_{AA}^\alpha(q) + \phi_\beta S_{AA}^\beta(q) \tag{64}$$

$$S_{BB}(q) = \phi_\alpha S_{BB}^\alpha(q) + \phi_\beta S_{BB}^\beta(q) \tag{65}$$

$$S_{AB}(q) = \phi_\alpha S_{AB}^\alpha(q) + \phi_\beta S_{AB}^\beta(q) \tag{66}$$

where ϕ_α and ϕ_β denote the fractions of α and β in the $(\alpha + \beta)$ mixture, respectively $(\phi_\alpha + \phi_\beta = 1)$. $S_{AA}^\alpha(q)$ expresses the A–A correlation in the α block copolymer and similar notations apply for the other pair correlations. These formulae can be readily extended to a binary mixture of a homopolymer and a diblock copolymer [77,85,86].

6. Multicomponent Mixtures (A/B/C Mixtures)

The above RPA formula is for a two-component system, i.e., all the constituents comprise A- and B-type segments, and therefore only one interaction parameter (χ_{AB}) is in principle required to describe the interactions in the system. Here, the RPA theory is extended to the multicomponent mixtures, where more than two interaction parameters are required to describe the interactions in the system [87–90]. The elastic static scattering function,

$\mathbf{S}(q)$, can be given in matrix form from the following relationship obtained by Benoit et al. [91], and Akcasu and Tombakoglu [92]:

$$\mathbf{S}^{-1}(q) = \mathbf{S}_0^{-1}(q) + \mathbf{V}(q) \tag{67}$$

and thus

$$\mathbf{S}(q) = [\mathbf{E} + \mathbf{S}_0(q)\mathbf{V}(q)]^{-1}\mathbf{S}_0(q) \tag{68}$$

where $\mathbf{S}_0(q)$ and $\mathbf{V}(q)$ are the matrices for the ideal (Gaussian chain) structure factors and interactions, respectively. $\mathbf{E}$ is the unit matrix. For an incompressible mixture of A/B/C, the C component is assumed to be the reference, then the components of $\mathbf{S}_0(q)$ can be formulated as:

$$S_{\mathrm{AA}}^0(q) = N_{\mathrm{A}}\phi_{\mathrm{A}}v_{\mathrm{A}}\overline{g_{\mathrm{D}}(x_{\mathrm{A}})} \tag{69}$$

$$S_{\mathrm{BB}}^0(q) = N_{\mathrm{B}}\phi_{\mathrm{B}}v_{\mathrm{B}}\overline{g_{\mathrm{D}}(x_{\mathrm{B}})} \tag{70}$$

$$S_{\mathrm{AB}}^0(q) = 0 \tag{71}$$

where ϕ_{B} denotes the volume fraction of the B component. It is convenient here to demonstrate the result for A–B diblock copolymer/C homopolymer mixtures. For this purpose, Eq. (71) should be replaced by the following equation due to the connectivity between A and B segments:

$$S_{\mathrm{AB}}^0(q) = (N_{\mathrm{A}}\phi_{\mathrm{A}}v_{\mathrm{A}}N_{\mathrm{B}}\phi_{\mathrm{B}}v_{\mathrm{B}})^{\frac{1}{2}}\overline{g_1(x_{\mathrm{A}})}\,\overline{g_1(x_{\mathrm{B}})} \tag{72}$$

The components of the interaction matrix $\mathbf{V}(q)$ are now expressed by:

$$V_{\mathrm{AA}}(q) = \frac{1}{S_{\mathrm{CC}}^0(q)} - \frac{2\chi_{\mathrm{AC}}}{v_0} \tag{73}$$

$$V_{\mathrm{BB}}(q) = \frac{1}{S_{\mathrm{CC}}^0(q)} - \frac{2\chi_{\mathrm{BC}}}{v_0} \tag{74}$$

$$V_{\mathrm{AB}}(q) = \frac{1}{S_{\mathrm{CC}}^0(q)} + \frac{\chi_{\mathrm{AB}}}{v_0} - \frac{\chi_{\mathrm{AC}}}{v_0} - \frac{\chi_{\mathrm{BC}}}{v_0} \tag{75}$$

with

$$S_{\mathrm{CC}}^0(q) = N_{\mathrm{C}}\phi_{\mathrm{C}}v_{\mathrm{C}}\overline{g_{\mathrm{D}}(x_{\mathrm{C}})} \tag{76}$$

where ϕ_{C} denotes the volume fraction of the C component. Thus, the bare structure factor for the C component (ideal chain) affects the actual structure factor through $\mathbf{V}(q)$. using Eq. (68), the components of the actual structure factor matrix $\mathbf{S}(q)$ are given by:

$$S_{\mathrm{AA}}(q) = \left[\frac{S_{\mathrm{AA}}^0(q)}{F(q)} + V_{\mathrm{BB}}(q)\right] \Big/ G(q) \tag{77}$$

$$S_{\mathrm{BB}}(q) = \left[\frac{S^0_{\mathrm{BB}}(q)}{F(q)} + V_{\mathrm{AA}}(q)\right] \Big/ G(q) \tag{78}$$

$$S_{\mathrm{AB}}(q) = \left[\frac{S^0_{\mathrm{AB}}(q)}{F(q)} - V_{\mathrm{AB}}(q)\right] \Big/ G(q) \tag{79}$$

where the denominators $F(q)$ and $G(q)$ are, respectively, given by:

$$F(q) = S^0_{\mathrm{AA}}(q) S^0_{\mathrm{BB}}(q) - \left\{S^0_{\mathrm{AB}}(q)\right\}^2 \tag{80}$$

and

$$G(q) = \left[\frac{S^0_{\mathrm{BB}}(q)}{F(q)} + V_{\mathrm{AA}}(q)\right]\left[\frac{S^0_{\mathrm{AA}}(q)}{F(q)} + V_{\mathrm{BB}}(q)\right] - \left[\frac{S^0_{\mathrm{AB}}(q)}{F(q)} - V_{\mathrm{AB}}(q)\right]^2 \tag{81}$$

Consequently, the SANS intensity $I(q)$ is given by:

$$I(q) = \mathbf{B}^{\mathrm{T}} \mathbf{S}(q) \mathbf{B} = B_{\mathrm{A}}^2 S_{\mathrm{AA}}(q) + 2 B_{\mathrm{A}} B_{\mathrm{B}} S_{\mathrm{AB}}(q) + B_{\mathrm{B}}^2 S_{\mathrm{BB}}(q) \tag{82}$$

where B_{K} is a component of the column vector $\mathbf{B}$ ($\mathbf{B}^{\mathrm{T}}$ is the corresponding row vector) and is defined as:

$$B_{\mathrm{K}} = \frac{b_{\mathrm{K}}}{v_{\mathrm{K}}} - \frac{b_{\mathrm{C}}}{v_{\mathrm{C}}} \qquad (\mathrm{K} = \mathrm{A \ or \ B}) \tag{83}$$

with scattering length b_{K}, and molar volume, v_{K}, for the K segment.

It is straightforward to apply this equation to a polymer blend solution (A/B in solvent) or a block copolymer solution (A–B in solvent). Occasionally, for easing experimental limitations, solvent is sometimes added to reduce the bare A–B interaction, especially in high molecular weight polymer blends or block copolymers [93]. In order to analyze the scattering intensity profile for these solutions within the framework of the binary RPA formalism, some unfavorable assumptions must be made. Among them, the requirement of an optical theta condition [94], or in other words, contrast matching to extinguish the scattering from A/solvent and B/solvent fluctuations, is particularly rigorous, and trivial deviations from the optical theta condition may cause nontrivial effects in the evaluation of the binary segmental interaction, χ_{AB}. Even if that condition is satisfied, the dilution approximation as expressed by [81,95,96]:

$$\chi_{\mathrm{app}} = \chi_{\mathrm{AB}} \cdot \phi_{\mathrm{p}} \tag{84}$$

is usually assumed for evaluating the bare interaction parameter, χ_{AB}, from the apparent interaction paramter, χ_{app}, obtained from the scattering experiments on solutions, where ϕ_p stands for the polymer volume fraction in the solution. However, it has been found that the dilution approximation is barely upheld for a block copolymer solution [97]. Therefore, the three-component RPA formalism must be used to extract χ_{AB} from the measurements of the A/B polymer blend solution or the A–B block copolymer solution. If the individual interactions of A and B components against the solvent molecules, i.e., χ_{AS} and χ_{BS} are known, then χ_{AB} can be directly evaluated from the measurements of those solutions without any assumptions.

7. Compressible RPA Formula

The number of the publications reporting on the pressure dependence of the miscibility has increased remarkably due to improvements in the experimental technique [41,98–101]. Although this is another important factor to hasten the development of the compressible RPA formalism, the formalism itself is necessary for better evaluation of the interaction parameter. The compressible RPA was first formulated by McMullen and Freed [102] using the Edwards path integral method. Later, the theory was extended by Tang and Freed [103] to derive the static structure factors for the compressible polymer blends and diblock copolymers. More recently, Bidkar and Sanchez [104] have independently derived the structure factor for compressible polymer mixtures starting with the multicomponent scattering function of Eq. (67) by using the Sanchez–Lacombe lattice fluid (LF) model [20,105]. The technical procedure to determine the free volume fraction and the intercomponent interaction cohesive energy (P^*_{AB}) is well demonstrated in the paper of Hammouda and Benmouna [101]. Here, the procedure is introduced using the final form of the compressible RPA formula. In Fig. 8, the procedural flow chart is shown.

First of all, the LF equation of state (EOS) is given as:

$$(1-f_0)^2 + \frac{P}{P^*} + \frac{T}{T^*}[\ln(f_0) + (1-f_0)] = 0 \qquad (85)$$

where f_0 denotes a free volume fraction. P^* and T^* are the cohesive energy densities (internal pressures) and characteristic temperatures, respectively, whose values for ordinary polymers are available in the literature [106]. For instance, $P^* = 355$ MPa and $T^* = 731$ K for DPS and $P^* = 353$ MPa and $T^* = 657$ K for PVME [100,101].

Let us introduce a free volume fraction, f_0, to an A/B polymer blend with volume fractions of ϕ_A and ϕ_B $(= 1 - \phi_A)$, respectively. In turn, the true volume fractions are determined to be $\phi_A(1-f_0)$ and $\phi_B(1-f_0)$,

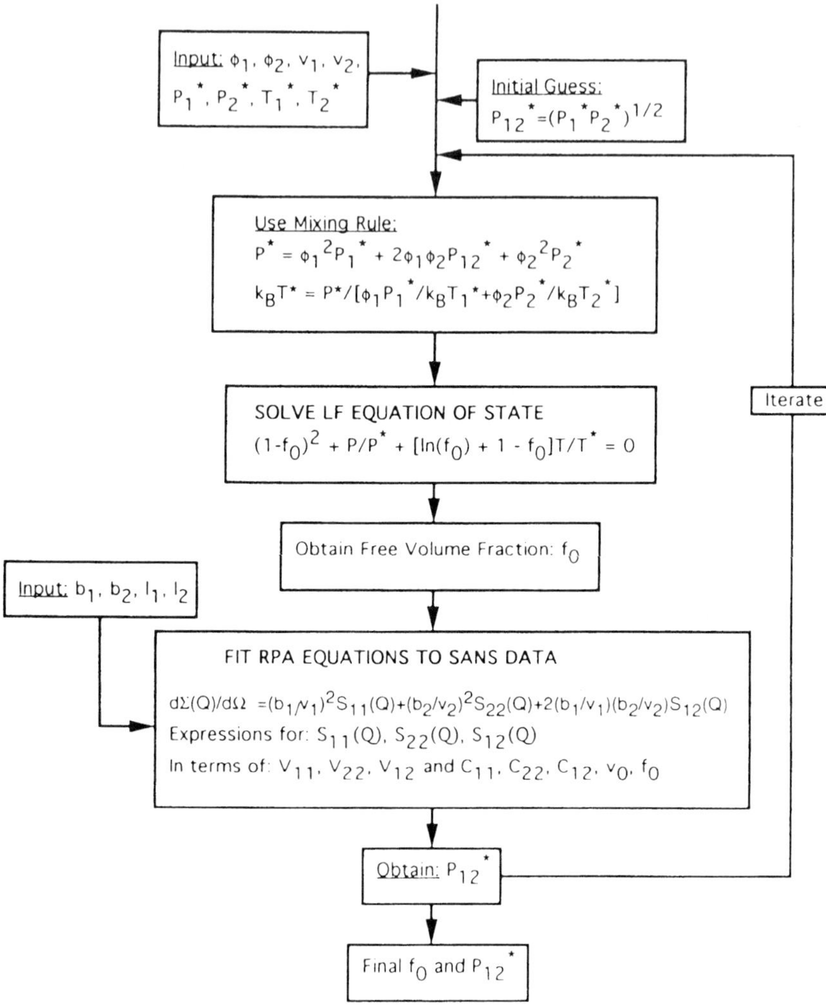

Figure 8 Procedural flow chart for the analysis of SANS profiles using the compressible RPA method. From Ref. 101.

respectively, for the A and B components. To avoid confusion, note here that characters "1" and "2" are used only in Fig. 8 to specify the components A and B. For polymer blends, the internal pressure P^*_{AB} is given by [101]:

$$P_{\mathrm{AB}}^* = -W_{\mathrm{AB}}^*/2v_0 \tag{86}$$

with W_{AB}^* and v_0 being the interaction energy between segments of type A and B, and a reference cell volume ($v_0 = \sqrt{v_{\mathrm{A}} v_{\mathrm{B}}}$), respectively. The FH interaction parameter, χ_{AB}, is related to those parameters by:

$$\chi_{\mathrm{AB}} = \frac{v_0}{k_{\mathrm{B}} T}(P_{\mathrm{A}}^* + P_{\mathrm{B}}^* - 2P_{\mathrm{AB}}^*) = \frac{1}{k_{\mathrm{B}} T}\left(W_{\mathrm{AB}} - \frac{W_{\mathrm{AA}} + W_{\mathrm{BB}}}{2}\right) \tag{87}$$

The fraction of free volume, f_0, is then obtained by solving the LF EOS with mixing rules [104] for P^* and T^*, respectively, as:

$$P^* = \phi_{\mathrm{A}}^2 P_{\mathrm{A}}^* + 2\phi_{\mathrm{A}}\phi_{\mathrm{B}} P_{\mathrm{AB}}^* + \phi_{\mathrm{B}}^2 P_{\mathrm{B}}^* \tag{88}$$

and

$$\frac{P^*}{k_{\mathrm{B}} T^*} = \frac{\phi_{\mathrm{A}} P_{\mathrm{A}}^*}{k_{\mathrm{B}} T_{\mathrm{A}}^*} + \frac{\phi_{\mathrm{B}} P_{\mathrm{B}}^*}{k_{\mathrm{B}} T_{\mathrm{B}}^*} = \frac{1}{v_{\mathrm{f}}} \tag{89}$$

where v_{f} denotes the volume of voids that constitute the free volume. P^* and T^* are the cohesive energy densities (internal pressures) and characteristic temperatures for homogeneous mixtures, respectively. These values are needed to solve the LF EOS (Eq. 85)). P_{K}^* and T_{K}^* denote the internal pressures and characteristic temperatures for pure components K (= A or B), respectively.

Practically, in order to solve the EOS to obtain f_0 (the fraction of the free volume), the parameter P_{AB}^* (or χ_{AB}) must be known. The value P_{AB}^* is evaluated to give the best fit of the scattering function to the experimentally measured profile. However, it is not straightforward because the value of f_0 is needed to calculate the scattering function for compressible polymer blends, as shown below. Therefore, solving the EOS and the parameter fitting should be conducted self-consistently. Using the multicomponent RPA formula, the scattering function is derived as [104]:

$$S(q) = \left(\frac{b_{\mathrm{A}}}{v_{\mathrm{A}}}\right)^2 S_{\mathrm{AA}}(q) + \left(\frac{b_{\mathrm{B}}}{v_{\mathrm{B}}}\right)^2 S_{\mathrm{BB}}(q) + 2\left(\frac{b_{\mathrm{A}}}{v_{\mathrm{A}}}\right)\left(\frac{b_{\mathrm{B}}}{v_{\mathrm{B}}}\right) S_{\mathrm{AB}}(q) \tag{90}$$

with

$$S_{\mathrm{AA}}(q) = \left[S_{\mathrm{AA}}^0(q)\left(1 + V_{\mathrm{BB}}(q) S_{\mathrm{BB}}^0(q)\right)\right]/H(q) \tag{91}$$

$$S_{\mathrm{BB}}(q) = \left[S_{\mathrm{BB}}^0(q)\left(1 + V_{\mathrm{AA}}(q) S_{\mathrm{AA}}^0(q)\right)\right]/H(q) \tag{92}$$

$$S_{\mathrm{AB}}(q) = -S_{\mathrm{AA}}^0(q) V_{\mathrm{AB}}(q) S_{\mathrm{BB}}^0(q)/H(q) \tag{93}$$

where

$$H(q) = \left(1 + V_{\mathrm{AA}}(q)S^0_{\mathrm{AA}}(q)\right)\left(1 + V_{\mathrm{BB}}(q)S^0_{\mathrm{BB}}(q)\right) - V^2_{\mathrm{AB}}(q)S^0_{\mathrm{AA}}(q)S^0_{\mathrm{BB}}(q) \tag{94}$$

and

$$V_{\mathrm{AA}}(q) = \frac{1}{v_{\mathrm{f}}f_0} - \frac{2P^*_A}{k_{\mathrm{B}}T} + C_{\mathrm{AA}} \tag{95}$$

$$V_{\mathrm{BB}}(q) = \frac{1}{v_{\mathrm{f}}f_0} - \frac{2P^*_{\mathrm{B}}}{k_{\mathrm{B}}T} + C_{\mathrm{BB}} \tag{96}$$

$$V_{\mathrm{AB}}(q) = \frac{1}{v_{\mathrm{f}}f_0} - \frac{2P^*_{\mathrm{AB}}}{K_{\mathrm{B}}T} + C_{\mathrm{AB}} \tag{97}$$

The voids only contribute through the excluded volumes, V_{ij} ($i, j = \mathrm{A}$ or B). The ideal chain structure factors, $S^0_{\mathrm{AA}}(q)$ and $S^0_{\mathrm{BB}}(q)$, are given by the following equations with the Debye function, $g_{\mathrm{D}}(x_{\mathrm{K}})$, as:

$$S^0_{\mathrm{AA}}(q) = N_{\mathrm{A}}\phi_A(1 - f_0)v_{\mathrm{A}}g_{\mathrm{D}}(x_{\mathrm{A}}) \tag{98}$$

$$S^0_{\mathrm{BB}}(q) = N_{\mathrm{B}}\phi_{\mathrm{B}}(1 - f_0)v_{\mathrm{B}}g_{\mathrm{D}}(x_{\mathrm{B}}) \tag{99}$$

Finally, the coefficients C_{ij} ($i, j = \mathrm{A}$ or B) in Eqs. (95)–(97) designate contributions from the composition dependence of the average void size, v_{f}. These are given by:

$$C_{\mathrm{AA}} = -2\phi_{\mathrm{B}}u(f_0, \{P^*_{\mathrm{K}}\}, \{T^*_{\mathrm{K}}\}) \tag{100}$$

$$C_{\mathrm{BB}} = 2\phi_{\mathrm{A}}u(f_0, \{P^*_{\mathrm{K}}\}, \{T^*_{\mathrm{K}}\}) \tag{101}$$

$$C_{\mathrm{AB}} = (\phi_{\mathrm{A}} - \phi_{\mathrm{B}})u(f_0, \{P^*_{\mathrm{K}}\}, \{T^*_{\mathrm{K}}\}) \tag{102}$$

with

$$u(f_0, \{P^*_{\mathrm{K}}\}, \{T^*_{\mathrm{K}}\}) = \frac{\ln(f_0) + (1 - f_0)}{(1 - f_0)^2}\left(\frac{P^*_{\mathrm{A}}}{k_{\mathrm{B}}T^*_{\mathrm{A}}} - \frac{P^*_{\mathrm{B}}}{k_{\mathrm{B}}T^*_{\mathrm{B}}}\right) \quad (k = A \text{ or } B) \tag{103}$$

Note that the procedure to self-consistently solve the LF EOS is started using an initial guess of $P^*_{\mathrm{AB}} = \sqrt{P^*_{\mathrm{A}}P^*_{\mathrm{B}}}$ [101].

C. Discussion on Phenomenological χ_{AB} and Miscibility Control

In the framework of the original FH lattice theory, the interaction parameter, χ_{AB}, is introduced to describe the enthalpic contribution to the free energy upon mixing, as defined by Eq. (2), and evaluated from the cohesive energy density through Eq. (4). However, it has been shown experimentally that χ_{AB} depends on other parameters, e.g., the molecular weight, composition, the distance between two interacting segments, and the pressure. For a multicomponent polymeric system comprising more complicated chain structures, such phenomenological aspects of χ_{AB} have also been reported, for example, the effects of the arm number of star-block polystyrene-*b*-polyisoprene copolymers [79], the effects of the homopolymer composition in binary blends of SI/homoPS [86], the effects of the chain architecture for block copolymers comprising the same chemical species with polystyrene-*b*-polyisoprene diblocks and polyisoprene-*b*-polystyrene-*b*-polyisoprene triblocks [107], the topological effects on the miscibility where cyclic chains of polycarbonate (PC) are found to be more miscible with PS than linear chains of PC [108], the effects of crosslinks on the miscibility (see Chapter 5), and others. This section concerns those effects, especially isotope labeling effects, composition and molecular weight dependences, pressure dependence, and microstructure effects. Furthermore, a method of miscibility control using random copolymers is discussed.

1. Isotope Labeling Effects of χ_{AB}

As already stated, the SANS technique is one of the most powerful methods and is widely used at present for the studies of multicomponent systems, structural analyses, miscibility, and critical phenomena. The reason why the SANS technique is favorably used is ascribed to the advantages of deuterium labeling. As is well known, the scattering length of hydrogen is very different from that of deuterium [70]. Therefore, substitution of hydrogen by deuterium causes large differences in the scattering length while the thermodynamic properties are trivially affected by the substitution. This advantage is utilized for measurements of the single chain dimensions of a polymer [66–68], where a very small amount of the deuterated polymer, which has a sufficient contrast for the neutrons, is mixed into the hydrogenous polymer matrix comprising the same chemical species excepting the deuterium substitution. It can be said that the deuterated molecules "shine" when exposed to neutrons. However, a repulsive interaction between deuterated and hydrogenous components has been confirmed experimentally [36,37,69,109–116]. This is negligibly small on the segmental level (i.e., a very small value of χ_{HD}) but may cause phase separation in high molecular weight polymer blends [117] since the net interaction is expressed by a product of χ_{HD} and the degree of polymerization. The concept that deuterium labeling may induce phase separation was quite controversial [118]. Moreover, the general effects of the deuterium labeling are not straightforward [87,119–121]; sometimes it induces a favorable interaction, for instance in PS/PVME blends [87,119,120]. In this section, the effects of deuterium labeling on the miscibility and the phase diagram are summarized.

First, let us focus on the isotopic polymer mixtures comprised of deuterated and hydrogenous polymers of the same chemical species. Bates and coworkers [109,110] first reported small but positive values of χ_{HD} for both DPB/HPB (DPB, HPB; deuterated and hydrogenous polybutadiene, respectively) and DPS/HPS (hydrogenous polystyrene) [110] pairs. For such athermal pairs, the origin of the positive interaction was attributed to the reduction of the carbon–hydrogen bond length due to zero-point vibrations and the anharmonicity of the interatomic potential [110]. This reduction of the bond length results in a decrease of the segmental volume and the segment polarizability, which together generate a positive value of B ($\equiv \partial\chi_{AB}/\partial(1/T)$), which in turn results in UCST-type behavior. Sakurai et al. [69] have presented mostly negative values for χ_{HD} in the experimentally accessible temperature range (25–180°C) for DPB/HPB blends. However, they also expected a UCST-type phase diagram where the critical temperature was expected to be around −37°C. This temperature is far below the experimentally examined temperature range. The apparent discrepancy between the results of Bates and coworkers and those of Sakurai et al. has been interpreted in terms of random copolymer effects. Namely, the PB samples used by the two different groups should be treated as random copolymers comprising different amounts of 1,2 and 1,4 geometric isomers. By taking this effect into account, the χ_{HD} value between the deuterated and hydrogenous segments of 1,2 as well as 1,4 isomers was estimated to be:

$$\chi_{HD} = 1.68 \times 10^{-4} + 0.39/T \tag{104}$$

where $\chi_{HD}$for 1,2 isomers and χ_{HD} for 1,4 isomers were assumed to be identical due to a lack of data available for quantitative estimates of the individual χ_{HD} parameters. The details of the analysis based on the random copolymer effects are described in Sec. IV. C. 4. From the result of Eq. (104), it was confirmed that the segmental interaction is small but positive, which induces phase separation for high molecular weight polymer blends with a UCST-type phase diagram. Deuterated poly(methyl methacrylate)/hydrogenous poly(methyl methacrylate) (DPMMA/HPMMA), as studied by Hopkinson et al. [37], also showed a UCST-type phase diagram.

Next, we consider the effects of deuterium labeling on LCST-type phase behavior. One of the most thoroughly investigated pairs is the blend of DPS and PVME. Fig. 9 shows the phase diagrams for the three particular systems, DPS/PVME, DPS/HPS/PVME, and HPS/PVME, as determined by laser light cloud-point measurements [120]. The characteristics of those polymers are $M_w = 1.19 \times 10^5$, $M_w/M_n = 1.05$ (DPS), $M_w = 1.00 \times 10^5$, $M_w/M_n = 1.05$ (HPS), and $M_w = 9.9 \times 10^4$, $M_w/M_n =$

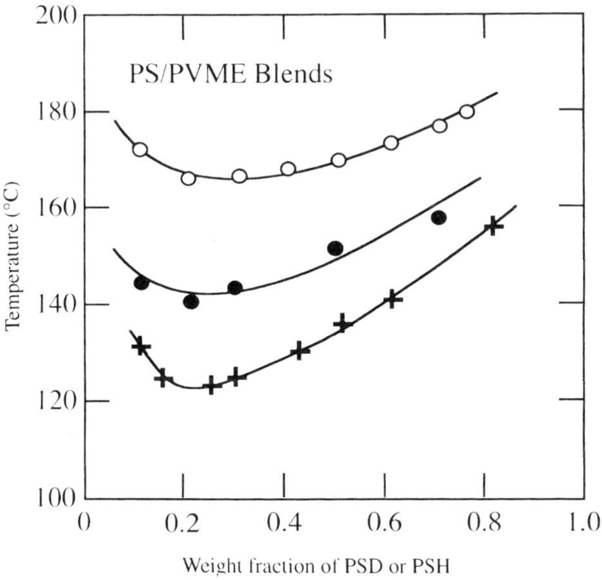

Figure 9 Phase diagram showing a plot of the cloud-point temperature versus weight fraction of PS(= DPS + HPS) for three sets of mixtures; DPS/PVME (open circle), DPS/HPS/PVME (closed circle), and HPS/PVME (cross). DPS: $M_w = 1.19 \times 10^5$, $M_w/M_n = 1.05$; HPS: $M_w = 1.00 \times 10^5$, $M_w/M_n = 1.05$; and PVME: $M_w = 9.9 \times 10^4$, $M_w/M_n = 2.13$. From Ref. 120.

2.13 (PVME). Although the molecular weight of DPS is slightly larger than that of HPS, it is obvious that the deuterium labeling enhances miscibility. Hammouda et al. [87] have experimentally determined the temperature dependence of χ_{AB} for both DPS/PVME and HPS/PVME by conducting SANS measurements on ternary mixtures of DPS/HPS/PVME then analyzing the data in terms of the multicomponent RPA procedure (see Sec. IV. B. 6). The results are shown in Fig. 10 in a plot of χ_{AB} versus the inverse of the absolute temperature ($1/T$). It is clear that the χ_{AB} values decrease upon deuterium substitution. The intercept values are found to be affected, while the slope of the line is invariant. From this observation, it is interesting to note that only the entropic contribution to χ_{AB} appears to be affected by the substitution.

Finally, the recent work by Russell [121] is noteworthy. Russell examined the effects of cross-deuteration (the so-called "switching effect") as well as dual deuteration for PMMA/PS segmental interaction. Fig. 11 shows

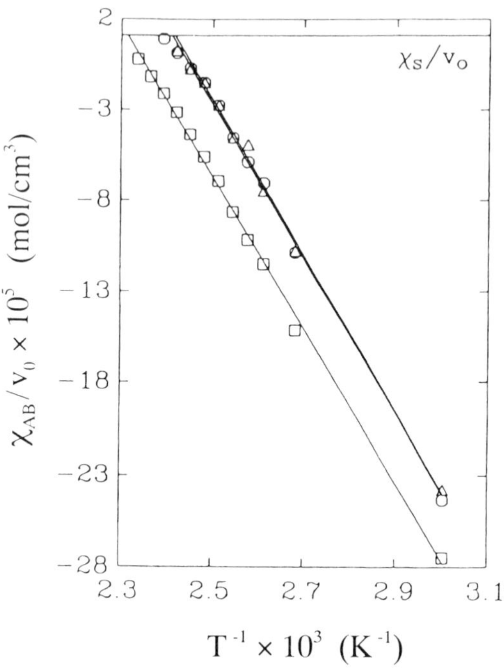

Figure 10 Plots of χ_{AB}/v_0 versus $1/T$ for DPS/PVME (□: from the binary mixture 48/52 by volume) and HPS/PVME (△ and ○) pairs. The χ_{AB} values for the HPS/PVME pair are obtained by analyzing the SANS data of a ternary mixture of DPS/HPS/PVME (△: 36/51/13; ○: 24/50/26 by volume) with the multicomponent RPA procedure. DPS: $M_w = 1.95 \times 10^5$, $M_w/M_n = 1.02$; HPS: $M_w = 1.90 \times 10^5$, $M_w/M_n = 1.04$; PVME: $M_w = 1.59 \times 10^5$, $M_w/M_n = 1.30$. From Ref. 87.

three sets of χ_{AB} parameters for HPMMA/DPS, DPMMA/HPS, and DPMMA/DPS as a function of $1/T$. The values were determined by analyzing the SANS profiles of the three diblock copolymers of PMMA-*b*-PS in the disordered state using the RPA. The χ_{AB} for HPMMA/HPS, shown in Fig. 11 by the broken line, was estimated from a separate study [122]. From these results, a very interesting conclusion on the effects of deuterium substitution can be drawn. First of all, when the substitution is performed in only one component, the repulsive interaction is dramatically increased, almost irrespective of which component is substituted. On the other hand, when both components are substituted, although the repulsive interaction is enhanced compared with the undeuterated pair, it is remarkably reduced compared with the case of one-component substitution. Again, only the entropic part of the χ_{AB} parameter is affected.

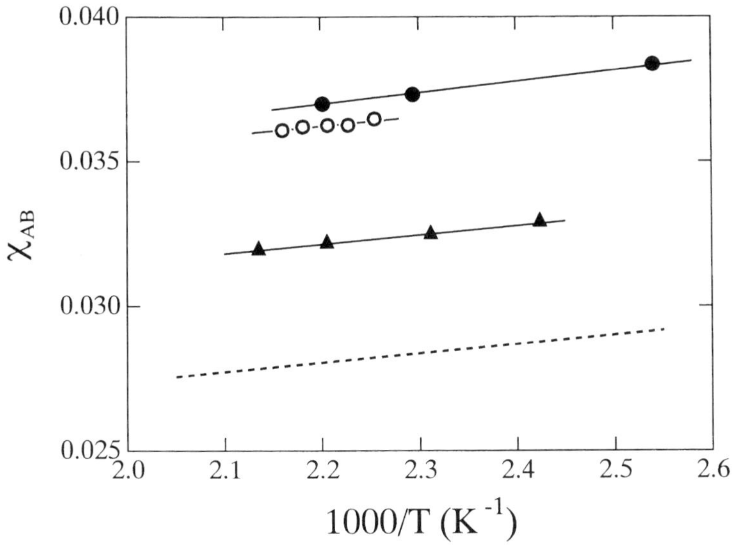

Figure 11 Plots of χ_{AB} versus $1/T$ for DPS-*block*-HPMMA (closed circles), HPS-*block*-DPMMA (open circles), and DPS-*block*-DPMMA (closed triangles) as analyzed by SAXS. From Ref. 121.

2. Composition and Molecular Weight Dependencies of χ_{AB}

The composition dependence of the interaction parameter, $\chi_{AB}(\phi_A)$, has been exclusively studied using the SANS technique with the RPA procedure for data analysis [25–27,36,37,69,113,116,123–125]. Some blend systems show a parabolic dependence while others show a linear dependence of χ_{AB} on concentration. For the parabolic dependence, both upward ($\partial^2 \chi_{AB}(\phi_A)/\partial\phi_A^2 > 0$) and downward ($\partial^2 \chi_{AB}(\phi_A)/\partial\phi_A^2 < 0$) curvatures have been found, depending on the blend system. Moreover, some blend systems (ethylene-but-1-ene copolymer, i.e., hydrogenated PB) are reported to show a parabolic shape with upward curvature changing gradually into a linear dependence with increasing temperature. Although the phenomenological results seem to be in a "chaotic" situation, some of them are reasonably understood. First of all, with respect to equation-of-state theories and from the analogy to a low molecular weight fluid mixture, it is well understood that $\chi_{AB}(\phi_A)$ shows a linear dependence for blends comprised of components with a large asymmetry in the free volumes, e.g., PS/PVME with $\partial\chi_{AB}(\phi_{DPS})/\partial\phi_{DPS} < 0$ [20,105]. Note that this type of mixture generally exhibits an LCST phase behavior. Dudowicz and Freed (126) have also reproduced the experimentally obtained composition dependence by using the lattice cluster theory [61], which is an extended lattice theory. However, recently, Hammouda and Bauer [100] have reported that the composition dependence is maintained under high pressure and concluded that the effect of the compressibility is not responsible for the composition dependence of χ_{AB}. Therefore, even for the classical PS/PVME blends, the origin of the composition dependence is still unclear.

The parabolic upward type ($\partial^2 \chi_{AB}(\phi_A)/\partial\phi_A^2 > 0$) dependence has been found for most isotopic mixtures [113,116,123]. As mentioned above, an isotopic mixture can be characterized in terms of the value of χ_{HD} which is very small but positive in UCST-type phase behavior. Bates et al. [113] have studied the composition and molecular weight dependences of the χ_{HD} parameter for isotopic blends of poly(ethylethylene) (PEE) and poly(vinylethylene) (PVE). They have obtained $\chi_{HD}(\phi_D, N)$ as depicted in Fig. 12. Note here that χ for the ordinate stands for the interaction parameter determined from scattering experiments, which is identical to χ_{sc} used in Sec. II. B. Moreover, ϕ_D for the abscissa denotes volume fraction of the deuterated component. It is obvious that the composition dependence of χ_{HD} is parabolic upward with respect to ϕ_D and that χ_{HD} decreases with an increase of N (PVE12; PVE1(d)/PVE2(h), PVE13; PVE1(d)/PVE3(h), where $N_n = 1270$, 1690, and 791 for PVE1, PVE2, and PVE3, respectively). Bates et al. have also compared their experimental results with a theoretical prediction by Muthukumar [127], which was developed by including a monomer concentration fluctuation effect and was originally applied to a polymer solution. According to their formalism, the corrected FH free energy of mixing is:

$$\frac{\Delta F_{\text{mix}}}{k_B T} = \frac{\phi_A}{N_A}\ln\phi_A + \frac{(1-\phi_A)}{N_B}\ln(1-\phi_A) + \phi_A(1-\phi_A)\chi_{FH} + (24\pi\xi^3)^{-1} \tag{105}$$

Here, χ_{FH} is the original FH interaction parameter (see Sec. II. B). Using the relationship of Eqs. (20) or (21), the interaction parameter determined from scattering experiments, χ_{sc}, can be written as:

$$\chi_{sc}(\phi_A, N) = \chi_{FH} - \frac{27}{4\pi}(\chi_s - \chi_{FH})^{1/2} \left\{\frac{3}{2}\left[\frac{1}{N_B(1-\phi_A)^2} - \frac{1}{N_A\phi_A^2}\right]\left[\phi_A^{1/2}(1-\phi_A)^{3/2} - (1-\phi_A)^{1/2}\phi_A^{3/2}\right]\right.$$

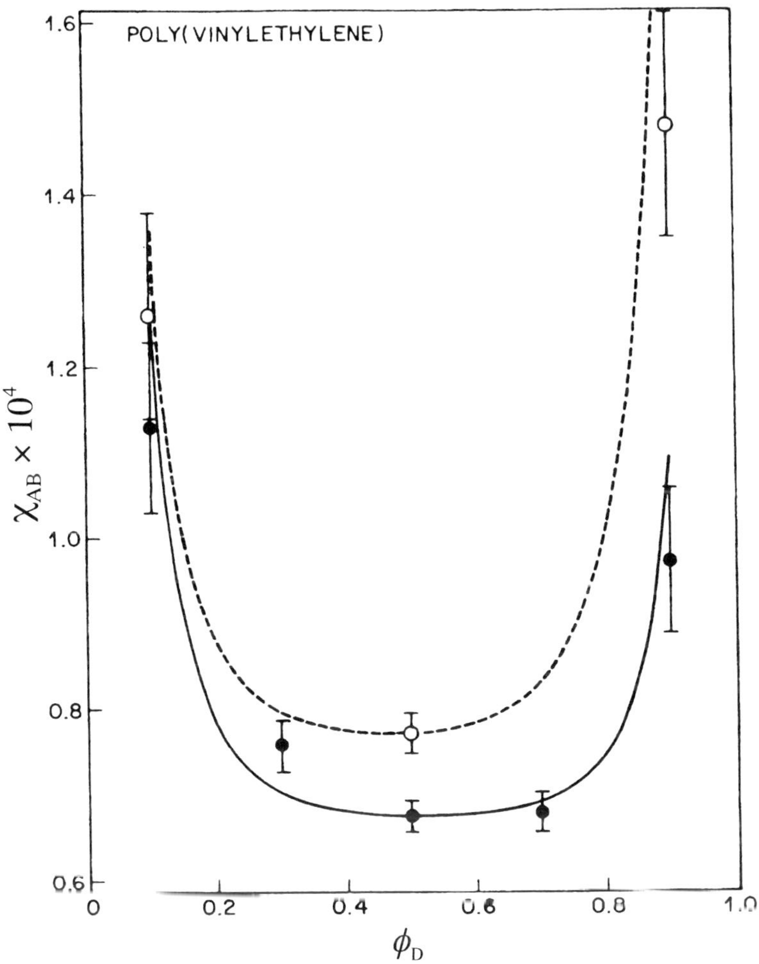

Figure 12 Composition dependence of the interaction parameter for two data sets of isotopic mixtures of poly(vinylethylene) (PVE). Note here that χ for the ordinate stands for the interaction parameter determined from scattering experiments, which is identical to χ_{00} used in Sec. II. B. ϕ_D for the abscissa denotes volume fraction of the deuterated component. The closed and open circles are for blend systems PVE12 (having larger molecular weights) and PVE13 (having smaller molecular weights), respectively. PVE12 designates the PVE1(deuterated)/PVE2(protonated) blend and PVE13 stands for the PVE1(deuterated)/PVE3(protonated) blend, where $N_n = 1270$, 1690, and 791 for PVE1, PVE2, and PVE3, respectively. The solid and broken curves are calculated using the theoretical consideration as expressed by Eq. (106) based on χ_{FH} determined at $\phi_D = 0.5$. Here, χ_{FH} is the original FH interaction parameter (see Sec. II. B). From Ref. 113.

$$+\frac{1}{2}(\chi_s - \chi_{FH})\Big[\phi_A^{-1/2}(1-\phi_A)^{3/2}$$
$$+(1-\phi_A)^{-1/2}\phi_A^{3/2} - 6\phi_A^{1/2}(1-\phi_A)^{1/2}\Big]$$
$$+\left[\frac{1}{N_A\phi_A^3} + \frac{1}{N_B(1-\phi_A)^3}\right]\phi_A^{3/2}(1-\phi_A)^{3/2}$$
$$+\frac{\phi_A^{3/2}(1-\phi_A)^{3/2}}{8(\chi_s - \chi_{FH})}\left[\frac{1}{N_B(1-\phi_A)^2} - \frac{1}{N_A\phi_A^2}\right]^2\Bigg\} \tag{106}$$

Fig. 12 also shows the curves based on Eq. (106) fitted to the data points. Thus, the theory describes both the compositional and the molecular weight dependences of interaction parameter well.

Sariban and Binder [128] have shown a similr tendency for the ϕ_A dependence of χ_{AB} (parabolic upward curvature) with the χ_{AB} values decreasing with increasing chain length (molecular weight), as revealed by a Monte Carlo simulation. The PRISM (Polymer Reference Interaction Site Model) theory by Schweizer and Curro [129], which is an integral equation theory, has demonstrated the same tendency [130] where the effects of compressibility are taken into account by that theory. The compressible RPA formalism developed by Bidkar and Sanchez [104], which is discussed in Sec. IV. B. 7, also turned out to be useful in explaining the composition dependence of χ_{AB} determined from the incompressible RPA formalism. As shown in Fig. 13, the effects of compressibility account reasonably for the parabolic upward shape of the $\chi_{AB}(\phi_A)$ for the isotopic mixture of polyethylene (PE) [104,116]. In contrast, the data for DPS/HPS blends show the parabolic *downward* curvature.

$$\frac{\partial^2 \chi_{AB}(\phi_D)}{\partial \phi_D^2} < 0$$

This result is interpreted by the lattice cluster theory [131].

The molecular weight dependence of χ_{AB} has been discussed exclusively in terms of scaling relationships, although it is unclear whether the universal scaling law is valid for $\chi_{AB}(N)$. The scaling relations, however, are experimentally verified for some blends in the restricted range of the molecular weights. Mori et al. [125] have recently reported the molecular weight dependence of χ_{AB} as:

$$\chi_{AB} \sim N_n^{-0.5} \qquad (10^2 \le N_n \le 10^3) \tag{107}$$

by the analysis of SAXS results for polystyrene-*b*-polyisoprene (SI) diblock copolymers using the RPA analysis. They have further commented that the

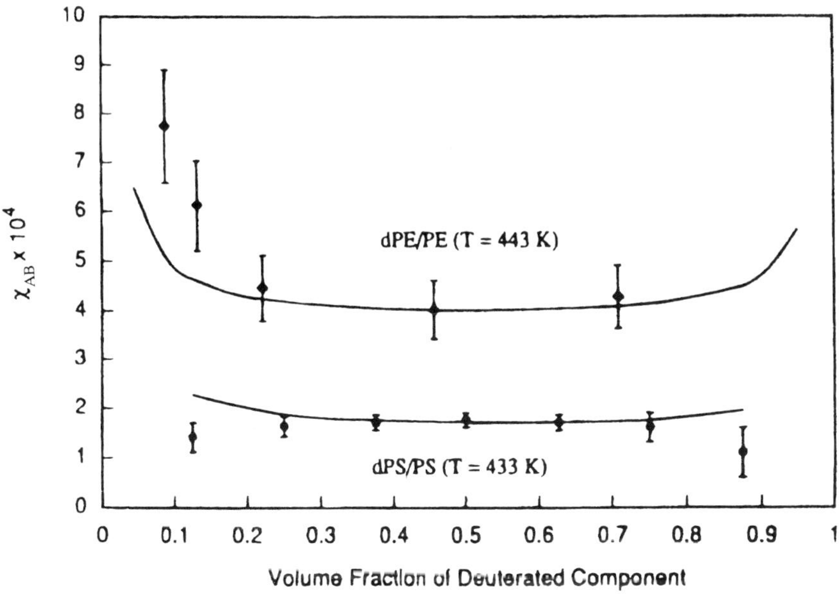

Figure 13 Comparison between the theoretical and experimental results for the composition dependence of the interaction parameter, χ_{AB}, for two data sets of isotopic mixtures of polyethylene (PE) and polystyrene (PS). The solid lines are calculated by using the compressible RPA theory of Bidkar and Sanchez [104]. From Ref. 104.

theoretical consideration by Dudowicz and Freed [132] using the lattice cluster theory for PEP-*b*-PEE leads to the following relationship:

$$\chi_{AB} \sim N_n^{-0.7} \qquad (10^2 \leq N_n \leq 10^3) \tag{108}$$

in the relevant range of molecular weights ($10^2 \leq N_n \leq 10^3$), although the theoretical results themselves do not show a simple scaling relation. In fact, their results showed a trivial molecular weight dependency in the limit of high molecular weight and the apparent scaling relationship for $10^3 \leq N_n \leq 10^4$ can be approximately given by

$$\chi_{AB} \sim N_n^{-0.1} \qquad (10^3 \leq N_n \leq 10^4) \tag{109}$$

Schweizer and Curro [133] have obtained the same result as Eq. (107) ($\chi_{AB} \sim N^{-0.5}$) using the PRISM theory, which is in good accord with the results of Mori et al. However, Yethiraj and Schweizer recently corrected the result to give $T_c \sim N^1$ even for a low molecular weight, where T_c denotes the critical temperature [134]. The correction is due to the change of molecular closures used in the PRISM theory. This result ($T_c \sim N^1$) agrees with the experimental result of Gehlsen et al. [135]. From Eq. (12), $\chi_c \sim N^{-1}$ is deduced for $N_A = N_B = N$. If the temperature dependence of χ_{AB} can be simply given by $\chi_{AB} = B/T$, then $T_c \sim BN^1$ is evaluated. Consequently, the term B turns out to be independent of N by comparing the relation $T_c \sim BN^1$ with the result $T_c \sim N^1$. This indicates, in turn, that the χ_{AB} parameter is independent of the molecular weight. Although both the applicability of the scaling relation for $\chi_{AB}(N)$ and its exponent remain open questions for future work, the simple conclusion at present is that the χ_{AB} parameter decreases with increasing molecular weight.

3. Pressure Dependence

Applying pressure may give a new perspective on understanding the thermodynamics of polymer blends, besides having importance in technological applications. Recently, SANS and SAXS studies were conducted to reveal the pressure dependence of χ_{AB}. Phenomenologically, only the entropic term of χ_{AB} (i.e., $A[\equiv \partial(T\chi_{AB})/\partial T]$) is altered with pressure, while the enthalpic term of χ_{AB} (i.e., $B[\equiv \partial\chi_{AB}/\partial(1/T)]$) does not change. Note that the χ_{AB} parameter discussed here is evaluated using ordinary incompressible RPA theories. Such results are to be expected and have been predicted by EOS theories. Janssen et al. [98] have examined the prediction of the Flory–Orwoll–Vrij (FOV) equation of state [136] by comparing SANS results for DPS/PVME and DPS/PPMS (poly(phenylmethyl siloxane)) blends. Generally, the agreement between the FOV EOS and the SANS results was found to be unsatisfactory. Hammouda and Bauer [100] have found in DPS/PVME and DPS/poly(butyl methacrylate) (PBMA) blends that the enthalpic part of the interaction parameter shows pressure dependence, as well as the entropic part. For those blends, the value of χ_{AB} decreases with increasing pressure.

For a better understanding, the interaction parameter or the intercomponent interaction cohesive energy (P^*_{AB}), which is the measure of the degree of segmental miscibility (see Eq. (87)), should be evaluated without assuming incompressibility, since the χ_{AB} values obtained using the ordinary incompressible RPA theory are affected by the free volume fraction in the sample (i.e., the compressibility). As introduced in Sec. IV. B. 7 (compressible RPA formula), some compressible RPA theories are available at present. In this paragraph, the results obtained using the compressible RPA analysis are discussed with respect to the pressure dependence of χ_{AB}. For DPS/PVME blends, Hammouda and Benmouna [101] have obtained values of P^*_{12} and the free volume fraction (f_0) as analyzed by the compressible RPA theory of Bidkar and Sanchez [104]. It was found that the free volume fraction decreases with pressure and increases with temperature. The free volume fraction also shows a very weak linear dependence on the DPS volume fraction ($\phi_{\rm DPS}$) as $\partial f_0/\partial\phi_{\rm DPS} < 0$. As for the interaction cohesive density, $P^*_{\rm AB}$, it was found to be almost independent of pressure. This is very much in contrast with the results obtained using the incompressible RPA procedure. Thus, for this particular case, the compressible RPA analysis seems to work well. Another notable result has been reported by Hajduk et al. [41] for SI diblock copolymers under pressure studied by SAXS using synchrotron radiation. For analyses of the SAXS data, they also used compressible RPA formulae developed originally by Tang and Freed [103]. However, the interaction parameter, $\chi_{\rm AB}$, evaluated without assuming incompressibility shows unambiguous linear dependence with pressure ($\partial\chi_{\rm AB}/\partial P > 0$). Moreover, they found that the radius of gyration, R_g, also increases with increasing pressure. Although no reasonable interpretation is given in their paper, the result of Hajduk et al. may suggest that pressure enhances the chain stretching of diblock copolymers. Note here that chain stretching of block copolymers was observed in the vicinity of the order–disorder transition (ODT) under ordinary pressure [152]. They also observed an increase in the SAXS peak intensity with increasing pressure, which indicates that the SI in the disordered state becomes less miscible with application of the pressure. It is noted that the SAXS peak observed in the disordered state of the diblock copolymers is ascribed to the correlation hole effect (as demonstrated in Fig. 7). It is important for future studies to further probe the effects of pressure on miscibility.

4. Microstructure Effects of $\chi_{\rm AB}$ and Miscibility Control Using Random Copolymers

The phenomenological $\chi_{\rm AB}$ parameter as evaluated using the incompressible RPA theory shows dependences on various experimental factors, as discussed in the previous sections. This section concerns the miscibility in connection with the structures of the polymer chains or the monomers. Namely, we describe the effects of microstructure in polydiene rubbers [27,69,137–140], and polyolefins (hydrogenated (saturated) polydiene) [115], and tacticity effects on the $\chi_{\rm AB}$ parameter for PS/PVME [33].

Here, microstructure effects in polydiene rubber blends are discussed as an example of such a structure–miscibility relationship. Some very interesting examples are DPB/HPB blends. Polydiene rubbers such as DPB and HPB are comprised of geometrical isomers of trans-1,4, cis-1,4, and 1,2 (vinyl) units. As a result of the miscibility evaluation using the incompressible RPA analysis for the SANS data, it was found that the interaction parameter, $\chi_{\rm AB}$, depends strongly on the microstructure [69,139]. Fig. 14

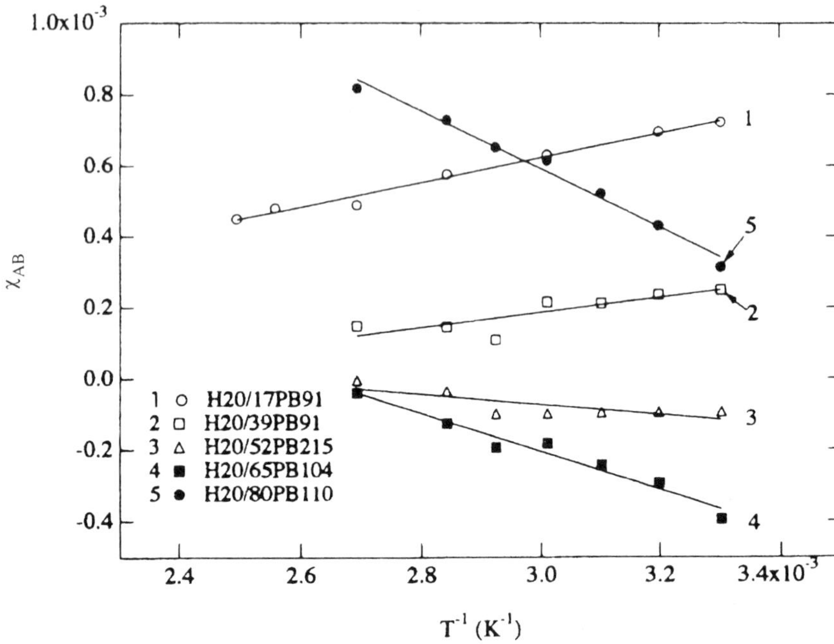

Figure 14 Plot of χ_{AB} versus the inverse absolute temperature for five sets of DPB/HPB blend samples, where the five HPB samples with various 1,2 (vinyl) contents are blended with the identical DPB sample with 15.6% of 1,2 content. The contents of 1,2 unit in the HPB samples are 17, 39, 52.2, 64.9, and 80.4% for the data set numbers 1 (open circle), 2 (open square), 3 (open triangle), 4 (closed square), and 5 (closed circle), respectively. From Ref. 139.

shows the effective (phenomenological) χ_{AB} as a function of the inverse absolute temperature for five sets of DPB/HPB blend samples, where the five different HPB samples having various 1,2 (vinyl) contents are blended with the identical DPB sample having 15.6% of 1,2 content. Not only the χ_{AB} value itself but also the temperature dependence of the χ_{AB} are affected by the microstructure. First of all, the isotopic pair with matching 1,2 content in the DPB and HPB (No. 1 pair) exhibited UCST phase behavior, as expected from the pure deuteration of effects discussed previously in Sec. IV. C. 1. Although the No. 2 sample (39% of 1,2 content in HPB) also shows UCST behavior, the other pairs surprisingly exhibit LCST behaviors. The slopes of the χ_{AB} versus $1/T$ plots systematically decreases with increasing 1,2 content in HPB, and, further, the sign of the slope changed (from UCST to LCST behavior). Generally, LCST phase behavior is explained by the EOS (effect of free volume) coupled with specific interactions [141], for instance for PS/PVME blends. For the LCST-type DPB/HPB blends, however, the EOS explanation is not valid, since the isotopic mixture definitely shows UCST behavior. Alternatively, LCST behavior can be explained by the random copolymer effects [137]. According to the random copolymer theory [142–144], the effective interaction between two random copolymers A_xB_{1-x} and C_yD_{1-y} is given by:

$$\chi_{\text{blend}} = \chi_1 - \chi_2 \tag{110}$$

with

$$\chi_1 = xy\chi_{AC} + x(1-y)\chi_{AD} + (1-x)y\chi_{BC} + (1-x)(1-y)\chi_{BD} \tag{111}$$

and

$$\chi_2 = x(1-x)\chi_{AB} + y(1-y)\chi_{CD} \tag{112}$$

where χ_2 denotes the sum of the intramolecular interactions. The interesting conclusion which is drawn from this theory is that the effective interaction χ_{blend} is considerably reduced when the intramolecular interaction (χ_2) is large. In other words, a random copolymer comprised of two kinds of monomers with strong repulsive interaction has good miscibility with other polymers. This effect is well known as the "miscibility window" effect [142–144]. When 1,2(D), 1,4(D), 1,2(H), and 1,4(H) isomers are chosen as the A, B, C, and D components, then the effective interaction for any pair of DPB/HPB can be estimated by Eqs. (110)–(112), if the microstructure content in those samples is known. Conversely, the six fundamental interactions $\chi_{1,2D/1,4D}$, $\chi_{1,2D/1,2H}$, $\chi_{1,2D/1,4H}$, $\chi_{1,4D/1,2H}$, $\chi_{1,4D/1,4H}$, and $\chi_{1,2H/1,4H}$ can be evaluated by consistently solving the simultaneus equations with satisfactory sets of experimentally determined effective interaction parameters for the known microstructures, x and y. In fact, the temperature dependence of χ_{HD} given by Eq. (104) in Sec. IV. C. 1 is evaluated by a similar procedure with simplified equations after making some assumptions [69]. Using the random copolymer effects, the LCST behavior can be explained as follows, in spite of the fact that only UCST-type fundamental interactions are present [137]. Fig. 15 schematically illustrates the temperature dependence of the interaction, where χ_1 and χ_2 denote the inter- and intramolecular interactions as defined by Eqs. (111) and (112), respectively. Note that both χ_1 and χ_2 decrease with temperature (the UCST type). If the relationship of χ_1 and χ_2 is such as depicted, then the effective interaction (χ_{blend}) as determined by $\chi_1 - \chi_2$ becomes an LCST type, i.e., χ_{blend} increases with an increase of temperature. For the particular case of DPB/HPB, the random copolymer effect gives a possible explanation of their LCST behavior.

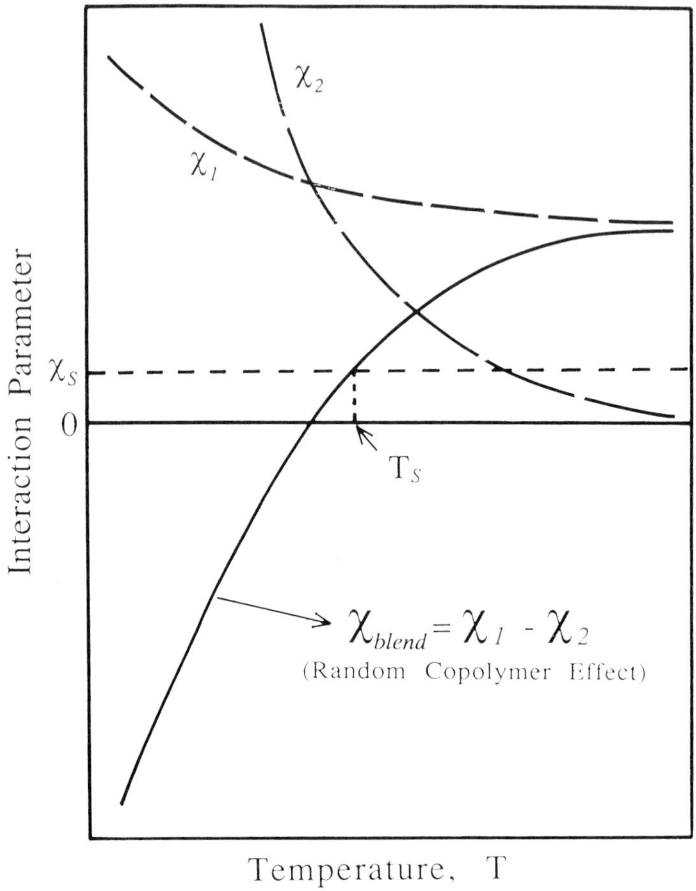

Figure 15 Schematic illustration of the temperature dependence of the interaction parameter, where χ_1 and χ_2 denote the inter- and intramolecular interactions as defined by Eqs. (111) and (112), respectively. Note that both χ_1 and χ_2 decrease with temperature (UCST behavior). If the relationship between χ_1 and χ_2 is such as depicted, then the effective interaction (χ_{blend}) as determined by $\chi_1 - \chi_2$ becomes LCST type, i.e., χ_{blend} increases with an increase of temperature. From Ref. 137.

The random copolymer theory provides a sophisticated method to control miscibility through random copolymerization. As mentioned above, not only the degree of miscibility but also the temperature dependence of the phase behavior, including the type of phase behavior (UCST or LCST) are controllable. One of the most interesting examples is a poly(styrene-*co*- butadiene) (SBR). The importance of SBR is known empirically in the rubber industry, especially for control of the miscibility and that of the glass transition temperature [145]. Sakurai et al. [138] have demonstrated, using the random copolymer theory, that by controlling the microstructure of the butadiene units and the styrene content in SBR, the changes in miscibility are not straightforward for SBR/PB blends. There are five fundamental segment units in the SBR/PB blends, namely, styrene, 1,4-, and 1,2-units in SBR, and additionally, 1,4- and 1,2-units in PB. Extending the random copolymer theory to the five-component system, the miscibility can be predicted for any combination of those segmental units (microstructures). The complicated miscibility behavior caused by the microstructures can never be understood by the simple empirical rule for miscibility in SBR/PB blends.

V. SUMMARY AND FUTURE PERSPECTIVES

Recent progress in the understanding of the critical phenomena in binary polymer mixtures is summarized in the first half of this chapter. It has been generally recognized that polymeric mixtures exhibit a crossover from mean-field to 3D-Ising critical behavior in the vicinity of the critical point. Although the mean-field picture can be applicable to high molecular weight polymers, the crossover also exists in the region very close to the critical point. However, the extent of deviation from the mean-field behavior becomes negligible as the molecular weight increases. The Ginzburg criterion has been introduced to discuss quantitatively the crossover in terms of the universal Ginzburg number which depends on the molecular weight. However, precise definition for the universal Ginzburg number remains a subject of controversy. Further studies are desired to examine the character of the critical behavior in polymer mixtures by changing various experimental conditions (e.g., microstructure, pressure, etc.). The Lifshitz behavior in ternary blends of A, B homopolymers with the symmetric A–B diblock copolymers also deserves further work.

The progress in evaluation methods for the miscibility of polymer blends and other polymeric systems has been summarized in the second half of this chapter. We focused intensively on recent results obtained using scattering techniques as evaluated by RPA theory. The validity of the evaluated χ_{AB} parameter mainly depends on the reliability of the theoretical scattering function used. Therefore, the RPA theory should be modified to include factors deviated from ideal conditions. Some modifications for the effects of polydispersity and the difference in segmental volumes were discussed. The multicomponent and the compressible RPA formulae were presented as well. The multicomponent RPA has been especially important in examining the effects of partial deuteration in SANS studies

[30,31,146–148] and the effect of nonrandomness in mixtures containing statistically random copolymers. The multicomponent RPA is also useful for systems containing solvent (polymer A/polymer B/solvent C), because most polymeric systems are immiscible in the bulk state. Solvents reduce the repulsive interaction and induce miscibility of the polymers in the presence of solvent. Finally, the compressible RPA theory is necessary for mixtures where the change in the free volume fraction upon mixing plays a very important role (e.g., a strong composition dependence of χ_{AB}) in order to evaluate the χ_{AB} and to predict the miscibility behavior without assuming incompressibility.

The RPA theory has been extended to calculate the scattering functions for any type of chain architecture (e.g., multiblock, graft, starburst, etc.) along with corrections for such effects. Generally, most theoretical considerations stay within the framework of the FH lattice theory. Corrections for deviations from the mean-field approximation and the framework of the FH lattice theory have not yet been addressed satisfactorily. Future work will hopefully be developed to consider these effects and to overcome these problems. As for the correction for compressibility effects, compressible RPA methods are promising, although only a small number of experimental studies have been conducted. In the near future, the analysis scheme using the compressible RPA will become easier for the evaluation of the interaction parameter value, which will in turn encourage the experimentalists to recognize the importance of the compressibility more than at present. Recently, three-component copolymers such as the A–B–C-type triblock copolymers (see Chapter 4 in this book) and ABC three-arm starpolymers [149,150] have been successfully synthesized. It is expected that these copolymers will play an important role in the understanding of structure–property relationships and that they undergo novel types of phase transitions [151]. In order to understand the phase behavior in such systems, the multicomponent RPA should be applied and this should result in progress toward a more thorough understanding of the structure–miscibility relationship.

The phenomenological χ_{AB} parameter has been shown to depend on various experimental parameters including pressure. For any experimental parameters, the dependence is not necessarily monotonic and is not universal, either. Namely, in some blend systems χ_{AB} increases with the given experimental parameter, but in other systems χ_{AB} decreases. Since no unified explanation has been given, continuing studies are required.

Finally, it should be noted that nonrandomness in a statistical random copolymer, which may exist in real samples, may cause nontrivial effects in the evaluation of the miscibility (e.g., the separation of the isotope labeling effect of χ_{AB} from microstructure effects [69,115,139]) and the miscibility prediction in random copolymer blends. Moreover, the possible composition dependence of χ_{AB} and the effects of the free volume (compressibility) should be included in the random copolymer theory for improved reliability of the theory.

ACKNOWLEDGMENT

Dr Alan I. Nakatani at the National Institute of Standards and Technology is gratefully acknowledged for critical proofreading and his valuable comments.

REFERENCES

1. DR Paul, S Newman, ed. Polymer Blends. New York: Academic Press, 1970.
2. O Olabisi, LM Robeson, MT Shaw. Polymer-Polymer Miscibility. New York: Academic press, 1979.
3. LA Utracki. Polymer Alloys and Blends—Thermodynamics and Rheology. Munich: Carl Hanser Verlag, 1989.
4. JW Cahn, JE Hilliard, J Chem Phys 28:258, 1958.
   JW Cahn. J Chem Phys 30:1121, 1959.
   JW Cahn, JE Hilliard. J Chem Phys 31:688, 1959.
5. CC Han, AZ Akcasu. Ann Rev Phys Chem 43:61, 1992.
6. PJ Flory. Principles of Polymer Chemistry. Ithaca, NY: Cornell University Press, 1953.
7. PG de Gennes. Scaling Concepts in Polymer Physics. Ithaca, NY: Cornell University Press, 1979.
8. HE Stanley. Introduction to Phase Transitions and Critical Phenomena. Oxford: Clarendon press, 1971.
9. A Maconnachie, RP Kambour, DM White, S Rostami, DJ Walsh. Macromolecules 17:2645, 1984.
10. JH Hildebrand, RL Scott. The Solubility of Non-Electrolytes. 3rd ed. Princeton, NJ: Van-Nostrand-Reinhold, 1950 (reprinted by New York: Dover Publications, 1964).
11. EA Grulke. In: J Brandrup, EH Immergut, ed. Polymer Handbook. 3rd ed. New York: Wiley, 1989, pp VII/519.
12. R Koningsveld, LA Kleintjens. Macromolecules 4:637, 1971.
13. H Fujita. Polymer Solutions. Amsterdam: Elsevier, 1990, pp 282–324.
14. KF Freed, AI Pesci. J Chem Phys 87:7342, 1987.
15. T Nishi, TK Kwei, Polymer 16:285, 1975.
16. D Patterson. Polym Eng Sci 22:64, 1982.
17. B Hammouda, BJ Bauer, TP Russell. Macromolecules 27:2357, 1994.
18. PJ Flory, RA Orwoll, A Vrij. J Am Chem Soc 86:3507, 3515, 1964.
    J Am Chem Soc 87:1833, 1965.
    A Abe, PJ Flory. J Am Chem Soc 87:1838, 1965.
    BE Eichinger, PJ Flory. Trans Faraday Soc 65:2035, 1968.

19. I Prigogine. The Molecular Theory of Solutions. Amsterdam: North-Holland Publishing Company, 1957.
20. RH Lacombe, IC Sanchez. J Phys Chem 80:2568, 1976.
21. S Rostami, DJ Walsh. Macromolecules 17:315, 1984.
IC Sanchez. Macromolecules 17:967, 1984.
DJ Walsh, S Rostami. Macromolecules 18:216, 1985.
S Rostami, DJ Walsh. Macromolecules 18:1228, 1985.
DJ Walsh, GT Dee, JL Halary, JM Ubiche, M Millequant, J Lesec, L Monnerie. Macromolecules 22:3395, 1989.
HW Kammer, T Inoue, T Ougizawa. Polymer 30:888, (1989).
DJ Walsh, WW Graessley, S Datta, DJ Lohse, LJ Fetters. Macromolecules 25:5236, 1992.
J Kressler, N Higashida, K Shimomai, T Inoue, T Ougizawa. Macromolecules 27:2448, 1994.
22. G Hadziioannou, RS Stein. Macromolecules 17:567, 1984; Macromolecules 17:1059, 1984.
CT Murray, JW Gilmer, RS Stein. Macromolecules 18:996, 1985.
H Ito, TP Russell, GD Wignall. Macromolecules 20:2213, 1987.
PE Tomlins, JS Higgins. Macromolecules 21:425, 1988.
23. P Debye. J Appl Phys 15:338, 1946.
24. IC Sanchez. Polymer 30:471, 1989.
25. M Shibayama, H Yang, RS Stein, CC Han. Macromolecules 18:2179, 1985.
26. CC Han, M Okada, Y Muroga, FL McCrackin, BJ Bauer, Q Tran-Cong. Polym Eng Sci 26:3, 1986.
CC Han, M Okada, Y Muroga, BJ Bauer, Q Tran-Cong. Polym Eng Sci 26:1208, 1986.
CC Han, BJ Bauer, JC Clark, Y Muroga, M Matsushita, M Okada, Q Tran-Cong, T Chang, IC Sanchez. Polymer 29:2002, 1988.
27. H Hasegawa, S Sakurai, M Takenaka, T Hashimoto, CC Han. Macromolecules 24:1813, 1991.
28. PG de Gennes. J Phys (Paris) Lett 38:L441, 1977.
JF Joanny. J Phys A 11:L117, 1978.
CC Han. In: M Nagasawa, ed. Molecular Conformation and Dynamics of Macromolecules in Condensed Systems. New York: Elsevier, 1988.
29. H Hayashi, F Hamada, A Nakajima. Macromolecules 7:959, 1974; Macromolecules 9:543, 1976; Makromol Chem 178:827, 1977; Colloid Polym Sci 256:651, 1978.
30. JD Tanzer, B Crist Jr. Macromolecules 18:1291, 1985.
31. JC Nicholson, TM Finerman, B Crist. Polymer 31:2287, 1990.
32. FS Bates, JH Rosedale, P Stepanek, TP Lodge, P Wiltzius, GH Fredrickson, RP Hjelm, Jr. Phys Rev Lett 65:1893, 1990.
33. G Beaucage and RS Stein. Macromolecules 26:1617, 1993. It is noteworthy that sufficiently wide q range is available with $0.08 < q < 1.8\,\mathrm{nm}^{-1}$ at the LANSCE facility.

34. D Schwahn, K Mortensen, T Springer, H Yee-Madeira, R Thomas. J Chem Phys 87:6078, 1987.
35. D Schwahn, H Takeno, L Willner, H Hasegawa, H Jinnai, T Hashimoto, M Imai. Phys Rev Lett 73:3427, 1994.
36. D Schwahn, K Hahn, J Streib, T Springer. J Chem Phys 93:8383, 1990.
37. I Hopkinson, FT Kiff, RW Richards, SM King, H Munro. Polymer 35:1722, 1994.
38. B Chu, Q Ying, K Linliu, P Xie, T Gao, Y Li, T Nose, M Okada. Proceedings of the American Chemical Society Division of Polymer Material Science and Engineering 65:209, 1991; Macromolecules 25:7382, 1992.
39. Q Tran-Cong, R Kawakubo, S Sakurai. Polymer 35:1236, 1994.
40. C Riekel, P Boesecke, O Diat, P Engstroem, M Kocsis, S Lequien, A Snigirev, I Snigireva, K Toth. J Phys IV Coll C8 (France) 3:403, 1993.
A Snigirev, I Snigireva, C Riekel, A Miller, L Wess, T Wess. J Phys IV Coll C8 France 3:443, 1993.
41. DA Hajduk, P Urayama, SM Gruner, S Erramilli, RA Register, K Brister, LJ Fetters. Macromolecules 28:7148, 1995.
42. AJ Ryan, CW Macosko, W Bras. Macromolecules 25:6277, 1992.
43. K Binder. Adv Polym Sci 112:181, 1994.
44. D Schwahn, K Mortensen, H Yee-Madeira. Phys Rev Lett 58:1544, 1987.
45. S Janssen, D Schwahn, T Springer. Phys Rev Lett 68:3180, 1992.
46. DW Hair, EK Hobbie, AI Nakatani, CC Han. J Chem Phys 96:9133, 1992.
47. EK Hobbie, L Reed, CC Huang, CC Han. Phys Rev E 48:1579, 1993.
48. LS Ornstein, F Zernike. Proc Acad Sci Amsterdam 17:793, 1914.
49. K Binder. J Chem Phys 79:6387, 1983.
50. D Schwahn, T Schmackers, K Mortensen. Phys Rev E 52:R1288, 1995.
51. J Dudowicz, M Lifschitz, KF Freed, JF Douglas. J Chem Phys 99:4804, 1993.
52. M Lifschitz, J Dudowicz, KF Freed. J Chem Phys 100:3957, 1994.
53. FS Bates, W Maurer, TP Lodge, MF Schulz, MW Matsen, K Almdal, K Mortensen. Phys Rev Lett 75:4429, 1995.
54. M Teubner, R Strey. J Chem Phys 87:3195, 1987.
55. CC Han. Private communication, to be published.
56. RM Hornreich, M Luban, S Shtrikman. Phys Rev Lett 35:1678, 1975.
R Holyst, M Schick. J Chem Phys 96:7728, 1992.
57. D Schwahn, K Mortensen, S Janssen. Phys Rev Lett 73:1452, 1994.
58. ME Fisher. Phys Rev 176:257, 1968.
ME Fisher, PE Scesney. Phys Rev B 2:825, 1970.
D Broseta, L Leibler, JF Joanny. Macromolecules 20:1935, 1987.
DW Hair, EK Hobbie, JF Douglas, CC Han. Phys Rev Lett 68:2476, 1992.
H Yajima, DW Hair, AI Nakatani, JF Douglas, CC Han. Phys Rev B 47: 12 268, 1993.
59. JW Barlow, DR Paul. Ann Rev Mater Sci 11:299, 1981.
JF Harris, DR Paul, JW Barlow. Polym Eng Sci 23:676, 1983.
60. AI Pesci, KF Freed. J Chem Phys 90:2017, 1989.
61. KF Freed, J Dudowicz. Trends Polym Sci 3:248, 1995.

62. KW Foreman, KF Freed. J Chem Phys 102:4663, 1995.
63. AM Nemirovsky, MG Bawendi, KF Freed. J Chem Phys 87:7272, 1987. AI Pesci, KF Freed. J Chem Phys 90:2003, 1989.
64. EK Hobbie, CC Han. J Chem Phys 105:738, 1996.
65. BH Zimm. J Chem Phys 16:1093, 1948.
66. JS Higgins. In: G Kostorz, ed. Treatise on Materials Science and Technology, Vol. 15, Neutron Scattering. New York: Academic Press, 1979.
67. S Sakurai, H Hasegawa, T Hashimoto, CC Han. Polym Comm 31:99, 1990.
68. RM Briber, BJ Bauer, B Hammouda. J Chem Phys 101:2592, 1994.
69. S Sakurai, H Hasegawa, T Hashimoto, IG Hargis, SL Aggarwal, CC Han. Macromolecules 23:451, 1990.
70. G Kostorz, SW Lovesey. In:G Kostorz, ed. Treatise on Materials Science and Technology, Vol. 15, Neutron Scattering. New York: Academic Press, 1979, pp 1–67.
71. H Benoit, EW Fischer, HG Zachmann. Polymer 30:379, 1989.
72. TP Russell, LJ Fetters, JC Clark, BJ Bauer, CC Han, Macromolecules 23:654, 1990.
73. B Hammouda. J Polym Sci Part B, Polym Phys Ed 30:1387, 1992; Adv Polym Sci 106:87, 1993.
74. G Hadziioannou, H Benoit, W Tang, K Shull, CC Han. Polymer 33:4677, 1992.
75. B Hammouda. J Chem Phys 99:9182, 1993.
76. B Hammouda. J Chem Phys 98:3439, 1993.
77. K Mori, H Tanaka, T Hashimoto. Macromolecules 20:381, 1987.
78. Y Ijichi, T Hashimoto. Polym Comm 29:135, 1988.
79. Y Ijichi, T Hashimoto, LJ Fetters. Macromolecules 22:2817, 1989.
80. LH Peebles Jr. Molecular Weight Distributions in Polymers. New York: Interscience, 1970.
81. S Sakurai, K Mori, A Okawara, K Kimishima, T Hashimoto. Macromolecules, 25:2679, 1992.
82. FS Bates, MA Hartney. Macromolecules 18:2478, 1985; Macromolecules 19:2892, 1986.
83. L Leibler. Macromolecules 13:1602, 1980.
84. S Sakurai, S Nomura. Polymer, 38:4103, 1997.
85. H Tanaka, T Hashimoto. Polym Comm 29:212, 1988.
86. H Tanaka, T Hashimoto. Macromolecules 24:5398, 1991.
87. B Hammouda, RM Briber, BJ Bauer. Polymer 33:1785, 1992.
88. JK Kim, K Kimishima, T Hashimoto. Macromolecules 26:125, 1993.
89. NP Balsara, SV Jonnalagadda, CC Lin, CC Han, R Krishnamoorti. J Chem Phys 99:10 011, 1993.
90. CC Lin, SV Jonnalagadda, NP Balsara, CC Han, R Krishnamoorti. Macromolecules 29:661, 1996.
91. H Benoit, W Wu, M Benmouna, B Bauer, A Lapp. Macromolecules 18:986, 1985.
92. AZ Akcasu, M Tombakoglu. Macromolecules 23:607, 1990.

93. T Hashimoto, M Shibayama, H Kawai. Macromolecules 16:1093, 1983.
94. T Fukuda, M Nagata, H Inagaki. Macromolecules 17:548, 1984; Macromolecules 19:1411, 1986.
    T Hashimoto, K Mori. Macromolecules 23:5347, 1990.
95. E Helfand, Y Tagami. J Chem Phys 56:3592, 1972.
96. A Onuki, T Hashimoto. Macromolecules 22:879, 1989.
97. JG Connell, RW Richards. Macromolecules 23:1766, 1990.
    TP Lodge, C Pan, X Jin, Z Liu, J Zhao, WW Maurer, FS Bates. J Polym Sci Part B Polym Phys Ed 33:2289, 1995.
98. S Janssen, D Schwahn, K Mortensen, T Springer. Macromolecules 26:5587, 1993.
99. S Janssen, D Schwahn, T Springer, K Mortensen. Macromolecules 28:2555, 1995.
100. B Hammouda, BJ Bauer. Macromolecules 28:4505, 1995.
101. B Hammouda, M Benmouna. J Polym Sci Part B Polym Phys Ed 33:2359, 1995.
102. WE McMullen, KF Freed. Macromolecules 23:255, 1990.
103. H Tang, KF Freed. Macromolecules 24:958, 1991
104. UR Bidkar, IC Sanchez. Macromolecules 28:3963, 1995.
105. IC Sanchez, RH Lacombe. Macromolecules 11:1145, 1978.
106. IC Sanchez, CG Panayiotou. In: S. Sandler, ed. Models for Thermodynamic and Phase Equilibria Calculations. New York: Marcel Dekker, 1993.
107. CC Lin, SV Jonnalagadda, PK Kesani, HJ Dai, NP Balsara. Macromolecules 27:7769, 1994.
108. WL Nachlis, JT Bendler, RP Kambour, WJ MacKnight. Macromolecules 28:7869, 1995.
109. FS Bates, GD Wignall, WC Koehler. Phys Rev Lett 55:2425, 1985.
110. FS Bates, GD Wignall. Phys Rev Lett 57:1429, 1986.
111. FS Bates, GD Wignall. Macromolecules 19:932, 1986.
112. FS Bates, SB Dierker, GD Wignall. Macromolecules 19:1938, 1986.
113. FS Bates, M Muthukumar, GD Wignall, LJ Fetters. J Chem Phys 89:535, 1988.
114. FS Bates, LJ Fetters, GD Wignall. Macromolecules 21:1086, 1988.
115. J Rhee, B Crist. J Chem Phys 98:4174, 1993.
116. JD Londono, AH Narten, GD Wignall, KG Honnell, ET Hsieh, TW Johnson, FS Bates. Macromolecules 27:2864, 1994.
117. P Wiltzius, FS Bates, WR Heffner. Phys Rev Lett 60:1538, 1988.
118. EL Atkin, LA Kleintjens, R Koningsveld, LJ Fetters. Makromol Chem 185:377, 1984.
     JL Lin, D Rigby, R-J Roe. Macromolecules 18:1609, 1985.
     H Yang, RS Stein, CC Han, BJ Bauer, EJ Kramer. Polym Comm 27:132, 1986.
119. H Yang, G Hadziioannou, RS Stein. J Polym Sci Polym Phys Ed 21:159, 1983.
120. H Yang, M Shibayama, RS Stein, N Shimizu, T Hashimoto. Macromolecules 19:1667, 1986.
121. TP Russell. Macromolecules 26:5819, 1993.
122. TA Collaghan, DK Paul. Macromolecules 26:2439, 1993.

123. R Krishnamoorti, WW Graessley, NP Balsara, DJ Lohse. J Chem Phys 100:3894, 1994.
124. YW Cheung, RS Stein. Macromolecules 27:2512, 1994.
125. K Mori, A Okawara, T Hashimoto. J Chem Phys 104:7765, 1996.
126. J Dudowicz, KF Freed. Macromolecules 24:5112, 1991.
127. M Muthukumar. J Chem Phys 85:4722, 1986.
128. A Sariban, K Binder. J Chem Phys 86:5859, 1987.
129. KS Schweizer, JG Curro. Adv Polym Sci 116:319, 1994.
130. KS Schweizer, JG Curro. J Chem Phys 91:5059, 1989.
131. J Dudowicz, KF Freed, M Lifschitz. Macromolecules 27:5387, 1994.
132. J Dudowicz, KF Freed. Macromolecules 26:213, 1993.
133. KS Schweizer, JG Curro. Chem Phys 149:105, 1990.
JG Curro, KS Schweizer. Macromolecules 24:6736, 1991.
134. A Yethiraj, KS Schweizer. J Chem Phys 97:5927, 1992.
135. MD Gehlsen, JH Rosedale, FS Bates, GD Wignall, L Hansen, K Almdal. Phys Rev Lett 68:2452, 1992.
136. PJ Flory, RA Orwoll, A Vrij. J Am Chem Soc 86:3507, 1964; J Am Chem Soc 86:3515, 1964.
137. S Sakurai, H Jinnai, H Hasegawa, T Hashimoto, CC Han. Macromolecules 24:4839, 1991.
138. S Sakurai, T Izumitani, H Hasegawa, T Hashimoto, CC Han. Macromolecules 24:4844, 1991.
139. H Jinnai, H Hasegawa, T Hashimoto, CC Han. Macromolecules 25:6078, 1992.
140. RN Thudium, CC Han. Macromolecules 29:2143, 1996.
141. D Patterson, Macromolecules 2:672, 1969.
142. RP Kambour, JT Bendler, RC Bopp. Macromolecules 16:753, 1983.
143. G ten Brinke, FE Karasz, WJ MacKnight. Macromolecules 16:1827, 1983.
144. DR Paul, JW Barlow. Polymer 25:487, 1984.
145. AK Bhowmick, HL Stephens, ed., Handbook of Elastomers, New York: Marcel Dekker, 1988.
146. B Crist, JC Nicholson. Polymer 35:1846, 1994.
147. NP Balsara, DJ Lohse, WW Graessley, R Krishnamoorti. J Chem Phys 100:3905, 1994.
148. GC Reichart, RA Register, WW Graessley, R Krishnamoorti, DJ Lohse. Macromolecules 28:8862, 1995.
149. N Hadjichristidis, H Iatrou, SK Behal, JJ Chludzinski, MM Disko, RT Garner, KS Liang, DJ Lohse, ST Milner. Macromolecules 26:5812, 1993.
150. T Fujimoto, H Zhang, T Kazama, Y Isono, H Hasegawa, T Hashimoto. Polymer 33:2208, 1992.
151. C Neumann, V Abetz, R Stadler. Polym Bull 36:43, 1996.
152. K Almdal, JH Rosedale, FS Bates, GD Wignall, GH Fredrickson. Phys Rev Lett 65:1112, 1990.

4

Block and Graft Copolymers

Yushu Matsushita
University of Tokyo, Tokyo, Japan

I. INTRODUCTION

Equilibrium multiphase structures of block and graft copolymers and their stability are introduced and discussed in this chapter.

Since molecular weight and composition distributions sensitively affect the structures of these polymers in condensed systems, it is very important to use samples with well-defined structures to obtain definite conclusions. As such molecules can be in general synthesized by anionic polymerizations [1,2], preparation methods of well-defined samples based on anionic polymerizations will be briefly described in this chapter first.

Block and graft copolymers show microphase-separated structures (MSSs) in condensed systems by intramolecular phase separation if polymer components are immiscible with each other [3]. The magnitude of the repulsive forces between immiscible components rules the critical condition for phase separation of the systems because MSSs can be caused by those forces. The critical condition is primarily discussed in the second section [4]. The system is obviously in the ordered state in microphase-separated bulk. If external conditons have been changed to lower the magnitude of the interaction parameter, so that it starts to mix, the system is in the disordered state. This phenomenon is called order–disorder transition (ODT), [5,6]. ODT is briefly mentioned in the second part of the second section.

It is easily understood that the shape of the polymer/polymer interfaces varies depending on the relative chain lengths of the component polymers [7]. These morphological changes with composition of several copolymers with diferent geometrical features such as the AB and the

BAB two-component block copolymers, the ABC three component triblock copolymers and the star-branched ABB graft copolymers are introduced in Sec. II. C.

Characterization of the surface structure is very important in the understanding of the morphologies of complex polymer systems [8]. Hence, in Sec. IV, the features of polymer–polymer interfaces are reviewed, focusing on the curvatures of their structures. First, the characteristics of the curvatures of polymer–polymer interfaces are classified. Three-dimensionally bicontinuous and tricontinuous structures have only been found quite recently for several soft materials including polymers [9–11]. The surface characteristics of these structures is described in Sec. IV. B. Recently, moreover, since bicontinuous structures have also been found for homopolymer–block copolymer blends, experimental results on these blend systems are introduced in Sec. IV. C [12].

Block polymer chains must be deformed in intramolecularly phase-separated bulk from the unperturbed Gaussian chain in the amorphous bulk state. Since the theoretical treatments in the strong segregation regime are relatively easy [13,14], the predictions on domain sizes in this regime are discussed and the predicted sizes are compared with the experimental results in Sec. V. A.

The molecular information of polymers in condensed systems can be directly obtained by neutron scattering experiments. [15], therefore, scattering theory for condensed systems will be described and the related "contrast-matching" techniques will be introduced in Sec. V. B. Subsequently, the experimental observations on several different polymer systems are compared. Finally, the results on the molecular information are combined with those on structural information and are discussed.

Polymer chains with higher molecular weights in many cases have crystalline regions in the bulk state because of the stabilization by the crystallization energy. However, only the amorphous polymers will be dealt with in this chapter to simplify the discussions.

II. PREPARATION OF BLOCK AND GRAFT COPOLYMERS WITH WELL-DEFINED STRUCTURES

A. Preparation of Block and Graft Copolymers

The anionic polymerization method is normally adopted to prepare block and graft copolymers with definite structures because this method seems to be the most suitable of several polymerization methods for synthesizing copolymers without contaminations [1]. Polymerization reactions are usually carried out either under vacuum or inert gas atmosphere because water, carbon dioxide, and other tiny amount of chemicals in the air can easily attack the propagating "living" carbanions during the monomer addition reaction and terminate the polymerization reactions.

The most standard procedures for preparing block copolymers are the sequential monomer addition methods [2]. If one tries to synthesize an AB diblock copolymer by this method, the monomers a and b have to be properly polymerized under the same experimental conditions, i.e., the monomers should be polymerized by the same initiator in the same solvent because procedures of switching the polymerization conditions in the course of the reactions may normally be accompanied by some difficulties. The schematic of synthesizing AB diblock copolymers is shown in Scheme I. In this scheme RM denotes an initator where R is an alkyl group and M is normally an alkali metal. ABC triblock copolymers and $(AB)_n$ alternating multiblock copolymers have been synthesized by essentially the same techniques so far and many copolymers with well-defined structures are known. ABA two-component triblock copolymers can be also obtained by this method, but two-step reactions initiated by bifunctional initiators may be more common, which is also shown in Scheme II. Here RM_2 also denotes an initiator having an alkyl group with two alkali metal atoms.

Coupling reactions have to be adopted to prepare copolymers with more complicated structures, such as star-shaped branched copolymers [16].

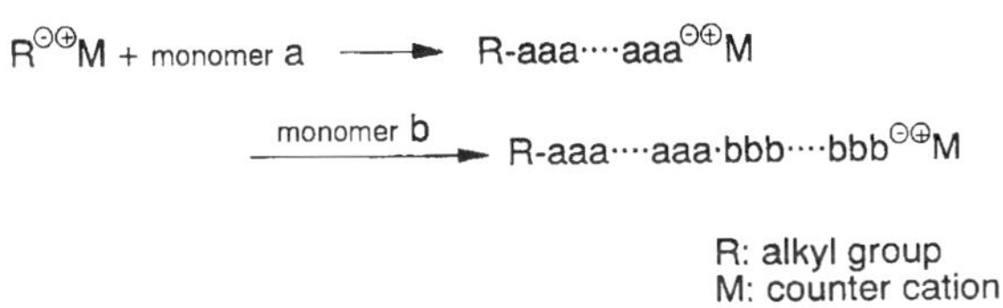

Scheme I

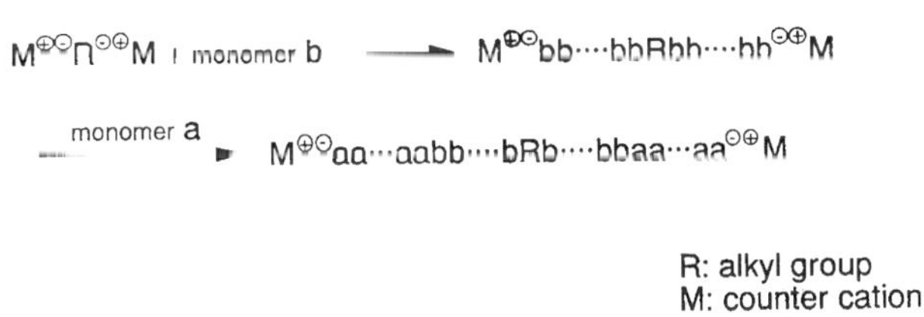

Scheme II

One of the most representative is the $(AB)_n$-type star-shaped block copolymers, where n is the number of the arms. This polymer can be obtained by the coupling reactions among the linear AB diblock copolymer chains with living ends and n-functional coupling agent. The basic idea of synthesizing a star-shaped copolymer is shown in Scheme III. The compound RX_n denotes the coupling agent, where X is a halogen atom and R is an alkyl or silylalkyl group. The most well-known polymers of this type synthesized so far are $(SI)_n$-type two-component copolymers with n being 4, 8, 12, and 18 [17], where the SI arm is a styrene-isoprene (SI) diblock copolymer polymerized from butyllithium. Chlorinated silylalkyl compounds are normally used as coupling agents.

The other example of the branched polymer is the graft copolymer of the AB_n type [18,19]. This type of polymer can be considered as the basic unit of general graft copolymers. Scheme IV represents the coupling reaction among one A polymer chain with bifunctional compounds on one end and two B living polymer chains, thus producing the AB_2 graft copolymers.

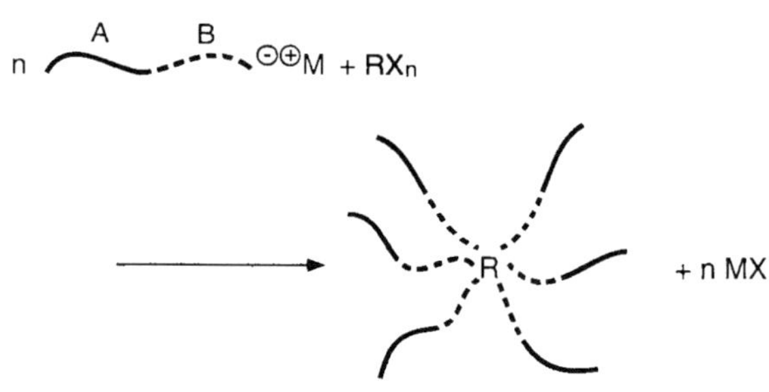

Scheme III

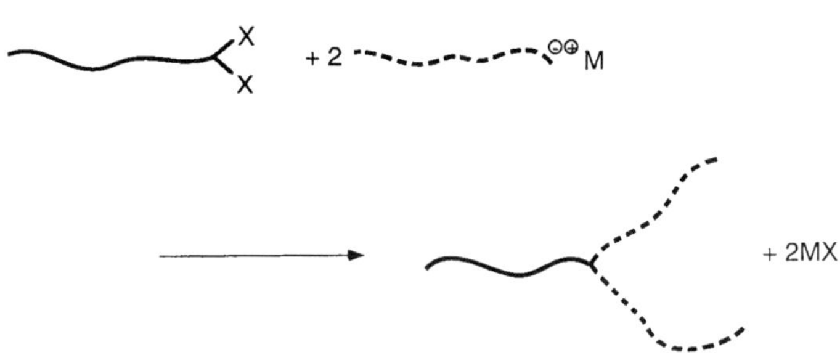

Scheme IV

Another method to produce graft copolymers is the polymerization of macromonomers [20]. However, it is in general not so easy to produce polymers with well-defined structures by this method since polymerization reactions of macromonomers, in general, do not proceed anionically under control, but they do radically.

B. Characterization Methods of Block and Graft Copolymers

Concise molecular characterization of polymers from the determination of the molecular weight and the molecular weight distributions is very important and has to be done carefully prior to the study of structures and properties of multiphase polymers. Also the compositions of copolymers are sometimes sensitive to the structural transitions or the physical properties of these polymers.

The molecular weight of a polymer is the most basic and important piece of molecular information. One, therefore, must know the molecular weight precisely in advance of any other research activities in most cases, so experimental techniques by which the absolute molecular weight can be obtained should be used. Vapor pressure osmometry ($M_n < 10^4$) [21] and membrane osmometry ($10^4 < M_n < 2 \times 10^5$) [22] have been used to measure the number-averaged molecular weight of samples, M_n, with relatively low molecular weight. We do not have to pay any attention to the detailed nature of the polymer, even when the samples are copolymers, because this method measures the difference in chemical potentials between the pure solvent and the polymer solutions, of which the magnitude is proportional to the number of molecules per unit volume in the solution. Light scattering ($M_w > 10^5$) [23] has been commonly used to measure the weight-averaged molecular weight, M_w, of polymers with fairly high molecular weight. M_w of copolymers, however, cannot be measured easily by light scattering because in general the refractive index increment of polymer A, $(\partial n/\partial c)_{A-C}$, and polymer B, $(\partial n/\partial c)_{B-C}$, in solvent C are different from each other.

In these days the molecular weight distributions of polymers including copolymers are measured by gel-permeation chromatography (GPC) [24] or size exclusion chromatography (SEC)* in most cases. This method is very simple and easy and shows high reproducibility though it is a relative method and requires "standard samples" having a known molecular weight M to calibrate the chromatograms of the samples.

*GPC and SEC are essentially the same method using analytical columns with crosslinked polymer gels which exhibit the size exclusion effect.

Conventional spectroscopy methods, such as ultraviolet (UV) and Infrared (IR) spectroscopies are still being used for determining compositions of polymers in addition to the elemental analysis. Pyrolysis-gas chromatography is a new reliable method to measure compositions of copolymers [25].

III. STABILITY OF MICROPHASE-SEPARATED STRUCTURES

A. Formation of Microphases

Block copolymer chains diluted with good solvents behave like homopolymer chains in a dilute solution of good solvent owing to the solvation of many solvent molecules on the polymer chains, which prevents incompatible segments from contacting [26,27]. This situation is schematically shown in Fig. 1(a). When the concentration of the polymer increases, the repulsion force between unlike segments is generated gradually and the block copolymer starts to phase separate intramolecularly, and at the same time chains of the same kind gather together. Eventually A-rich and B-rich regions are produced spontaneously as shown in Fig. 1(b) [28]. This tendency becomes more striking if the polymer concentration goes higher, and finally A- and B-domains are formed when solvent molecules vanish completely. Here the A-chain and B-chain in the same molecule cannot separate beyond the chain length as A and B are connected. Consequently, block chains are aligned in the order ···A–BB–AA–B···, with the junction points staying on the interfacial region as shown in Fig. 1(c). The repeating distance D, which is called the domain spacing, corresponds to the length of two molecules. The

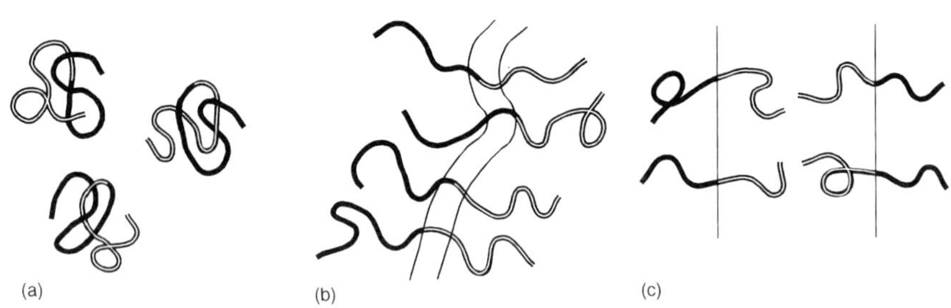

Figure 1 Chain conformations of block copolymers at various states; (a) in dilute solution; (b) in concentrated solution; (c) in microphase-separated bulk.

very periodic structure reflecting intramolecular phase separation is the microphase-separated structure (MSS).

There are two ways to obtain a MSS by experiment. One is to obtain thin films by solvent-casting from dilute solutions of sample specimens on flat surfaces such as mercury or glass Petri dishes with particularly smooth surfaces [8,29]. This method is based on the idea of utilizing the solvent power of a common good solvent which even gives the uniformly expanded coils for incompatible block copolymers in dilute solutions. This procedure needs slow evaporation of solvent, normally for several days [29], to avoid as much as possible the introduction of the artificial kinetic effects or non-equilibrium factors. Further, thin films are usually annealed at high tempertures well above the glass transition temperatures of the component polymers employed under vacuum for several days to remove residual undesirable nonequilibrium factors.

The other method to produce films possessing equilibrium MSSs is sample molding under vacuum [30]. In principle, this method could be a more suitable one to obtain equilibrium structures compared with solvent-casting if it can be started from a disordered state at much higher or lower temperatures than ODT temperatures. However, in many cases we have to be careful about the sample degradation owing to heat, and for this reason, this latter method is not so commonly adopted at least in the academic field.

B. Order–Disorder Transition

Leibler predicted the phase diagram in the weak segregation regime, [4] where the positive interaction between block A and B is very weak, as shown in Fig. 2, assuming that the deformation of the block chain due to the segregation is negligible. This figure illustrates the following well-known critical condition where MSS is favored over random-mixing for block chains with a volume fraction of component A, ϕ_A, of 0.5.

$$\chi_{AB}N > 10.5 \tag{1}$$

In Eq. (1), χ_{AB} is the Flory–Huggins interaction parameter between component polymers and N is the degree of polymerization of the whole block chain. If a block copolymer satisfies the condition in Eq. (1), the system is in the ordered state. Alternatively, if χ_{AB} becomes very small and/or N becomes small enough not to satisfy the condition in Eq. (1), block chains tend to be mixed randomly, being in the disordered state. This phase transition is called the order–disorder transition (ODT) of the block copolymer in the weak segregation regime. This phenomenon was clearly

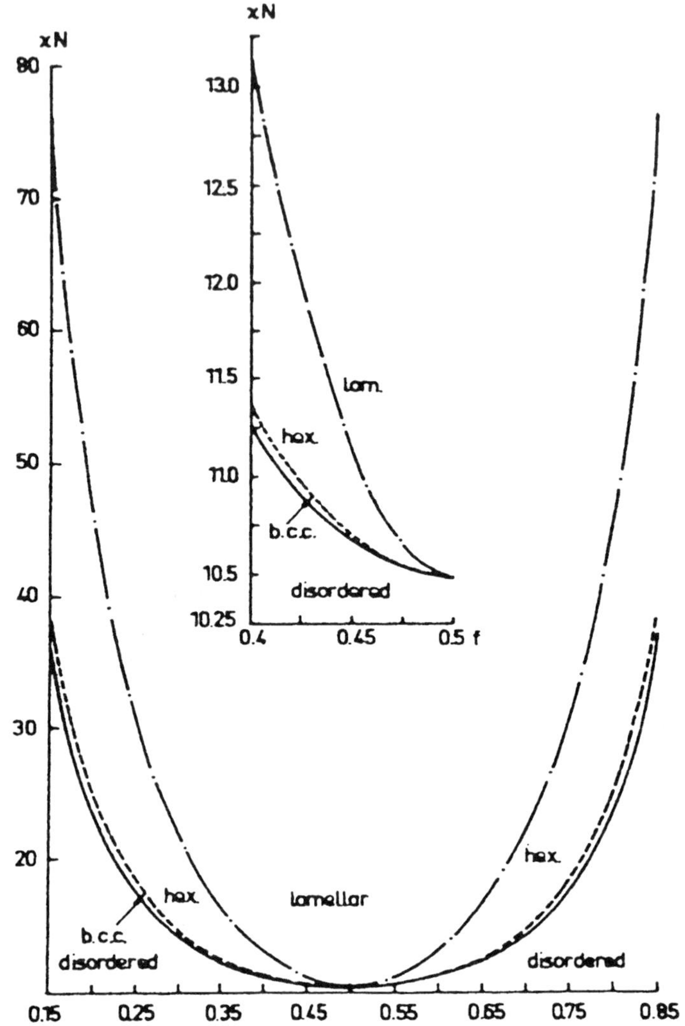

Figure 2 Phase diagram of AB diblock copolymers from Ref. 4. "lam.," "hex.," and "b.c.c." denote alternating lamellar structure, hexagonal cylinders, and spherical microdomains with body-centered cubic arrangements, respectively.

observed by experiments on a styrene–isoprene diblock and star-shaped block copolymers [5].

Moreover, Fredrickson and Helfand proposed the fluctuation theory [6] where diblock coils are deformed because of the composition fluctuation even at temperatures well above or below the ODT temperature, and this theory was supported by the experiments on poly(ethylene-propylene)–poly(ethyl ethylene) (PEP–PEE) diblock copolymers [31,32].

In the remaining part of the chapter, however, we will only deal with copolymers in the strong segregation regime where positive interactions between component polymers are strong enough for the systems to have MSS with very narrow polymer/polymer interphases.

C. Morphological Changes with Composition

It is obvious from the discussion of the previous section that the symmetric block copolymers form flat interfaces as is shown in Fig. 3(a), when the volume fractions of component A and B are the same. However, what happens if the composition deviates greatly from 50/50 for A/B? Fig. 3(b) expresses the imaginary case where asymmetric polymers, i.e., the volume fraction of the minor component is much lower than that of the major one, would form a flat interface. Actually this is not realized because the conformational entropy loss of the major component is too high to form the flat interface, so that the chain of the major component tends to expand along the direction parallel to the interface to gain conformational entropy under the condition that the segment densities of both block chains have to be kept constant and must be the same as those of bulk densities of homopolymers. Consequently, the interface becomes concave toward the major component

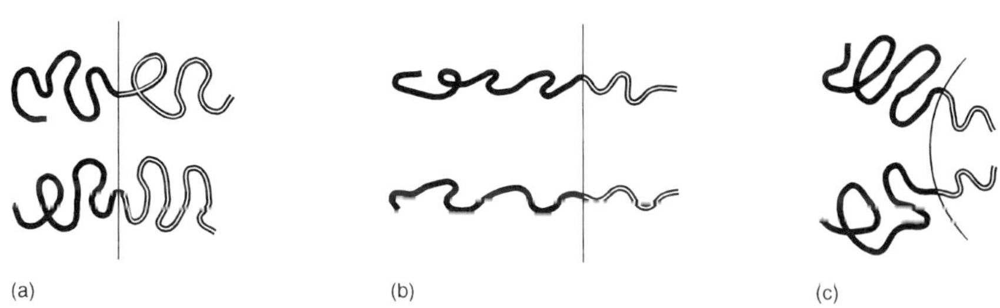

Figure 3 Schematics of chain conformation at the microphase-separated state; (a) stable flat interface from copolymers whose volume fraction of polymer A (black chain), ϕ_A, is comparable to that of polymer B (white chain), ϕ_B; (b) an unstable flat interface in the case of $\phi_A \gg \phi_B$, (c) a stable curved interface in the case of $\phi_A \gg \phi_B$

as shown in Fig. 3(c). This is the reason for curving the interface and this effect is more apparent if the volume fraction of the minor component becomes even smaller.

Fig. 4 schematically expresses the morphology changes with compositions [7]. Fig. 4(a) shows the spherical microdomains of the minor component A (white) in a matrix of B (black), while Fig. 4(b) represents the cylindrical microdomains of A in a matrix of B. Fig. 4(c) is the alternating lamellar structure of two components, which corresponds to the copolymer whose ϕ_A and ϕ_B are not very different. These three are well-established structures and have been observed for several copolymer samples experimentally [3,8,33]. When ϕ_A is higher than ϕ_B, inverse cylindrical and inverse spherical structures have been reported. Fig. 5 shows typical transmission electron micrograph (TEM) images of three different morphological structures from styrene–isoprene (SI) diblock copolymers. The dark portions show the polyisoprene microdomains as sample specimens were stained with osmium tetroxide. It was also reported that styrene-2-vinylpyridine (SP) diblock copolymers show qualitatively the same morphology change as SI copolymers except that the compositions where morphological transition occur are recognizably different from each other [34].

Fairly recently, two kinds of bicontinuous structures (BS), where the minor as well as the major components form three-dimensionally continuous network domains, were observed for several two-component systems. One is the double-diamond structure called the tetrapod-network structure which was discovered for strongly segregated SI diblock copolymers [35]. Essentially the same structure but called the ordered-bicontinuous double-diamond (OBDD) structure was observed for $(SI)_n$-type star-shaped block copolymers [10,36].

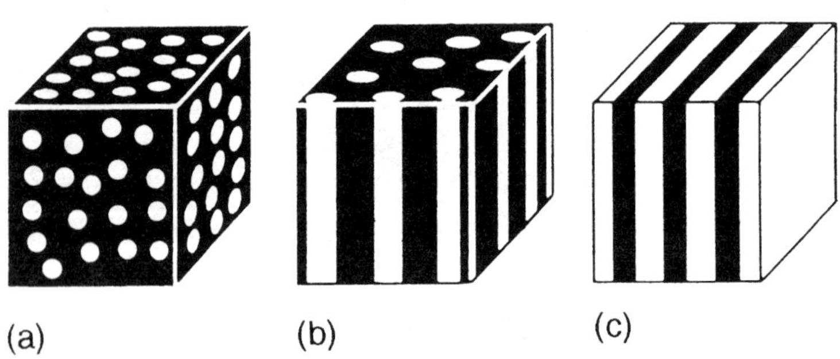

Figure 4 Three well-known MSS of AB diblock copolymers: (a) spherical structure; (b) cylindrical structure; (c) alternating lamellar structure.

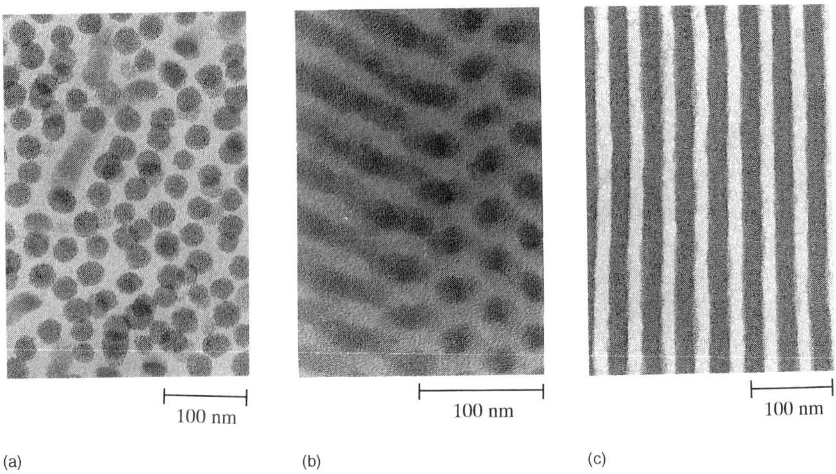

Figure 5 Examples of TEM images. Samples are styrene-isoprene diblock copolymers whose polyisoprene volume fractions are 0.18, 0.24, and 0.52 for (a), (b), and (c), respectively.

A schematic drawing of the tetrapod network structure is shown in Fig. 6(a) and the corresponding TEM image is compared in Fig. 6(b). Fig. 6(a) expresses a unit cell of this structure. It includes eight tetrapods, where two sets of four tetrapods are connected to each other, though only one of each of them is drawn in this figure to avoid confusion. Actually the minor component, polyisoprene, whose volume fraction is 0.34 forms two kinds of interwoven diamond network in a matrix of polystyrene phase as revealed in Fig. 6(b), where the polyisoprene phase is stained with osmium tetroxide. The double-diamond lattice is one of the cubic lattices and its symmetry belongs to the *Pn*3*m* crystallographic space group [10].

The other cubic lattice is the one derived from the "Gyroid," which was arithmetically discovered and named by Schoen [37]. This structure was observed for a weakly segregated SI [38] and well-annealed $(SI)_n$ star-shaped block copolymers [39]. Fig. 7(a) is the schematic drawing of the framework of the Gyroid structure and Fig. 7(b) is the corresponding TEM image observed for an SI diblock copolymer. This structure is also a cubic lattice and it has the symmetry of an *Ia*3*d* space group. It is actually impossible to distinguish OBDD and Gyroid-derived structure from TEM images alone. Small-angle X-ray scattering data from the specimens can give us a little more definite evidence so that the structure can be assigned to a particular type, but again it is very difficult to distinguish the two clearly at present. There is no definite conclusion as

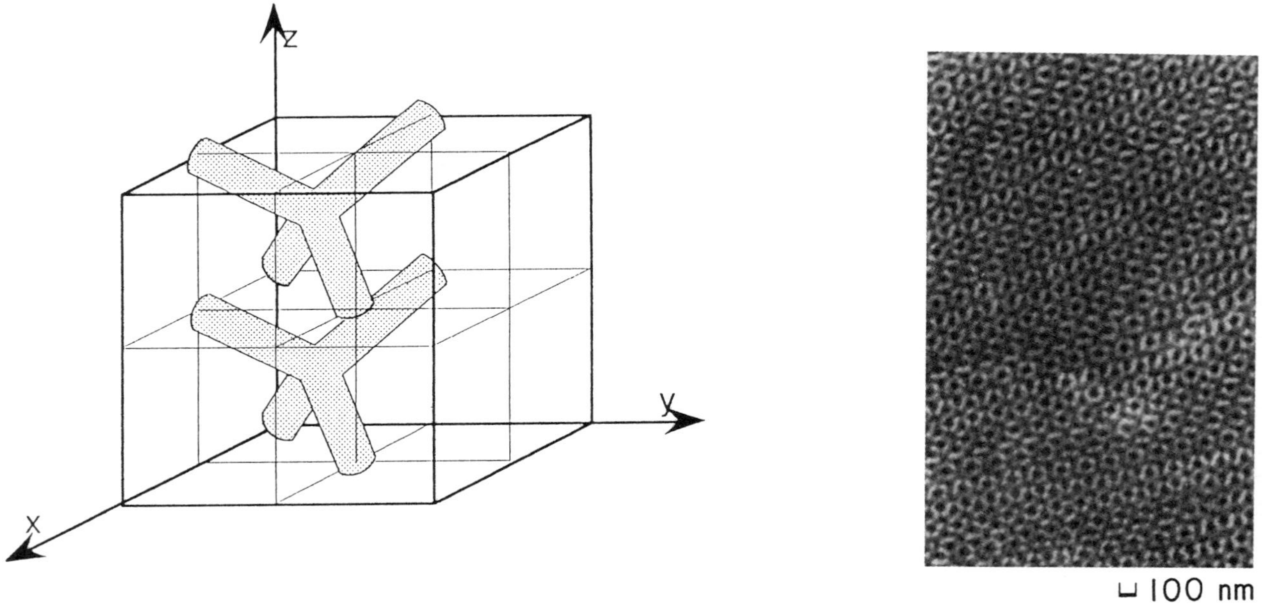

(a) (b)

Figure 6 The tetrapod-network structure. (a) A unit cell of the double-diamond framework from the minor component. (b) A TEM image obtained from an SI diblock copolymer, whose ϕ_S is 0.66. This figure is from Ref. 35.

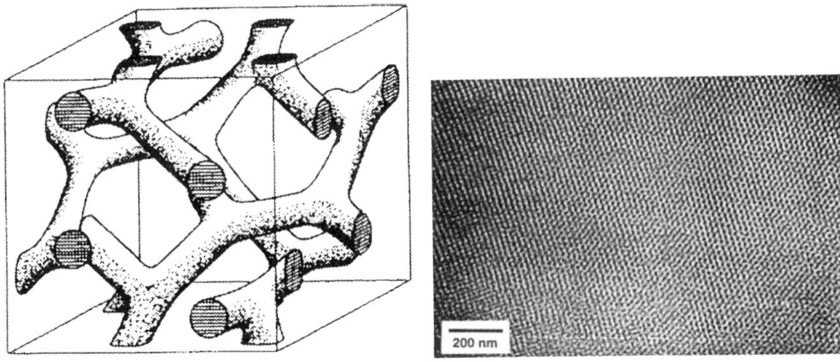

(a) (b)

Figure 7 The family of the "Gyroid" structure. (a) A schematic representation of the double network derived from the "Gyroid" surface. (b) A TEM image of an SI diblock copolymer whose ϕ_S is 0.33. These figures are from Ref. 38.

to which structure is more stable. Bicontinuous and tricontinuous structures will be described in the next section in some more detail, focusing on the surface structures.

Morpholoical change with composition for ABA two-component [34], ABC three-component triblock copolymers [11] and also ABB graft copolymers [40,41] have also been investigated. Their phase diagrams are compared in Fig. 8. It is apparent from this figure that no essential differences are recognized between AB (SP) diblock and ABA (PSP) triblock copolymers, while the phase diagrams for ABB (SPP) graft copolymers are markedly different from that of an AB diblock, reflecting the difference in architectures. So far, no bicontinuous structures have been observed as for ABB graft copolymers probably because of the lack of experimental data; in reality, the white portions on the figure are the regions where experiments have not yet been carried out. Further very peculiar MSSs and hence phase diagrams, which are fairly different from the former three, have been observed for ABC triblock copolymers.

Fig. 9 compares TEM images of isoprene-styrene-2-vinylpyridine (ISP) triblock copolymers. These polymers possess the molecular characteristics of keeping the volume fractions of two end block polymers constant, i.e. $\phi_I = \phi_P$, so that the molecules are symmetric in terms of volume fractions. Fig. 9(a) shows a spherical structure with two kinds of spheres from the I and P domains (black and gray spheres in the TEM image). It is very hard to figure out the arrangement of spheres just from TEM images, but diffraction experiments using small-angle X ray scattering (SAXS) reveal

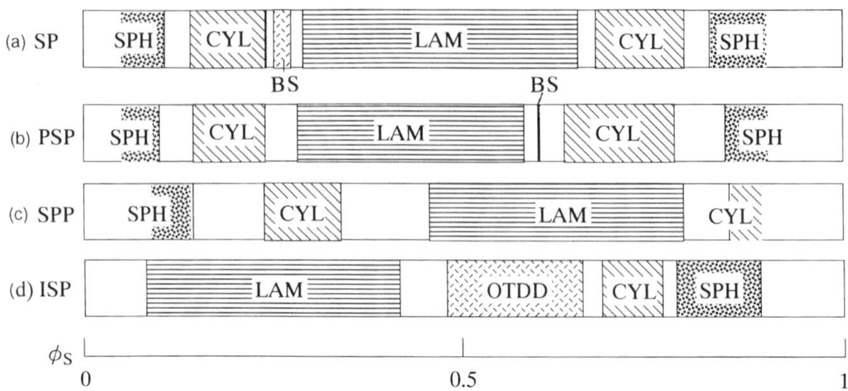

Figure 8 A comparison of the phase diagrams of copolymers with four different architectures. SPH, CYL, BS, and LAM denote spherical, cylindrical, bicontinuous, and lamellar structures, respectively, while OTDD in the diagram for ISP expresses ordered-tricontinuous double-diamond structure. Samples are: (a) styrene(S)-2-vinylpyridine(P), labeled as SP, diblock copolymers; (b) PSP triblock copolymers; (c) SPP graft copolymers; (d) isoprene(I)-styrene(S)-2-vinylpyridine(P), labeled as ISP, triblock copolymers.

that the spheres are packed in body-centered cubic arrangements where I spheres stay at the body-center of the eight neighboring P spheres and vice versa. The resulting structure is the CsCl-type one, reflecting the symmetry of the molecule [42].

Fig. 9(b) shows a cylindrical structure. It is again very difficult to determine the exact arrangements of cylinders from TEM images alone. SAXS experiments verified the mutually tetragonally packed configuration of the cylinders for this structure [42], which is in contrast to the well-known hexagonal packing patterns for diblock copolymer systems [33].

The morphological feature of the ISP triblock copolymers with $\phi_I = \phi_P$ is highlighted in Fig. 9(c) [43]. This figure looks quite complicated compared with the others, but it turns out that this is a periodically triply-continuous structure where the two end block polymers I and P form equivalent frameworks which are mutually interwoven in a matrix of the center block S.

Finally Fig. 9(d) shows a very simple lamellar structure. The black, white, and gray stripes represent I, S, and P lamellae. Looking at this TEM carefully, one can notice that there are twice as many white stripes as the other two, representing the order of the block chain connections. In short, the center block, polystyrene, has to have a bridge-conformation connecting

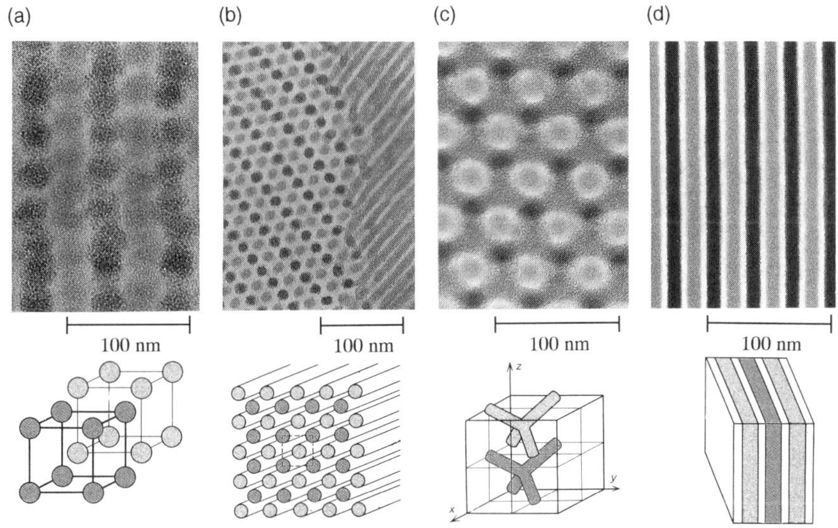

Figure 9 Four kinds of MSS of ISP triblock copolymers (top) and their schematic models (bottom) [42]. (a) A spherical structure of the C_sCl type. Volume factions of I, S, and P blocks of the sample are 0.08, 0.84, and 0.08, respectively. (b) A tetragonally-packed cylindrical structure (0.14-0.68-0.18). (c) An ordered tricontinuous structure (0.26-0.48-0.26). (d) A three-phase four-layer lamellar structure (0.36-0.31-0.33).

two lamellae because I and P chains are incompatible with each other, whereas the situation for end block chains of this ISP molecule is almost the same as that for a diblock molecule. Consequently, the repeating unit of microdomain structures should include one –ISP– plus another S domain, that is, –I–S–P–S– four layers. One can notice again that the ISPS unit includes two ISP molecules in the direction perpendicular to the lamellar interface just as for diblock copolymers. This structure, therefore, has been named the three-phase four-layer lamellar structure.

If we look for the common feature of the four structures in Fig. 9, one notices that the microdomains from I and P domains are interchangeable, though the dimensions of the four structures are different. This structural feature is nothing but the "superlattice structure" which is known for metal alloys and other multicomponent materials. Thus, we can summarize the features of a series of the ABC triblock copolymer with the condition $\phi_A = \phi_C$ as conforming to the superlattice structure [42].

IV. THE FEATURES OF POLYMER–POLYMER INTERFACES

A. Curvatures of Polymer–Polymer Interfaces

As it was described in the previous section, the shape of the polymer/polymer interface changes with its composition, and accordingly, the curvature of the interface depends on the morphology. For example, the curvatures of alternating lamellar structures are zero if we ignore local fluctuations or waviness of the compositions at the surfaces, while those of spherical and cylindrical structures are positive and constant. Then, what kind of surfaces and curvatures do bicontinuous structures have? It is meaningful to adopt the idea of mean curvature [45], and some quantitative discussion will be required to make the point clearer.

Consider now the features of a surface from an arithmetic viewpoint. The mean curvature is defined as the arithmetic mean, $(k_1 + k_2)/2$, of two principal curvatures, where k_1 and k_2 are the maximum and minimum curvature along two orthogonal directions going through certain points under consideration. The feature of the surface can be characterized by the sign of the product, k_1k_2, of these curvatures, which is called Gaussian curvature. A particular surface whose Gaussian curvature is negative is schematically shown in Fig. 10 and the characteristics of surfaces known for the equilibrium morphology of block copolymers are summarized in Table 1. A remarkable feature of the surface for a bicontinuous structure is its minus sign in Gaussian curvature. Moreover, the mean curvature, $(k_1 + k_2)/2$, must be constant for this structure because the system is considered to be unstable if the local mean curvature is significantly different from place to place. This may be the requirement of a constant-mean-curvature surface for equilibrium morphology of block copolymers [46]. The surface whose mean curvature is exactly zero is called the "minimal

Table 1 Mean Curvature and Gaussian Curvature for Equilibrium Microphase-Separated Structures of Block Copolymers

Structure	Spherical structure	Cylindrical structure	Bicontinuous structure	Lamellar structure
Mean curvature $(k_1 + k_2)/2$	Plus	Plus	Constant	Zero
Gaussian curvature k_1k_2	Plus	Zero	Minus	Zero

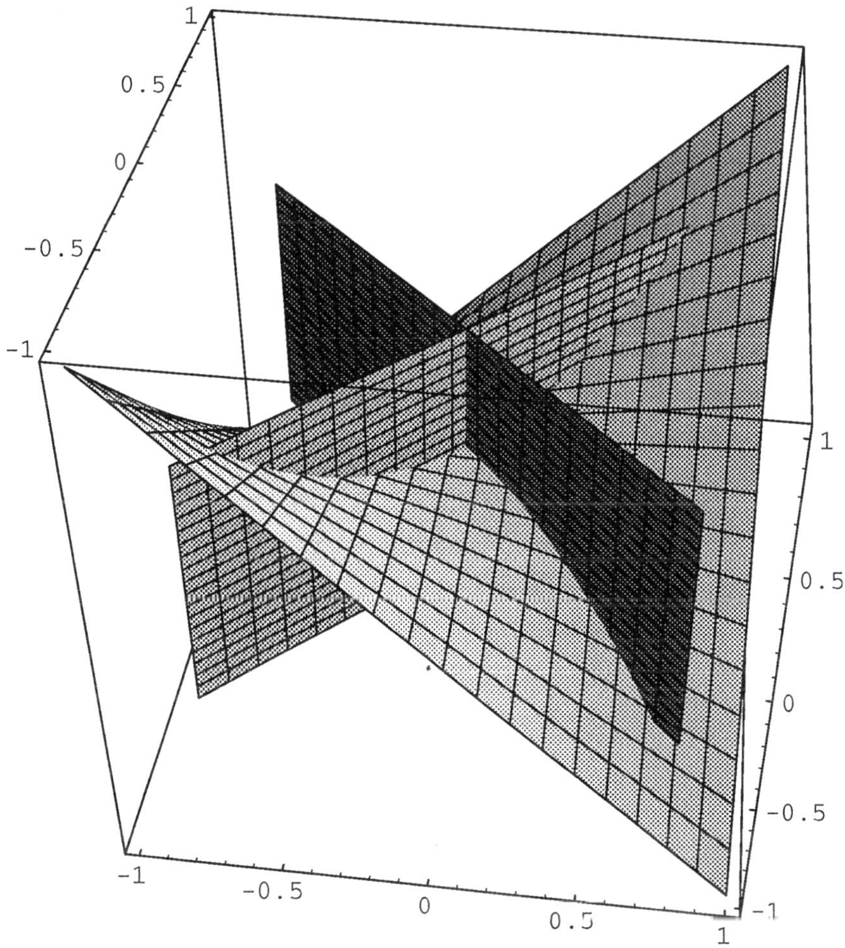

Figure 10 A surface with negative Gaussian curvature.

surface." Many minimal surfaces, such as Schwarz's D-surface [44], Scherk's first surface [8] and Schoen's G-surface [37] have already been discovered mathematically. They can be strictly expressed by the Weierstrass function* [47] but in general very large calculations are

---

*(x, y, z) coordinates of the minimal surfaces in $O - xyz$ Cartesian coordinates can be expressed by the Weierstrass function $R(\omega)$ as: $x = R_e \int (1 - \omega^2) R(\omega)\, d\omega$; $y = -I_m \int (1 + \omega^2) R(\omega)\, d\omega$; $z = R_e \int 2\omega R(\omega)\, d\omega$, where R_e and I_m denote the real and the imaginary parts, respectively.

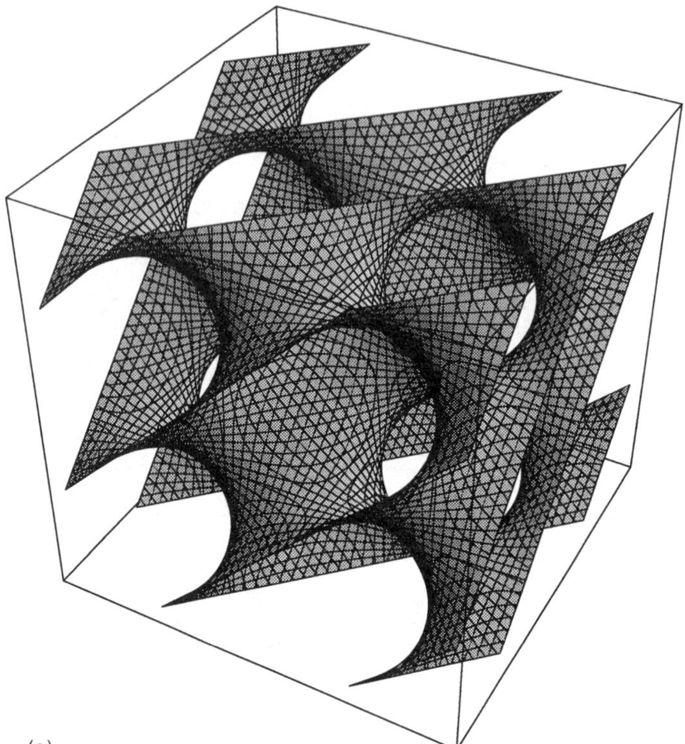

(a)

Figure 11 Computer-generated wireframe models for minimal surfaces: (a) Schwarz's D-surface; (b) Schoen's G-surface.

required to obtain the "real" minimal surfaces; therefore, we normally use the approximate equations to describe the surfaces. The calculated three-dimensional pictures for D-surfaces and G-surfaces thus obtained, for example, are compared as shown by the wire frame models in Fig. 11.*

*The approximate equations for a D-surface and a G-surface are: $\cos x \cos y \cos z + \cos x \sin y \sin z + \sin x \cos y \sin z + \sin x \sin y \cos z = 0$ (D-surface); $\sin x \cos y + \sin y \cos z + \sin x \cos z = 0$ (G-surface).

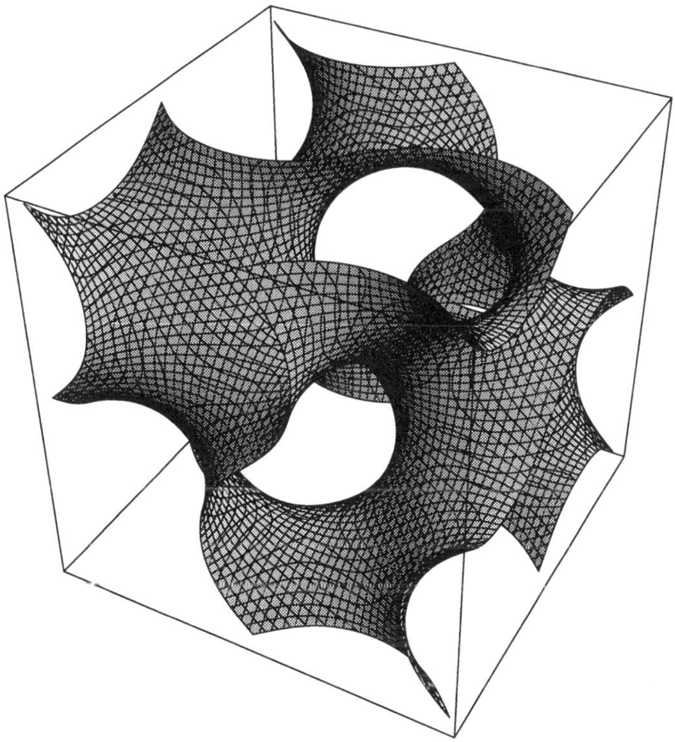

(b)

B. Surface Characteristics of Bicontinuous and Tricontinuous Structures

If we apply the idea of these imaginary minimal surfaces to our real polymer–polymer interfaces, we have to pay attention to the volume fraction first. Each one of the minimal surfaces strictly divides the space into two equivolume regions [44]; however, real polymers never have minimal surfaces but instead have apparently lamellar phases if the volume fractions of A chains in AB diblock copolymers are 0.5. Then, the idea of parallel surfaces, which are produced by adding layers of uniform thickness on both sides of the particular minimal surface, has been introduced to adjust the difference in volume fractions between imaginary and real systems [48]. Chemical junctions of block chains are placed on the two different surfaces setting the minimal surfaces as planes of symmetry in the parallel surface model, as shown in Fig. 12. The resulting two surfaces are interwoven and their Gaussian curvatures are both negative and the magnitudes are the same. One should notice from Fig. 12 that the actual microdomain thickness is never constant in bicontinuous or tricontinuous structures.

Several attempts have been made to prove that the surface structures of block copolymers have the character of this modeled structure by comparing TEM images and computer-simulated images [36,38]. Fig. 13 compares the computer-simulated image, assuming double-diamond frames formed by cylindrical struts for the [111] projection within the thickness of one repeat unit and a TEM image observed for a strongly segregated $(SI)_8$-type star-shaped block copolymer [36]. Fig. 14, on the other hand, compares a computer-simulated image assuming the parallel surfaces derived from the G-surface and a TEM image observed for a weakly segregated SI diblock copolymer [38]. It seems that the TEM and the computer-generated images are quite similar to each other for both cases in Figs. 13 and 14.

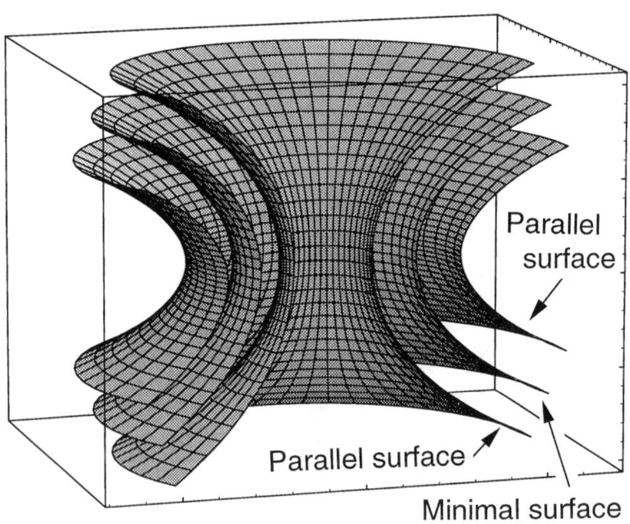

Figure 12 Spatial relations in between a minimal surface and two equivalent parallel surfaces which are produced by shifting the surface normal to the original minimal surface toward both sides. The Catenoid surface is chosen as an example of the minimal surfaces in this figure because this surface is relatively simple.

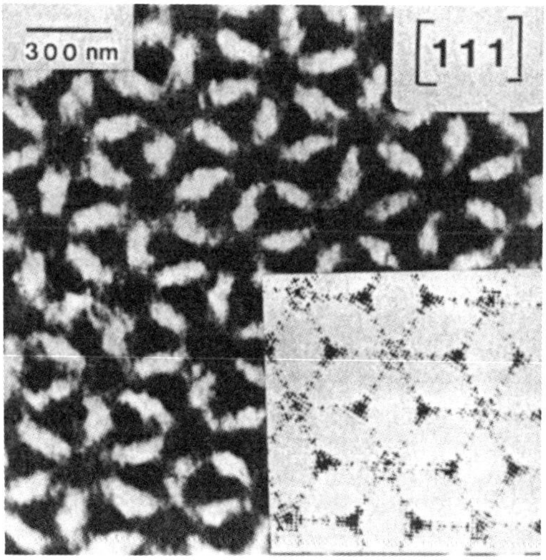

Figure 13 A family of D-surfaces. The bright field TEM image is the [111] projection of the double-diamond structure. Sample: $(SI)_8$ star-shaped block copolymer with eight arms whose ϕ_S is 0.27. The right lower corner shows a computer-generated image assuming double-diamond frameworks formed by cylindrical struts. This figure is from Ref. 36.

In general, however, these attempts do not necessarily lead to definite conclusions because the procedure does not give necessary and sufficient conditions in addition to the facts that both TEM and computer-generated images are very sensitive to the angle of observation, the sample thickness, and the contrast between microdomains of component polymers.

C. Block Copolymer/Homopolymer Blends

If homopolymers of the same components as those of block copolymers are mixed with parent block copolymers, the resulting blends show the full variety of MSSs depending on homo/block molecular weight ratios and also blend ratios. Here we assume an AB diblock copolymer with an alternating lamellar structure possessing a flat surface as shown in Fig. 15(a). If homo-A is mixed with AB and if the molecular weight of homo-A is much lower than that of block A of AB, homo-A can dissolve into the A-microdomain uniformly so that it eventually bends the interface concave to the A-microdomain as schematically shown in Fig. 15(b). On the other hand, when A and B homopolymers both with smaller molecular weights than those of

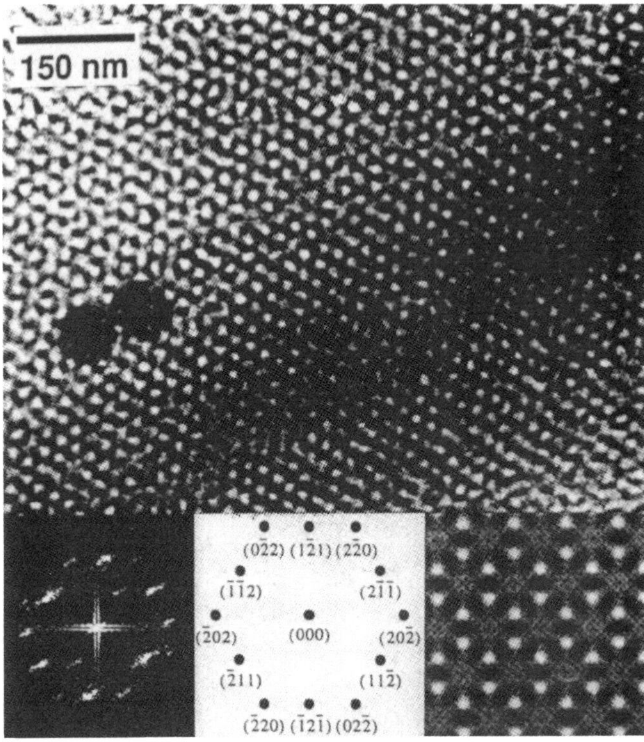

Figure 14 A family of G-surfaces. A bright field TEM image is the [111] projection of the "Gyroid" structure. Sample: SI diblock copolymer, ϕ_S; 0.33. The insets on the left show the Fourier-transformed image and its indexed pattern displaying six (211) and six (022) reflections. The smaller figure in the right lower corner is a computer-generated image assuming twin parallel surfaces to the G-surface. This figure is from Ref. 38.

the blocks are mixed with the same amount of AB, the shape of the interface does not change though the neighboring chemical junction points of the block copolymer are pulled apart because of the solubility of the homopolymer molecules as shown in Fig. 15(c).

Alternatively, let us consider the case of homo-A with molecular weight comparable to that of block A is mixed with AB. Apparently the homo-A can be involved in the microdomain structure formed by the block copolymer, but both polymers are weakly segregated from each other and consequently the homopolymer chain is localized and its distribution peak is at the center of the lamellar microdomain as shown in Fig. 15(d). The average distance between junction points of block chains is not changed and hence the A

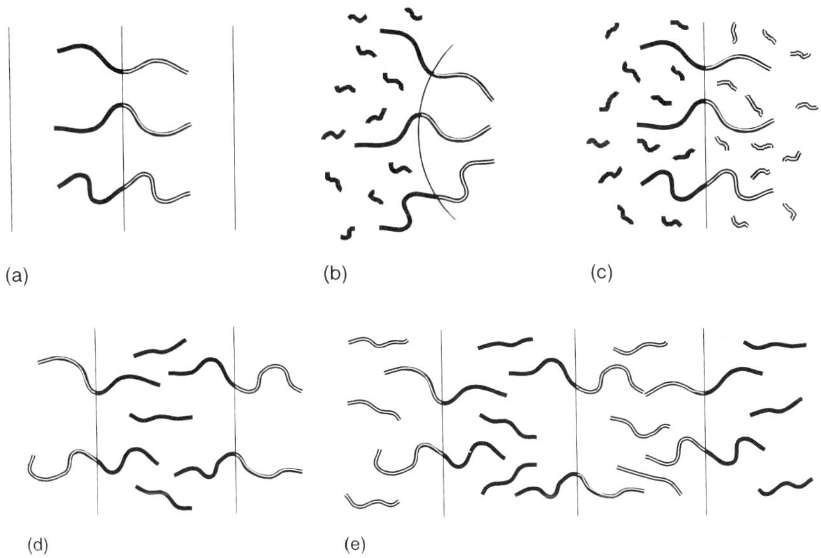

Figure 15 A schematic representation of the manner of mixing of block copolymer/homopolymer blends. The black chains are for polymer A, while the white chains are for polymer B. (a) The pure block copolymer. (b) The AB/B binary blend, the molecular weight of homo-A is much smaller than that of block-A. (c) The AB/A/B ternary blend, the molecular weights of homo-A, homo-B are much smaller than those of corresponding block chains. (d) The AB/A binary blend, the molecular weight of homo-A is comparable to block A. (e) AB/A/B ternary blend, the molecular weights of homo-A and -B are comparable to those of block A and B.

lamellar thickness is expanded leaving the B lamellar thickness unchanged in this case. The situation is essentially the same as for A-homo/AB/B-homo blends, though both A and B lamellar thicknesses are expanded in the latter case as shown in Fig. 15(e). If the molecular weight of the homopolymer added is sufficiently larger than that of the block chain, macrophase separation between the homopolymer and the block polymer is favored.

Summarizing the above considerations, it seems that the homopolymers dissolved in microdomains have two different roles in the course of domain formations. One role is that of a diluent for block polymer chains and the other is that of having a space-filling effect. These two roles help to control the morphologies of block copolymers and their blends, especially to produce bicontinuous structures. In short, the domain thicknesses are not uniform in any known bicontinuous structures, while they are constant for three well-known structures, that is, spherical, cylindrical, and alternating lamellar structures. This nonuniformity of the domain thickness is one of the reasons for the difficulty for pure block copolymers to form bicontinuous structures because the block chains have to be in different environments at different places, which leads to the nonuniformity of the conformational entropy. This is an unfavorable condition for the system to form bicontinuous morphologies. Blended homopolymers can behave as space-fillers and they can also change the curvatures of the interfaces. Thus, homo/block blend systems can more easily lead to the formation of a bicontinuous structure than pure block copolymers. In fact, Hasegawa et al. [48] have found the P-surface for an SI/S blend system and the tD-surface for an SIS/homopolymer, while Winey et al. [12] reported that the ordered bicontinuous double-diamond (OBDD) structures were created for SI/S blends from a block copolymer with lamellar structure.

V. CHAIN CONFORMATIONS OF BLOCK COPOLYMERS IN CONDENSED SYSTEMS

A. Microdomain Sizes

As was described in Sec. III, immiscible polymers mutually prevent each other from forming their own microdomains in bulk. Obviously the sizes of the repeating units of microdomains depend on the chain dimensions of polymers. It is relatively easy to predict the domain sizes by theoretical considerations in the strong segregation limit regime. Among several conceivable factors, two forces or energies play the dominant roles in forming equilibrium structures, therefore, the contributions from the other minor factors are omitted here.

One of these forces is the repulsive force generated between two different polymer species due to the segregation power. This force can be understood as the surface tension. Block chains have to be elongated along the direction perpendicular to microdomain interfaces to keep segment density constant within microdomains. Fig. 16(a) expresses the surface tension which acts to narrow the interfacial area and so causes the distance between adjacent chemical junction points to decrease. The interfacial free energy G_i per chain can be described as

$$G_i \propto N/r \tag{2}$$

where N is the degree of polymerization of the whole molecule and r is the chain dimension of the block polymer along the direction perpendicular to the lamellar interface. Alternatively, in general, once a polymer chain is elongated in a certain direction, the chain loses conformational entropy so

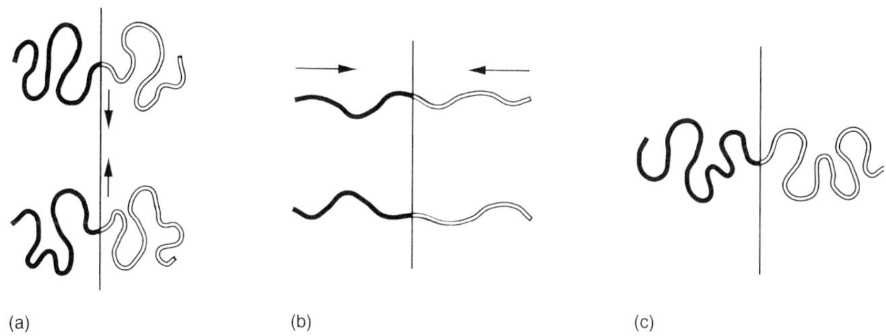

Figure 16 The forces acting on intramolecularly phase-separated chains: (a) the surface repulsive force; (b) The elastic force; (c) the balance between the two forces.

that an elastic force is generated to gain entropy. This phenomenon is known as the rubber elasticity of a polymer chain and is shown in Fig. 16(b). Thus the other contribution to the total free energy comes from the elastic force. The corresponding elastic energy G_e per chain can be written as

$$G_e \propto r^2/Nb^2 \tag{3}$$

where b is the segment length of the polymer.* The free energies in Eqs. (2) and (3) cause the forces acting in opposite directions and they counterbalance each other, so that the block chain is in the intermediate state as shown in Fig. 16(c).

The chain dimension r in the direction perpendicular to the lamellar interface should be proportional to the thickness of the microdomain, and hence proportional to the domain spacing D, i.e., $r \propto D$. Using this condition, the total free energy to form the microdomain structure $G(= G_i + G_e)$ per chain is

$$G \propto (N/D + D^2/Nb^2) \tag{4}$$

We can easily obtain the equilibrium D value by applying the condition, $\partial G/\partial D = 0$, because G should have a minimal value when the system is thermodynamically in equilibrium. Thus we have the well-known relationship between D and N [49,50].

$$D \propto N^{2/3} \tag{5}$$

*The segment length of polymer A and that of polymer B are different from each other in the real system, however, we put $b_A = b_A = b$ for simplicity according to Helfand and Wasserman [14].

Comparing Eq. (5) and the relationship of the radius of gyration (R_g) and the degree of polymerization for an amorphous polymer chain in bulk, i.e., $R_g \propto N^{1/2}$ [51,52], one can see that block chains are elongated along the direction perpendicular to the lamellar interface.

Several experimental results are known. If we write the D versus N relationship in a more general expression as

$$D \propto N^{\nu} \tag{6}$$

most of the experiments seem to indicate that the exponents ν in Eq. (6) are larger than 1/2 and quite close to 2/3 [30,53–55]. One set of experimental results is given in Fig. 17, showing the data for SP diblock and also for PSP triblock copolymers [55]. The data for the latter are actually plotted against half the molecular weight of each polymer because a triblock copolymer fills just one repeating unit while a diblock copolymer fills half of the repeating

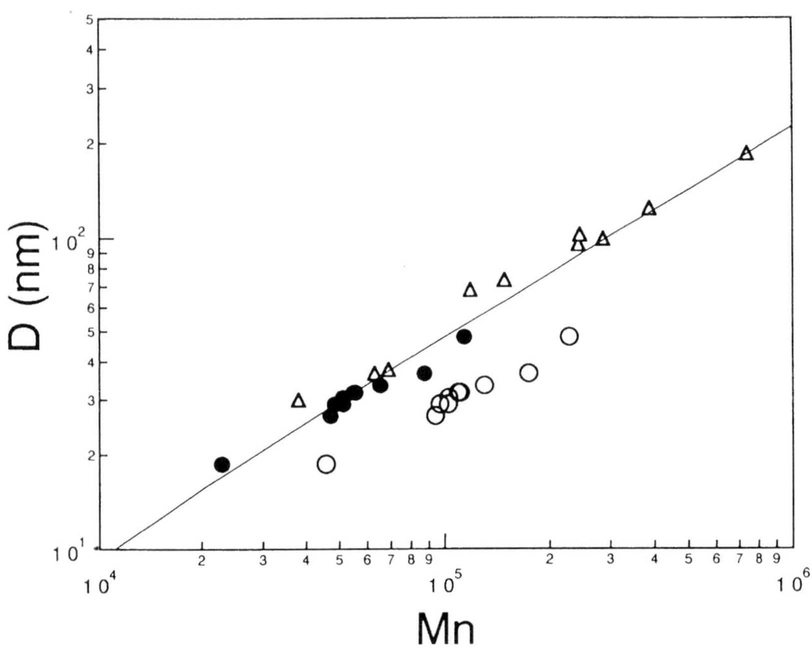

Figure 17 Domain spacing D versus the number-averaged molecular weight of whole block chains on a double logarithmic scale. The open circles are the data of PSP triblock copolymers while the filled circles are the same data but plotted against $M/2$ instead of M. The triangles are the data of SP diblock copolymers (cf. Ref. 54). This figure is from Ref. 55.

unit. The solid line in Fig. 17 is drawn by the least-squares method for data points of diblock copolymers and its slope is about 0.64. One can easily recognize from this figure that the experimental results well support the theoretical prediction, i.e., the relation $D \propto N^{2/3}$.

B. Single Chain Dimensions of Block Polymers in Bulk

It has been recognized both theoretically and experimentally that block chains are expanded at least toward the direction normal to the lamellar interface. Then what is the dimension of the deformed chain parallel to the interface? No theories are known which predict the dimension of block chains along this parallel direction. Small-angle neutron scattering (SANS) experiments only enable us to measure the dimensions of block polymers even in the microphase-separated phase. In principle, molecular scattering theory for condensed polymer systems can be extensively applied to these two-phase systems [56,57].

Firstly, let us consider homopolymer systems containing labeled and unlabeled species, ssuming that:

1. the system is incompressible and
2. labeled and unlabeled species are identical, namely they have the same degree of polymerization and there is no thermodynamic interaction between the two species.

The coherent scattering intensity $I(q)$ for this "two-component" system can simply be described by

$$I(q) \propto N X_D (1 - X_D)(\beta_D - \beta_H)^2 P(q) \tag{7}$$

Here $P(q)$ is the particle scattering function from which one can derive information on chain conformation, β_D and β_H are the coherent scattering length densities of the deuterated and the hydrogenated polymers, respectively, and X_D is the volume fraction of the deuterated chain. Eq. (7) is obtained by adopting the random phase approximation (RPA) derived by de Gennes [15]. More details are described in Chapter 3 of this book. From Eq. (7), one should recognize that $P(q)$ can be obtained from mixtures of labeled and unlabeled polymers with "any" volume fractions. This treatment is totally different from those for dilute solutions, in which extrapolation of the data to infinite dilution is always required. The possibility of labeling is the eminent advantage when neutron scattering experiments are applied to condensed polymer systems.

Next, we will consider microphase-separated block copolymer systems. Of course we need to mix deuterium-labeled A_DB block copolymers and their unlabeled counterparts A_HB with the same chain lengths and compositions. Applying the idea of RPA to this system, the coherent scattering intensity for these mixtures can be described by

$$I(q) \propto X_D(1 - X_D)(\beta_{AD} - \beta_{AH})^2 P(q) + [X_D\beta_{AD} + (1 - X_D)\beta_{AH} - \beta_B]^2/(\beta_B - \beta_{AH})^2 Q(q) \tag{8}$$

where $Q(q)$ is the interparticle scattering function and β_{AD}, β_{AH}, and β_B are the coherent scattering length densities for polymer species A_D, A_H, and B, respectively [29,58]. The second term in Eq. (8) is actually the contribution from the phase structure, and this term is apparently undesirable when obtaining $P(q)$.

There are two ways to suppress this contribution. One is to use the blend samples which satisfy the condition

$$X_D\beta_{AD} + (1 - X_D)\beta_{AH} = \beta_B \tag{9}$$

so that the second term in Eq. (8) vanishes. This operation is called the "contrast-matching" method and it is peculiar to neutron scattering. Fig. 18 schematically shows the relationship among samples, neutron beams, and detector planes in the case of alternating lamellar structure. The degree of contrast-matching can apparently be estimated by evaluating the magnitude of diffractions if neutrons are incident parallel to the lamellar interface, which is called the edge-view as shown in Fig. 18. In fact, Fig. 19 compares the variation of the diffraction patterns with the mixing ratios of labeled and unlabeled block copolymers. The samples are again a PSP triblock copolymer and its labeled counterpart and the data from a two-dimensional detector are averaged along the horizontal axis (cf. Fig. 18), i.e., within sectors at $\psi = 0^\circ \pm 5^\circ$ and $\psi = 180^\circ \pm 5^\circ$. The values along the third axis in Fig. 19 are the weight percentages of the labeled polymer. We can recognize that contrast matching is achieved when the labeled polymer is mixed at around 10%, while the calculated matching point which satisfies the condition in Eq. (9) is 10.7%.

The other method is to choose a scattering geometry in which we do not observe diffractions from the phase structure by using samples with highly oriented structures [59]. This corresponds, for example, to choosing the through-view geometry where the neutrons are irradiated perpendicular to the lamellar interface (cf. Fig. 18). If the coherent scattering intensity, $I(q)$, free from diffraction intensities is obtained from an A_DB/A_HB block copolymer blend, we can estimate the dimension of an A block by applying the modified Guinier's expression to $I(q)$ as:

$$I(q) = I_0 \exp(-q^2 R_{g,k}^2) \tag{10}$$

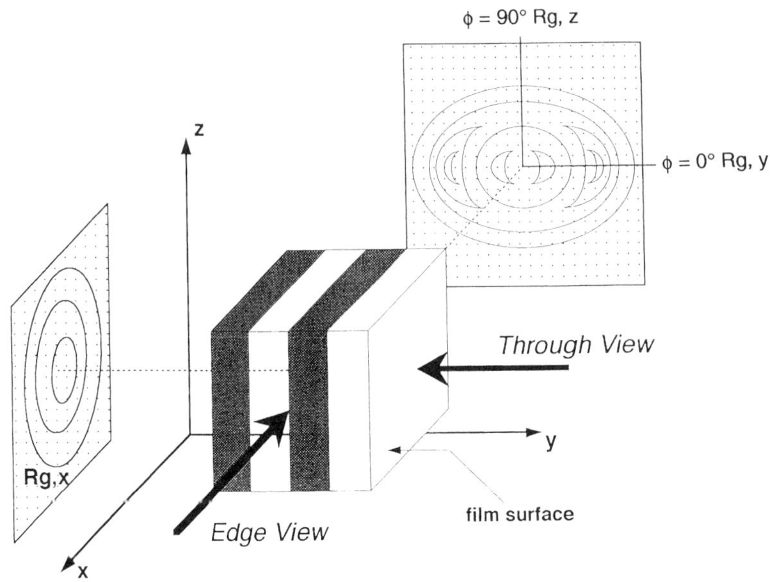

Figure 18 Mutual relations among sample specimen, observation views, and a two-dimensional detector for SANS experiments. This figure is from Ref. 55.

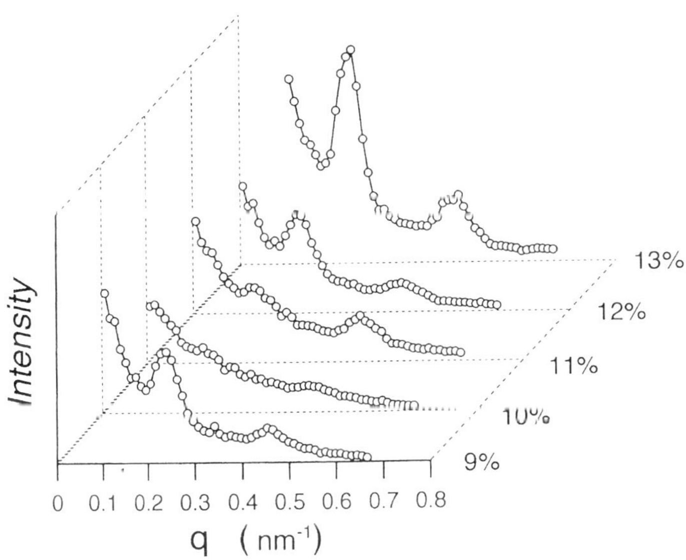

Figure 19 Variation of the edge-view coherent scattering intensity with different blend ratios. This figure is from Ref. 55.

where $R_{g,k}$ is the component of the radius of gyration along the $k(= x, y$, or $z)$ axis.

$R_{g,k}$ values for polystyrenes of SP diblock copolymers thus obtained are plotted against molecular weight in Fig. 20. The $R_{g,x}$–M relationship was estimated to be $R_{g,x} = 0.0289M^{0.43}$ (nm), and that for $R_{g,y}$–M to be $R_{g,y} = 0.00453M^{0.64}$ (nm) [60]. The latter relationship was obtained by drawing a straight line with slope 0.64, which is the value of the exponent for the D–M relationship evaluated in Fig. 17, going through one data point. Obviously, $R_{g,x} = R_{g}$, is smaller than $R_{g,y} = R_{g} \perp$ if M is larger than 10^4.

By using these $R_{g,k}$ values, the volume occupied by one molecule can be calculated as:

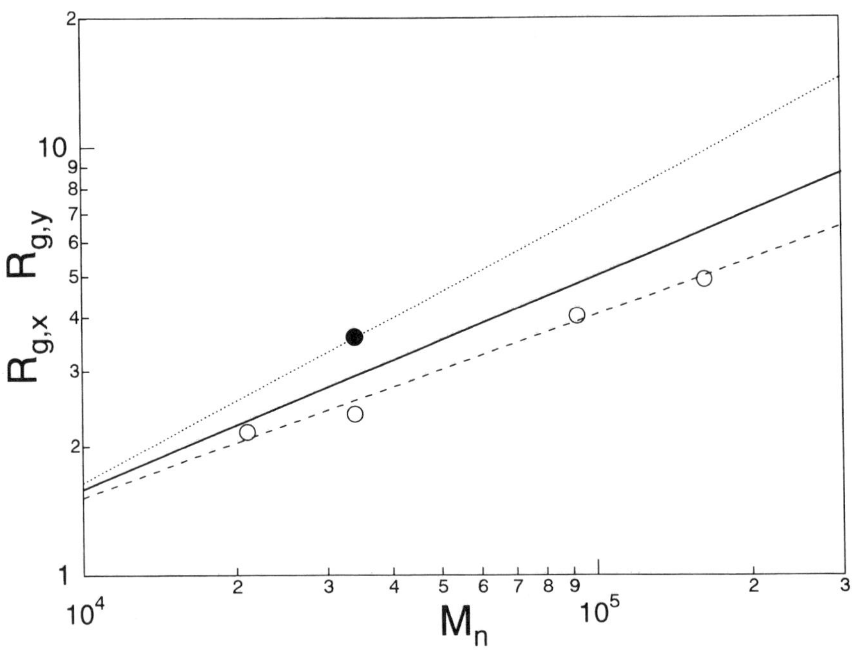

Figure 20 Double-logarithmic plots of the one-dimensional radius of gyration versus the molecular weight of polystyrene blocks of SP diblock copolymers. The open circles are for $R_{g,x}$, and the filled one is for $R_{g,y}$. The solid line expresses the empirical equation, $R_{g,k0} = 0.0165M^{1/2}$ (nm) for the unperturbed polystyrene chains in bulk, [60] whereas the broken line and the dotted line denote the relationships $R_{g,x} = 0.0289M^{0.43}$ (nm) and $R_{g,y} = 0.00453M^{0.64}$ (nm), respectively [60] (see text). This figure is from Ref. 60.

$$R_{g,x}R_{g,y}R_{g,z} = R_{g,x}^2 R_{g,y} = 4.0 \times 10^{-6} M^{1.5}\,(\mathrm{nm}^3) \tag{11}$$

The volume occupied by the unperturbed chain in free space is estimated by using $R_{g,k0}$ as [60]:

$$R_{g,k0}^3 = (0.0165\ M^{1/2})^3 = 4.5 \times 10^{-6} M^{1.5}\,(\mathrm{nm}^3) \tag{12}$$

Comparing Eqs. (11) and (12), we reach the important conclusion that a block chain of a diblock copolymer in a confined space is deformed so as not to change its volume [59–61], as is schematically shown in Fig. 21. This conclusion is confirmed to hold also for the center block of ABA and ABC triblock copolymers [55,62,63]. Therefore, we may consider this conclusion as the general feature of block chains in bulk irrespective of their architectures.

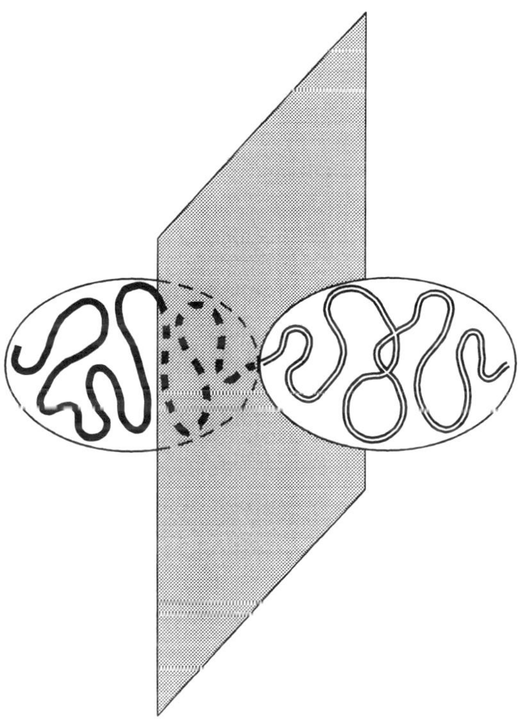

Figure 21 A schematic drawing of the conformation of an AB diblock copolymer chain in bulk. Both black (A) and white (B) chains can be conceived to preserve the volumes of the unperturbed chains in free space.

VI. SUMMARY AND PERSPECTIVES

Microphase-separated structures of several block and graft copolymers have been clarified at the molecular level. The principal conclusions are as follows.

1. The compositions where morphological transitions occur are significantly different among copolymers depending on the architectures of the copolymers.
2. Polymer/polymer interfaces always show the common feature of their "constant-mean-curvature."
3. Block polymer chains in microphase-separated structures are shrunk in the direction parallel to the lamellar interface to compensate for the elongation in the perpendicular direction.

The fundamentals of the morphology control of these copolymers have been consolidated from the results of scientific investigations. However, the following unsolved problems for bicontinuous and tricontinuous structures are still open.

1. Are bicontinuous structures from two-component block copolymers really the equilibrium structures?
2. Can the characteristics of surfaces be uniquely assigned for each particular copolymer system?
3. What are the density fluctuation profiles along the surfaces?

REFERENCES

1. M Szwarc. In: Carbanions Living Polymers and Electron Transfer Processes. New York: Wiley, 1968.
2. JE McGrath, ed., ACS Symposium Series. In: Anionic Polymerization; Kinetics, Mechanisms and Synthesis, Washington, DC: American Chemical Society, 1981.
3. BRM Gallot. Adv Polym Sci 29:85, 1978.
4. L Leibler. Macromolecules 13:1602, 1980.
5. T Hashimoto, Y Ijichi, LJ Fetters. J Chem Phys 89:2463, 1988.
6. GH Fredrickson, E Helfand. J Chem Phys 87:697, 1987.
7. GE Molau. In: SL Aggarwal, ed., Block Polymers. New York: Plenum Press, 1970.
8. EL Thomas, DM Anderson, CS Henkee, D Hoffman. Nature 334:598, 1988.
9. LE Schriven. Nature 263:123, 1976.
10. DB Alward, DJ Kinning, EL Thomas, LJ Fetters. Macromolecules 19:215, 1986.

11. Y Mogi, H Kotsuji, Y Kaneko, K Mori, Y Matsushita, I Noda. Macromolecules 25:5408, 1992.
12. KI Winey, EL Thomas, LJ Fetters. Macromolecules 25:422, 1992.
13. DJ Meier. J Polym Sci C 26:81, 1969.
14. E Helfand, ZR Wasserman. Macromolecules 9:879, 1976.
15. PG de Gennes. In: Scaling Concepts in Polymer Physics. Ithaca, NY: Cornell University Press, 1979.
16. S Bywater. Adv Polym Sci 30:89, 1979.
17. N Hadjichristidis, LJ Fetters. Macromolecules 13:191, 1980.
18. H Iatrou, N Hadjichristidis. Macromolecules 25:4649, 1992.
19. Y Matsushita, J Watanabe, F Katano, Y Yoshida, I Noda. Polymer 37:321, 1996.
20. PF Rempp, E Franta. Adv Polym Sci 58:1, 1984.
21. S Arichi, MY Pedram, JMG Cowie. Eur Polym J 15:125, 1979.
22. L Utracki, R Simha. J Phys Chem 67:1052, 1963.
23. H Inagaki, H Suzuki, M Fujii, T Matsuo. J Phys Chem 70:1718, 1966.
24. AL Spatorico, B Coulter. J Polym Sci Polym Phys Ed 11:1139, 1973.
25. S Tsuge, T Kobayashi, T Nagaya, T Takeuchi. J Anal Appl Pyrolysis 1:133, 1979.
26. T Tanaka, M Omoto, H Inagaki. Macromolecules 12:146, 1979.
27. Y Matsushita, K Shimizu, I Noda, T Chang, CC Han. Polymer 33:2412, 1992.
28. M Shibayama, H Hasegawa, T Hashimoto, H Kawai. Macromolecules 16:1427, 1983.
29. Y Matsushita, Y Nakao, R Saguchi, K Mori, H Choshi, Y Muroga, I Noda, M Nagasawa, T Chang, CJ Glinka, CC Han. Macromolecules 21:1802, 1988.
30. G Hadziioannou, A Skoulios. Macromolecules 15:258, 1982.
31. FS Bates, JH Rosedale, GH Fredrickson, CJ Glinka. Phys Rev Lett 61:2229, 1988.
32. JH Rosedale, FS Bates, K Almdal, K Mortensen, GD Wignall. Macromolecules 28:1429, 1995.
33. RW Richards, JL Thomason. Polymer 22:581, 1981.
34. Y Matsushita. J Phys Soc Jpn. 65 (Suppl. A):119, 1996.
35. H Hasegawa, H Tanaka, K Yamasaki, T Hashimoto. Macromolecules 20:1651, 1987.
36. EL Thomas, DB Alward, DJ Kinning, DC Martin, DL Handlin Jr, LJ Fetters. Macromolecules 19:2197, 1986.
37. AH Schoen. NASA Tech. Note TN D-5541, 1970.
38. DA Hajduk, PE Harper, SM Grunner, CC Honeker, G Kim, EL Thomas, LJ Fetters. Macromolecules 27:4063, 1994.
39. DA Hajduk, PE Harper, SM Grunner, CC Honeker, EL Thomas, LJ Fetters. Macromolecules 28:2570, 1995.
40. N Hadjichristidis, H Iatrou, SK Behal, JJ Chludzinski, MM Disko, RT Garner, KS Liang, DJ Lohse, SJ Milner. Macromolecules 26:5812, 1993.
41. Y Matsushita, H Momose, Y Yoshida, I Noda. Polymer 38:149, 1997.

42. Y Mogi, M Nomura, H Kotsuji, K Ohnishi, Y Matsushita, I Noda. Macromolecules 27:6755, 1994.
43. Y. Mogi, K Mori, Y Matsushita, I Noda. Macromolecules 25:5412, 1992.
44. HA Schwarz. Gesammelte Mathematische Abhandlungen Berlin: Springer, 1890.
45. A Gray. In: Modern Differential Geometry of Curves and Surfaces. Boca Raton: CRC Press, 1993.
46. DM Anderson, EL Thomas. Macromolecules 21:3221, 1988.
47. S Andersson, ST Hyde, HG von Schnering. Z Kristallogr 168:1, 1984.
48. H Hasegawa, Y Nishikawa, S Koizumi, T Hashimoto, ST Hyde. In: H Aoki, ed. Advanced Materials '93, II/A: Biomaterials, Organic and Intelligent Materials. (Trans Mat Res Soc Jpn, Vol. 15A. Amsterdam: Elsevier, 1994.
49. AV Semenov. Sov Phys JETP (Engl. Transl.) 61:733, 1985.
50. T Ohta, K Kawasaki. Macromolecules 19:2621, 1986.
51. PJ Flory. In: Principles of Polymer Chemistry. Ithaca, NY: Cornell University Press, 1953.
52. J Higgins, RS Stein. J Appl Cryst 11:346, 1978.
53. T Hashimoto, M Shibayama, H Kawai. Macromolecules 13:1237, 1980.
54. Y Matsushita, K Mori, R Saguchi, Y Nakao, I Noda, M Nagasawa. Macromolecules 23:4313, 1990.
55. Y Matsushita, M Nomura, J Watanabe, Y Mogi, I Noda. Macromolecules 28:6007, 1995.
56. SN Jahshan, GC Summerfield. J Polym Sci Polym Phys Ed 18:1859, 1980.
57. JT Koberstein. J Polym Sci Polym Phys Ed 20:593, 1982.
58. X Quan, JT Koberstein. J Polym Sci Part B Polym Phys Ed 25:1381, 1987.
59. H Hasegawa, T Hashimoto, H Kawai, TP Lodge, EJ Amis, CC Han. Macromolecules 18:67, 1985.
60. Y Matsushita, K Mori, Y Mogi, R Saguchi, I Noda, M Nagasawa, T Chang, CJ Glinka, CC Han. Macromolecules 23:4317, 1990.
61. G Hadjiioannou, C Picot, A Skoulios, M-L Ionescu, A Mathis, R Duplessix, Y Gallot, J-P Lingelser. Macromolecules 15:263, 1982.
62. X Quan, I Gancarz, JT Koberstein, GD Wignall. J Polym Sci Part B Polym Phys Ed 25:641, 1987.
63. Y Mogi, K Mori, H Kotsuji, Y Matsushita, I Noda, CC Han. Macromolecules 26:5169, 1993.

5

Phase Separation and Morphology of Chemically Reacting Polymer Blends

Qui Tran-Cong
Kyoto Institute of Technology, Kyoto, Japan

I. OUTLINE

Over the past decades, the fact that most polymers cannot mix with each other to form homogenous mixtures at the segmental scale has been the driving force for a large number of studies on multiphase polymeric materials. It has been known for a long time that physical properties such as toughness, flowability, transparency, weather resistence, and so on of these phase-separated polymers are dictated by their morphologies [1]. As a consequence, a variety of chemical as well as physical methods has been extensively exploited to control the morphology of immiscible polymer mixtures. From the chemistry aspects, the sizes of the disperse phases resulting from the phase separation of two immiscible polymers can be modified and controlled by chemically linking two different chains to form a block copolymer via living anionic polymerization [2]. The chain lengths and thus the domain sizes can be well regulated by these chemical techniques [3]. However, due to the limit of the reactivity, block copolymers with very high molecular weights cannot be produced. Consequently, the regular morphology resulting from the microphase separation of block copolymers is limited in the nanometer scale. Practically, the morphology of polymer blends in the micrometer scale is controlled by stirring the mixtures of immiscible components, and therefore, the domain sizes can be regulated by adjusting the stirring rates. The former methods based on living polymerization provide ways to control the sizes and the regularity of the morphology whereas the latter contributes mainly to the regulation of domain sizes. There are also a growing number of studies on controlling the morphology of polymer mixtures by applying a shear field [4]. Since the details of the static and dynamical aspects of the phase separation of polymer mixtures as well as of microphase separation of block copolymers are carefully described in Chapters 1, 2, and 4 of this book, only phase separation of reactive polymer blends is described in this chapter to avoid repetition. Furthermore, there have been many textbooks and monographs discussing the phase behavior of thermally reacting blends [5], these results will be only briefly mentioned in this chapter in order to highlight the advantages as well as the disadvantages of photochemically and thermally activated chemical reactions. The scope of this chapter is therefore to emphasize the phase-separation phenomena induced by photochemical reactions and particularly their relations with the pattern formation processes in chemically reacting systems.

II. PHASE SEPARATION IN REACTIVE POLYMER BLENDS

A. Polymer Blends with Thermally Activated Reactions

Reactive polymers are widely utilized for various practical purposes such as the modification of surface tension between different phases via which morphology and physical properties of multiphase polymeric materials can be modified and controlled. These techniques are termed reactive compatibilization [6,7]. Depending upon the types of chemical reactions performed in the blends, these processes can be classified into three main categories [8].

1. *Reactions between functionalized polymers.* The polymer components either contain reactive repeating units or are functionalized with appropriately reactive side groups which are capable of reacting with other component in the mixture to form block and/or graft copolymers. The products of these reactions are used to modify the interfacial tensions between different phases, and therefore, the morphology can be controlled to provide multiphase polymeric materials with desirable physical properties. Transesterification reactions between polyesters such as polycarbonate and polybutylene terephthalate are examples in this category [9]. As the particular case of reacting polymer blends, ionic bondings or electrostatic interactions can also be used for compatibilization [10,11].

2. *Reaction via additives of low molecular weight compounds.* By incorporating additives of low molecular weight compounds which can induce chemical reactions to form copolymers or networks between polymer chains, the morphology of the immiscible blends can be modified and

stabilized. An example for the former case is adding p-toluene sulfonic acid to the blends of polyethylene terephthalate and polyamide 6, 6, [12]. Adding peroxide compounds to crosslink the blends of polyethylene and polypropylene is an example for the latter case [13].

3. *Reacting systems involving polymerization.* This is the case encountered in the preparation process of interpenetrating polymer networks (IPNs) [14]. Contrary to the first two cases given above, the system here is initially inside the one-phase region and undergoes phase separation upon reacting. The mixtures are thermodynamically unstable due to the network formation and phase separation takes place during the crosslinking process. The morphology obtained here is the result of the competition between the reaction and phase separation. The latter dominates the former in most cases and the resulting morphology is the random multiphase structure.

In order to compare these results with those described later for photo-crosslinking, typical experiments performed with thermally activated reactions are briefly summarized in the next section.

1. Phase Separation in Thermally Crosslinked Polymer Blends

For a given binary polymer blend, phase separation is generally favored when one component is crosslinked because of the increase in the free energy of mixing. Theoretical studies on this subject were initiated by de Gennes [15], continued by Binder and Frisch and were extended recently by Binder and coworkers [16]. From the experimental viewpoint, small-angle neutron scattering (SANS) can provide a powerful technique to investigate the critical phenomena associated with the crosslinking reactions. Bauer and Briber have utilized SANS to examine the phase behavior of mixtures composed of deuterated polystyrene (PSD) networks with varying crosslinking densities swollen in linear poly(vinyl methyl ether) (PVME) [17], PSD/PVME blends with three different types of crosslink A–A, B–B, and A–B simultaneously induced by γ-ray irradiation [18], and mixtures of protonated polystyrene (PSH) and PSD blends crosslinked with divinyl benzene [19]. The general tendency is that the blends become thermodynamically unstable and undergo phase separations when one polymer component is crosslinked. Particularly, for the blend containing PSH chains trapped in the PSD crosslinked networks, the scattering intensity at zero-angle, $S(0)$, and the correlation length ξ of concentration fluctuations obtained from the Ornstein–Zernike (Oz) plot both diverge with increasing crosslinking density, indicating that the blends are destablized upon crosslinking. On the other hand, it was also found that the PS/PVME blend becomes more miscible when all the three different types of crosslinking reactions A–A, B–B, and A–B were performed with γ-ray irradiation as shown in Fig. 1

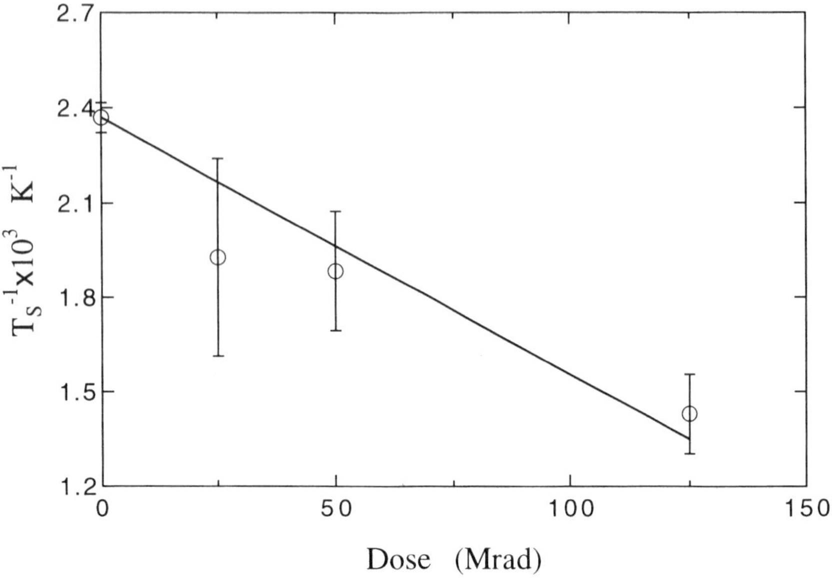

Figure 1 Effects of radiation dose on the spinodal temperature of PSD/PVME blends. (Redrawn from Ref. 18, with permission of the American Chemical Society.)

where the spinodal line shifts to high temperatures as radiation dose increases [18]. Recently, it was also demonstrated by SANS that a blend of perdeuterated polybutadiene (DPB) and protonated polybutadiene does not undergo phase separation and remains miscible when crosslinked with peroxides. The effects of the concentration fluctuations at the crosslinking temperatures seems to play a determining role in the scattering function of the crosslinked blends [20]. These works provide a molecular basis for the understanding of the influence of crosslinking reactions on the phase behavior of the miscible blends.

2. Phase Separation in Epoxy Resins Prepared by Curing Reactions

Epoxy resins are widely used in most rubber-toughened plastics. The materials are prepared from mixtures of prepolymers bearing epoxy end-groups such as diglycidine ether and liquid rubbers with reactive terminated groups such as carbonyls or amines [21]. The two components react with each other and form networks. During the curing time, the sample becomes cloudy because phase separation takes place. This process has been examined by Inoue and coworkers [22,23], and also by Lee and Kyu [24] using light scattering. The phase separation in these curing systems is determined by the competitions between phase separation triggered by the network formation and the curing reaction. By changing the curing rates via changing the curing agents, catalysts, and curing temperatures, three different structures, i.e., spherical domain structures, bimodal domain structures and interconnected globule structures, all developing from co-continuous structures of the spinodal decomposition, were observed [23]. The last morphology was obtained when the curing effects became comparable to the phase separation. As a result, the structures developing in the late stage of the spinodal decomposition process were frozen. Physical properties such as vibrational damping efficiency, the stress–strain relationship, and the modulus of the resins with these particular morphologies were also examined. Furthermore, the scaling law proposed by Furukawa [25] for the late stage of the spinodal decomposition process was also used to analyze the phase separation kinetics of these cured expoxies [24].

3. Phase Separation in Polymerizing Systems

Phase separation associated with polymerization is encountered in many aspects of applied polymer science. When the crosslinking reaction is also involved in the polymerization process, one obtains IPNs, which is an important field of applied polymer science. Since there have been a number of monographs [14,26] as well as reviews [27] on this subject, the IPNs prepared by conventional methods will not be mentioned in this chapter. Instead, recent experimental work on a *non*crosslinked systems will be summarized in this section. Phase separation accompanying the polymerization of 2-chlorostyrene in polystyrene containing di-*n*-butyl methacrylate as a platicizer was recently examined by Okada and coworkers using light scattering and electron microscopy [28]. It was found, in this work, that the time-evolution of the structures and the scattering intensity during the phase separation can be generally divided into three different regimes characterized by the power laws $q_m \approx (t - t_0)^{-\alpha}$ and $I_m \approx (t - t_0)^{\beta}$ where I_m and q_m are respectively the maximum scattering intensity and the corresponding magnitude of the scattering vector. The exponents α and β change from one regime to another and t_0 is the induction time of the phase-separation process. A series of pictures showing the morphology obtained by electron microscope in the late stage of phase separation indicates the existence of the droplet-deposition-and-coalescence phenomena where the droplets develop to form co-continuous domains as reaction time increases. On the other hand, phase separation associated with polymerization is also encountered in the so-called "frontal polymerization," the techniques which were first initiated by Merzhanov and Borovinskaya for preparation of ceramic and intermetallic compounds [29] and were applied to free-radical polymerization by Chechilo and Enikolopyan [30]. Recently, these techniques have been intensively developed for polymer processing by Khan and Pojman [31]. This polymerization technique takes advantage of the positive feedback loop driven by the heat released in a local region of the reaction. Because the thermal diffusion is much faster than the mass diffusion of monomers as well as polymers, once initiated, the polymerizing front is produced and propagates by the heat of the reaction which in turn triggers the polymerization reaction. This process keeps going on until the monomer is completely consumed. When two or more monomers are used in this polymerization procedure, phase separation between the resulting polymers takes place along the propagation direction of the reaction heat. At this stage, hydrodynamic instabilities such as Rayleigh–Taylor instabilities and the so-called spinal mode arise from the couplings between the heat of the reaction and the mass diffusion of polymers. It has been found that a semi-IPN produced by this frontal polymerization has a morphology more uniform than those obtained by the conventional isothermal polymerization. The techniques have been suggested for the preparation of novel IPNs and molecular composites with reduced energy costs and reduced waste production [32].

4. Phase Separation in Polymer Blends with Transesterification

Transesterification reactions are common among polyesters and have been utilized not only to control the miscibility of polymer blends but also to optimize the morphology developing during processing. Generally, the exchanges between functional groups (esters) take place first, resulting in block copolymers of the two polymer components. As the reaction proceeds, the copolymers become more and more random, and eventually the reactiong blend becomes miscible. The most studied systems are the transesterification of commonly used polyesters such as bisphenol A polycarbonate/poly(butylene terephthalate) (PC/PBT) blends, PC/poly(ethylene terephthalate) (PC/PET) and PBT/PET. Furthermore, the blends between copolyesters such as p-hydroxybenzoic acid-PET copolyester (PHB–PET), hydroxybenzoic acid–hydroxynaphthoic acid copolyester (Vectra), and other commonly known polyesters such as PET, PBT, and PC have been intensively investigated. The conclusions about the achievement of miscibility between two polymers undergoing transesterification were confused for a while because of the influence of catalyst residuals on the reactions. Several examples for the improvements of the physical properties due to the transesterification of polyester blends have been reported. These include the appearance of the maxima in tear strength and the improvement of moduli for these polyester blends with the reaction time. Most recent work on the miscibility and physical properties of polymer blends with transesterification can be found in the review articles by Kotliar [33], and by Porter and Wang [34]. The effects of transesterification on the miscibility of polymer blends have been examined by conventional characterization techniques such as differential thermal analysis, differential scanning calorimetry and dynamic mechanical analysis. Recently, more sophisticated techniques such as nuclear magnetic resonance (NMR) have been employed to monitor the transesterification reactions [35,36]. On the other hand, the direct observation of the changes in morphology accompanying the reaction in polyester blends has been carried out recently with light scattering in PET/PBT blends [37], small-angle neutron scattering (SANS) for PC/PET blends [38], and atomic force microscopy (AFM) for thin films of PC/poly(methyl methacrylate) (PMMA) blends [39]. These scattering techniques provide not only direct information on the development of the interfacial thickness of the copolymer layers formed upon transesterification, but also knowledge of the competition between reaction and phase separation at different temperatures. These results are useful for the morphology control of polyster/polyester blends undergoing transesterification during the processing.

B. Polymer Blends with Photochemical Reactions

Compared to thermally activated chemical reactions such as curing and transesterification described in the previous section, photochemical reactions have a number of advantages and disadvantages as summarized in Table 1. In general, photochemical reactions can be initiated as well as terminated under arbitrary conditions independent of the thermodynamics (P,V,T) of the blends. Furthermore, these reactions are also highly selective because the reacting sides on the polymer chains can be specified by using the labeling method. As illustrated in Fig. 2 for example, the three types of reactions A–A, B–B, and A–B between the two polymer chains A and B in a mixture, in principle, can be performed independently by appropriately selecting the reactions between two photoreactive species (a) and (b) with the help of chemical synthesis. Furthermore, by using photochemical reactions, the spatial distribution of the reaction in the sample can also be controlled by using polarized light, provided that the crosslinker has a well-defined absorption transition moment either in the ultraviolet or the visible region [40]. The experimental results obtained by photo-crosslinking binary blends inside the unstable (spinodal) and miscible (one-phase) regions are summarized in this chapter. Multiphase materials with modu-

Table 1 Comparison Between Photochemically and Thermally Activated Crosslinking Reactions

Photo-crosslinking reaction	Thermally activated crosslinking reaction
Depending on mobilities because most of the reactions are diffusion-controlled.	Depending strongly on temperature via the half-life of radicals generated by crosslinkers.
The reaction can be initiated and terminated rapidly and independently from thermodynamical variables *P*, *V*, *T* of the mixtures.	Initiation and termination of the reaction depend strongly on *P*, *V*, *T*, of the system under consideration.
High selectivity of the reaction due to the labeling techniques. Spatial distribution of the reaction can also be regulated and controlled by using polarized light.	In most cases, the crosslinkers are dispersed in the mixtures. The selectivity of the reactions is not high due to the difference in chemical affinity between the crosslinker and the polymer components.
In general, reaction can proceed without adding catalyst.	Requirements of catalysts to proceed the reaction and therefore introducing extra component into the mixtures.
Not suitable for thick and/or opaque samples.	Performable with thick and opaque samples.
Limited by complicated organic synthesis required for the labeling reactions.	Simple procedure of crosslinking because chemical labeling is generally not necessary.

lated structures were obtained in the former case whereas various highly ordered phases were observed in the latter.

1. Photo-Crosslinking Reactions Initiated Inside the Unstable Region

According to the phase-separation mechanism of polymer blends described precisely in Chapter 2, as soon as the sample enters the unstable region, the concentration fluctuations develop and grow with time. Initially there exist, in the mixture, various fluctuations of different wavelengths. However, under a given experimental condition, selection processes among different modes take place. In a closed system, the mode with the greatest growth rate will dominate the morphological evolution as predicted by linear stability

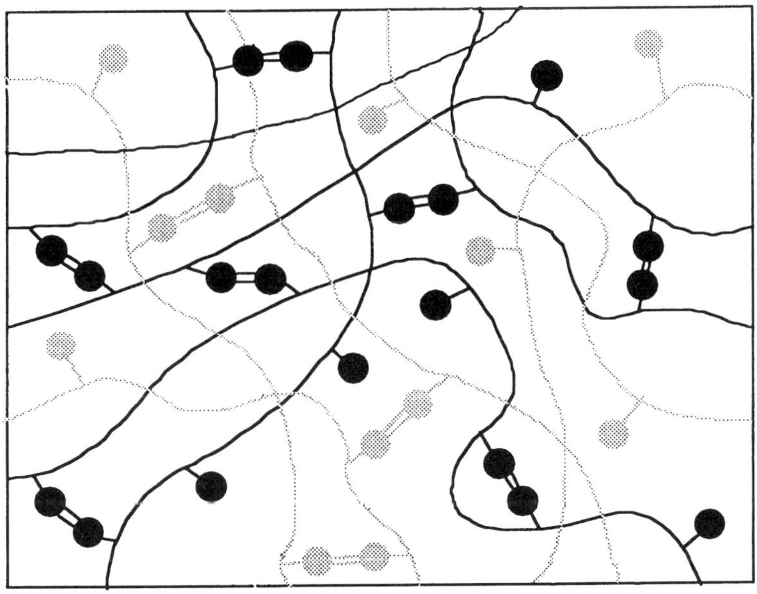

Figure 2 A schematic presentation of photo-crosslinkable polymer blends (—) polymer A; (- - -) polymer B; (●,○) photo-crosslinkers.

analysis [41,42]. As a result, the morphology of the blend in this initial stage of phase separation is characterized by the interconnecting structures with the length scale (ξ) given by [43]:

$$\xi^2 = \frac{4\pi^2}{9} R_0^2 \left(\frac{\chi_s}{\chi - \chi_s} \right) \tag{1}$$

where R_0 is the unperturbed chain dimension, and χ and χ_s are respectively the binary interaction parameters of the blend at an arbitrary and the spinodal temperatures.

These interconnecting structures become coarsened with time, gradually losing their connectivity and eventually result in random two-phase structures. Multiphase polymeric materials with controllable periodic structures can be obtained if one can suddenly freeze the diffusion of polymer chains during this time-evolution process of the phase separation. In principle, it is possible to obtain such polymeric materials by quenching a blend undergoing spinodal decomposition from the unstable region into its glassy region. However, this procedure is not efficient because the phase separation restarts as temperatures increase, particularly in the temperature range higher than the glass transition temperature (T_g) of the sample. By taking advantages of *in situ* photo-crosslinking, polymers with permanent modulated structures in the micrometer range were obtained.

Photodimerization of anthracene derivatives chemically attached to one polymer component in binary polymer blends has been used for these purposes. The spatially modulated structures developing during the spinodal decomposition process of the following binary blends polystyrene/poly(2-chlorostyrene) (PS/P2CS) [44,45], polystyrene/poly(vinyl methyl ether) (PS/PVME) [46], poly (2-chlorostyrene)/poly(vinyl methyl ether) (P2CS/PVME) [47] were efficiently frozen by irradiation with light. As an example, shown in Fig. 3 is the morphology of anthracene-labeled P2CS/PVME (60/40) photo-crosslinked at the jump depth 2°C above the spinodal temperature of this particular composition. The morphology of the irradiated sample reveals interconnecting structure which are in contrast with the random two-phase structure obtained from a blend with the same composition having similar thermal history and without irradiation. These results indicate that the modulated structures developing during the spinodal decomposition process were frozen by the photodimerization of anthracenes attached on P2CS chains. The resulting morphology crucially depends on the reaction rates and the growth rate of the spinodal decomposition process. The former is determined by the irradiation intensity and the label content of the P2CS

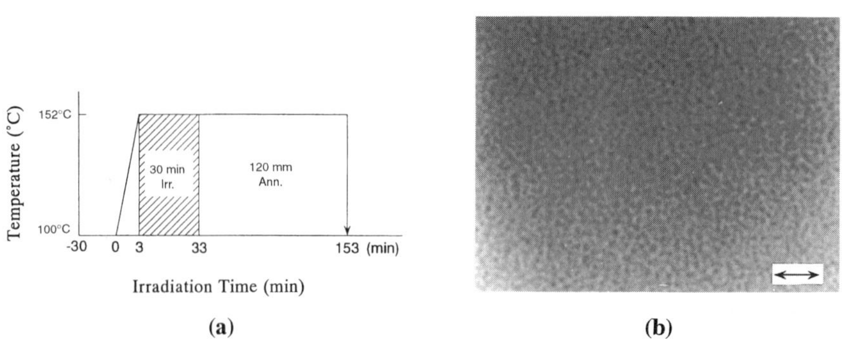

Figure 3 Temperature-jump diagram (a) and the spinodal structures (b) of a poly(2-chlorostyrene)/poly(vinyl methyl ether) (P2CS/PVME)(60/40) blend obtained after irradiation over 30 min and subsequently annealed in the dark for 120 min at 152°C. Photo-crosslinking was performed with anthracene labeled on P2CS chains and the jump depth is 2°C. The scale corresponds to 5 μm. (With permission from the American Chemical Society.)

chains whereas the latter strongly depends on the experimental temperatures. It was found that for a given label content and an experimental temperature, the characteristic length scales of the structures decrease with increasing irradiation intensity. However, the morphology becomes unchanged as the light intensity reaches a threshold [46]. Most of the spatially modulated structures obtained by photo-crosslinking reactions as described in this work are under nonequilibrium conditions just after irradiation. Upon annealing the reacted blend in the dark at the crosslinking temperature, these phase-separated structures relax toward equilibrium which is determined by the network structures formed by irradiation. The time-evolution of this relaxational process can be followed by phase-contrast optical microscopy and subsequently analyzed by digital image analysis. As an example, Fig. 4 shows the one-dimensional (1D) power spectra extracted from the two-dimensional (2D) fast Fourier transform (FFT) data of the morphologies of a P2CS/PVME (60/40) blend annealed over different time intervals after being photo-crosslinked over 30 min at 152°C. Under this crosslinking condition, it was estimated from ultraviolet–visible spectroscopy that half of the anthracenes attached to the P2CS chains were reacted during the 30 min of irradiation. The crosslinking density corresponding to this particular irradiation condition is one crosslinking junction per 300 repeating monomer units. Upon annealing in the dark, these interconnecting structures initially grow with time and eventually approach an equilibrium value of about 2 μm. Before reaching the equilibrium length scale, the coarsening process can be described by the power law $\xi \propto t^{\alpha}$ where t is the annealing time and ξ is the characteristic length scale of the morphology calculated from the maximum frequency $q_{\max} = 2\pi/\xi$ of the 2D-FFT power spectra. It was found that for most cases, the exponent α is approximately 1/4 which is somewhat smaller than the value $\alpha = 1/3$ predicted by the Lifshitz–Slyozov [48] and Wagner [49] theories probably due to the presence of the cross-linking junctions. These results suggest that multi-phase polymeric materials with controllable modulated structures in the micrometer range can be designed by photo-crosslinking polymer blends undergoing a spinodal decomposition process. Examination of the viscoelastic behavior of the photo-crosslinked P2CS/PVME blends with varying length scales shows that the dynamic modulus E' increases with the characteristic length scale ξ. On the other hand, the temperature corresponding to the maximum of the loss modulus, E'', is located in the temperature range between those of the miscible and the two-phase blends. Furthermore, as ξ increases, these peaks slightly shift toward the side of higher temperature [47]. The temperature dependence of the tan δ of these crosslinked blends is very broad, representing the typical characteristics of IPNs. Similar behavior of E'' and tan δ was also obtained by dielectric relaxation for PS/P2CS blends photo-crosslinked

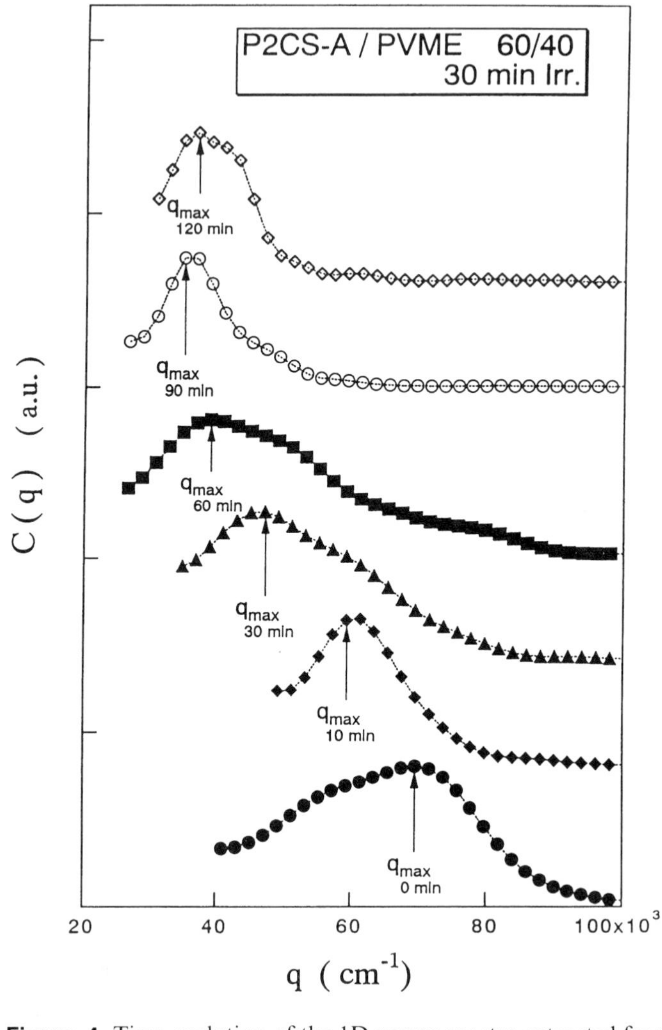

Figure 4 Time-evolution of the 1D power spectra extracted from the 2D-FFT data of a P2CS/PVME(60/40) annealed at 152°C after crosslinking for 30 min at the same temperature. The numbers shown in the figure indicate the annealing time.

in the spinodal region [44,45]. It is well known that for phase-separated polymer blends at equilibrium, the dispersed phases with sizes in the micrometer scale play a significant role in mechanical properties such as toughness, impact resistance strength, etc. [1]. The blends with the spatially modulated structures obtained by photo-crosslinking described above are different from the polymers with dispersed phases because the concentration profiles at the interfaces between two phases are much broader in the former. As mentioned earlier, the final morphology of these reacting blends is determined by the competitions between the phase separation and the crosslinking reactions. Therefore, it is crucial to monitor the crosslinking kinetics not only to obtain the precise crosslinking densities at a given irradiation time but also to understand the mechanism of the morphology resulting from the reaction. The details of the reaction kinetics are summarized in Sec. II. C. below.

2. Photo-Crosslinking Reactions Initiated Inside the Miscible Region

For most cases, polymer blends photo-crosslinked in the miscible region become thermodynamically unstable and undergo phase separation. Compared to thermally activated reactions utilized for the preparation of IPNs described in the previous section, *in situ* photo-crosslinking of miscible blends in the one-phase region can provide an efficient way to control the morphology. This is because the diffusion of polymer chains in the bulk state is much slower than in solutions and also because the photochemical reactions can be promptly initiated by irradiation, facilitating the manipulation of the morphology of the blends undergoing phase separation. Furthermore, the reactions can be accelerated by increasing the label contents via copolymerization with appropriate monomers such as chloromethyl styrene (CMS) bearing reactive groups for subsequent labeling reactions. However, precautions should be taken because in some particular cases, the high label contents lead to a considerable decrease in the blend miscibility [50]. Since the reactivity ratio of CMS (0.72) is close to that of styrene (1.31) [51], the copolymers synthesized from these two monomers are almost random and have no microphase separation as examined by small-angle X-ray scattering (SAXS) [52]. In general, the morphology emerging from the crosslinked blends is the result of the competition between the phase separation and the crosslinking reaction. The latter acts as a long-range effect in comparison to the former because the composition is not conserved as the crosslinking reaction proceeds. Physical and chemical systems with competition between antagonistic interactions which occurs over different length scales have recently attracted the attention of many theorists as well as experimentalists because of the universality observed in a wide variety of ordering phenomena in condensed matter ranging from ferrofluids under a magnetic field [53] to Langmuir monolayers at an air–water interface [54]. The characteristic length scales of these structures, known as modulated phases, are determined by the strength of this competition and can vary from nanometer in the ripple phase of vesicles [55] to centimeters in the Rayleigh–Bénard instabilities [56]. In polymer science, block copolymers are typical materials exhibiting modulated phases which originate from the competition between the short-range attraction and long-range repulsion between immiscible polymer components forced to be linked on the same chain (see Chapter 1 of this book and also Refs. 57–60).

In general, for polymers photo-crosslinked in the one-phase (miscible) region, the mixture becomes unstable and phase separation starts as soon as the new coexistence curve reaches the experimental temperatures. Due to the presence of the forming networks, the morphology is not only controlled by the competition between the reactions and the phase separation but is also affected by the elasticity building up during the reactions. Such effects on reacting polymer blends have been theoretically discussed [15,16,61]. There are two extreme cases for polymer blends photo-crosslinked in the one-phase region: one is close to the glass transition temperature (T_g) of the blends and the other is in the vicinity of the coexistence curves. The former corresponds to the "shallow" quench condition because the phase boundary reaches the experimental temperatures at high crosslinking densities (γ). In the other case, the reacted blend undergoes phase separation at low crosslinking densities because the experimental temperature is located close to the coexistence curve. Miscible semi-IPNs were apparently obtained with anthracene-labeled poly(styrene-co-chloromethyl styrene)/poly(vinyl methyl ether) (PSCS/PVME) (50/50) blends photo-crosslinked at 30°C above T_g and 40.5°C below the cloud point of the blend as revealed by SAXS [50]. However, these crosslinked blends undergo microphase separation upon annealing over a long time at high temperatures. Namely, the PVME chains were kinetically frozen in the PSCS networks at the reaction temperatures. Upon increasing temperatures after crosslinking, PVME chains are activated and move away from the PSCS networks, initiating the phase separation. Because their diffusion is restricted by the PSCS networks, only microphase domains at about 100 Å were observed by SAXS. On the other hand, miscible IPNs are obtainable when both the polymer components are simultaneously crosslinked. Such data have been reported based on dynamic mechanical relaxation of simultaneously crosslinked full-IPNs [62] where the diffusion of both polymer chains is completely suppressed by the forming networks. In the intermediate range of temperatures located between the glass transition temperatures and the cloud points of the blend, a variety of ordered structures was observed. These ordering processes are determined by the rates of the crosslinking reactions and the phase separation.

(a) Spinodal Structures. Spatially modulated structures are obtained when the photo-crosslinking reactions were performed in the vicinity of the coexistence curve. The range of temperatures for these structures to be observable depends on the crosslinking kinetics and thus sensitively changes with the label contents and intensity of the light source. For PSCS bearing one anthracene per 44 repeating monomers units crosslinked with a conventional Hg–Xe lamp (3 $mJ/cm^{-2}s^{-1}$), these structures were observed within 20°C from the coexistence curve. These modulated structures appear when the new phase boundary of the reacting blends crosses the experimental temperatures under the low crosslinking densities (less than four junctions per chain) which correspond to the "deep" quench conditions. Namely, the mixtures quickly pass through the metastable region and then enter the unstable region where the spindodal decomposition process is initiated. The phase separation in this particular case proceeds until the chain diffusion is frozen by the crosslinking reaction. Thus the bi-continuous morphology gradually develops with irradiation time and eventually reaches the stationary state where the polymer chains of both components are kinetically locked by the forming networks. As an example, the morphology of a PSCS/PVME (20/80) blend photo-crosslinked at 110°C over 55 min is shown in Fig. 5. The structures grow with the reaction time in the early stage of the crosslinking reaction and eventually become unchanged after 55 min of irradiation. Further analysis

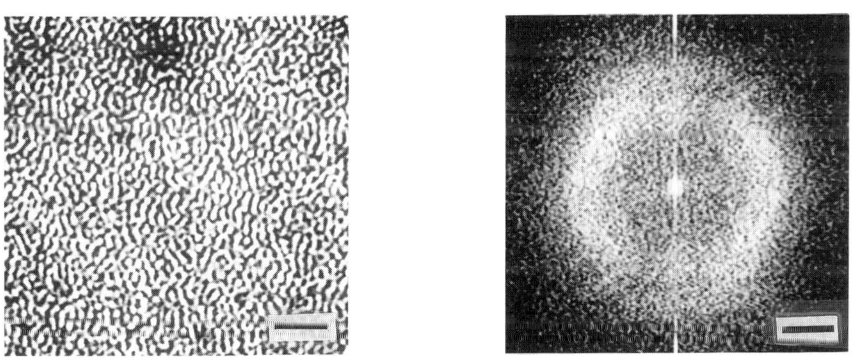

(a) (b)

Figure 5 Interconnecting morphology (a) and the corresponding 2D-FFT power spectra of PSCS/PVME(20/80) photo-crosslinked at 110°C over 55 min. These structures are at equilibrium and are unchanged upon further irradiation. The scales are 10 μm for (a) and $2 \times 10^4\ cm^{-1}$ for (b).

shows that the time-evolution of the phase separation under irradiation can be described by the following scaling law [63,64]:

$$\xi(t) = t^{\alpha} G(x) \tag{2}$$

and

$$x = kt \tag{3}$$

where $\xi(t)$ is the characteristic length scale of the structures at the reaction time t, α is the growth exponent characterizing the phase separation kinetics and k is the reaction rate.

Since the reaction rate k is almost constant in the early stage of irradiation, the proposed scaling function $G(x)$ can be obtained experimentally by rewriting Eq. (2) as:

$$G(x) = \frac{\xi(t)}{t^{\alpha}} \tag{4}$$

Because all the quantities in the right-hand side of (4) are obtainable independently of the experiments, the scaling function $G(x)$ can be experimentally verified for phase separation kinetics in chemically reacting systems. It was found that the growth of these structures can be well described by the scaling law (2) with the exponent α varying from 1/3 to 1/5 [65].

(b) Labyrinthine Structures. These structures were obtained when the blends with lower compositions of PSCS [PSCS/PVME (20/80)] were photo-crosslinked at relatively low temperatures, i.e., between 80°C and 100°C. Upon irradiation, the blends undergo phase separation when the crosslinking density reaches a critical value (γ_c) which depends strongly on the reaction temperature. Generally, under a constant irradiance, the higher the reaction temperature is, the earlier the phase separation takes place. The morphology is initially isotropic and gradually becomes anisotropic as the reaction time increases. Eventually, the structures are frozen at longer irradiation time because both the PSCS and PVME chains are kinetically trapped in the forming networks. For the crosslinking reaction performed at 80°C, phase separation of the reacting blends becomes observable at 45 min after irradiation (about 15% of anthracenes were consumed). During the first 120 min of the reaction, the contrast of the morphology becomes stronger with the reaction time whereas the sizes of the interconnecting domains are almost unchanged, suggesting that the reacting blend might pass through the Cahn-Hilliard linear region [42]. Shown in Fig. 6 is an example of the above-mentioned *isotropic-anisotropic* transition obtained with a PSCS/PVME (20/80) blend photo crosslinked at 80° [66]. This structural anisotropy partially originates from the inhomogeneity of the crosslinking reaction. As described below,

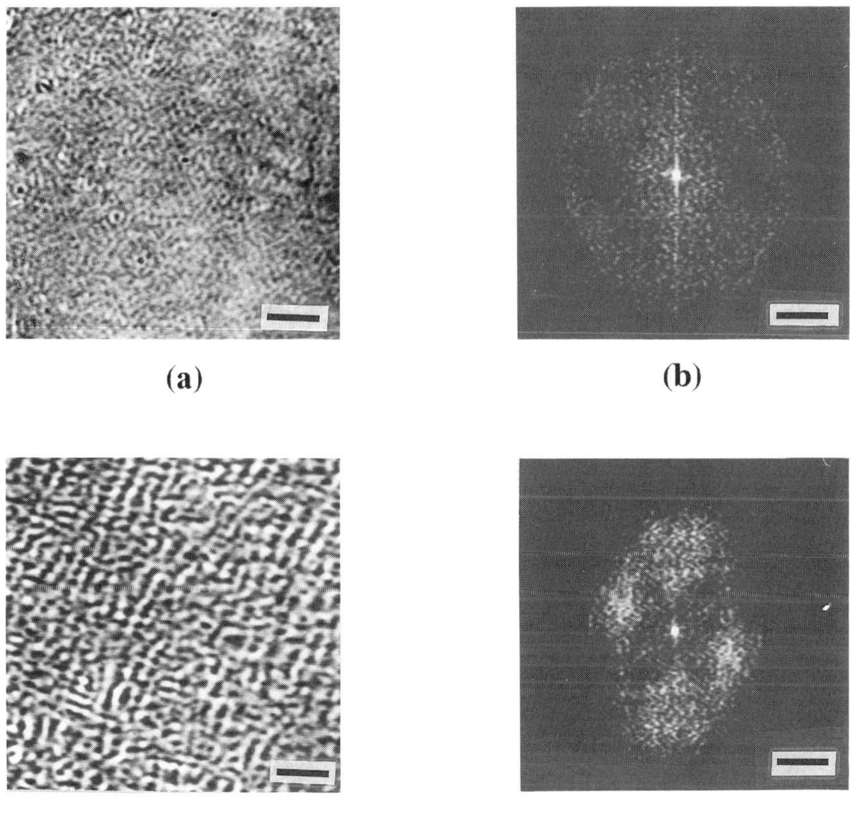

Figure 6 Isotropic–anisotropic transition observed in PSCS/PVME(20/80) photo-crosslinked at 80°C. (a) 60 min; (c) 240 min after irradiation; (b) and (d) are the corresponding 2D-FFT power spectra. The scales are 10 μm and 2×10^4 cm^{-1}.

the photo-crosslinking kinetics of PSCS chains in the blend is fairly complicated and cannot be simply described by the mean-field kinetics [67]. At the early stage of irradiation, the reaction approximately proceeds with a constant rate which is determined by the diffusion-controlled mechanism of the photodimerization of anthracenes [68]. However, as the reaction proceeds to some extent, the viscosity around anthracenes increases and the number of adjacent photo-crosslinkers decreases. These processes have two effects: one is slowing down the reaction and the other is building up an inhomogeneous network in the sample. As a result, there exists in the reacting blend a local gradient of crosslinking density from which an elastic stress field is built up. This sort of elastic stress has often been observed during the course of epoxy curing [69]. As the reaction proceeds, the newly formed coexistence curve approaches the experimental temperatures (the point A in Fig. 7) and at the same time the concentration fluctuations are amplified because the reacting blend becomes less stable compared with the early stage of the reaction. This thermodynamical instability, in turn, accelerates the reaction because of the condensation effects taking place inside the PSCS-rich region. As a consequence, a positive feedback loop driven by concentration fluctuations might be built up in the blends at the late stage of the reaction. Under the influence of this feedback process, the local anisotropy develops globally and grows to the micrometer (mesoscopic) scale as depicted in Fig. 6.

(c) Nucleation-Assisted Spinodal Decomposition Process. These particular processes are observed when mixtures with high PSCS compositions are photo-crosslinked at temperatures lower than the conditions to obtain the target patterns (described below) and the spinodal structures. Initially, small nuclei start appearing under phase-

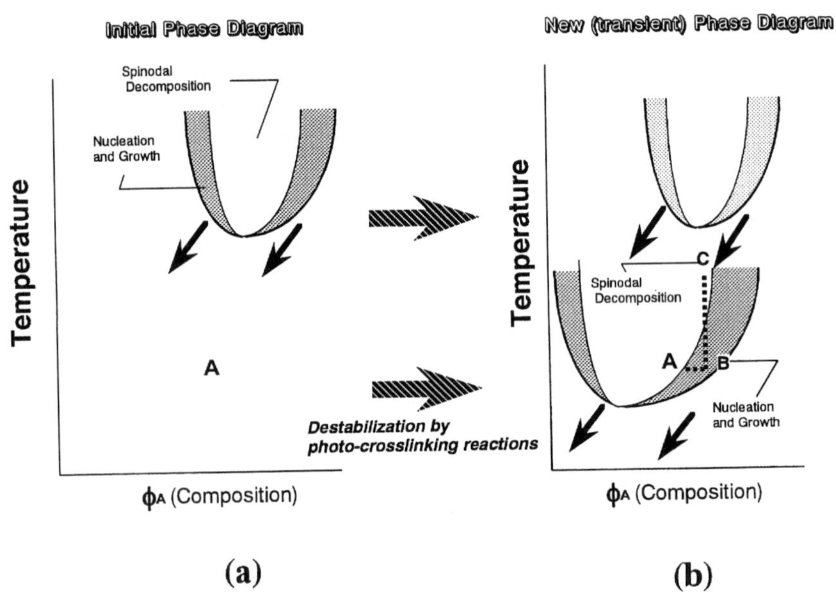

Figure 7 Proposed mechanism for the nucleation-assisted spinodal decomposition-process observed in photo-crosslinked blends. (With permission from the American Chemical Society.)

contrast optical microscopy at 30 min after irradiation. These nuclei grow slowly with irradiation time whereas the interconnecting structures start emerging in the region between these nuclei at 120 min of the reaction. As the reaction time increases, these interconnecting structures become coarsened and eventually freeze at the late stage of irradiation. The time-evolution of this nucleation-assisted spinodal decomposition process was monitored by using digital image analysis. The results were illustrated in Fig. 8 where the characteristic length scales ξ are plotted versus the irradiation time on a double-logarithmic scale. It was found that the evolution process in this particular case can be well expressed by the scaling law given by Eq. (4) with $\alpha = 1/2$. This exponent has been found in systems with a non-conserved order parameter [70]. The nonconservation probably comes from the competition between nucleation and spinodal

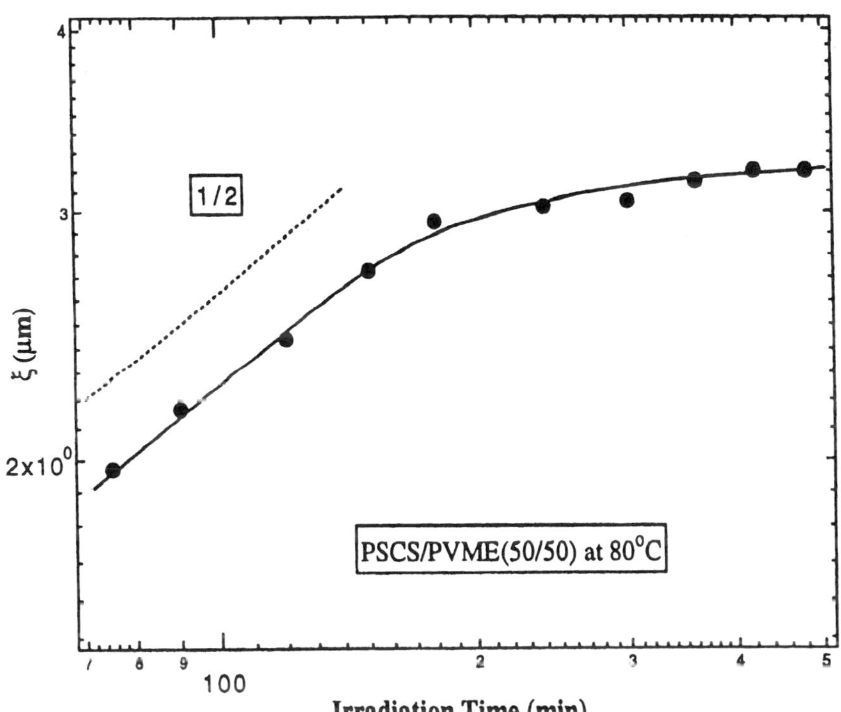

Figure 8 Time evolution of the interconnecting structures observed in the nucleation-assisted spinodal decomposition process of a PSCS/PVME(50/50) blend photo-crosslinked at 80°C.

decomposition. The mechanism is schematically illustrated in Fig. 7. In general, depending upon the crosslinking kinetics, the coexistence curve can pass through point A in various different manners. For most cases, point A enters the unstable region via the metastable region where the nucleation-and-growth process takes place. As shown in Fig. 7(b), upon irradiation, the phase boundary shifts toward the lower temperature side and eventually crosses the experimental temperature at A. This is equivalent to the situation where point A was brought by the crosslinking reaction from low to higher temperatures. If the system stays in the binodal region for a certain period of time so that the nucleation-and-growth process can proceed with stable nuclei, and then enters the unstable region (at point C) by successive crosslinking reactions, so-called nucleation-assisted spinodal decomposition takes place. Namely, these structures are formed by the nucleation-and-growth process subsequently followed by the spinodal decomposition. These morphologies are similar to those obtained recently by computer simulation for the spinodal decomposition process in the presence of seeds [71,72].

(d) Concentric (Target) Patterns. Upon irradiation at higher temperatures, i.e., between 90°C and 110°C, concentric patterns appear. These peculiar structures obtained with a PSCS/PVME (50/50) blend become observable under optical phase-contrast microscopy after 60 min of irradiation. Under this condition, the crosslinking density at the time where the sample begins to become cloudy, as about 14 junctions per chain as estimated by ultraviolet–visible spectroscopy. The sizes of the target layers do not remarkably change whereas their contrast becomes stronger with the irradiation time. For longer reaction times, the layers are disconnected, resulting in quite regular spherical domains as shown in Fig. 9 with average sizes of about 3μm which are responsible for the higher-ordered diffraction patterns obtained by 2D light scattering [73]. The cross-section of the irradiated sample shows broken lamellae with the direction perpendicular to the surface of the sample, indicating that the concentric patterns shown in Fig. 9 are not spherical but rather have cylindrical symmetry. One of the reasons for this peculiar pattern is that the crosslinking reaction proceeds with a front moving from above to below due to the Lambert–Beer law. Several experiments have been carried out in an attempt to clarify the formation mechanism of the target patterns such as interference of irradiating light by the growing nuclei [74]. However, the mechanism is not yet clearly understood and it seems that the interference of the irradiation light, even though a possible factor, is not the only cause for the formation of these target patterns. Further experiments such as triggering the spinodal decomposition in binary blends containing

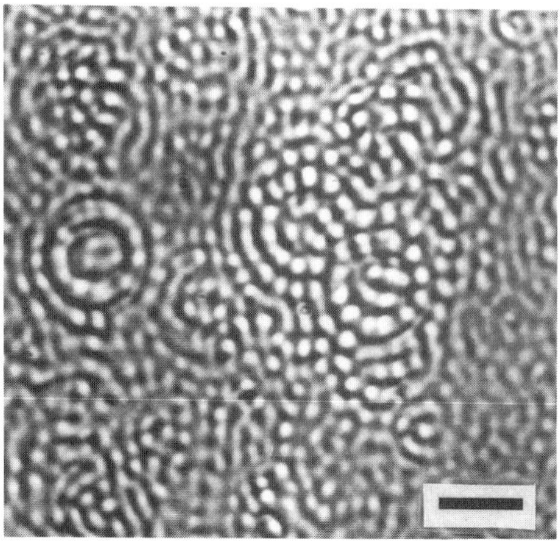

Figure 9 Target patterns observed for a PSCS/PVME(50/50) blend with high label content (120 anthracenes per chain) photo-crosslinked at 90°C for 600 min. The scale is 10 μm.

impurities dispersed as nuclei in the blend might provide some insights into the relevant mechanism.

The morphologies emerging from the competitions between the phase separation and the crosslinking reactions in the miscible region of a binary polymer blend are summarized in Fig. 10. It is worth noting that this morphological phase diagram should be sensitive to external conditions such as light intensity and is therefore changeable with the label content of the polymers.

C. Kinetics of *In Situ* Photo-Crosslinking Reactions of Polymer Blends

The kinetics of crosslinking reactions plays a very important role in the determination of the final morphologies of reactive blends. These phase structures determine the physical properties, particularly the viscoelastic and mechanical properties of multiphase polymers prepared by chemical reactions. There exists a number of experimental methods which have been utilized to directly monitor the crosslinking reactions in polymer systems. The most widely used methods are spectroscopic techniques such as Fourier transform infrared (FT-IR) [75–77] or FT-Raman [78], and differential scanning calorimetry (DSC) [79–82] because most of the reactions involved in the curing processes are exothermic. It has been shown that,

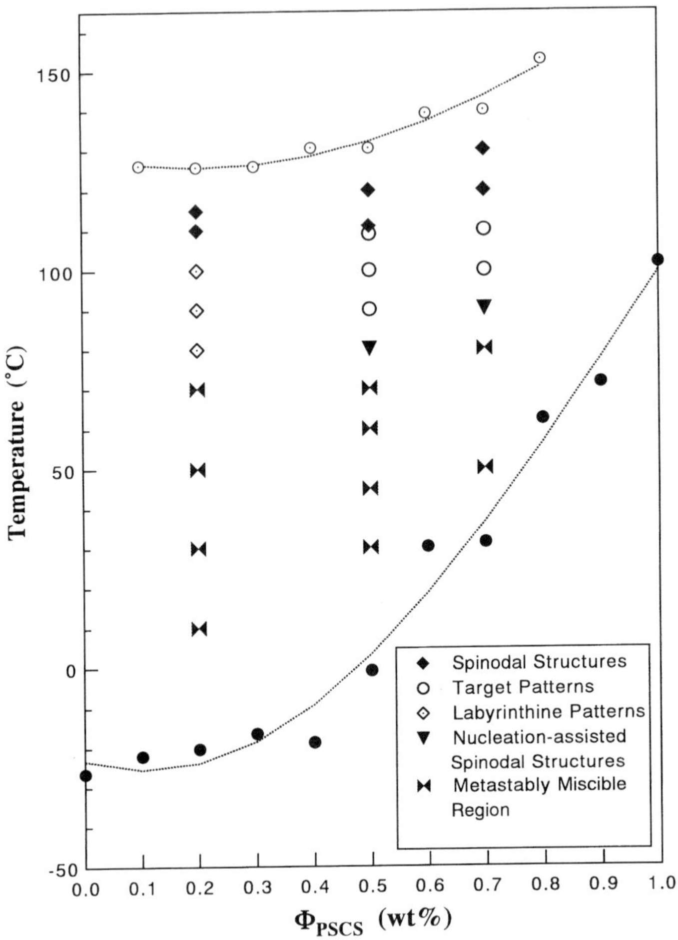

Figure 10 Morphological phase diagram of PSCS/PVME blends photo-crosslinked in the one-phase region. PSCS ($M_w = 260{,}000$, 65 anthracenes per chain), PVME ($M_w = 96{,}000$). Light source: 500 W Hg–Xe lamp. Irradiance: 3.0 mJ cm^{-2} s^{-1} at 365 nm. (⊙) and (●) are the cloud point and the glass transition temperature. (With permission from the American Chemical Society.)

for most cases, the reactions do not proceed homogeneously during the preparation of IPNs or expoxy resins, but instead exhibit autocatalytic behavior due the heat generated by the formation of free radicals [79]. This exothermicity accelerates the reaction until the viscosity of the systems increases, slowing down the polymer mobilities.

For anthracene-labeled polymer blends, the photodimerization, known as a diffusion-controlled reaction, can be monitored by following the changes in absorbance of anthracenes under irradiation by using ultraviolet–visible spectroscopy. In polymer blends, the reaction approximately follows the mean-field kinetics at low conversion. However, as the conversion increases the non-mean-field behavior becomes noticeable. In order to describe the reaction kinetics, the reaction yield or conversion $p(t)$and the crosslinking density $\gamma(t)$ are defined as follows:

$$p(t) = \frac{OD_0 - OD(t)}{OD_0} \tag{5}$$

$$\gamma(t) = \tfrac{1}{2}\theta p(t) \tag{6}$$

where OD_0 and $OD(t)$ are the absorbances of anthracenes measured at an arbitrary wavelength (typically at 410 nm) at the irradiation times $t = 0$ and t, respectively. θ is the label content, i.e., the average number of anthracene per chain.

1. Kinetics of Photo-Crosslinking Inside the Spinodal Region

As described above, the kinetics of photo-crosslinking reactions can be monitored after each time interval of irradiation by using ultraviolet–visible spectroscopy. However, one of the difficulties associated with these experiments is that the mixtures become cloudy during the reactions because of phase separation. Therefore, the extent to which the reaction still proceeds after the phase separation takes place is not clearly known. In an attempt to correlate the crosslinking kinetics with the resulting morphology of the blends, the changes in concentrations of the photo-crosslinker anthracenes with irradiation time for the P2CS/PVME blends shown in Fig. 3 were monitored by using ultra-violet–visible spectroscopy. The correction for the effects of cloudiness on the absorbance of anthracenes was performed by using the transmission data of the reacted blends monitored at two different wavelengths. One wavelength is located inside and the other was outside of the absorption range of anthracene. By this correction, it was found that for the P2CS/PVME (60/40) blends photo-crosslinked under the conditions indicated in Fig. 3, the reaction approximately follows the mean-field kinetics:

$$\frac{dC_A(t)}{dt} = -kC_A^2(t) \tag{7}$$

where $C_A(t)$ is the concentration of anthracene at the irradiation time t and k is the reaction rate. By defining the initial concentration of anthracene as

$C_A(0)$, the solution of Eq. (7) can be expressed in terms of the reaction yield $p(t)$ as:

$$\frac{1}{(1-p)} = kC_A(0)t \qquad \text{with} \quad p(t) = \frac{C_A(0) - C_A(t)}{C_A(0)} \tag{8}$$

Eq. (8) provides a test for the crosslinking reaction kinetics [83]. It was found that within 30 min of irradiation, $1/(1-p)$ varies linearly with irradiation time. Therefore, the crosslinking under the condition shown in Fig. 3 can be regarded as a homogeneous reaction. As described in the next section, for the higher label content and the crosslinking reaction performed over a longer time, the kinetics generally exhibits the non-mean-field behavior and can be well expressed by the Kohlrausch–Williams–Watts (KWW) empirical function.

2. Kinetics of Photo-Crosslinking in the One-Phase Region

Generally, in the temperature region far away from the cloud point T_{cl} of the blend, the reaction proceeds homogeneously in the early stage because anthracene can diffuse and encounter other anthracene molecules when the concentration is still high. However, at long irradiation time, the concentration of crosslinkers decreases and the viscosity of the blend increases. Under this circumstance, the reaction proceeds much slower than in the early stage. As a result, the kinetics can be well described by the stretched exponential function of time (KWW) [84,85]:

$$OD(t) = (OD_0 - B)\exp[-(k_0 t)^{\beta}] + B \tag{9}$$

where k_0 and β are the average reaction rate and the inhomogeneity index of the reaction. The baseline B was introduced in order to compensate for the fact that the photo-crosslinkers cannot completely react under irradiation. It was found that for all the photo-crosslinking reactions carried out in the temperature range close to T_g, i.e., "shallow" quench conditions, the photo-crosslinking process can be well expressed by the above equation. An example is shown in Fig. 11 for an anthracene-labeled PSCS/PVME (50/50) crosslinked at several temperatures far away from the cloud point (127.4°C) of the miscible blend [67]. Under these conditions, the crosslinking density is so high that the coexistence curve of the reacted blend cannot reach the experimental temperature (T_{exp}) due to the strong kinetic effects coming from the formation of the PSCS networks. The resulting *semi*-IPNs are therefore metastably miscible. On the other hand, the blends undergo phase separation when crosslinked at higher temperatures. Under these particular crosslinking conditions, it was found that the reaction exhibits autocatalytic behavior driven by concentration fluctuations. Namely, when

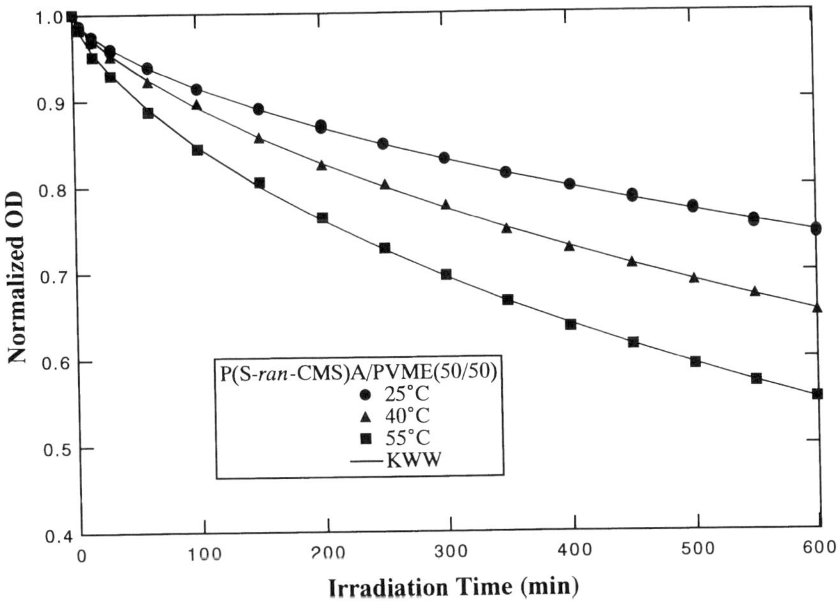

Figure 11 Photo-crosslinking kinetics of a PSCS/PVME(50/50) blend observed at different temperatures. The solid curves were obtained by fitting the experimental data to Eq. (11) using a nonlinear least-squares method with regression. The exponent β is in the range 0.66–0.71.

the sample is crosslinked at a temperature close to the phase boundary, the coexistence curve approaches T_{exp} while the concentration fluctuations are amplified. These fluctuations associated with the moving phase boundary become more remarkable as the new phase boundary becomes closer to T_{exp} (point A in Fig. 7). These concentration fluctuations can be observed directly by SANS [86] or SAXS [87]. During this period, there are several phenomena occurring simultaneously:

1. increase in viscosity due to the network formation;
2. consumption of the photo-crosslinkers by the reaction;
3. decrease in volume of the sample due to the crosslinking reaction [69];
4. increase in the concentration of photo-crosslinkers in the PSCS-rich regions built-up by concentration fluctuations (condensation effects).

Depending upon the irradiation condition and the blend composition, the condensation effects induced by the concentration fluctuations might lead to the formation of a positive feedback loop in the reacting systems. The crosslinking reactions are thus accelerated by this autocatalytic feedback process until the sample becomes cloudy. An example of such a thermodynamics-driven feedback loop has been observed when PSCS/PVME (20/80) and (50/50) blends were photo-crosslinked at 80°C. In the initial stage of crosslink, the average reaction rate (dp/dt) decreases because of the consumption of anthracenes and insignificant effects of concentration fluctuations. As the phase boundary approaches the experimental temperatures, T_{exp}, the reaction is accelerated. As a consequence, (dp/dt) passes through a maximum before the coexistence curve of the blend crosses the T_{exp} point. The fact that this autocatalytic behavior was not observed with the photo-crosslinking reaction of PSCS homopolymers, suggests that this autocatalysm is driven by concentration fluctuations and has a thermodynamic origin [88]. Autocatalytic behavior of crosslinking reactions has been observed in a large number of chemically curing systems as well as in the synthesis of IPNs where the formation of free radicals is involved [76,79,89,90]. The origin of these autocatalytic feedbacks is known as the heat generated by the radical formation which in turn accelerates the reaction. The relevant kinetics has been a central subject of studies on curing as well as crosslinking reactions in IPNs [91,92]. On the other hand, the feedback observed in photo-crosslinked blends is actually driven by thermodynamics and should be discriminated from one driven by the heat released by polymerization or curing reactions described above. Moreover, the presence of this fluctuation-driven feedback loop might be responsible for the spatial coherence emerging in the morphology of photo-crosslinked blends. An example can be seen in the morphology shown in Fig. 6(c) where the orientation of the lamella structures extends over 100 μm.

In summary, the kinetics of the *in situ* photo-crosslinking reactions in the miscible region of polymer blends cannot be described by the mean-field kinetics and generally can be expressed by the (KWW) function at temperatures far from the coexistence curve (shallow quench conditions). As the newly formed coexistence curve of the blend approaches the experimental temperature, the crosslinking reaction proceeds in an autocatalytic fashion driven by concentration fluctuations before phase separation takes place. This autocatalytic feedback might be the origin of the spatially coherent structures emerging in the morphology of these photo-crosslinked blends.

D. Reaction-Induced Phase Separation as a Pattern-Formation Process

Phase separation induced by chemical reactions, particularly by reversible reactions, is a fascinating critical phenomenon because it is an example of systems with competing interactions at different length scales. Here, the chemical reaction acts as a long-range effect whereas phase separation plays the role of a short-range interaction. For chemically reacting polymer mixtures, theoretical works, at the present time, are not yet completely developed because of the complexity coming from the inhomogeneity of the polymer reactions and the iteration of the effects of the chemical reactions on both the static (via modification of the free energy of mixing) as well as the kinetic (via modification of diffusion) aspects of the phase separation. Therefore, most of the currently available models are dealing with mixtures of low molecular weight compounds instead of polymers [93]. In general, the early stage of reaction-induced phase separation is analyzed by linear stability analysis to obtain the dominating modes and the range of unstable modes characterizing the phenomena. In the late stage where the amplitudes of concentration fluctuations are getting larger, the linear stability analysis is no longer available and most of the equations of motion are analytically intractable. Instead, computer simulation becomes a powerful tool to predict the time-evolution of the systems.

1. Recent Analytical Theories and Computer Simulation for Chemically Reacting Systems

The modified Cahn–Hilliard equation for chemically reacting binary mixtures with conserved order parameter can be written in terms of the free energy functional [93]:

$$\frac{\partial\phi(r,t)}{\partial t} = M\nabla^2\frac{\delta F\{\phi(r,t)\}}{\delta\phi(r,t)} + g(\phi) \tag{10}$$

where ϕ, M, and $F\{\phi(r,t)\}$ are respectively the composition, the relative mobility coefficient and the free energy functional of the mixture. $g(\phi)$ is the reaction term containing the appropriate chemical kinetics.

The free energy functional $F\{\phi(r,t)\}$ can be taken as the Cahn-Hilliard free energy:

$$F\{\phi(r,t)\} = \int dr\{f(\phi) + \kappa[\nabla\phi(r,t)]^2\} \tag{11}$$

The first term in the right-hand side of Eq. (11) is the bulk free energy of mixing of the mixture. Depending upon the system under consideration, $f(\phi)$ can be replaced by either the Flory–Huggins mean field free energy in the case of polymer blends or by the Ginzburg–Landau free energy for small molecule mixtures. The second term of the right-hand side of Eq. (11), the so-called square gradient term, was first introduced by Cahn and Hilliard to express the contribution of the fluctuations to the free energy of mixing. κ is the coefficient related to the interaction range of the two components in the mixtures.

The reaction term $g(\phi)$ is introduced into the Cahn–Hilliard equation in order to take into account the influence of chemical reactions on the stability of the mixtures. This type of formulation was introduced for the first time by Huberman to formulate the critical phenomena associated with chemical reactions containing autocatalytic feedback steps [94]. Depending upon the reaction kinetics involved in the phenomena, $g(\phi)$ can have various forms. The simplest expression for the dispersion relation between the growth rate $\omega(q)$ and the wavelength λ of the unstable modes in Fourier space q (where $q = 2\pi/\lambda$) for reacting mixtures is:

$$\omega(q) = -K - M\left(\frac{\partial^2 f}{\partial\phi^2}\right)q^2 - 2\kappa M q^4 \tag{12}$$

Here q is the wavenumber. The constant K varies with the reaction kinetics and is a function of the reaction rates involved in the chemical process. It is worth noting that in the absence of chemical reactions, K vanishes and the growth rate $\omega(q)$ reduces to the well-known Cahn–Hilliard equation for the amplification factor. $\omega(q)$ is sketched with q in Fig. 12. Due to the presence of K, the fluctuations with large wavelengths in the vicinity of $q \to 0$ (the so-called soft-modes) are affected by the chemical reactions. If $K > 0$, the above analysis indicates that the growth of all the unstable modes satisfying the condition $0 < q < q_{c_1}$ is suppressed and the emerging patterns are stabilized by the chemical reactions. This primary prediction is very important not only for the mode-selection process in chemical systems undergoing critical phenomena, but also for the morphology control of multiphase materials using chemical reactions. The effects of the following reaction kinetics on the phase separation of binary mixtures have been examined.

(a) Autocatalytic Reaction [94].

$$\mathrm{A + X \xrightarrow{k_A} 2X} \tag{13}$$

$$\mathrm{X + B \xrightarrow{k_B} C} \tag{14}$$

The intermediate species X is autocatalytically produced in Eq. (13) and consumed to give the product C in Eq. (14). By incorporating the above

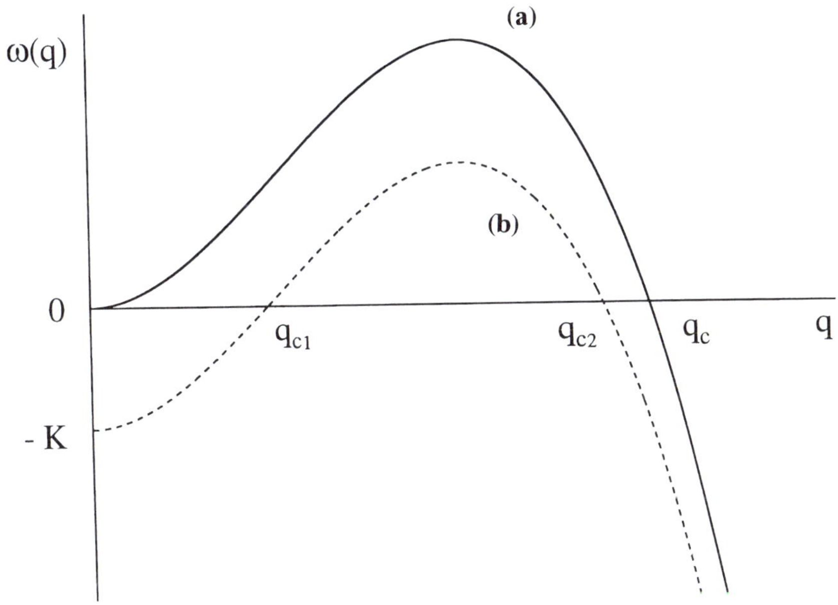

Figure 12 A schematic presentation of the dispersion relation given by Eq. (12): (a) $K = 0$, without chemical reaction where the mode at $q = 0$ is marginally stable (the Cahn–Hilliard case); (b) $K \neq 0$, in the presence of chemical reactions.

chemical kinetics into the modified Cahn–Hilliard equation (10), the reaction term $g(\phi)$ corresponding to this particular case is:

$$g(\phi) = k_1\phi - k_A\phi^2 \tag{15}$$

where ϕ is the concentration of X and $k_1 = k_A\phi_A(0) - k_B\phi_B$. Here, $\phi_A(0)$ and ϕ_B are the concentrations of the species A at time 0 and of the species B at time t, respectively. From the dispersion relation of the modified Cahn–Hilliard equation with the reaction term given by Eq. (15), the intercept K is:

$$K = k_A\phi_A(0) - k_B\phi_B - 2k_A\phi_0 \tag{16}$$

where ϕ_0 is the concentration of X at equilibrium. As a result, the stability of the soft modes are determined by the rate constants k_A and k_B.

(b) Reversible First-Order Reaction. The morphology of a binary mixture A/B undergoing spinodal decomposition with simultaneous exchange reaction:

$$\mathrm{A} \underset{k_2}{\overset{k_1}{\rightleftharpoons}} \mathrm{B}$$

was investigated recently by computer simulation [95]. The modified Cahn-Hilliard equation for this particular case is:

$$\frac{\partial\phi}{\partial t} = M\nabla^2 \frac{\delta F\{\phi(r,t)\}}{\delta\phi} - k_1\phi + k_2(1-\phi) \tag{17}$$

with

$$g(\phi) = -k_1\phi + k_2(1-\phi)$$

where $\phi(r,t)$, k_1 and k_2 are respectively the composition of the component A and the rates of the forward and backward reactions.

Linear stability analysis of Eq. (17) gives the dispersion relation indicated in Eq. (12) with:

$$K = k_1 + k_2 \tag{18}$$

For $k_1 = k_2 = k$, the dispersion relation becomes simpler:

$$\omega(q) = -M\left(\frac{\partial^2 f}{\partial\phi^2}\right)q^2 - 2\kappa Mq^4 - 2k \tag{19}$$

As seen previously with the case of Eq. (12), the last term in the right-hand side of Eq. (19) implies the suppression of the phase separation process by the reaction.

The time-evolution of the average length scales of the emerging structures has been shown to follow the scaling law given by Eq. (2). The average size ξ of the domains grows with the reaction time in the early stage and subsequently saturates at a certain equilibrium length scale ξ_{eq} which can be scaled with the reaction rate (k) like $\xi_{eq} \approx (1/k)^\alpha$ [95]. It was found by computer simulation [96] that the exponent α approaches the Lifshitz–Slyosov–Wagner exponent (1/3) [48,49] for the small reaction rates. Recently, the effects of reaction rates on the phase separation kinetics were also investigated by computer simulation [97]. The behavior of the phase separation for small reaction rates obtained previously in [96] was confirmed whereas the exponent α was found to be 1/4 for the case of high rates.

On the other hand, the effects of reversible reaction $\mathrm{A} \rightleftharpoons \mathrm{B}$ on the two-phase morphology of polymer mixtures have also recently been investigated by computer simulation [98]. By changing the initial compositions of the mixtures and the two rate constants k_1 and k_2, the morphological transitions (from random two-phase structures to concentric patterns, disk-like type and/or interconnection structures) have been observed.

(c) Relation with Microphase Separation in Block Copolymers. When the order parameter $\phi(r, t)$ is taken as $-1 < \phi(r, t) < 1$, Eq. (10) combined with the condition $k_1 = k_2 = K$ (or $k_b = 2k$) has the simpler form [99]:

$$\frac{\partial \phi}{\partial t} = M\nabla^2 \frac{\delta F\{\phi(r, t)\}}{\delta \phi} - k_b \phi \tag{20}$$

It was pointed out that by incorporating the linear term with respect to the order parameter ϕ into the free energy functional $F\{\phi(r, t)\}$, Eq. (20) can be written in the following form [99–101]:

$$\frac{\partial \phi}{\partial t} = M\nabla^2 \frac{\delta \mathcal{J}\{\phi(r, t)\}}{\delta \phi} - k_b \phi \tag{21}$$

with

$$\mathcal{J}\{\phi(r, t)\} = F\{\phi(r, t)\} + \frac{k_b}{2M} \int\int G(r, r')\phi(r, t)\phi(r', t)\, dr\, dr' \tag{22}$$

Here $G(r, r')$ is the Green's function satisfying the following Poisson equation $\nabla^2 G(r, r') = -\delta(r - r')$. The free energy functional $\mathcal{J}\{\phi(r, t)\}$ in Eq. (22) contains two terms expressing two antagonistic interactions competing over different length scales: phase separation as a short-range interaction and the chemical reaction as a long-range effect. Eq. (22) also resembles the free energy functional used to describe the microphase separation of block copolymers where the second term expresses the long-range repulsive interactions and the first term represents the short-range attractive interactions. The correspondence between the free energy functionals of chemically reacting polymer mixtures and of microphase separation can be seen by comparing Eq. (22) with the free energy functional of diblock copolymers [102,103]:

$$F\{\phi\} = \int dr[f(\phi) + \kappa(\nabla\phi)^2] + \frac{\alpha}{2}\int\int dr\, dr' G(r, r')\phi(r)\phi(r') \tag{23}$$

where:

$$\alpha = \frac{3}{[Nf(1 - f)]^2} \tag{24}$$

and N is the polymerization index and f is the block ratio.

The analogy between the macrophase separation of chemically reacting polymer mixtures and the microphase separation of block copolymers can be elucidated by comparing Eq. (22) with Eq. (23). It turns out that:

$$\frac{\alpha}{2} = \frac{k_b}{2M} \tag{25}$$

Eq. (25) implies that by changing the reaction rate k_b and the mobility M independently, the phase morphology similar to those observed in the microphase separation of block copolymers can be obtained, in principle, with reacting polymer blends.

(d) Phase Separation Associated with Heterogeneous Catalytic Reactions [104]. Heterogeneous catalytic reactions exhibit the behavior of systems far from equilibrium and have been extensively investigated. It is now established that chemical systems with these particular reactions have rich varieties of dynamics [105]. To model the phase transition phenomena at the surfaces of catalysts such as platinum (Pt) arising from the adsorption and desorption of gas molecules such as carbon monoxide (CO), Verdasca et al. have recently incorporated adsorption–desorption kinetics into the modified Cahn–Hilliard equation (10) with the free energy functional given by Eq. (11) [104]. The reaction term $g(\phi)$ in this case is:

$$g(\phi) = psk_{\mathrm{ad}}(1-\phi) - k_{\mathrm{des}}\phi^2 \tag{26}$$

where p is the gas pressure, s is the coefficient of stickness, and k_{ad} and k_{des} are respectively the adsorption and desorption rate constant of gas at the surface.

The modified Cahn–Hilliard equation with the reaction term given by Eq. (26) can be analytically solved by linear stability analysis to obtain the dispersion relation:

$$\omega(q) = \left(\frac{sk_{\mathrm{ad}}}{k_{\mathrm{des}}}\right)p + 2\phi_{\mathrm{s}} - M\left(\frac{\partial^2 F}{\partial \phi^2}\right)_{\phi_{\mathrm{s}}} q^2 - 2M\kappa q^4 \tag{27}$$

The first two terms in the right-hand side of (27) come from the contribution of the reaction and the last two terms are the conventional Cahn–Hilliard equation. Here ϕ_{s} is the composition at the steady state.

Eq. (27) again indicates the suppression of the long wavelength instability ($q \to 0$) due to the chemical reactions. It has been shown that in the weak segregation limit, the morphology obtained by numerical calculation of the modified Cahn–Hilliard equation (10) with $g(\phi)$ given by Eq. (26) is either hexagonal or striped (lamellar) structures. On the other hand, in the strong segregation limit, it was found that the dynamics cannot be described by a single length and seems to display the characteristics of multilength scales. It was also found that there are, at least, two length scales dominating the dynamics of the process: the average sizes of the phase domains and the averages of reciprocal inter-droplet distance.

2. On the Effects of Thermodynamics on the Reaction Kinetics in the Phase Separation of Chemically Reacting Polymer Mixtures

In Eq. (20), the reaction rate k_b was introduced into the Cahn–Hilliard equation as a constant via the reaction term $g(\phi)$. In general, the reaction rate depends strongly on the reaction environments, particularly for diffusion-controlled bimolecular reactions such as photodimerization of anthracenes [106]. Therefore, it is natural to assume that the reaction term $g(\phi)$ is, in general, a function of the thermodynamics of the mixtures. Such effects on the phase separation of chemically reacting mixtures have been pointed out recently by Lefever and coworkers [107]. By taking into account the effects of thermodynamics on chemical reactions, these authors have shown that the presence of the cut-off q_{c_1} on the side of low wavenumbers (long wavelengths) in the dispersion curve illustrated in Fig. 12 is a special case of reaction-driven phase separation under thermodynamically open conditions. In general, the phase separation is instead not stabilized or even accelerated by the presence of chemical reactions if the chemical kinetics is coupled to the thermodynamics [108]. As mentioned in Section II. C. 2 it was experimentally found that the photo-crosslining kinetics of PSCS chains in PSCS/PVME mixtures was remarkably affected by the concentration fluctuations as the newly formed coexistence curve of the reacting mixtures closely approached the experimental temperatures (T_{exp}) [88]. Namely, the concentration fluctuations are amplified as the new phase boundary is located in the vicinity of T_{exp}. Under this circumstance, in the PSCS-rich region, the local concentration of the photo-crosslinkers increases, accelerating the crosslinking reactions. This step, in turn, enhances the instability of the blends and thus amplifies further the concentration fluctuations. These processes are repeated under irradiation until the blends undergo phase separation. The presence of this positive feedback loop is also responsible for the emergence of spatially coherent structures in the micrometer scale observed previously with photo-crosslinked PSCS/PVME blends [73]. On the other hand, reversible photoisomerization, i.e., a noncrosslinking reaction, of stilbene derivatives has been used to induce phase separation of stilbene-labeled polystyrene/poly(vinyl methyl ether) (PSCS/PVME) mixtures [109]. In this case, the blend becomes unstable and undergoes phase separation by switching the *trans* → *cis* isomerization of the stilbene moieties chemically labeled on polystyrene chains by irradiation with ultraviolet light. This process corresponds to releasing pressure applied to the PSCS component of the blend. Contrary to the case of crosslinking by photodimerization of anthracenes, the photoisomerization kinetics monitored by ultraviolet–visible spectroscopy under the "deep" quench (low-yield) conditions is almost unaffected by the presence of concentration fluctuations in the vicinity of T_{exp}. Furthermore, compared to the case where only temperature was used to induce phase separation, it was also found that the phase separation in the presence of photoisomerization of stilbene proceeds extremely slow. The Lifshitz–Slyozov–Wagner exponent α given in Eq. (2) is approximately in between 1/3 and 1/4 in the case of thermally inducing phase separation, whereas it is less than 0.1 in the case of irradiation [109]. These experimental results reveal the existence of the soft-modes suppression process which originates from the cut-off wavenumber k_{c_1} introduced by the reversible photoisomerization to the dispersion curve illustrated in Fig. 12. Though the number of experimental data is limited, it can be currently concluded that the validity of Eq. (20) for describing phase separation accompanied by chemical reactions depends on the reaction kinetics. In a general sense, for diffusion-controlled reactions such as photodimerization of anthracene shown here, the contribution of thermodynamical fluctuations to the reaction kinetics should be taken into account as pointed out recently by Lefever and coworkers [107,108]. In other words, the reaction rate k is not a constant and should depend on the chemical potential of the reacting mixture. However, for unimolecular reactions such as the *trans* ⇔ *cis* photoisomerization of stilbene described above, the effects of thermodynamics might not be remarkable. In this case, Eq. (10) can be used to describe the phase-separation phenomena in chemically reacting systems. In conclusion, the results of reaction kinetics obtained with photodimerization of anthracene and photoisomerization of stilbene suggest that there exist some correlations between the wavelengths of concentration fluctuations (L_f) and the effective range (L_r) of the diffusion-controlled reaction. The effects of thermodynamics on the reaction will be significant if $L_f \leq L_r$, i.e., the case of crosslinking reactions (an intermolecular reaction). On the other hand, if $L_f \geq L_r$, i.e., the case of photoisomerization of stilbene (an intramolecular reaction), the effects of thermodynamics would be insignificant.

III. CONCLUDING REMARKS AND PERSPECTIVES

1. As mentioned earlier, chemically reacting polymer mixtures are widely encountered in polymer engineering and science through the preparation of practically useful materials such as IPNs or curing epoxy resins. However, with the conventional methods of preparation used so far, phase separation overcomes the crosslinking reactions, and in most cases, the phase equilibrium is achieved before the competition sets in. As a consequence, the resulting morphology is the random two-phase structure. In

order to efficiently control the morphology of multiphase polymeric materials, it is necessary to perform experiments with fast reactions. This fact was pointed out for the first time in the pioneering work of Huberman in 1976 [94]. Photochemical reactions are suitable candidates for these purposes because they can be quickly triggered and terminated independently of the variables P, V, and T. Therefore, photochemical reactions can provide great opportunities to induce and control the morphology of polymer blends. Furthermore, the reacting sites can be specified and selected by the labeling techniques. An example for this selection is that the reactions A–A and/or B–B between the two component polymers A and B in the blends can be independently performed by irradiating the samples with two different wavelengths for two different photo-crosslinkers. On the other hand, the applicability of photochemical reactions is limited by the sample thickness and also by the cloudiness of the sample as phase separation takes place.

2. From the viewpoint of the pattern selection process, with proper choice, photochemical reactions can provide not only the way to modify the cut off wavelengths at low q, but also the shape of the dispersion curve, i.e., $\omega(q)$ versus q, particularly when there exists a strong coupling between order parameters via their nonlinear reaction terms. Furthermore, the symmetry of the free energy with respect to the order parameter of the reacting mixtures can also be altered by changing the order of the reactions.

For crosslinking reactions, the elastic stress associated with the network formation participates in the late stage of the reaction as a long-range interaction, making the problem more interesting and complex. From this particular aspect, there is an analogy between the patterns formed by the crosslinking-induced phase separation, and the pattern formation arising from the reaction diffusion mechanism in developmental biology [110]. In order to explain all the morphologies obtained by photo-crosslinking reaction, it is necessary to incorporate into the modified Cahn–Hilliard equation the contributions of elastic stress associated with the reaction.

3. From the thermodynamics viewpoint, in order to explain the emergence of spatially coherent structures in photo-crosslinked blends, it is necessary to take into account the contributions of the feedback process driven by concentration fluctuations in the Cahn-Hilliard equation.

In summary, phase separation induced by chemical reactions is a complicated and interesting problem cutting across many scientific disciplines. Elucidation of these fundamental problems will not only help to design polymer materials with novel morphology, but can also provide important information for establishing some rules about the pattern selection processes in polymeric systems.

ACKNOWLEDGMENTS

The works described in this chapter are based on the collaborations with my former graduate students at the Department of Polymer Science & Engineering, Kyoto Institute of Technology, Kyoto, Japan during the period 1993–1996: Akira Imagawa (Sekisui Chem. Co., Osaka), Toshikazu Tamai (Dynik Inc., Shiga), Asuka Harada (DaiNippon Ink & Chem. Inc., Osaka) and Naoki Kawazoe (Kyoto University), Takashi Ohta and Katsunari Kataoka (both currently with the Department of Polymer Science and Engineering, Kyoto Institute of Technology). The financial support from the Ministry of Education, Science and Cultures, Japan during these years is greatly acknowledged. Finally, we thank Dr Barry J. Bauer (Polymers Division, National Institute of Standards and Technology, Gaithersburg, Maryland) for his critical reading of the manuscript.

REFERENCES

1. For example, see CK Riew, ed. Rubber-Toughened Plastics. Vols I and II Adv Chem Ser Nos 222 and 252, Washington, DC: American Chemical Society, 1987 and 1996.
2. M Szwarc. Carbanions, Living Polymers and Electron-Transfer Processes, New York: Wiley, 1968.
3. M Morton, LJ Fetters. Anionic Polymerizations and Block Copolymers. In: CE Schildknecht, I Skeist, ed. Polymerization Processes. New York: Wiley Interscience, 1977. See also Y Matsushita, Chapter 4 of this book for recent developments.
4. See AI Nakatani, CC Han. Chapter 7 of this book.
5. DR Paul, LH Sperling, ed. Multicomponent Polymer Materials. Adv Chem Ser No 211, Washington, DC: American Chemical Society, 1986.
   LA Utracki. Polymer Alloys and Blends. Munich: Carl Hanser Verlag, 1989.
6. JL Lee. Reaction Injection Molding. In: G Allen and JC Bevington, ed. Comprehensive Polymer Science. Oxford: Pergamon Press, 1989, vol 7, pp 379–426.
   SB Brown, CM Orlando. Reactive Extrusion. In: Encyclopedia of Polymer Science and Engineering. New York: Wiley, 1988, vol 14, pp 169–189.
7. VJ Triacca, S Ziaee, JW Barlow, H Keskkula, DR Paul. Polymer 32:1401, 1991.
8. M Xanthos, SS Dagli. Polym Eng Sci 31:929, 1991.
9. J Devaux, P Godard, JP Mercier. Polym Eng Sci 22:229, 1987.
10. JC Golba, GT Seeger. Plast Eng 43:3, 57, 1984.
11. Z Zhou, A Eisenberg. J Polym Sci Polym Phys Ed 21:595, 1983.
    M Hara, A Eisenberg. Macromolecules 17:1335, 1984.

12. LZ Pillon, LA Utracki. Polym Eng Sci 24:1300, 1984.
13. NY Chiu, SJ Fang. J Appl Polym Sci 30:1473, 1985.
14. For example, see LH Sperling. Interpenetrating Polymer Networks and Related Materials. New York: Plenum Press, 1981.
15. PG de Gennes. J Phy Lett (France) 40:69, 1979.
16. K Binder, HL Frisch. J Chem Phys 81:2126, 1984.
S Stepanow, M Schulz, K Binder. J Phys II (France) 4:819, 1994.
17. BJ Bauer, RM Briber, CC Han. Macromolecules 22:940, 1989.
18. RM Briber, BJ Bauer. Macromolecules 21:3296, 1988.
19. RM Briber, BJ Bauer. Macromolecules 24:1899, 1991.
20. H Jinnai, H Hasegawa, T Hashimoto, RM Briber, CC Han. Macromolecules 26:182, 1993.
21. K Dusek. Adv Polym Sci 78:1, 1986.
22. K Yamanaka, T Inoue. Polymer 30:662, 1989.
23. K Yamanaka, Y Takagi, T Inoue. Polymer 30: 1839, 1989.
24. H-S Lee, T Kyu. Macromolecules 23:459, 1990.
25. H Furukawa. Adv Phys 34:703, 1985.
26. See, for example, D Klempner, LH Sperling, L Utracki, ed. Interpenetrating Polymer Networks. Adv Chem Ser No 239, Washington DC: American Chemical Society, 1994.
27. R Bansil and G Liao. Trends Polym Sci 5:146, 1997; LW Barrett, LH Sperling. Trends Polym Sci 1:45, 1993 and references therein.
28. M Okada, K Fujimoto, T Nose. Macromolecules 28:1795, 1995.
29. AG Merzhanov, IP Borovinskaya. Dokl Akad Nauk SSSR 204:336, 1972.
30. NM Chechilo, NS Enikolopyan. Dokl Phys chem 221:392, 1975.
31. AM Khan, JA Pojman. Trends Polym Sci 4:253, 1996 and references cited therein.
32. JA Pojman, VM Ilyashenko, AM Khan. J Chem Soc Faraday Trans 92:2825, 1996.
33. AM Kotliar. J Polym Sci Macromol rev 16:367, 1981.
34. RS Porter, L-H Wang. Polymer 33:2019, 1992.
35. P Godard, JM Dekoninck, V Pevlesaver, J Devaux. J Polym Sci Polym Chem Ed 24:3315, 1986.
36. G van der Velden, G Kolfschoten-Smitsamans, A Veermans. Polymer Commun 28:169, 1987.
37. A Nakai, T Shiwaku, H Hawegawa, T Hashimoto. Macromolecules 19:3008, 1986.
38. M Motowoka, H Jinnai, T Hashimoto, Y Qui, CC Han. J Chem Phys 99:2095, 1993.
H Yoon, Y Feng, Y Qiu, CC Han. J Polym Sci Polym Phys 32:1485, 1994.
39. A Karim, JF Douglass, SK Satija, APY Wong, CC Han, preprints.
40. Q Tran-Cong. Polarization-Induced Photochromic Reactions in Polymer Solids. In: JC Salamone, ed. Polymeric Materials Encyclopedia. Boca Raton: CRC Press, 1996, vol 7, pp 5128–5136.

41. G Nicolis, I Prigogine. Self-Organization in Nonequilibrium Systems. New York: Wiley, 1977, Chapters 6, 7.
42. JW Cahn. Acta metall 9:795, 1961; J Chem Phys 42:93, 1965.
43. PG de Gennes, J Chem Phys 72:4756, 1980.
P Pincus. J Chem Phys 75:1996, 1981.
44. Q Tran-Cong, T Nagaki, T Nakagawa, O Yano, T Soen. Macromolecules 22:2720, 1989.
45. Q Tran-Cong, T Nagaki, O Yano, T Soen. Macromolecules 24:1505, 1991.
46. Q Tran-Cong, M Imura, T Soen, M Shibayama. Polym Eng Sci 33:772, 1993.
47. A Imagawa, Q Tran-Cong. Macromolecules 28:8388, 1995.
48. IM Lifshitz, VV Slyozov. J Phys Chem Solids 19:35, 1961.
49. CZ Wagner. Z Electrochem 65:581, 1961.
50. T Tamai, A Imagawa, Q Tran-Cong. Macromolecules 27:7486, 1994.
51. S Kondo, T Otsuka, K Ogura, K Tsuda. J Macromol Sci Chem A13:767, 1979.
52. A Harada, Q Tran-Cong. unpublished results.
53. RE Rosensweig. Ferrohydrodynamics. Cambridge: Cambridge University Press, 1985.
54. RM Weis, HM McConnell. Nature 310:47, 1984.
55. J Zasadzinski, MB Schneider. J Phys (France) 48:2001, 1987.
56. E Bodenschatz, JR de Bruyn, G Ahlers, DS Cannel. Phys Rev Lett 67:3078, 1991. For a review, see M Seul, D Andelman. Science 267:476, 1995.
57. N Goldenfeld. Lectures on Phase Transitions and the Renormalization Group. New York: Addison-Wesley, 1992.
58. C Sagui, RC Desai. Phys Rev E 49:2225, 1994.
59. P Borckmans, G Dewel, A De Wit, D Walgraef. Turing Bifurcations and Pattern Selection. In: R Kapral, K Showalter, ed. Chemical Waves and Patterns, Dordrecht: Kluwer Academic Publishers, 1995.
60. FS Bates, GH Fredrickson. Ann Rev Phys Chem 41:525, 1990.
61. H Nakazawa, K Sekimoto. J Chem Phys 104:1675, 1996.
62. S Singh, HL Frisch, H Ghiradella. Macromolecules 23:375, 1990.
63. SC Glotzer, MF Gyure, F Sciortino, A Coniglio, EH Stanley. Phys Rev Lett 70:3275, 1993.
64. SC Glotzer, MF Gyure, F Sciortino, A Coniglio, EH Stanley. Phys Rev E 49:247, 1994.
65. A Harada, Q Tran-Cong. Macromolecules 29:4801, 1996.
66. A Harada, Q Tran-Cong. Macromolecules 30:1643, 1997.
67. K Kataoka, A Harada, O Urakawa, Q Tran-Cong. J Polym Sci Polym Phys (in press), 1997.
68. See, for example, AA Frost, RG Pearson. Kinetics and Mechanism. New York: Wiley, 1961, Chapter 11.
69. J Lange, S Toll, J-A Manson, A Hult. Polymer 36:3135, 1995.
70. M Allen, JW Cahn. Acta Metall 27:1085, 1979.
71. H Furukawa. J Phys Soc Jpn 63:3744, 1994.
72. H Furukawa. J Phys Soc Jpn 63:3919, 1994.

73. Q Tran-Cong, A Harada. Phys Rev Lett 76:1162, 1996.
74. A Harada, Q Tran-Cong. Bussei Kenkyu 63:581, 1994.
75. A Hakiki, JE Herz, G Beinert. Polymer 33:4575, 1992.
76. YS Lipatov, GM Semenovitch, SI Skiba, LV Karabanova, LM Sergeeva. Polymer 33:361, 1993.
YS Lipatov, TT Alekseeva, VF Rosovitsky, NV Babkina. Polymer 33:610, 1993.
77. K Moussa, C Dekker. J Polym Sci Polym Chem 31:2633, 1993.
78. M Younes, S Wartewig, D Lellinger, B Strehmel, V Strehmel. Polymer 35:5269, 1994.
79. JM Barton. Adv Polym Sci 72:111, 1985.
80. P Cassagnau, M Bert, V Verney, A Michel. Polymer 34:124, 1993.
81. J-P Chen, YD Lee. Polymer 36:55, 1995.
82. HL Lin, TL Yu. Polymer 37:581, 1996.
83. N Kawazoe, A Imagawa, Q Tran-Cong. J Appl Polym Sci, (in press).
84. TV Ramakrishnan, M Raj Lakshmi ed. Non-Debye Relaxation in Condensed Matter. Singapore: World Scientific, 1986.
85. R Richert, A. Blumen, Eds. Disorder Effects on Relaxational Processes. New York: Springer, 1994.
86. CC Han, BJ Bauer, JC Clark, Y Muroga, M Matsushita, M Okada, Q Tran-Cong, T Chang, IC Sanchez. Polymer 29:2002, 1988.
87. Q Tran-Cong, R Kawakubo, S Sakurai. Polymer 35:1236, 1994.
88. Q Tran-Cong, A Harada, K Kataoka, T Ohta, O Urakawa. Phys Rev E 55:R6340, 1997.
89. X Ramis, JM Salla. Polymer 36:3511, 1995.
90. CS Su, EM Woo. Polymer 36:2883, 1995.
91. ME Ryan, A Dutta. Polymer 20:203, 1972.
92. JM Kenny, J Appl Polym Sci 51:761, 1994.
93. For a simple and compact formulation of the Cahn–Hilliard equation, see the recent review by SC Glotzer. Computer Simulations of Spinodal Decomposition in Polymer Blends. In: D Stauffer, ed. Annual Reviews of Computational Physics. Singapore: World Scientific, 1995, vol 2, pp 1–46.
94. BA Huberman. J Chem Phys 65:2013, 1976.
95. SC Glotzer, EA Di Marzio, M Muthukumar. Phys Rev Lett 74:2034, 1995.
96. SC Glotzer, D Stauffer, N Jan. Phys Rev Lett 72:4109, 1994.
97. JJ Christensen, K Elder, HC Fogedby. Phy Rev E 54:2212, 1996.
98. M Motoyama. J Phys Soc Jpn 65:1894, 1996.
99. SC Glotzer, A Coniglio. Phys Rev E 50:4241, 1994.
100. Y Oono, M Bahiana. Phys Rev Lett 61:1109, 1989.
M Bahiana, Y Oono. Phys Rev A: 41:6763, 1990.
101. F Liu, N Goldenfeld. Phys Rev A 39:4805, 1989.
102. L Leibler. Macromolecules 13:1602, 1980.
103. T Ohta, K Kawasaki. Macromolecules 19:2621, 1986.
104. J Verdasca, P Borckmans, G Dewel. Phys Rev E 52:4616, 1995.
105. R Imbihl, G Ertl. Chem Rev 95:697, 1995.

106. Q Tran-Cong, Y Ishida, A Tanaka, T Soen. Polym Bull 29:89, 1992.
107. R Lefever, D Carati, N Hassani. Phy Rev Lett 75:1674, 1995.
108. R Lefever, private communication; D Carati, R Lefever. Phys Rev E 56:3127, 1997.
109. Q Tran-Cong, T Ohta, O Urakawa. Phys Rev E 56:R59, 1997.
110. JD Murray, PK Maini, RT Tranquillo. Phys Repts 171:60, 1988.

6

Multiphases in Polymeric Gels

Mitsuhiro Shibayama
Kyoto Institute of Technology, Kyoto, Japan

I. INTRODUCTION

In the previous chapters, multicomponent polymeric systems have been discussed theoretically and experimentally, where the systems are mostly in the bulk state. Apart from those systems, we treat here highly diluted polymeric systems called gels. In this chapter various aspects of phases and related phenomena of gels are reviewed with an emphasis on the analogy between gels and multicomponent polymeric systems, e.g., polymer blends and block copolymers. Particularly, it is demonstrated that the Coulombic interaction plays an important role in the structure formation and generation of unique properties of gels. The concept of microphase separation in weakly charged polymer gels is introduced, and a classification of the macrophase separation, microphase separation, and volume phase transition is made by taking account of the chain connectivity and molecular interactions.

A gel is a three-dimensional "network" comprising slightly crosslinked long-chain molecules and a large amount of solvent [1] as schematically depicted in Fig. 1. It is worth listing here some of the characteristic features of gels as shown in Fig. 1 in order to understand comprehensively the physical properties of gels.

(a). A gel is a self-sustainable polymer network and a solvent container with a semipermeable boundary, across which the solvent molecules can move. Because of the network architecture, a gel is a self-sustainable material with finite shear modulus [2]. The nonzero shear modulus is a characteristic of a gel which distinguishes it from a semi-dilute solution of

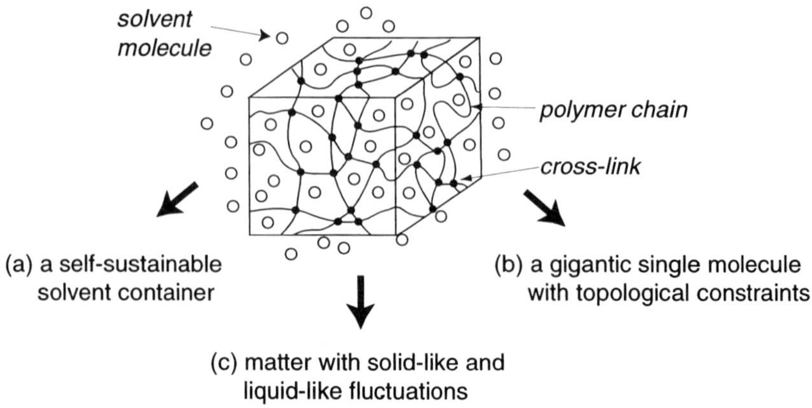

Figure 1 A schematic representation of a gel and its characteristic features.

macromolecules, although both consist of a large number of solvent molecules and polymer chains. The solvent is trapped in the gel network by an osmotic balance between the inside and outside of the gel. Therefore, a gel is an open system at equilibrium with the surrounding solvent. Gels can be not only a solvent container but also a chemical reactor, such as, a dry cell (battery) and the biological cell.

(b). A piece of gel is a gigantic single molecule in which subchains, i.e., the polymer chains between crosslinks, are connected with each other. This architecture leads to a strong cooperativity in gels, and undergoes swelling or shrinking together as a piece. A gel can also be treated as an isolated system where local thermodynamics applies. This type of treatment is necessary for small gels where the surface of gels contributes to the phase behavior.

(c). A gel possesses both liquid-like and solid-like fluctuations because of permanent crosslinking. The former are time-dependent thermal fluctuations and the latter are static frozen inhomogeneities resulting from topological constraints, ranging from nanometers to micrometers or larger [3]. The inhomogeneity in gels has been called crosslink inhomogeneity [4–7].

Phases in gels are interesting subjects for scientific discussion. In the field of physical gels where crosslinks connect polymer chains temporarily in the gels, there are two phases, i.e., the sol and gel phases. In the case of chemical gels crosslinked with chemical bonds, a swollen gel (in a swollen phase) collapses to a shrunken gel (in a shrunken phase) by reducing the solubility. This can be regarded as a macroscopic manifestation of the coil–globule transition of polymer solutions [8,9]. A gel immersed in a poor solvent becomes turbid and undergoes phase separation, resulting in the appearance of polymer-rich and polymer-poor phases [10]. This is analogous to the phase separation of polymer blends. Furthermore, if a gel has both attractive and repulsive interactions, a microphase separation takes place in the gel because a macroscopic phase separation is inhibited due to the presence of crosslinks [11]. Some gels have multistable phases at a given thermodynamic condition. The number of the phases can be more than ten [12]. In this particular case, the gel cannot be specified by one order parameter, but it requires another type of order parameter, e.g., such as the hydrogen bonding density.

In this chapter, a review of recent studies on the structure and properties of gels is presented with the keywords of "phases" and/or "fluctuations." We will not deal with physical gels and gelation kinetics. Readers who are interested in the structure and physical properties of thermoreversible physical gels may consult the book of Guenet [13]. Although a continuous transition from a swollen gel to a collapsed gel or vice versa is usually observed, some gels undergo a discrete transition and the volume is changed by a multiple of a few tens or a few hundreds. This is called the "volume phase transition" of gels. The volume phase transition was predicted by Dušek and Patterson [14] and was first observed by Tanaka [15] in acrylamide gel in a mixed solvent of acetone and water. This finding initiated an enormous number of studies on the volume phase transition of gels and environment sensitive gels [16,17]. The volume phase transition can be triggered by an infinitesimal change of one of the environmental variables, such as temperature, solvent composition, or pH [16,17]. The transition has been discussed mostly on the basis of thermodynamics without the concept of fluctuations [14,15,18–24] although the concentration fluctuations are of particular importance to describe the transitions and phases of gels as pointed out by Onuki [25]. Since concentration fluctuations play a major role in the phase transition of gels, we will focus on the concentration fluctuations, both static (frozen) inhomogeneities and dynamic thermal fluctations. In order to describe these fluctuations, we often use the spectrum of the concentration fluctuations i.e., the structure factor in Fourier space. This introduces a characteristic length scale of concentration fluctuations. Because these fluctuations lead to phase separation, we define here two types of phase separation; one is macrophase separation and the other is microphase separation. In the former case, the characteristic length scale can vary from segment order to infinity depending on thermodynamic and kinetic conditions. However, in the latter case, it is strongly dependent on the molecular architecture and is limited by the size of the constituent molecules. Thus, a microphase separation is characterized by the appearance of a scattering maximum in the structure factor.

This chapter is organized as follows: In Sec. II, examples of volume phase transition are shown and a thermodynamic interpretation is given. We mostly treat environment-sensitive gels because it is easy to control the phases of the gel by the thermodynamic variables, such as Flory's interaction parameter (via temperature or solvent composition) and ionic strength. It will become clear that the antagonistic interactions, particularly the Coulombic and hydrophobic interactions, are important to initiate a discrete volume phase transition. Therefore, in Sec. III, the role of charges on a gel network will be discussed with an emphasis on the possible formation of microphase separation. Sec. IV provides the relationship between the volume phase transition and the macrophase separation, followed by a discussion of related properties. In Sec. V, a short summary will be given which addresses the future direction of studies on polymer gels.

II. VOLUME PHASE TRANSITION AND THERMODYNAMICS

A. Phenomenological View of Volume Phase Transition

As introduced in Sec. I, a volume phase transition can be triggered by various kinds of external stimuli. Variations of solvent composition have been commonly employed to initiate discrete volume phase transitions of gels [15,18,26–29]. However, temperature variation has been extensively employed by many workers [30–38] since it is capable of a variety of applications, such as in drug delivery [39–44], functional immobilized biocatalysts [45], and actuators [46]. In most of these cases, poly(N-isopropylacrylamide) gels (abbreviated hereafter as NIPA gel) have been employed because NIPA gels undergo a swollen-to-shrunken transition at 33.6–33.8°C in an aqueous medium [30]. Photo-sensitive [47–50], electric-field sensitive [51–53], stress-sensitive [48,54–56], and pressure-sensitive gels [57,58] have also been investigated with a wide range of applications. However, even in these cases of environment-sensitive gels, temperature-sensitive gels are often employed and one of these stimuli is used as a supplemental trigger of the phase transition. Therefore, the volume phase transition of these gels can be theoretically treated within the same framework of temperature sensitive gels by taking account of the additional contribution to the free energy of gels. Hence, we now discuss the outline of the thermodynamics of the volume phase transition of gels.

B. Thermodynamic Theory of Volume Phase Transition

1. Classical Thermodynamics

(a) Neutral Gels. The thermodynamic theory for equilibrium swelling has been discussed for many decades. One can find a prototype of the swelling equilibrium in the textbook written by Flory [1]. The free energy of a gel is given as a sum of the mixing free energy term, ΔF_{mix}, and the entropy elasticity term, ΔF_{el}, as [1]:

$$\Delta F = \Delta F_{\text{mix}} + \Delta F_{\text{el}} \tag{1}$$

$$\Delta F_{\text{mix}} = k_{\text{B}}T[n_s \ln(1-\phi) + \chi n_s \phi] \tag{2}$$

$$\Delta F_{\text{el}} = \frac{nk_{\text{B}}T}{2}\left[3\left(\frac{\phi_0}{\phi}\right)^{2/3} - 3 - B\ln\left(\frac{\phi_0}{\phi}\right)\right] \tag{3}$$

where $k_{\text{B}}T$ is the Boltzmann energy, n_{s} and n are the numbers of solvent molecules and effective polymer chains in the system, respectively, χ is the Flory's interaction parameter, and ϕ and ϕ_0 are the volume fraction of the network polymer at measurement and at the reference state, respective. B is a controversial coefficient depending on the model [59], which we set to be unity here according to Flory's treatment [1]. By differentiating ΔF with respect to n_{s}, the chemical potential of the solvent, $\Delta\mu$, is obtained, which is related to the osmotic pressure of the gel,

$$-\frac{\Delta\mu}{V_{\text{s}}} \equiv \Pi = \Pi_{\text{mix}} + \Pi_{\text{el}} \tag{4}$$

where V_{s} is the molar volume of the solvent and

$$\Pi_{\text{mix}} = -\frac{k_{\text{B}}T}{V_{\text{s}}}[\phi + \ln(1-\phi) + \chi\phi^2],$$

$$\Pi_{\text{el}} = \nu k_{\text{B}}T\left[\frac{1}{2}\left(\frac{\phi}{\phi_0}\right) - \left(\frac{\phi}{\phi_0}\right)^{1/3}\right] \tag{5}$$

where ν is the number of the effective polymer chains per unit volume ($\nu \equiv n/V_0$, with V_0 being the total volume of the system). By letting $\Pi = 0$, one gets the equilibrium swelling ratio, ϕ_0/ϕ (or the polymer volume fraction at equilibrium, ϕ). By increasing χ, the gel undergoes a shrinking transition. Theory predicts that the transition is continuous as far as χ changes continuously. Therefore, this is regarded as a second-order transition. However, as was found by Tanaka, a weakly charged gel undergoes a discrete volume phase transition from a swollen state to a shrunken state and vice versa by a slight modification of the solvent nature, such as an infinitesimal change of solvent composition or temperature [15].

(b) Charged Gels. In the case of charged gels, Eqs. (1) and (4) are modified by adding the counter-ion free energy term, i.e., the Donnan potential term, ΔF_{ion} (or Π_{ion}) which is written in the following form,

$$\Delta F = \Delta F_{\text{mix}} + \Delta F_{\text{el}} + \Delta F_{\text{ion}} \tag{6}$$

and

$$\Pi = \Pi_{\text{mix}} + \Pi_{\text{el}} + \Pi_{\text{ion}} \tag{6a}$$

Tanaka and coworkers used Π_{ion} as the counter-ion free energy given by

$$\Pi_{\text{ion}} = \nu n_{\text{ion}} k_{\text{B}} T(\phi/\phi_0) \tag{7}$$

where n_{ion} is the number of counter-ions per subchains (i.e., polymer chains between crosslinks). Hence the net osmotic pressure is given by

$$\frac{\Pi}{k_{\text{B}}T} = -\frac{1}{V_{\text{s}}}[\phi + \ln(1-\phi) + \chi\phi^2] + \nu\left[\left(\frac{1}{2} + n_{\text{ion}}\right)\left(\frac{\phi}{\phi_0}\right) - \left(\frac{\phi}{\phi_0}\right)^{1/3}\right] \tag{8}$$

This type of equation has been frequently used to account for the discrete volume phase transition of gels [15,18,19].

(c) pH and Salt Concentration Dependence. Eq. (8) does not take into account the charge balance between the outside and the inside of the gel. The contribution of the imbalance of the charges is known as the Donnan potential, which is given by [1,21,38,60–65]

$$\Pi_{\text{ion}} = -\frac{\Delta\mu_{\text{ion}}}{V_{\text{s}}} \approx k_{\text{B}} T \Delta C \tag{9}$$

where ΔC is the difference of the mobile ion concentrations between the inside and outside of the gel and is given by

$$\Delta C = (C_+ + C_-) - (C'_+ + C'_-) \tag{10}$$

If the polymer chain is charged negatively, the following relations hold [66],

$$
\begin{aligned}
z_+C_+ &= z_-C_- + z_p C_p && \text{(inside gel)} \\
z_+C'_+ &= z_-C'_- && \text{(outside gel)} \\
\gamma_\pm^2 C_+^{z_-} C_-^{z_+} &= \gamma_\pm'^2 C_+'^{z_-} C_-'^{z_+} && \text{(osmotic balance)}
\end{aligned}
\tag{11}
$$

where z_i and C_i are the valency and the concentrations of positive ($i = +$), negative ions ($i = -$) and polymers ($i = p$), respectively, inside the gel (without the prime) and outside the gel (with the prime). By letting the activity coefficient gamma $\gamma_\pm = 1$, and considering monovalent electrolytes, Eq. (11) simplifies to

$$
\begin{aligned}
C_+ &= C_- + z_p C_p \\
C'_+ &= C'_- \equiv I \\
C_+C_- &= C'_+C'_- = I^2
\end{aligned}
\tag{12}
$$

In the case of poly(N-isopropylacrylamide-co-acrylic acid) (NIPA/AAc) copolymer gels, ΔC is simply given by

$$
\Delta C \approx \sqrt{[\mathrm{COO^-}] + 4I^2} - 2I \tag{13}
$$

where $[\mathrm{COO^-}]$ is the concentration of carboxylate ions. Fig. 2(a) shows the linear swelling ratio, d/d_0, $(= (\phi_0/\phi)^{1/3})$, of NIPA/AAc gels (open circles) as a function of pH and that calculated with the above theory (solid curve) [38], where d and d_0 are the diameters of the gel at observation and at preparation, respectively. The pH dependence of the corresponding noncharged gel, i.e., the NIPA homopolymer gel, is also given (open squares), which is independent of pH as expected (dashed line). The pH dependence of d/d_0 for NIPA/AAc gel seems to be well reproduced by the theory. The deviations at intermediate pHs are due to an underestimation of the crosslink density since the contribution of entanglements was not taken into account in the theory. Fig. 2(b) shows the pH dependence of Π_{mix}, Π_{el}, and Π_{ion}, for the case of NIPA/AAc gels. This example shows that the Donnan potential (Π_{ion}) is the major component responsible for gel swelling, and Π_{mix} and Π_{el} terms act as compensators to keep a finite size of the gel. It is worth noting here that an additional effect, such as, intersegmental complexation (i.e., a complexation at the segment order) may need to be taken into account to describe the swelling behavior of NIPA/AAc gels. This deduction is based on the recent report by Staikos et al. who discussed interpolymer complexation between poly-NIPA and poly-AAc in aqueous solution [67]. However, we ignore these kinds of additional interactions arising from the choice of the pair of monomers when discussing the thermodynamics of gels.

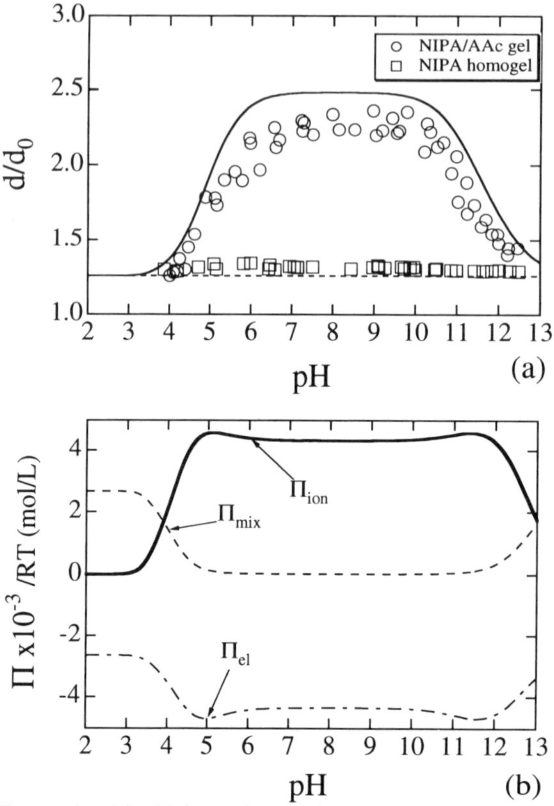

Figure 2 (a) pH dependence of the linear swelling ratio, d/d_0 (the ratio between the gel diameters at observation d, and at preparation, d_0) for NIPA/AAc copolymer gel and NIPA homopolymer gel. (Source: (Ref. 38) *J Chem Phys* 105:4358, 1996. Copyright: American Institute of Physics.) (b) pH dependence of the mixing Π_{mix}, elastic Π_{el}, and ionic terms Π_{ion}, contributing to the net osmotic pressure, $\Pi_{mix} + \Pi_{el} + \Pi_{ion}$.

(d) Uniaxial Stretching. If a piece of gel is uniaxially stretched with a force f, does the gel favor the swollen or shrunken state? The elastic term of the free energy, ΔF_{el}, for a gel stretched from length L_0 to L, is given by [54,68,69]

$$\Delta F_{el} = \frac{\nu k_B T}{2}\left[(\alpha_{//}^2 + 2\alpha_{\perp}^2) - 3 - \ln\left(\frac{\phi_0}{\phi}\right)\right] - f(L - L_0) \tag{14}$$

where $\alpha_{//}$ and $\alpha_{\perp}$ are the extension ratios along and perpendicular to the stretching direction, respectively, and

$$\alpha_{//} = L/L_0 \qquad \phi_0/\phi = \alpha_{//}\alpha_{\perp}^2 \tag{15}$$

The answer of the above question is given by Hirotsu and Onuki [54]. Fig. 3 shows the temperature dependence of the linear swelling ratio of a cylind-

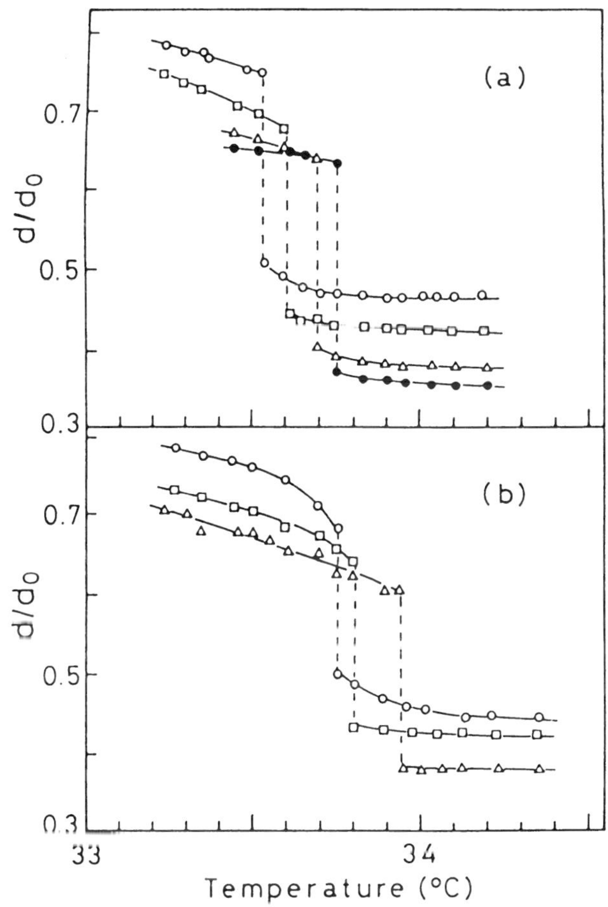

Figure 3 The temperature dependence of the linear swelling ratio of a cylindrical NIPA gel, d/d_0, subject to uniaxial stretching with several values of load along the cylinder axis on (a) cooling and (b) heating. Values of tension when $d = d_0$ are (in 10^8 dyn/cm^2); ○:0, □:2.4, △:4.7, ●:7.1. (Source: (Ref. 54) *J Phys Soc Jpn* 58.1508, 1989. Copyright: Japan Physical Society.)

rical NIPA gel, d/d_0, subject to uniaxial stretching with several values of load along the cylinder axis. This shows that there are two phases, such as the swollen (low temperatures) and shrunken (high temperatures) phases, and the volume phase transition temperature increases with increasing load. Therefore, the gel favors a swollen phase when it is stretched. A hysteresis between (a) cooling and (b) heating processes is also seen. A similar result was obtained by Suzuki and his coworkers [48–50].

As described above, the thermodynamics of the volume phase transition of gels has been described in the context of Flory theory although it sometimes requires further assumptions, such as very high crosslink density [19] or discrete temperature dependence of χ at the transition [70], or strong ϕ dependence of χ [71,72]. These examples indicate incompleteness of the theoretical treatment; e.g., (1) by the employment of Gaussian statistics for entropy elasticity; (2) assuming affine deformation; (3) ignoring the fact that the volume fraction of polymer in the shrunken state, ϕ_{sh}, is much less than unity (e.g., $\phi_{\mathrm{sh}} \approx 0.65$) [37] in spite of the theoretical prediction being $\phi_{\mathrm{sh}} = 1$, etc. These problems have been treated by Marchetti et al. [73,74], Prausnitz and coworkers [75,76], Grosberg and Nechaev [77], and Moerkerke et al. [78]. In the following section, we briefly describe some of these theories.

2. Modified Equation of State Theory for Gels

Marchetti et al. [73,74] employed a lattice-fluid theory proposed by Sanchez and Lacombe [79,80] to explain the pressure effect of gel swelling. The chemical potential of the solvent, μ_{s}, is given by

$$\frac{\mu_{\mathrm{s}}}{v_{\mathrm{s}}^* k_{\mathrm{B}} T} = \frac{1}{v_{\mathrm{s}}^*}\left[\ln \phi_{\mathrm{s}} + \left(1 - \frac{v_{\mathrm{s}}^*}{v_0^*}\right)\phi_{\mathrm{h}} + \phi_{\mathrm{g}}\right] + \frac{\nu \phi_{\mathrm{g}}^0}{V_{\mathrm{g}}^*}\left(\frac{\phi_{\mathrm{g}}}{\phi_{\mathrm{g}}^0}\right)^{1/3} + (\chi'_{\mathrm{sg}}\phi_{\mathrm{g}} + \chi'_{\mathrm{s0}}\phi_{\mathrm{h}})(\phi_{\mathrm{g}} + \phi_{\mathrm{h}}) - \chi'_{\mathrm{gh}}\phi_{\mathrm{g}}\phi_{\mathrm{h}} \tag{16}$$

where v_0^* and v_{s}^* are the volumes of the lattice site and the solvent molecule; ϕ_{h}, ϕ_{s}, and ϕ_{g} are the lattice site fractions of holes (i.e., vacancies), solvent, and gel; and $\nu\phi_{\mathrm{g}}^0/V_{\mathrm{g}}^*$ is the density of tetrafunctional crosslinks per unit volume of gels at syntheses. The interaction parameters per lattice site, χ'_{sh}, χ'_{gh}, and χ'_{sg}, which are purely energetic in origin, are given by

$$\chi'_{\mathrm{sh}} = P_{\mathrm{s}}^*/k_{\mathrm{B}}T \qquad \chi'_{\mathrm{gh}} = P_{\mathrm{g}}^*/k_{\mathrm{B}}T$$
$$\chi'_{\mathrm{sg}} = \frac{1}{k_{\mathrm{B}}T}\left[(\sqrt{P_{\mathrm{s}}^*} - \sqrt{P_{\mathrm{g}}^*})^2 + 2Z_{\mathrm{sg}}\sqrt{P_{\mathrm{s}}^* P_{\mathrm{g}}^*}\right] \tag{17}$$

where P_s^* and P_g^* are the cohesive energy densities of the solvent and gel, respectively, and Z_{sg} is a correction parameter characterizing the mixture. The first term of the right-hand side of Eq. (16) is entropic in origin, arising from mixing three components, i.e., holes, solvent, and gel. The second and third terms denote the elastic and energetic terms, respectively. Though there are many parameters, most of them are evaluated by independent experiments. This theory predicts an increase of transition temperature, T_c with increasing hydrostatic pressure, which has been experimentally observed by Kato and coworkers [57,58].

Prausnitz and coworkers generalized the Flory theory by taking account of (1) nonrandom mixing, such as hydrogen bonding, and (2) the phantom network nature of the gel [75,76]. This theory satisfactorily reproduces their experimental results on acrylamide/[(methacrylamido)propyl]trimethylammonium chloride (MAPTAC) charged gels. Grosberg and Nechaev [77] considered topological constraints of network chains and proposed a concept of crumpled globules for the shrunken state, which accounts for $\phi_{sh} \ll 1$.

Onuki [25,68] introduced a Ginzburg–Landau-type concentration gradient term, ΔF_{inh}, to the free energy function of gels having the following form,

$$\Delta F_{inh} = \frac{k_B T}{2} \int \eta |\nabla\phi|^2 \, dV \tag{18}$$

where V is the volume of the system and η is a constant of the order of $1/(\phi v^{1/3})$ and v is the volume of one solvent molecule. This leads to an additional contribution to the net free energy of the gel. This gradient term becomes important when the structure factor of gels is discussed at the transition temperature. It should be noted here that the assumption of the linear additivity of ΔF_{mix} and ΔF_{el}, employed in all the theories described above, is criticized by Deam and Edwards [81].

3. Multiple Phases

In the preceding sections, we discussed two phases, i.e., the swollen and shrunken phases. However, some gels exhibit more than two phases. Annaka and Tanaka [12] have shown the existence of multiple phases in copolymer gels made of acrylic acid (AAc) and MAPTAC. This gel is a polyampholyte gel, and AAc and MAPTAC act as the anionic and cationic constituents, respectively. The number of phases is strongly dependent on pH and the composition and the gel volume changes discretely among the phases like a flip-flop. In the case of AAc/MAPTAC (460 mM/240 mM), the number of phases is more than seven in a moderate pH region, e.g., pH 6 to

8. Note that the numbers in the parenthesis denote the monomer concentrations at preparation. These phases are dependent on the history of pH treatment. In the case of MAPTAC-rich or AAc-rich copolymers, the number of discrete phases decreases and both homopolymers show no discrete transition. There is no doubt that at least two opposite interactions, such as the Coulombic attractive and repulsive interactions, are necessary for volume phase transition. In addition to these, it is deduced that hydrogen-bond interaction may be necessary to induce transitions among these multiple phases. These kinds of multiple phases are also observed in AAc/styrene copolymer gels [82], interpenetrating networks of MAPTAC and AAc, polypeptide gels [83], and protein gels [84]. Theoretically, multiple phases may be understood as a result of competition between various factors. However, it is clear that the mean-field picture with the order parameter of $\Delta\phi$ does not predict the presence of so many phases and a novel theory is required. Though Annaka and Tanaka proposed to introduce another order parameter, such as the hydrogen bonding density [83], the appearance of multiple phases has not been theoretically understood yet. It should be noted that such multiple phases seem to be characteristic of gels since the intrinsic viscosity, a measure of the size of the individual polymer dimension, of the corresponding AAc/MAPTAC copolymer solution does not exhibit such a discrete change with pH [85].

III. MICROPHASE SEPARATION AND ITS MICROSCOPIC VIEW

A. Phenomenology

A discrete volume phase transition was accidentally observed in an acrylamide gel in which the amide groups underwent spontaneous partial ionization by aging [16]. This is why thermodynamic theories of the volume phase transition have been developed with an emphasis of the contribution of charges as discussed in the previous section. The theory predicts a shift of the transition temperature, T_c, and an increase of the difference of the volume fractions, $\Delta\phi = \phi_{sh} - \phi_{sw}$, at T_c with increasing ionization of gels [15,17–19]. Fig. 4 shows a micrograph of a cylindrical NIPA/AAc (668 mM/32 mM) gel at T_c in water. Interestingly, both swollen and shrunken phases coexist at this temperature with the volume fractions, ϕ_{sh} and ϕ_{sw}. ϕ_{sh} and ϕ_{sw} are calculated from the diameters of the swollen and shrunken phases, i.e., $\phi_{sw} \equiv (d_0/d_{sw})^3\phi_0$ and $\phi_{sh} \equiv (d_0/d_{sh})^3\phi_0$ where d_{sw} and d_{sh} are the diameters of the gel at the swollen and the shrunken states, respectively, as shown in the figure. This coexistent structure lasts more than a week unless the temperature is changed. Though Hirotsu reported that this type of

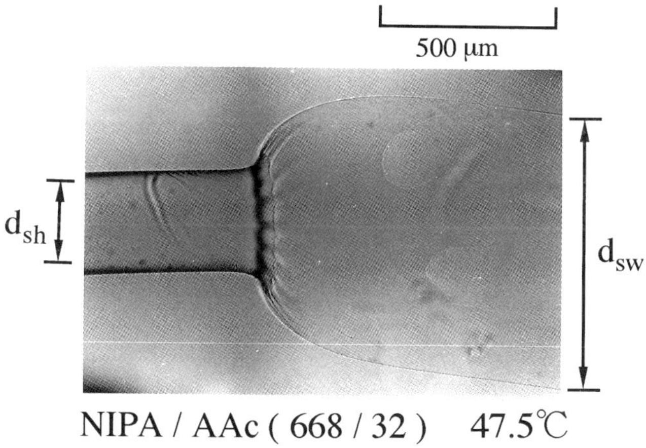

Figure 4 Optical micrograph of a cylindrical NIPA/AAc(668/32) gel at T_c in water. Swollen and shrunken phases coexist. Though this coexistent structure is stable, a slow kinetics leads to either shrinking or swelling dependent on the temperature. The diameters of the swollen phase and shrunken phase are denoted as ϕ_{sw} and ϕ_{sh}, respectively.

coexistence is thermodynamically stable [86], our recent investigation indicates that the coexistence is a transient state during an extremely slow swelling or shrinking process of the order of a few hundred hours [87]. Note that $\Delta\phi$ is an order parameter and is nonzero at the transition for a discrete volume phase transition. A discrete transition is observed even in a homopolymer gel made of NIPA, provided that the crosslink density is low enough [31,87]. However, in many cases, the presence of charges is required to generate a discrete volume phase transition.

Let us consider a gel containing chargeable groups. By immersing it in water, it may swell more than a nonchargeable gel because of ionization of the chargeable groups. However, if water becomes a poor solvent to the gel, it tends to shrink. In this case, strong concentration fluctuations may be generated owing to the antagonistic forces, i.e., the interplay between the short range tendency toward immiscibility and the large-scale homogenization due to the requirement of electroneutrality [88]. In order to describe such concentration fluctuations of matter, a scattering technique is one of the best means as will be discussed below.

A small-angle neutron scattering (SANS) experiment on polymer gels was conducted by Plestil et al. [89] on poly(N, N′-diethylacrylamide-co-sodium methacrylate) networks. They found a clear scattering maximum in the structure factor. The scattered intensity increases with temperature.

Shibyama and coworkers studied the structure factors of a series of NIPA homopolymer gels and NIPA/AAc copolymer gels as a function of temperature and polymer concentration. Fig. 5 shows the SANS intensity function, $I(q)$, of (a) NIPA homopolymer gels [17,36] and (b) NIPA/AAc(668/32) gels [11,17] in deuterated water at various temperatures. In both cases, $I(q)$ is a monotonous decreasing function with q at temperatures below the Θ temperature of poly-NIPA aqueous solutions ($\approx 31^\circ$C) [73,90]. By approaching Θ, however, $I(q)$ of NIPA homopolymer gels increases at $q = 0$ and diverges at $T_c = 34.6^\circ$C [91]. On the other hand, the $I(q)$ values of NIPA/AAc copolymer gels have a distinct peak at $q \approx 0.02$ Å^{-1} at $T > \Theta$. The presence of a scattering maximum indicates an occurrence of microphase separation with a well-defined length scale above this temperature. It should be noted here that the intensity change at a characteristic q vector is more than a hundred fold during a variation of temperature of about 25°C.

Fig. 6 shows that the corresponding swelling–shrinking curve of NIPA/AAc gels in deuterated water. A shrinking curve of neutral gel (NIPA homopolymer gel) is shown by the open squares. The NIPA homopolymer gel undergoes a continuous volume phase transition at T_c. At T_c,

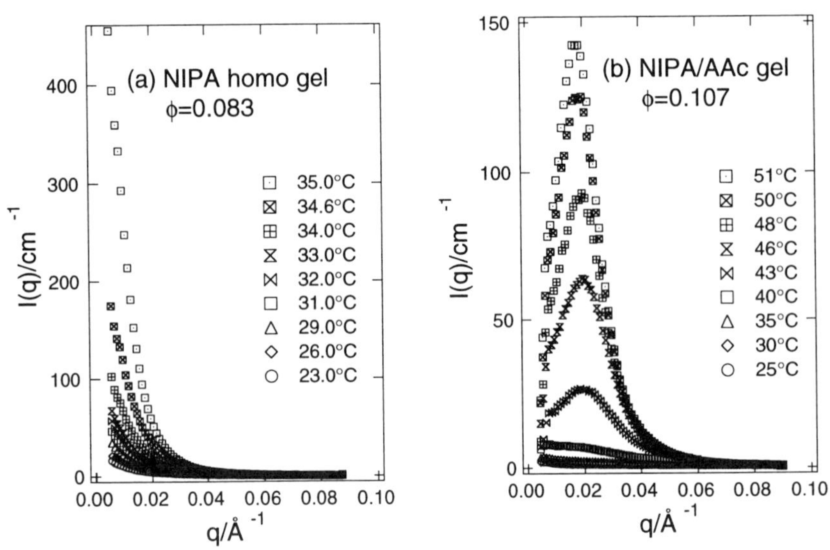

Figure 5 SANS intensity functions for (a) NIPA homopolymer gels and (b) NIPA/AAc (668 mM/32 mM) copolymer gels. The numbers in parenthesis indicate the monomer concentrations at preparation. The gel volume fractions, ϕ, at observation are 0.083 and 0.107, respectively. (Source: (Ref. 17) *Adv Polym Sci* 109:1, 1993. Copyright: Springer Verlag.)

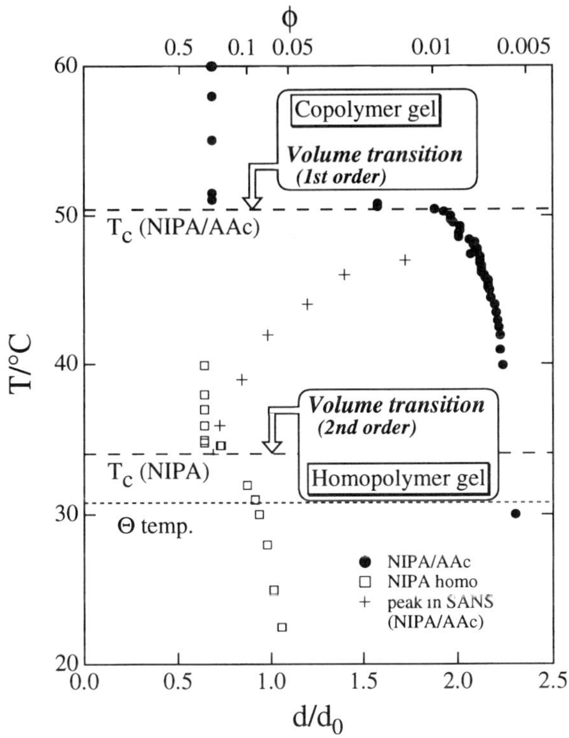

Figure 6 Swelling–shrinking curves of NIPA homopolymer gel (open squares) and NIPA/AAc (solid circles). The temperatures, T_{peak}, at which a scattering maximum appears for isochore gels are indicated with crosses. The corresponding polymer volume fraction, ϕ, is given in the top abscissa. A microphase separated structure is expected above T_{peak}. (Source: (Ref. 11) *J Chem Phys* 97:6842, 1992. Copyright: American Institute of Physics.)

both $I(q = 0)$ and the correlation length diverge (Fig. 5(a)). In the case of NIPA/AAc gels, on the other hand, the transition is discontinuous and T_c is much higher than Θ as reported by Hirotsu et al. [19]. On the basis of the SANS results (Fig. 5), the temperatures (T_{peak}) at which a scattering maximum appears in the SANS curves of NIPA/AAc gels are superimposed on the shrinking curve (filled circles) with crosses (Fig. 6). The corresponding volume fraction, ϕ, is given in the top abscissa. This indicates the presence of a microphase separated structure in NIPA/AAc gels above T_{peak}. Two important findings here are as follows: (1) though the gel is still in a swollen phase macroscopically at $\Theta < T < T_c$, the NIPA/AAc gels undergo a microphase separation at $T > T_{peak} > \Theta$; (2) even in the shrunken phase,

ϕ is far below unity. This experimental evidence clearly suggests that the physical meaning of the macroscopic volume phase transition has to be re-examined. Now, we focus on the concentration fluctuations near T_c, discuss the static structure factor of weakly charged gels on the basis of a Debye–Hückel theory of polyelectrolyte solutions, and predict the possibility of "microphase separation" in gels.

B. Static Structure Factor of Polyelectrolyte Solutions

First of all, we review a Debye–Hückel-type theory for polyelectrolyte solutions developed by Borue and Erukhimovich [92] (BE) and discuss the characteristic features of weakly charged polymer gels. Let us assume that the polyelectrolytes in a solvent are sufficiently uniform so as to apply the random phase approximation. The free energy of the system consists of the free energy of a basic system F_b and of the Coulombic interaction energy U_{cl},

$$F(\{\rho_i(\mathbf{r})\}) = F_b(\{\rho_i(\mathbf{r})\}) + U_{cl}(\{\rho_i(\mathbf{r})\}) \tag{19}$$

where $\rho_i(\mathbf{r})$ is the deviation of the density of particles of type i from its corresponding equilibrium value. The term "basic" means the case where all of the Coulombic interactions are turned off. It is more convenient to use the Fourier transform of $\rho_i(\mathbf{r})$ so as to discuss the stability of the system. The Coulombic interaction energy U_{cl}, is given by

$$U_{cl}(\{\rho\}) = \frac{2\pi e^2}{\varepsilon}\int \frac{z_i\rho_i(\mathbf{q})z_j\rho_j(-\mathbf{q})}{q^2}\frac{d^3q}{(2\pi)^3} \tag{20}$$

where z_i is the valency of ions of type i, e is the elementary charge, ε is the dielectric constant, and $\mathbf{q}$ is the scattering vector. By expanding F_b in $\rho_i(\mathbf{q})$ and employing the first nonvanishing term, one gets

$$F_b(\{\rho\}) = F_0 + \frac{k_BT}{2}\int \tilde{G}_{ij}^{-1}(\mathbf{q})\rho_i(\mathbf{q})\rho_j(-\mathbf{q})\frac{d^3q}{(2\pi)^3} \tag{21}$$

where F_0 is the equilibrium density of the free energy of the basic polymer system. $\tilde{G}_{ij}^{-1}(\mathbf{q})$ is the matrix of correlation functions of the basic polymer system

$$\tilde{G}_{ij}^{-1}(\mathbf{q}) = \langle\rho_i(\mathbf{q})\rho_j(-\mathbf{q})\rangle_b/V \tag{22}$$

where $\langle\rangle_b$ means averaging over the basic polymer system having the volume V. From Eqs. (19) to (22), the inverse of the correlation function

$$G_{ij}^{-1}(\mathbf{q}) = \tilde{G}_{ij}^{-1}(\mathbf{q}) + \frac{4\pi e^2}{\varepsilon k_B T q^2} z_i z_j \tag{23}$$

is obtained. This leads to

$$G_{ij}(\mathbf{q}) = \tilde{G}_{ij}(\mathbf{q}) - \frac{4\pi l_B \tilde{G}_{im} z_m \tilde{G}_{in} z_n}{q^2 + \kappa^2(\mathbf{q})} \tag{24}$$

where l_B is the Bjerrum length defined by

$$l_B = \frac{e^2}{4\pi\varepsilon k_B T} \tag{25}$$

which is about 7 Å in water at 25°C and $\kappa(q)$ is the q-dependent inverse Debye screening length. If the basic system contains no polymers, $\kappa(q)$ becomes independent of q and Eq. (24) is reduced to the famous Debye–Hückel equation. The relationship between l_B and κ^{-1} is given by

$$\kappa^{-1} = (8\pi N_A l_B I)^{-0.5} \tag{26}$$

where N_A is the Avogadro's number and I is the ionic strength of the system having ions of concentration C_i and valency z_i,

$$I = \frac{1}{2}\sum_i C_i z_i^2 \tag{27}$$

The structure factor calculated from Eq. (24) has the following form,

$$S(x) = S^0(x)\left[1 - \frac{S_0(x)}{S^0(x) + (x^2 + s)}\right] \tag{28}$$

where x and s are the reduced scattering vector and the reduced salt concentration, respectively, as defined by

$$x = r_0 q \qquad s \equiv \kappa^2 r_0^2 \tag{29}$$

The parameter r_0 is the characteristic screening length of the Coulombic interaction by ideal Gaussian chains. $S^0(x)$ is the structure factor for basic polymer systems, i.e., that for noncharged polymer solutions, which is written as

$$S^0(x) = \frac{1}{x^2 + t} \tag{30}$$

where t is the reduced temperature related to the Flory's interaction parameter χ as defined by

$$t \simeq 12\left(\frac{r_0}{a}\right)^2 (1 - 2\chi) \tag{31}$$

where a is the segment length of the chain molecules. The parameter t is related to the correlation length, ξ, as follows,

$$\xi = r_0/\sqrt{t} \qquad \text{(for } t > 0\text{)} \tag{32}$$

The second term in the parenthesis of the right-hand side of Eq. (28) is the correction term of the structure factor due to the switching-on of the Coulombic interaction. By substituting Eq. (30) into Eq. (28), one obtains

$$S(x) = \frac{x^2 + s}{(x^2 + t)(x^2 + s) + 1} = \frac{1}{(x^2 + t) + 1/(x^2 + s)} \tag{33}$$

The same result as Eq. (33) was also obtained by Joanny and Leibler [93]. It is interesting to compare the behavior of $S(x)$ with other structure factors. As mentioned above, the structure factor of neutral (noncharged) polymer solutions is given by Eq. (30). By switching-on the Coulombic interaction, the additional term, $1/(x^2 + s)$ appears in the denominator as depicted in Eq. (33). de Gennes calculated the structure factor of polymer blends which are crosslinked in the miscible state and observed in the weakly incompatible region [94]. It has the following form,

$$I(q) = \frac{1}{(a^2/24)q^2 + \frac{1}{2}(\chi_s - \chi) + C/q^2} \sim \frac{1}{x^2 + t + C'/x^2} \equiv S_B(x) \tag{34}$$

where χ_s is the Flory interaction parameter at the spinodal and C is a coefficient of internal rigidity and is given by

$$C \simeq \frac{36}{N^2 a^2} \tag{35}$$

and $x \equiv aq$ and $t \equiv 12(\chi_s - \chi)$. The quantity Na^2 is the mean square end-to-end distance of a chain of N units.

A similar functional form was also derived by Ohta and Kawasaki [95] for the structure factor of microphase separated block copolymers. They modified Leibler's theory of diblock copolymers in the weak segregation limit [96] to the case of the strong segregation limit and obtained the following equation

$$I(q) = \frac{1}{A(f)q^2 + B(f) + C(f)/q^2} \tag{36}$$

where

$$A(f) = \frac{R_g^2}{2N_B f(1-f)} \qquad B(f) \approx \frac{1}{2N_B f^2(1-f)^2} - 2\chi$$
$$C(f) = \frac{3}{2N_B f^2(1-f)^2 R_g^2} \tag{37}$$

The parameters, N_B, f, and R_g are the degree of polymerization, the fraction of one component, and the radius of gyration of the block copolymer, respectively. The term $B(f)$ stands for the segment-order short-range concentration fluctuations and $C(f)$ represents the long-range interaction due to the chemical bond connecting the two components. In the case of polyelectrolyte solutions, the long-range interaction term of the structure factor is further modified from C/q^2 to $C/(q^2 + D)$, i.e.,

$$I(q) = \frac{1}{Aq^2 + B + C/(q^2 + D)} \tag{38}$$

as has been already seen in Eq. (33), where D is a constant independent of q. Fig. 7 shows characteristic features of the three kinds of scattering functions $S^0(x)$, $S_B(x)$, and $S(x)$. $S^0(x)$ is a monotonic decreasing function of the reduced scattering vector, x. When the system approaches criticality, i.e., $t \to 0$, $S^0(x)$ diverges. In the case of crosslinked polymer blends or block copolymers, such divergence is inhibited in $S_B(x)$ due to the chain connectivity. On the other hand, a scattering maximum appears in a finite q value. For polyelectrolyte solutions, the behavior of $S(x)$ is similar to $S^0(x)$ when the solvent is good for the polyelectrolytes, i.e., $t \gg 0$. However, if the salt concentration, s, is low, the electrostatic screening effect is suppressed and the long-range Coulombic interaction becomes dominant. Hence, the polyelectrolyte chains prefer an expanded state. Such a situation is not attained if the solvent is poor ($t < 0$), resulting in microphase separation with a scattering maximum at a finite q. Fig. 8 shows the scattered intensity curves for poly(acrylic acid) aqueous solutions obtained by Schoessler et al. [7,97,98]. The solid curves are fitted ones on the basis of Eq. (33).

Fig. 9 shows the diagram of states proposed by Borue and Erukhimovich (BE diagram). The states are divided into eight different regimes. Note that it was originally divided into ten regimes by further dividing the phase VII to three regimes. This diagram well represents the physical picture of polyelectrolyte solutions as a function of the reduced

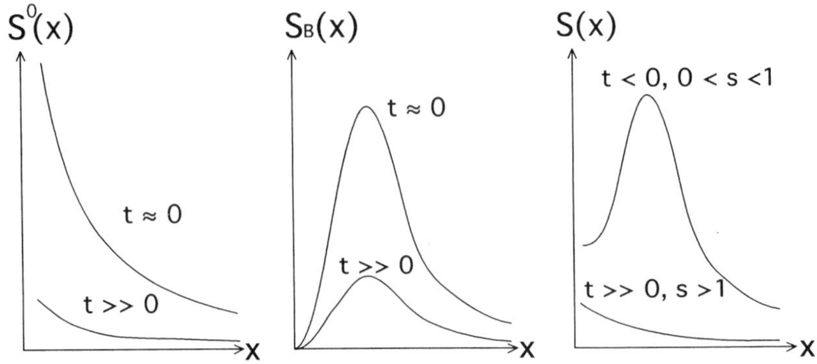

Figure 7 A schematic representation of the scattering functions of (a) neutral polymer solutions $S^0(x)$, (b) crosslinked polymer blends and block copolymers $S_B(x)$, and (c) polyelectrolyte solutions $S(x)$, where x, t, and s are the reduced scattering vector, the reduced temperature, and the reduced salt concentration, respectively.

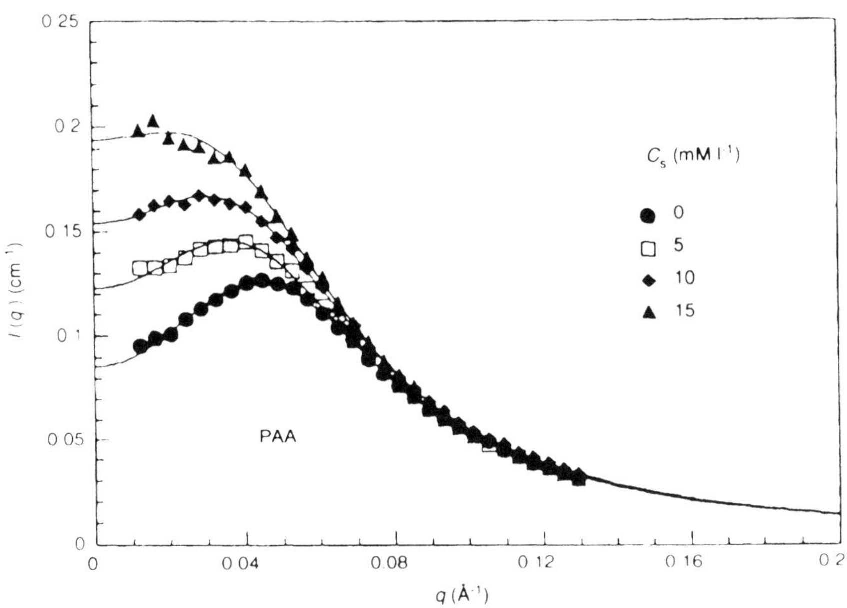

Figure 8 Scattered intensity curves for poly(acrylic acid) aqueous solutions. The solid curves are calculated scattered intensity functions from Eq. (33). (Source: (Ref. 98) *J Phys II (France)* 3: 573, 1993. Copyright: French Physical Society.)

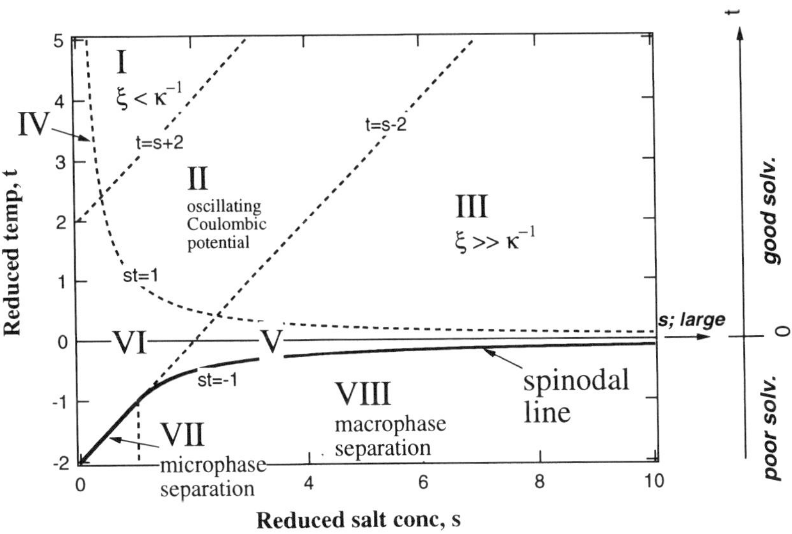

Figure 9 Borue–Erukhimovich diagram of states (BE map) for polyelectrolyte solutions. The diagram is divided into eight different regimes, I to VIII, depending on the reduced temperature, t, and the reduced salt concentration, s. The spinodal curve is shown as a thick curve. Note that the case of neutral polymer solutions correspond to $s \to \infty$.

temperature, t, and the reduced salt concentrations, s. It should be noted here that the noncharged polymer solutions correspond to the case $s \to \infty$, and the cases $t > 0$ and $t < 0$ indicate good and poor solvent regimes, respectively, as shown in the ordinate on the right-hand side. In regimes I to III, thermodynamic properties of the polyelectrolyte solution almost coincide with the Debye–Hückel ones. Regime I is the region where the Debye length κ^{-1} is larger than the Edwards screening length [3], ξ. Therefore, a strong polyelectrolyte effect appears in regime I. Regime III is, on the other hand, the high salt concentration regime, and a behavior similar to the noncharged polymer solutions is observed. Regime II is the region where two screening effects are competing and the Coulombic potential is oscillating. In regime V and VI, all excluded volume effects can be practically neglected, and the remaining Coulombic interaction becomes of great importance in regimes VI (poorly screened) and V (highly screened). In regime VII, both the repulsive interaction (Coulombic interaction) and the attractive (van der Waals and/or hydrophobic interactions) are strong enough to create a microphase-separated structure. However, in regime VIII, the repulsive interaction is highly screened by salt, resulting in a macrophase separation, similar to the case of noncharged polymer solutions. The boundary dividing phases VI–V and VII–VIII is the spinodal curve which is now a function of s. As already mentioned above, Borue and Erukhimovich divided regime VII to three regions depending on the symmetry of the microphases which would appear after microphase separation. These are lamellar, body centered cubic and two-dimensional hexagonal lattice phases. Dormidontova et al. discussed the stability of these phases and predicted that a triangular microdomain structure remains the most stable among microdomain phases of other symmetry [99]. Classification of the symmetry of the microdomain phases, however, has not been experimentally conducted yet. The importance of the two types of screening length, the excluded volume screening length (Edwards screening length [3]) ξ and the Debye screening length, κ^{-1}, is also discussed by Muthukumar as the "double screening" in polyelectrolyte solutions [100].

A more prominent scattering maximum in the structure factor than the case of poly(acrylic acid) aqueous solution was observed in temperature-sensitive weakly charged gels [38], and the corresponding solutions [101,102]. It is rather surprising that Eq. (33) well describes the structure factor of weakly charged gels as well as polyelectrolyte solutions, although the BE theory does not take into account the presence of crosslinks [38,102].

C. Microphase Separation in Weakly Charged Polymer Gels

A cruide theory for the structure factor of polyelectrolyte gels was proposed by Barrat et al. [103], which is a modification of the Joanny–Leibler theory [93] (similar to the BE theory). A comprehensive theory for weakly charged gels has recently been proposed by Rabin and Panyukov [104] (RP theory). This is a generalization of their theory on noncharged gels [105,106]. This theory deals with instantaneously crosslinked networks of Gaussian phantom chains with excluded volume. The structure factor of weakly charged gels, $S(q)$, was obtained by modifying the second virial coefficient w for noncharged gels to the cases of charged gels. The w parameter is now a function of q as follows,

$$w(q) = w_{\mathrm{NC}} + \frac{1}{s^{\mathrm{DH}}(q)} \tag{39}$$

where w_{NC} is the second virial coefficient for noncharged systems and $1/s^{DH}(q)$ is the inverse of the Debye–Hückel structure factor given by

$$\frac{1}{s^{DH}(q)} = \frac{4\pi l_B f^2}{\kappa^2 + q^2} \tag{40}$$

The parameter f here is the degree of ionization of the polymer chain, and κ^{-1} and l_B are the Debye length and the Bjerrum length as already defined in Eqs. (26) and (25), respectively. The explicit form of the structure factor, $S(q)$, which consists of the correlator of thermal fluctuations, $G(q)$, and the part of static density inhomogeneities, $C(q)$, is given by

$$S(q) = G(q) + C(q) \tag{41}$$

$$G(q) = \frac{a^{-3}\phi N g(q)}{1 + \hat{w}(q)g(q)} \tag{42}$$

$$C(q) = \frac{a^{-3}\phi N}{[1 + \hat{w}(q)g(q)]^2(1 + Q^2)^2}\left[6 + \frac{9}{\hat{w}_0 - 1 + (1/2)Q^2(\phi_0/\phi)^{2/3}\phi_0^{-1/4}}\right] \tag{43}$$

where N denotes the average degree of polymerization between crosslinks. ϕ_0 and ϕ are the volume fractions of polymer networks at preparation and at measurement, respectively. The dimensionless wavevector Q is defined by

$$Q = aN^{1/2}q \tag{44}$$

The function $g(q)$ is given as follows.

$$g(q) = \frac{1}{Q^2/2 + (4Q^2)^{-1} + 1} + \frac{2(\phi/\phi_0)^{2/3}\phi_0^{1/4}}{(1 + Q^2)^2} \tag{45}$$

The parameters $\hat{w}(q)$ and $\hat{w}_0$ are dimensionless interaction parameters in the final state (which is assumed to be in a Θ solvent or poor solvent) and in the state of preparation (in a good solvent), respectively, and are expressed by,

$$\hat{w}(q) = (1 - 2\chi + \phi)\phi N + \frac{\hat{l}_B f^2 \phi N^2}{Q^2 + \hat{l}_B(f + 2C_s)\phi N} \tag{46}$$

$$\hat{w}_0 = \phi_0^{5/4}N + \frac{\hat{l}_B f_0^2 \phi_0^{5/4} N^2}{Q^2(\phi_0/\phi)^{2/3} + \hat{l}_B f_0 \phi_0^{5/4} N} \tag{47}$$

where

$$\hat{l}_B = \frac{4\pi l_b}{a}$$

The important feature of the RP theory is that $S(q)$ of gels is described not only with the parameters at observation but also with those at preparation, e.g., ϕ_0, f_0, and $\hat{w}_0$. The RP theory predicts both macro- and microphase separations for weakly charged gels, which are now examined by comparing with SANS results [107]. This theory also predicts an anomalous cross-link density dependence of the structure factor. This aspect has recently been studied by Ikkai and Shibayama [108] and Shibayama et al. [109].

A microphase separation is induced not only by temperature but also by complexation with surfactants [110] or linear polymers [111]. A shrinking transition leads to a low permittivity state, which results from a high concentration of nonpolar polymer chains. In this case, some of the counterions may lose their mobility and condense on the corresponding co-ions and form ion pairs. Khokhlov and Kramarenko considered this case and predicted the presence of a new phase named the "supercollapsed state" of polymer gels [112]. The supercollapsed state is formed as a consequence of a positive feedback of the dielectric constant, i.e., the more collapsing, the lower is the dielectric constant. Thus ionomer multiples are progressively formed. Experimental evidence of this model was also given by the Moscow group [113,114]. Khokhlov et al. [115] also pointed out another important feature of microstructuring in the collapsed state of gels, which is the formation of glassy and kinetically frozen polymer-rich phases. This effect prevents the gel from a further collapsing. The formation of a supramolecular structure in polyelectrolyte gels in the presence of surfactants was also found in several systems, e.g., poly(diallyldimethylammonium chloride) gels with sodium alkylsulfate [116].

IV. MACROPHASE SEPARATION AND RELATED PROPERTIES

In Sec. II, we discussed the volume phase transition of gels without addressing the importance of phase separation. However, it should be realized that a volume phase transition is usually accompanied by phase separation. A macrophase separation is often observed even with the naked eye as a cloudy region formed in a gel undergoing a volume phase transition. In this section, physical properties of environment-sensitive gels are discussed in conjunction with macrophase and microphase transitions.

A. Mechanical Properties

The mechanical properties of gels are sensitive to volume phase transition and/or phase separation. As already discussed in Sec. II, the volume phase transition temperature shifts by stretching the gel. Hirotsu conducted a static deformation measurement of cylindrical NIPA gels immersed in water [72]. By loading the gel with a constant weight, the diameter and the length of the gel were measured as a function of temperature, from which the bulk modulus and Poisson ratio were evaluated. Interestingly, a softening of the gel occurred by increasing temperature and the Poisson ratio became negative at the transition temperature T_c. Accordingly, the bulk modulus had a discrete jump at T_c. This jump is successfully explained with the classical theory of the volume phase transition of gels as a result of a drastic change of the volume fraction of the network at T_c.

Shibayama et al. studied NIPA isochore gels and found a steep jump of the compressive modulus at T_c [117]. In this case, the gel was in the shrunken state and no volume change was allowed. Therefore, it is clear that the change of the modulus cannot be explained by a change of volume fraction. They interpreted this phenomenon with a concept of macrophase separation (or a formation of a "macro-network"). As a matter of fact, the gel turned opaque at T_c, above which the gel became very viscoelastic, like a sticky chewing gum. This example indicates the importance of structure change, i.e., macrophase separation, in studying the mechanical properties of gels.

B. Thermal Properties

Polymer gels consisting of monomers with strongly hydrophobic groups, such as NIPA, show a clear endotherm in a DSC (differential scanning calorimetry) curve [35,117,118]. Fig. 10 [117] shows a comparison of two DSC runs; one observed with (a) a sealed pan and the other with (b) an open pan. In the case of the latter, water evaporation during heating is allowed and the larger endotherm corresponds to the latent heat of water. On the other hand, water does not leak during a DSC run up to 60°C in the case of the sealed pan, which was confirmed by measuring the weights of the DSC pan before and after DSC runs. Therefore, it is clear that the endotherm for the open pan is solely due to dissociation of hydrophobic bonding. Shibayama et al. conducted a systematic study on two types of copolymer gels with DSC [37]. One is a copolymer gel consisting of NIPA and dimethyl acrylamide (NIPA/DMAA) which is a neutral gel and the DMAA reduces the hydrophobicity of NIPA. The other is also a copolymer gel of NIPA and acrylic acid (AAc) (NIPA/AAc), which is weakly charged due to ionization of AAc monomers. In the case of (a), the volume phase transition was continuous and T_c increased with increasing DMAA. However, as discussed in Secs. II and III, NIPA/AAc gels underwent a discrete volume phase transition with increasing temperature. It is interesting to compare the transition temperatures obtained by volumetry (i.e., swelling shrinking curve)

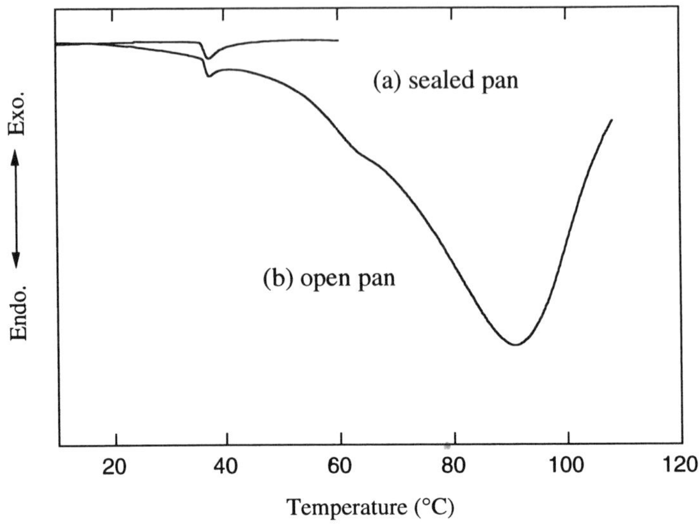

Figure 10 A comparison of two DSC runs of NIPA homopolymer gels: (a) with a sealed pan and (b) with an open pan. The large endotherm corresponds to the latent heat of water. By using a sealed pan, the enthalpy related to the dissociation of hydrophobic bonding is quantitatively evaluated. (Source: (Ref. 117) *Macromolecules* 27: 5060, 1994. Copyright: American Chemical Society.)

and by DSC. In the case of DSC, the transition temperature was defined to be the onset of the endothermic peak. The result is shown in Fig. 11. In the case of (a) NIPA/DMAA gels, both T_c values are almost identical and an increasing function with DMAA content. On the other hand, T_c of (b) NIPA/AAc gels obtained by DSC seems to be independent of AAc content while that obtained by volumetry increases almost linearly with AAc content. The comparison of the two systems clearly indicates that a microphase separation is involved in the thermal response of the NIPA/AAc gels because DSC is sensitive to the NIPA-rich domains and the NIPA domains are formed at $T > T_\Theta$. On the other hand, DMAA does not lead to a microphase separation because of lack of Coulombic interaction, and the thermal response is obtained within the framework of a random distribution of DMAA in the NIPA network (a mean-field). Thus, the discrepancy in T_c values in NIPA/AAc gels indicates again the presence of microphase separation.

DSC can also be used to monitor the structure relaxation of macrophase separated gels. Successive DSC runs of NIPA gels and of NIPA aqueous solutions aged in the phase separated state were carried out. The

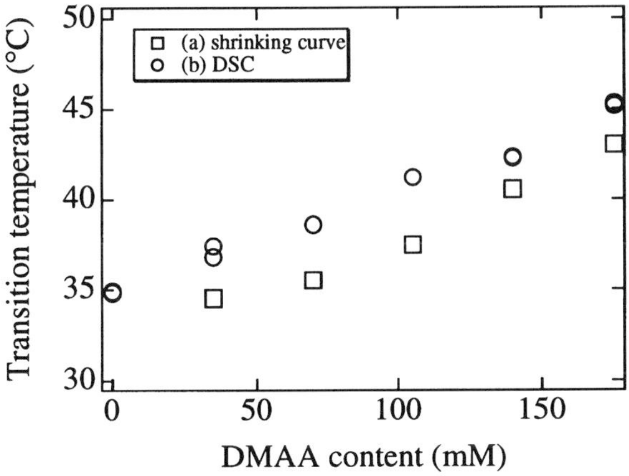

(a)

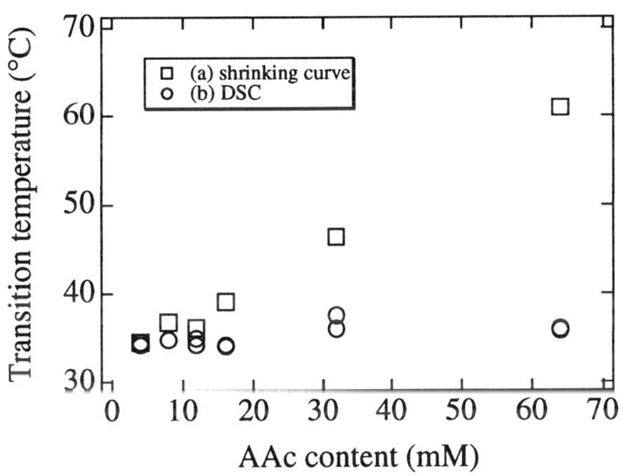

(b)

Figure 11 Transition temperatures of (a) NIPA/DMAA gels and (b) NIPA/AAc gels evaluated by volumetry (i.e., shrinking curve) and by DSC as a function of co-monomer concentration. (Source: (Ref. 37) *Macromolecules* 29:2019, 1996. Copyright: American Chemical Society.)

DSC thermogram of the NIPA gel was independent of aging times larger than 9 min. However, that of the NIPA solutions showed a strong dependence on the aging time (more than 2 h). This indicates a strong capability of recovery of gels [119] due to their cooperativity.

C. Optical and Light-Scattering Studies

As discussed in the previous section, formation of a microphase separated structure is studied with small-angle neutron (SANS) or small-angle X-ray scattering (SAXS). On the other hand, macrophase separation, which does not have a characteristic wavelength of concentration fluctuations, is commonly detected by turbidity and/or light-scattering measurements. Tanaka observed a divergence of scattered light from a polyacrylamide gel in water at -17°C [120].

A great advantage of light scattering is its capability of detecting the frozen inhomogeneities of gels. It is well known that the scattered intensity from a gel is strongly dependent on the position of the gel. However, its physical origin has not been well understood. Pusey and van Megen [121] proposed a theory of dynamic light scattering (DLS) by nonergodic media, such as glasses or gels, and stressed the importance of taking the ensemble average. A comprehensive study on DLS from gels was carried out by Joosten et al. [122]. They proposed a method to decompose the scattered intensity into the time-fluctuating (thermal fluctuations), I_F, and the frozen inhomogeneities. The intensity–intensity time correlation function for gels can be described with a single exponential function because of the cooperative nature of gels [2]. By plotting the apparent diffusion coefficient, D_A, as a function of the time-average scattered intensity, $\langle I \rangle_T$, one gets the cooperative diffusion coefficient, D, and the intensity component of the dynamic fluctuations, $\langle I_F \rangle_T$, by fitting with the following equation,

$$D_A = \frac{D}{2 - \langle I_F \rangle_T / \langle I \rangle_T} \tag{48}$$

Nishio and coworkers [123] conducted a preliminary work on a position-dependent light-scattering study on NIPA gels. Shibayama et al. studied the AAc content dependence of the thermal and frozen inhomogeneities of NIPA/AAc gels by DLS [124]. The results suggest the presence of two diffusive modes corresponding to the polymer-rich and polymer-poor phases. They modified Eq. (48) into the following form,

$$\frac{\langle I \rangle_T}{D_A} = \frac{2}{D} \langle I \rangle_T - \frac{\langle I_F \rangle_T}{D} \tag{49}$$

This allows one to conduct a linear regression analysis so as to evaluate D and $\langle I_F \rangle_T$. It should be noted here that the linear regression analysis with Eq. (49) has to be carried out with a weighting function of the form of $1/\langle I \rangle_T$ [125]. An example of the analyses is shown in Fig. 12 (a nonlinear regression with Eq. (48) and a linear regression analysis with Eq. (49) [124]. Li et al. [126] studied spinodal decomposition of NIPA gels by turbidity measurement, and found that the phase separation process consists of three stages; (1) the early stage where the Cahn–Hilliard–Cook theory [127,128] applies; (2) the association process of cascading network segments; (3) the frozen stage. Asnaghi et al. [129] conducted q-dependent static light-scattering experiments on polyacrylamide gels in polyethylene glycol and observed a spinodal-like growth of the scattered intensity with time. They found that the competition between spinodal decomposition and gelation can be profitably exploited for the fabrication of gels with exceptionally large pore size. This experimental evidence seems to support the model of the macro-

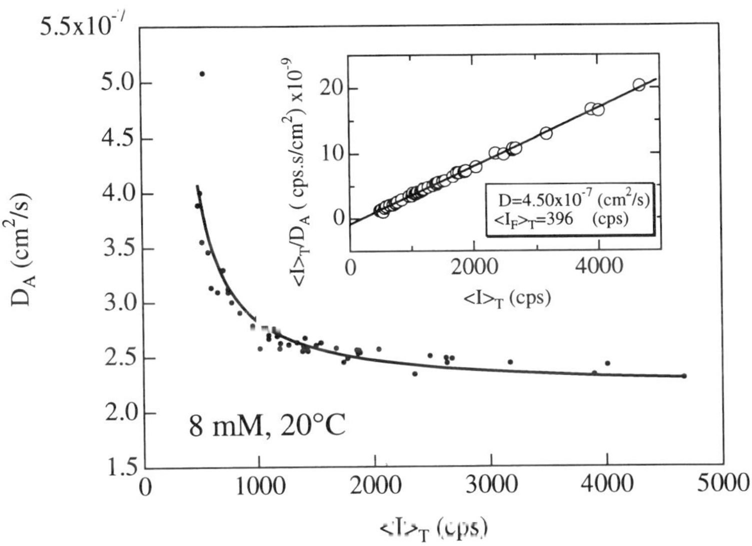

Figure 12 Evaluation of the cooperative diffusion coefficient, D, and the time-averaged fluctuating intensity component, $\langle I_F \rangle_T$, for NIPA/AAc (692 mM/8 mM) copolymer gels at 20°C. By plotting the apparent diffusion coefficient, D_A, versus the time-averaged scattered intensity, $\langle I \rangle_T$, D and $\langle I_F \rangle_T$ are obtained using Eq. (48) or Eq. (49) (the inset). (Source: (Ref. 124) *Macromolecules* 29. 6535, 1996. Copyright: American Chemical Society.)

network proposed by Shibayama et al. [117] in order to explain the mechanical transition of isochore gels. Micali et al. reported that a microphase separation takes place in a gel made of methyl-methacrylate cross-linked with ethylene-dimethacrylate [130]. The techniques they employed were depolarized Rayleigh, Brillouin, and elastic scattering. The Brillouin scattering data indicate the existence of well-defined solid-like and liquid-like islands with an extent of several hundred angstroms.

D. Spectroscopy

Nuclear magnetic resonance (NMR), infrared (IR) absorption, and ultraviolet (UV) absorption spectroscopies are useful techniques to study the local environment of gels undergoing a volume phase transition and/or phase separation. Tokuhiro et al. used NMR to study the volume phase transition of NIPA gels [131]. Shibayama et al. employed a total-attenuated IR technique to study the phase separation of NIPA gels and observed a noticeable shift of the CH_3 (on the side group) and CH_2 (on the main chain) and C = O absorption bands at T_c [117]. This indicates a drastic change of the local environment of the gel upon phase separation.

Ando and coworkers developed a ^{1}H-NMR imaging technique which allows one to visualize the spatial inhomogeneity in a gel [132]. They found that the spatial distributions of the ^{1}H spin density become lower and the molecular motion of water molecules in the gel becomes slower by compression. The response to an applied electric field was also studied [133].

Horie and coworkers studied the volume phase transition of poly(acrylamide) gels with fluorescent probes, such as probes carrying pyrenyl or dansyl groups [134,135]. They observed a shift of fluorescence peak to the shorter wavelength and an increase of the lifetime of the fluorescence. They concluded that the phase transition is accompanied by a change in the hydrophobic interaction of the main chains of the gels.

Zhou and Wu [136] studied the drying and swelling kinetics of ultrathin NIPA gel films with interferometry. The time evolution of the reflection spectra was monitored, from which the time dependence of the film thickness was evaluated. A strong temperature dependence of the drying rate was observed above and below Θ. This phenomenon is again strongly related to the phase separation.

E. Other Phenomena Related to the Phase Separation of Gels

In this subsection, we report on some of the interesting information from recent papers on gels.

1. Gels Having Unique Properties

A bi-gel, a bilayer of gel slab, was made by a sequential gel-slab preparation and its physical properties were investigated [137,138]. The bi-gel is composed of a temperature-sensitive gel laminated with a temperature-insensitive gel. Therefore, it bends upon a change of temperature like a bimetal. This type of gel may have a variety of applications, including thermo-sensitive actuators. Osada and Matsuda developed a shape memorizable gel. This gel consists of acrylic acid copolymerized with n-stearyl acrylate [139]. Below 25°C, the gel behaves as a hard plastic, but the gel becomes soft above 50°C. The Young's modulus changes by more than three orders of magnitude between these temperatures. The origin of the shape memory comes from the crystal-melt transition of the alkyl-side groups of gels.

2. Volume Phase Transition in Hydrocarbons

Annaka developed a gel which undergoes a volume phase transition in organic media [140]. The gel consists of poly(4-vinylpyridine) gel. The gel exhibited a discrete volume phase transition in benzene in the presence of tetrachloro-1,4-benzoquinone (acceptor) owing to donor–acceptor interaction.

3. Pattern Formation and Instability

Polymer gels undergoing shrinking or swelling transition sometimes show an interesting surface pattern [141–146]. Li et al. reported the temperature and time dependence of the surface pattern of NIPA gels [147] and pattern formation, e.g., hexagonal, grains, and bubbles, in acrylamide/sodium acrylate copolymer gels [148].

The dynamics of the volume phase transition has been theoretically treated from the viewpoint of thermodynamic instability. Yoshimura and Sekimoto considered the time variation of the instability of a gel triggered by changing the composition of the surrounding solution [149]. Tomari and Doi [150] studied the dynamics of the volume phase transition of a spherical gel by numerical calculation. They predicted the existence of an incubation period during which the whole gel is swollen only slightly, followed by the main process of a large volume change. Recent experimental results indicate that the volume phase transition of gels seems to be shape dependent [89]. The transition behavior of a cylindrical gel with a high aspect ratio is definitely different from that of a spherical gel. In this case, the local thermodynamics has to be applied by taking account of the surface free energy and the continuity of the gel. The theoretical consideration by Barriere and coworkers [151] may help in the understanding of this type of interesting phenomena.

V. CONCLUDING REMARKS AND FUTURE DIRECTIONS

The structure and properties of environment-sensitive gels have been discussed from the thermodynamic and microscopic points of view. It is emphasized that the well-known phenomenon of gels, the "volume phase transition," has to be discussed in connection with the fluctuations in the gel. In particular, the microphase separation is important in understanding the discrete volume phase transition. The microphase separation is formed as a result of the interplay of antagonistic interactions, such as hydrophobic and Coulombic interactions. It is unnecessary to mention that other interactions, such as hydrogen bonding, dipole–dipole interaction, and van der Waals interaction, are also very important to control the structure, phase, and stability of a gel. A combination of these interactions is significant to understand and to design multiple phases.

The kinetics of phase separation and of volume phase transition is another important issue in the understanding of gels, where the chain connectivity plays a major role as a cooperative phenomenon. Great advances in investigations of the inhomogeneity of gels have been made in the past decade. These interesting aspects are not extensively discussed in this chapter because of the space limitation. This will be treated in a review articel in the near future [152].

As introduced in Sec. I, gels have many aspects and are capable of a variety of applications. Environment-sensitive gels will be used in, for example, sensors, memory, actuators, and drug delivery systems, some of which have already been industrialized. Gels are used in the medical field as replacement and/or alternative materials for parts of the human body and/or organs owing to their softness and biocompatibility. The strong water absorbency of gels has been increasingly used in sanitary goods [153]. It is not just a dream to think that a barren desert could be turned into a fertile grain field by thermosensitive gels buried underground. Gels could supply water to the plants by shrinking in the daytime and retaining water in the night by swelling. Further understanding of gels will open the door to the development of biomimetic gels, for example in molecular recognition and motion, and commercialized high-performance and/or functionalized gels.

REFERENCES

1. P J Flory. Principles in Polymer Chemistry. Ithaca, NY: Cornell University Press, 1953.
2. T Tanaka, L O Hocker, G B. Benedek. J Chem Phys 59:5151, 1973.
3. P G de Gennes. Scaling Concepts in Polymer Physics. Ithaca, NY: Cornell University Press, 1979.
4. A Silberberg, J Elliassaf, A Katchalsky. J Polym Sci 23:259, 1957.
5. R S Stein. J Polym Sci B7:657, 1969.
6. F Bueche. J Coll Interf Sci 33:61, 1970.
7. J Bastide, S J Candau. In: J P Cohen Addad, ed. The Physical Properties of Polymeric Gels. New York: Wiley, 1996, pp. 145–308.
8. P G de Gennes. Phys Lett A38:339, 1972.
9. I M Lifshitz, A Yu Grosberg, A R Kokhlov. Rev Mod Phys 50:683, 1978.
10. A Koniglio, H E Stanley, W Klein. Phys Rev Lett 42:518, 1979.
11. M Shibayama, T Tanaka, C C Han. J Chem Phys 97:6842, 1992.
12. M Annaka, T Tanaka. Nature 355:430, 1992.
13. J-M Guenet. Thermoreversible Gelation of Polymers and Biopolymers. New York: Academic Press, 1992.
14. K Dušck, D Patterson. J Polym Sci A2, 6:1209, 1969
15. T Tanaka. Phys Rev Lett 40:820, 1978.
16. T Tanaka. Sci Am 244:110, 1981.
17. M Shibayama, T Tanaka. Adv Polym Sci 109:1, 1993.
18. T Tanaka. Polymer 20:1404, 1979.
19. S Hirotsu, Y Yoshikawa, T Tanaka. J Chem Phys 87:1392, 1987.
20. K Otake, H Inomata, M Konno, S Saito. J Chem Phys 91:1345, 1989.
21. H H Hooper, H P Baker, H W Blanch, J M Prausnitz. Macromolecules 23:1960, 1990.
22. M Marchetti, S Prager, E L Cussler. Macromolecules 23:1760, 1990.
23. M Marchetti, S Prager, E L Cussler. Macromolecules 23:3445, 1990.
24. A Khokhlov, S G Starodubtzev, V V Vasilevskaya. Adv Polym Sci 109:123, 1993.
R Moerkerke, R Koningsveld, H Berghmans, K Dusek, K Sorc. Macromolecules 28: 1103, 1995.
25. A Onuki. J Phys II (France) 2:45, 1992; Adv Polym Sci 109:63, 1993.
26. T Tanaka, D J Fillmore, S-T Sun, I Nishio, G Swislow, A Shah. Phys Rev Lett 45:1636, 1980.
27. T Katayama, A Ohta. Macromolecules 18:2781, 1985.
28. T Amiya, T Tanaka. Macromolecules 20:1162, 1987.
29. S Katayama, Y Takeshita, Y Akahori. Polymer 34:2677, 1993.
30. Y Hirokawa, T Tanaka, E S Matsuo. J Chem Phys 81:6379, 1984.
31. Y Li, T Tanaka. J Chem Phys 90:5161, 1989.
32. H Inomata, S Goto, S Saito. Macromolecules 23:4887, 1990.
33. H Inomata, S Goto, S Saito, Langmuir 8:1030, 1992.
34. K Otake, T Tsuji, M Konno, S Saito. J Chem Eng Jpn 21:443, 1988.

35. K Otake, H Inomata, M Konno, S Saito. Macromolecules 23:283, 1990.
36. M Shibayama, T Tanaka, C C Han. J Chem Phys 97:6829, 1992.
37. M Shibayama, S Mizutani, S Nomura, Macromolecules 29:2019, 1996.
38. M Shibayama, F Ikkai, S Inamoto, S Nomura, C C Han. J Chem Phys 105:4358, 1996.
39. A Hoffman. J Controlled Release 6:297, 1987.
40. Y H Bae, T Okano, S W Kim. J Controlled Release 9:271, 1989.
41. Y H Bae, T Okano, S W Kim. J Controlled Release 11:255, 1990.
42. Y H Bae, T Okano, S W Kim. J Polym Sci Polym Phys Ed 28:923, 1990.
43. T Okano. Adv Polym Sci 110:180, 1993.
44. R Yoshida, K Uchida, Y Kaneko, K Sakai, A Kikuchi, Y Sakurai, T Okano et al. Nature 374:240, 1995.
45. E Kokufuta. Adv Polym Sci 110:157, 1993.
46. M Suzuki, O Hirasa. Adv Polym Sci 110:241, 1993.
47. A Suzuki, T Tanaka. Nature 346:345, 1990.
48. A Suzuki. Adv Polym Sci 110:199, 1993.
49. A Suzuki, H Suzuki, O Sakashita, H Sakuyama. Phase Transitions 47:161, 1994.
50. A Suzuki, T Ishii, Y Maruyama. J Appl Phys 80:131, 1996.
51. Y Osada, K Umezawa, A Yamauchi. Makromol Chem 189:3859, 1988.
52. T Kurauchi, T Shiga, Y Hirose, A Okada. In: D DeRossi, K Kajiwara, Y Osada, A Yamauchi, ed. Polymer Gels. New York: Plenum Press, 1991, pp. 237–264.
53. M Doi, M Matsumoto, Y Hirose. Macromolecules 25:5504, 1992.
54. S Hirotsu, A Onuki. J Phys Soc Jpn 58:1508, 1989.
55. A Suzuki, S Kojima. J Chem Phys 101:10,003, 1994.
56. O Kondo, A Suzuki, T Shimazaki. J Interigent Mater Syst Struct 4:279, 1993.
57. E Kato, T Kitada. J Phys Soc Jpn 63:2831, 1994.
58. E Kato. J Chem Phys 101:468, 1994.
59. J E Mark, B Erman. Rubberlike Elasticity A Molecular Primer. New York: Wiley, 1988.
60. J Hasa, M Ilavsky, K Dusek. J Polym Sci Polym Phys Ed 13:253, 1975.
61. A Katchalsky, P Spitnik. J Polym Sci 2:432, 1947.
62. J Richa, T Tanaka. Macromolecules 17:2916, 1984.
63. M Seno, M L Lin, K Iwamoto. Colloid Polym Sci 269:873, 1991.
64. S T Moe, G Skjak-Braek, A Elgsaeter, O Smidstrod. Macromolecules 26:3589, 1993.
65. R A Siegel. Adv Polym Sci 109:233, 1993.
66. C Tanford. Physical Chemistry of Macromolecules. New York: Wiley, 1961.
67. G Staikos, G Bokias, K Karayanni. Polym Int 41:345, 1996.
68. A Onuki. J Phys Soc. Jpn 57:699, 1988.
69. A Onuki. J Phys Soc Jpn 57:1868, 1988.
70. S Hirotsu. J Phys Soc Jpn 56:233, 1987.
71. B Erman, P J Flory. Macromolecules 19:2342, 1986.
72. S Hirotsu. J Chem Phys 94:3949, 1991.

73. M Marchetti, S Prager, E L Cussler. Macromolecules 23:1760, 1990.
74. M Marchetti, S Prager, E L Cussler. Macromolecules 23:3445, 1990.
75. S Beltran, H H Hooper, H W Blanch, J M Prausnitz. J Chem Phys 92:2061, 1990.
76. H H Hooper, J P Baker, H W Blanch, J M Prausnitz. Macromolecules 23:1096, 1990.
77. A Yu Grosberg, S K Nechaev. Macromolecules 24:2789, 1991.
78. R Moerkerke, R Koningsveld, H Berghmans, K Dusek, K Solc. Macromolecules 28:1103, 1995.
79. I C Sanchez, R H Lacombe. J Phys Chem 80:2352, 1976.
80. R H Lacombe, I C Sanchez. J Phys Chem 80:2568, 1976.
81. R T Deam, S F Edwards. Phil Trans Roy Soc London Ser A 280:317, 1976.
82. M Annaka, D Berling, J Robert, T Tanaka. Macromolecules 26:3234, 1993.
83. M Annaka, T Tanaka. Phase Transitions 47:143, 1994.
84. M Annaka, T Tanaka. Physica A 204:40, 1994.
85. H Hirose, M Shibayama. In preparation.
86. S Hirotsu. Phase Transitions 47:183, 1994.
87. M Shibayama, H Hirose, Y Seutoh. To be submitted to Macromolecules.
88. A Yu Grosberg, A R Khokhlov. Statistical Physics of Macromolecules. New York: AIP Press, 1994.
89. J Plestil, Yu M Ostanevich, S Borbely, J Stejskal, M Ilavsky. Polymer Bull 17:465, 1987.
90. H G Shild. Prog Polym Sci 17:163, 1992.
91. Due to the isotope effect, the transition temperature of NIPA gels in D_2O is shifted by about 0.8°C from that of NIPA in H_2O (33.8°C). See Ref. 36.
92. V Borue, I Erukhimovich. Macromolecules 21:3240, 1988.
93. J F Joanny, L Leibler. J Phys France 51:545, 1991.
94. P G de Genne. J Phys Lett 40:69, 1979.
95. T Ohta, K Kawasaki. Macromolecules 19:2621, 1986.
96. L Leibler. Macromolecules 13:1620, 1980.
97. F Schosseler, A Moussaid, J P Munch, S J Candau. Macromolecules 24:225, 1991.
98. A Moussaid, F Schosseler, J P Munch, S J Candau. J Phys. II (France) 3:573, 1993.
99. E E Dormidontova, I Ya Erukhimovich, A R Khokhlov. Macromol Theory Simul 3:661, 1994.
100. M Muthukumar. J Chem Phys 105:5183, 1996.
101. M Shibayama, T Tanaka, C C Han. J Phys (France) IV Colloque C8:25, 1993.
102. M Shibayama, T Tanaka, C C Han. J Chem Phys 102:9392, 1995.
103. J-L Barrat, J-F Joanny, P Pincus. J Phys (France) 2:1531, 1992.
104. Y Rabin, S Panyukov. Macromolecules 30:301, 1997.
105. S Panyukov, Y Rabin. Phys. Rep 269:1, 1996.
106. S Panyukov, Y Rabin. Macromolecules 29:7960, 1996.
107. F Ikkai, M Shibayama. Submitted to Macromolecules.
108. F Ikkai, M Shibayama. Phys. Rev E 56:R51, 1997

109. M Shibayama, F Ikkai, Y Shiwa, Y Rabin. J Chem Phys 107:5227, 1997.
110. A R Khokhlov, E Yu Kramarenko, E E Makhaeva, S G Starodoubtsev. Macromol Theory Simul 1:105, 1992.
111. A R Khokhlov, E Yu Kramarenko. Macromol Theory Simul 2:169, 1992.
112. A R Khokhlov, E Yu Kramarenko. Macromol Theory Simul 3:45, 1994.
113. O E Philippova, T G Pieper, N L Sitnikova S G Starodoubtsev, A R Khokhlov, H G Kilian. Macromolecules 28:3925, 1995.
114. S G Starodoubtsev, A R Khokhlov, E L Sokolov, B Chu. Macromolecules 28:3930, 1995.
115. A R Khokhlov, O E Philippova, N L Sitnikova, S G Starodoubtsev. Faraday Discuss 101:125, 1995.
116. B Chu, F Yeh, E L Sokolov, S G Starodoubtsev, A R Khokhlov. Macromolecules 28:8447, 1995.
117. M Shibayama, M Morimoto, S Nomora. Macromolecules 27:5060, 1994.
118. H Inomata, K Nagahama, S Saito. Macromolecules 27:6459, 1994.
119. M Shibayama, Y Suetoh, S Nomora. Macromolecules 29:6966, 1996.
120. T Tanaka, S Ishiwata, C Ishimoto. Phys Rev Lett 38:771, 1977.
121. P N Pusey, W van Megen. Physica A 157:705, 1989.
122. J G H Joosten, J L McCarthy, P N Pusey. Macromolecules 21:6690, 1991.
123. Y Suzuki, K Nozaki, I Nishio. J Chem Phys 97:3808, 1992.
124. M Shibayama, Y Fujikawa, S Nomura. Macromolecules 29:6535, 1996.
125. M Shibayama, T Norisue, S Nomura. Macromolecules 29:8746, 1996.
126. Y Li, G Wang, Z Hu. Macromolecules 28:4194, 1995.
127. J W Cahn, J H Hilliard. J Chem Phys 28:258, 1958.
128. H E Cook. Acta Metall 18:297, 1970.
129. D Asnaghi, M Giglio, A Bossi, P G Righetti. J Chem Phys 102:9736, 1995.
130. N Micali, C Vasi, F Mallamace, R Bansil, S Pajevic, F Sciortion. Phys Rev E 48:4501, 1993.
131. T Tokuhiro, T Amiya, A Mamada, T Tanaka. Macromolecules 24:2936, 1991.
132. H Yasunaga, H Kurosu, I Ando, Y Li, G Wang, Z Hu. Macromolecules 25:6506, 1992.
133. T Shibuya, H Yasunaga, H Kurosu, I Ando. Macromolecules 28:4377, 1995.
134. Y Hu, K Horie, H Ushiki, F Tsunomori, Y Yamashita. Macromolecules 28:7324, 1992.
135. Y Hu, K Horie, H Ushiki, T Yamashita, F Tsunomori. Macromolecules 26:1761, 1993.
136. S Zhou, C Wu. Macromolecules 29:4998, 1996.
137. Z Hu, X Zhang, Y Li. Science 269:525, 1995.
138. X Zhang, Z Hu, Y Li. J Chem Phys 105:3794, 1996.
139. Y Osada, A Matsuda. Nature 376:219, 1995.
140. M Annaka, T Tanaka, Y Osada. Macromolecules 25:4826, 1992.
141. T Tanaka, S T Sun, Y Hirokawa, S Katayama, J Jucera, Y Hirose, T Amiya. Nature 325:796, 1987.
142. E Sato-Matuo, T Tanaka. J Chem Phys 89:1695, 1988.
143. E Sato-Matuo, T Tanaka. Nature 358:482, 1992.

144. A Onuki. J Phys Soc Jpn 57:703, 1988.
145. K Sekimoto, K Kawasaki. J Phys Soc Jpn 57:2597, 1988.
146. K Sekimoto, K Kawasaki. Physica A 154:384, 1989.
147. Y Li, C Li, Z Hu. J Chem Phys 100:4637, 1994.
148. C Li, Z Hu, Y Li. J Chem Phys 100:4645, 1994.
149. K Yoshimura, K Sekimoto. J Chem Phys 101:4407, 1994.
150. T Tomari, M Doi. Macromolecules 28:8334, 1995.
151. B Barriere, K Sekimoto, L Leibler. J Chem Phys 105:1735, 1996.
152. M Shibayama. Macromol Chem. In preparation.
153. Y-Q Zhang, T Tanaka, M Shibayama. Nature 360:142, 1992.

7

Shear Dependence of the Equilibrium and Kinetic Behavior of Multicomponent Systems

Alan I. Nakatani and Charles C. Han
National Institute of Standards and Technology, Gaithersburg, Maryland

I. INTRODUCTION

Equilibrium studies of the phase behavior of polymer blends and kinetic studies of the phase separation process have greatly improved our fundamental understanding of multicomponent polymeric materials. However, this improved understanding of polymer phase behavior in the commercially important arena of polymer processing has not been fully implemented. This is partly because the effect of shear rate, one of the principal processing variables, on the phase behavior and blend structure has not yet been studied in a systematic manner. Since morphological control is the ultimate goal of many manufacturing processes, it is essential that the influence of shear rate on polymer blend phase behavior be well characterized.

The first fundamental studies of phase separation utilized simple binary liquid mixtures, which display Ising-type behavior [1]. It has been recognized that critical fluctuations can be suppressed by shear flow, leading to a true shift of the critical temperature, T_c [2,3]. In the case of polymeric materials, however, which no longer display Ising-like behavior but mean-field behavior [4], it is not clear whether a true shift in T_c can be obtained by the application of shear flow. If a shift in T_c is observed with shear rate in polymer mixtures, studies of the basic mechanism of this shift and the underlying physics are warranted. In particular, the phase behavior under the influence of shear flow must be compared with the equilibrium phase diagram and other thermodynamic variables under quiescent conditions.

With the application of shear to a phase separated system, one intuitively expects that shear-induced mixing will occur in the sample. For example, a bottle containing phase-separated layers of oil and vinegar, when shaken vigorously, forms small, intermixed droplets and becomes more homogeneous. An understanding of the behavior of polymeric samples during shear has proven to be less intuitive. Early experiments on semidilute polymer solutions under shear demonstrated that for certain polymer–solvent pairs, shear-induced turbidity (phase separation) occurred [5–7]. These observations have spawned theoretical treatments [8,9] to explain this phenomenon and recently there have been a number of experimental studies exploring this behavior in more detail [10–12]. A review has been published by Rangel-Nafaile et al. on the stress-induced phase behavior of polymer solutions [13].

The behavior of a mixture of two homopolymers is the focal point of this chapter. In prior work on various two-component polymer blends, both shear-induced phase separation and shear-induced mixing behavior have been observed. One might argue that the observed differences in behavior are attributable to the initial state of the blend. Alternatively, some of these differences may be ascribed to the experimental technique and the length scale which is probed by the specific method. Therefore, the choice of instrumentation as well as the thermal and mechanical history of the sample play an important role in the design and interpretation of experiments. Larson has written an extensive review on flow-induced mixing and demixing, polymer migration in shearing flows, stress-induced diffusion, and flow effects on crystalline, liquid crystalline, and block copolymer ordering transitions [14] and Tirrell has also reviewed the phase behavior of flowing polymer mixtures [15]. The interested reader is referred to these reviews and the references cited therein for extensive background material and specific details of previous experimental work.

A wide variety of experimental methods are available for studying flow-induced structures in polymers. Light scattering has long been utilized in the characterization of polymers and there is an enormous amount of literature available on the subject. A book by Chu is a good source for reference material and contains a detailed discussion of light-scattering techniques [16]. Due to the incident wavelength, light-scattering techniques are capable of probing length scales on the order of 1 μm or larger. In order to probe smaller length scales, shorter wavelength probes must be utilized. Small-angle neutron scattering (SANS) and small-angle X-ray scattering are the methods of choice to examine size scales on the order of 10–1000 nm. Higgins and Benoit have addressed the applications of SANS to polymers [17]. Numerous instrumental devices have become available to examine flow-induced behavior in polymers and a review chapter by Nakatani discussed many recent developments in instrumentation [18]. Optical techniques have also been employed for studying different aspects of polymer orientation. Optical rheometry techniques have been reviewed by Fuller [19], while an in-depth coverage of flow birefringence techniques has been written by Janeschitz-Kriegl [20]. A monograph edited by Søndergaard and Lyngaae-Jørgensen contains general information and reviews on the rheo-physics of mutiphase systems [21].

In this chapter, we will describe and summarize recent progress in research on the phase separation behavior of multicomponent polymer samples. We will begin by reviewing some basic aspects of polymer blend phase behavior under quiescent conditions. Relevant theories will be discussed and related experimental data will be provided. We will also consider the phase-separation behavior of a two-component polymer blend with a block copolymer added as an interfacial modifying agent. Following the discussions of quiescent phase-separation behavior in the pure and modified blends, the behavior of these systems under the influence of a simple shear field will be examined. The experimental data have been obtained on a variety of different blend systems (mostly, polystyrene and polybutadiene) and will be utilized to address the following issues:

1. How is the phase behavior of a two-component polymer blend altered by the addition of a block copolymer modifier? What differences in behavior are observed with the application of shear?
2. How does shear flow suppress the composition fluctuations in the mean-field and Ising cases, as well as in the crossover region toward Ising-like behavior?
3. How is this suppression of fluctuations (in the miscible region) related to the T_c shift, and how does the measured suppression compare with the predictions developed for the Ising case by Onuki and coworkers [22–25]?
4. How does the behavior in the one-phase region compare with the phase behavior in the two-phase region under shear and can the observed differences be reconciled?

Both light scattering and SANS will be utilized as complementary techniques to approach these questions. The adequacy of techniques such as the invariance and correlation length traditionally used for analysis of scattering from phase-separated samples will be evaluated for the shear-rate dependence of the phase diagram, and compared with literature data. The steady shear light-scattering patterns formed in the two-phase region will be discussed as a function of quench depth and shear rate. Subsequently, various deformation sequences will be utilized to probe the dynamics of mixing and demixing under shear in comparison with the quiescent behavior. In addition, data from light-scattering instrumentation coupled with phase contrast optical microscopy (PCOM) and fluorescence microscopy (FM) to study the steady state and the kinetics after a "shear quench" in the two-phase region will also be presented. The fluorescence microscopy is specifically introduced for the measurement of the composition difference, $\Delta\phi$, in phase-separated blends.

II. QUIESCENT BEHAVIOR

A. Two-Component Polymer Blends

Much of the characterization of polymer blend equilibrium phase behavior has traditionally been conducted under quiescent conditions. The major emphasis of this work has been precise measurement of thermodynamic variables such as the free energy of mixing, the chemical potential, the temperature and composition dependence of the phase diagram, and in the mean-field case, the polymer–polymer interaction parameter, χ. Other studies have concentrated on the kinetics of the phase-separation process via spinodal decomposition. In the early stages of spinodal decomposition, the linearized theory of Cahn–Hilliard–Cook (CHC) [26–28] and some modified theories [29], which include nonlinearities due to q-mode coupling for polymers, have been used to analyze the data. However, only scaling arguments have been utilized to interpret the late stage phase-separation kinetics. We will begin this section with a discussion of polymer blend behavior in the one-phase region, the crossover from mean-field to Ising-like behavior, and finally the kinetics of phase separation.

SANS as an experimental technique for probing the structure of polymeric materials has been widely available since the 1970s. In conjunction with the evolution of experimental techniques, de Gennes [4] developed the random phase approximation (RPA) for calculating the static structure factor for polymer mixtures. This is a mean-field, self-consistent calculation, which gives the static structure factor, $S(q)$, for a mixture of polymers, A and B, as:

$$S(q) = k_n/\{[1/(\phi_A N_A v_A f(u_A))] + [1/(\phi_B N_B v_B f(u_B))] - 2\chi/v_0\} \tag{1}$$

where

$$k_n = N_0(a_A/v_A - a_B/v_B)^2 \tag{2}$$

with N_0 being Avogadro's number; a_i the scattering length per mole of monomer i; and v_0, and v_i the molar volumes of a reference unit cell and of the i-th monomer, respectively. The single-chain structure factor, $f(u_i)$, may be represented by the Debye function:

$$f(u_i) = (2/u_i^2)(\exp(-u_i) - 1 + u_i) \tag{3}$$

where $u_i = q^2 R_{gi}^2$, and q is the magnitude of the scattering wavevector

$$q \equiv |\mathbf{q}| = (4\pi/\lambda)\sin(\theta/2) \tag{4}$$

where λ is the wavelength of the incident beam, and θ is the scattering angle. R_{gi}, N_i, and ϕ_i are the radius of gyration, the degree of polymerization, and the volume fraction of polymer i, respectively.

In 1983, Herkt-Maetzky and Schelten [30] were the first to demonstrate for binary single-phase polymer mixtures that the correlation length, ξ, the susceptibility, $S(q = 0)$, and the interaction parameter, χ, can be obtained from the temperature dependence of experimental SANS data by Eq. (1). This procedure is the most commonly employed for the analysis of SANS data of binary polymer blends. Modified theories, which include more detailed molecular structures and correlations, have been discussed in the lattice cluster expansion theory (LCET) by Dudowicz and Freed [31–33], and PRISM theory by Schweizer and coworkers [34,35]. Also, the assumption of incompressibility used in the original de Gennes derivation has been reconsidered by Sanchez [36], Hammouda and Benmouna [37], and Taylor et al. [38]. In general, the main problem of these analysis techniques is the requirement of a specific model to fit a systematic set of extensive results. Hence, no simple, reliable analysis and feedback for experimental data is possible, even with the current computational availability. However, where specific models and experimental data coexist, agreement between theoretical predictions and experiments are quite good. For example, Dudowicz and Freed [31–33] have conducted a detailed analysis of results obtained by Han and coworkers on a blend of deuterated polystyrene (PSD) and poly(vinyl methyl ether) (PVME). When only an extensive series of experimental data with composition and temperature dependence is available for a given pair of polymers, the Flory–Huggins-type of free energy of mixing can still be empirically reconstructed through rigorous thermodynamic relations. This has been demonstrated by Han et al. [39], although the original meaning of the mean-field enthalpic interaction of χ parameter is obscured.

Another important question related to the mean-field approximation is the temperature crossover from the mean-field to the Ising region. Deviation from mean-field behavior has been observed by many groups [40–46]. Two discrepancies have been reported in terms of the mean field and RPA interpretations of the scattering results for polymer blends. In the first discrepancy, the crossover from mean-field to Ising behavior appears to occur at a temperature too distant from T_c. This gives an Ising region wider than the original prediction by de Gennes [47], Joanny [48], and Binder [49]. The second disagreement involves the bare correlation length, which, most notably for the PSD/PVME case, is too small compared to the mean-field prediction of $\xi_0 = (R_g^2/3)^{1/2}$ [39,50]. However, after a closer investigation of the experimental results, this discrepancy can be explained by an additional entropic contribution in the phenomenological representation of the χ parameter, such as $\chi = A/T + b$. Since this b-term can be negative (as in the case of PSD/PVME), it can change the magnitude of the coefficient of the squared order-parameter term in the Ginzberg–Landau representation [44] of the free energy expression:

$$f(\xi) = f_0 + \int d^3x((1/2)a\varepsilon\xi^2 + (1/4)u\xi^4 + (1/2)\kappa|\nabla\xi|^2) \tag{5}$$

with $\varepsilon = (1 - T_c/T)$ and $\xi(x) = \delta\phi(x)$, the order parameter. By comparing the Ginzberg–Landau free energy expression with the Flory–Huggins form of the free energy representation, one obtains an expression for a:

$$a = (k_B T_c/v)((N_A\phi_A)^{-1} + N_B^{-1}(1-\phi_A)^{-1} - 2b) \tag{6}$$

Since the bare correlation length, $\xi_0 = (\kappa/a)^{1/2}$ and

$$\kappa = (k_B T/v)(R_{g,A}^2/N_A\phi_A + R_{g,B}^2/N_B(1-\phi_A)) \tag{7}$$

this changes the bare correlation length by a ratio of:

$$((1/N_A\phi_A + 1/N_B(1-\phi_A))/(1/N_A\phi_A + 1/N_B(1-\phi_A) - 2b))^{1/2} \tag{8}$$

This changes the crossover temperature from mean-field to Ising response at

$$\begin{aligned}\varepsilon &= (1 - T_c/T_x)\\ &\sim v^2[N_A^{-1}\phi_A^{-3} + N_B^{-1}(1-\phi_A)^{-3}]^2/\{[(N_A\phi_A)^{-1}\\ &\qquad + N_B^{-1}(1-\phi_A)^{-1} - 2b]^4(\xi_0)^6\}\end{aligned} \tag{9}$$

Although these explanations rely upon a functional form for the χ parameter which is phenomenological, the arguments place the mean-field and RPA representations on a consistent basis.

The kinetics of the spinodal decomposition process have a direct bearing on the shear mixing/demixing behavior which will be described later in this chapter. We will first give a brief overview of spinodal decomposition studies of polymeric systems [49,51–53]. As a binary mixture is suddenly brought from its miscible state to an unstable state which is defined by

$(\partial^2 \Delta f / \partial \phi^2) \sim 0$, where Δf is the free energy of mixing, the system undergoes spontaneous phase separation. The current understanding of the dynamics of spinodal decomposition can be divided roughly into three stages: *early*, *intermediate*, and *late*.

1. The Early Stage

At the beginning of the early stage of spinodal decomposition near the critical composition, ϕ_c, the amplitude of the composition fluctuations, $|\phi(r, t) - \phi_c|$ is sufficiently small compared with the final equilibrium composition difference, $|\phi_A - \phi_B|$, and the kinetics can be described by the Cahn–Hilliard–Cook theory [26–28], as extended by de Gennes [52], Pincus [53], and Binder [49] for polymers. In this case, the time evolution of the structure factor, $S(q, t)$, after quenching into the spinodal region, can be described by:

$$\partial S(q, t)/\partial t = -2Mq^2\{[(\partial^2 \Delta f / \partial \phi^2)_0 + 2\kappa q^2]S(q, t)\} + 2Mk_B T q^2 \tag{10}$$

where M is the mobility defined as the proportionality constant in the relationship between the interdiffusion current density and the chemical potential gradient, and κ is an interfacial free energy coefficient defined as the proportionality constant in the relationship between the interfacial free energy density and the square of the concentration gradient. The last term is the thermal noise from the Cook [28] representation with an amplitude which is the same as that obtained from equilibrium fluctuation-dissipation theory.

The solution of Eq. (10) is given by:

$$S(q, t) = S_\infty + (S_0(q) - S_\infty(q))\exp(2R(q)t) \tag{11}$$

where $S_\infty(q)$ is the virtual structure factor:

$$S_\infty(q) = k_B T/((\partial^2 \Delta f / \partial \phi^2)_0 + 2\kappa q^2) \tag{12}$$

and

$$R(q) = -Mq^2((\partial^2 \Delta f / \partial \phi^2)_0 + 2\kappa q^2) \tag{13}$$

In general, the mobility, M, for polymers should be q-dependent. In the case of $qR_{gi} < 1$, M can be approximated as a constant, as is the case in most light-scattering studies. The expression for the interfacial free energy density, κ, was obtained by de Gennes from a comparison of the CHC linear theory and the RPA calculation of the static structure factor:

$$\kappa = (k_B T/36)(b_A^2/v_A \phi_A + b_B^2/v_B(1 - \phi_A)) \tag{14}$$

The interdiffusion coefficient, $D_{\text{int}} = M(\partial^2 \Delta f / \partial \phi^2)_0$, can be obtained from Eq. (13) as the intercept in a plot of $R(q)/q^2$ versus q^2 (often referred to as the Cahn plot). The value of D_{int} is negative for the case of fluctuation growth and positive for the case of diffusion.

The phase separation of polymeric systems [40,54–56] has been shown to be consistent with the linearized solution of the mean-field formulation of spinodal decomposition by CHC. In Fig. 1(a) [56], the time-dependent structure factor for a deuterated polycarbonate/poly(methyl methacrylate)

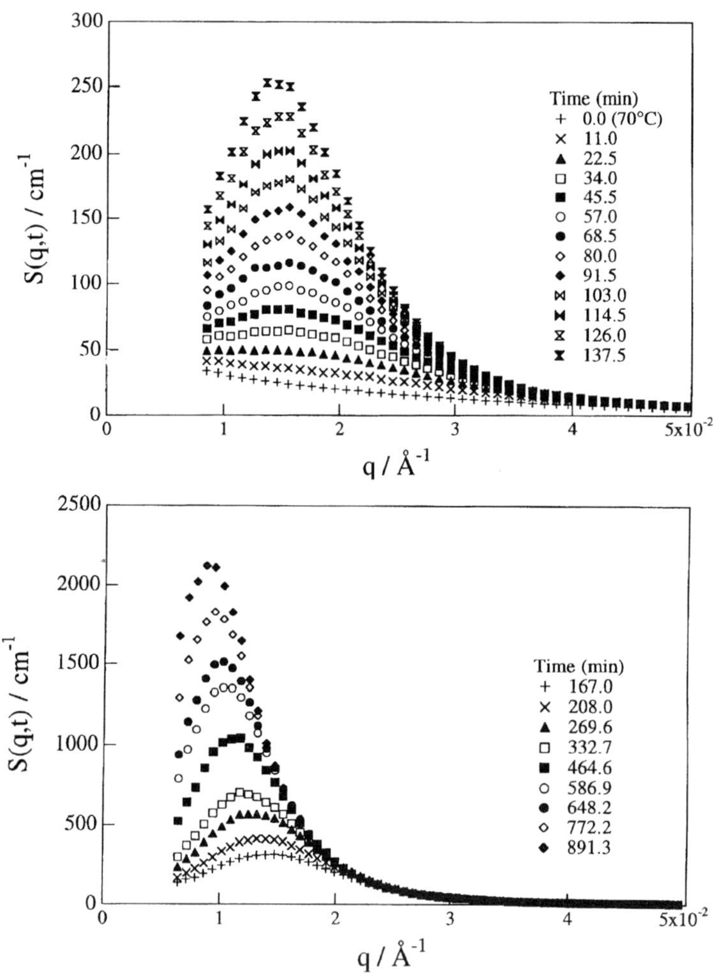

(a)

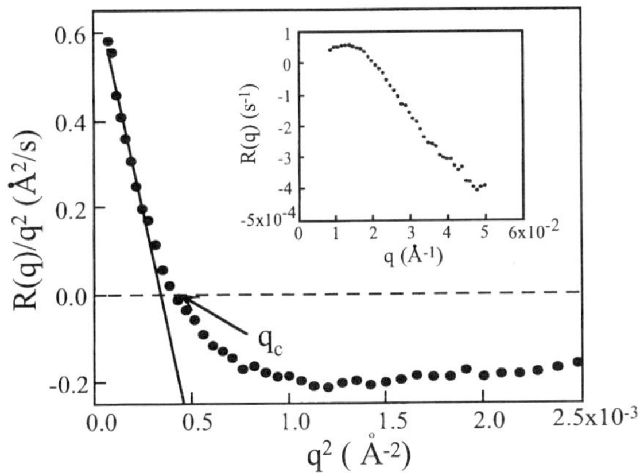

(b)

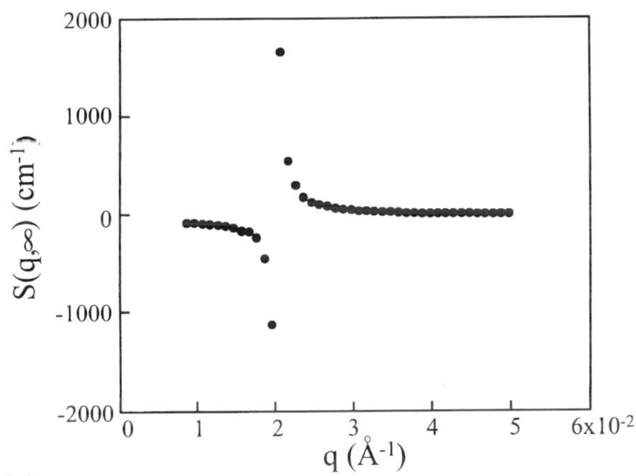

(c)

Figure 1 (a) The time-dependent structure factor for a 50:50 mass ratio deuterated polycarbonate/poly(methylmethacrylate) mixture undergoing phase separation in the spinodal region as measured by SANS. Top—early time; bottom—late time. Note the change in intensity scales. (b) Cahn plot to obtain the interdiffusion coefficient, D_{int}, and the virtual structure factor for the data shown in (a). The inset shows the rate of growth as a function of q. (c) Virtual structure factor obtained from (b) and Eqs. (12) and (13).

mixture undergoing phase separation in the spinodal region was measured by SANS. The initial growth rate of the concentration fluctuations can be followed from the intensity of the q-invariant peak position in the structure factor which can be analyzed according to Eq. (11). The rate of growth as a function of q, $R(q)$, the interdiffusion coefficient, D_{int}, and the virtual structure factor, $S_\infty(q)$, have been obtained according to Eqs. (12) and (13) as shown in Figs. 1(b) and 1(c) [56]. For q values greater than q_c ($q_c^2 = q_{max}^2/2$), $R(q)$ becomes negative and deviates from a straight line. This negative value of R reflects the diffusive motion of the individual chain segments instead of droplet coalescence, as dimensions smaller than the single chain dimensions are probed. The curvature in $R(q)$ can be modeled by different approaches for diffusive molecular motion [49,53].

2. The Intermediate Stage

As the amplitude of the concentration fluctuations continues to increase, the characteristic length scale also increases with time. In general, one observes a sharpening of the spinodal peak in the scattering structure factor and a gradual movement toward smaller q values. Because the amplitude of the concentration fluctuations is still changing during this time, the interfacial distribution is also continuously sharpening, and one does not expect to observe self-similar structures. However, a universal scaling of reduced size and time has been observed in polymeric systems [57,58]. In this scaling analysis, experimental results are reduced in a plot of the reduced wavenumber, Q_{max}, versus reduced time, τ, with:

$$\tau = t/t_c \tag{15a}$$

$$t_c = (Dq_{max}^2(t=0))^{-1} \tag{15b}$$

and

$$Q_m(\tau) = q_{max}(\tau)/q_{max}(t=0) \tag{15c}$$

Although this kind of scaling analysis may show some universal behavior, a direct analytical solution for the mode coupling calculation of the time-dependent scattering structure factor of binary liquids has been obtained by Langer et al. [59]. This has been extended to polymeric systems by Akcasu et al. [29].

3. The Late Stage

At the onset of the late stage of phase separation, the interface has developed into an equilibrium profile, and the compositions of the microdomains, which may be co-continuous, have reached their equilibrium coexistence compositions, ϕ_A and ϕ_B. As the system continuously coarsens in this stage, it probably follows a mechanism first suggested by Siggia [60]. Through hydrodynamic processes, the co-continuous structure is disrupted and coarsens with time in a self-similar fashion. Experimentally, one would expect to observe the spinodal decomposition peak intensity and position in the scattering structure factor change with time in the following manner:

$$q_{\max}(t) \sim (t)^{-\alpha} \tag{16a}$$

$$S(q_{\max}, t) \sim (t)^{\beta} \tag{16b}$$

and

$$\beta = d\alpha \tag{16c}$$

where d is the dimensionality ($d = 3$ for a volume filled three-dimensional system and $d < 3$ for a less dense, fractal structure). Rigorously, in this self-similar growth regime the scattering structure factor should also be self-similar:

$$S(q, t) = (q)^{-d} f(q/q_{\max}) \tag{16d}$$

$$= (q)^{-d} g(t/t_{\max}) \tag{16e}$$

In other words, plots of $S(q, t)q^d$ versus $q/q_{\max}$ for various t or of $S(q, t)q^d$ versus $t/t_{\max}$ for various q should yield a universal structure factor. For more details the interested reader is referred to Chapter 2 in this volume by Furukawa.

B. Phase Separation Kinetics with Block Copolymers

Another important topic in polymer blends is the effect of adding a third component as a compatibilizer/interfacial modifier. A detailed discussion on theories of phase segregation in copolymer and homopolymer mixtures is given by Olvera de la Cruz in Chapter 1 of this book. Technologically, the primary interest is to induce mixing and to retard or arrest further coarsening during processing and utilization. Fundamentally, equilibrium theories and simulation [61–63] have indicated that distribution of the block copolymer, A–B, is always at the immiscible interface of a phase-spearated A/B blend. However, static experiments [64] are not always in agreement with this prediction. The two major effects which need to be considered in the use of block copolymers for compatibilization are: (1) the thermodynamic effect of increasing the miscibility gap (lowering the T_c in the case of a UCST), and (2) the kinetic effects of enhanced droplet break up in a flow field as well as a slowing down of the phase separation and coarsening kinetics, presumably due to the modification of the interface and a lowering of the interfacial tension by the block copolymer. Prior experimental studies which examine the phase behavior of compatibilized blend systems have been summarized by Roe and Rigby [65] and by Hashimoto [54]. In this section, we will first discuss the influence of the block copolymer molecular weight* relative to the homopolymer on the partitioning and the shift of T_c in an A/B/A–B system relative to the A/B blend under quiescent conditions. Also, the kinetics of phase separation due to the influence of block copolymer at the interface will be discussed. The break up of droplets due to shear and the slowing down in the kinetics of the phase separation after a shear quench will be addressed in Sec. IV.

By using SANS [46] and light-scattering experiments, Sung and Han [66] have examined the effect of block copolymers on the phase behavior of a two-component polymer blend. A relatively low molecular weight blend of polystyrene ($M_w = 1 \times 10^3\,\mathrm{g\,mol^{-1}}$) and polybutadiene ($M_w = 5 \times 10^3\,\mathrm{g\,mole^{-1}}$) was utilized to demonstrate the phase diagram of the blend shifts downward with the addition of a PSD–polybutadiene (PB) diblock copolymer (M_w of each block $= 5 \times 10^3\,\mathrm{g\,mol^{-1}}$) as shown in Fig. 2(a) [66]. This depression shows a linear dependence with concentration in a fashion similar to low molecular weight diluents. The exact magnitude of the change in T_c (T_c of the pure blend $= 50.8^\circ\mathrm{C} \pm 0.4^\circ\mathrm{C}$)† depends on the concentration and the molecular weight of the block copolymer added. The T_c shift is shown in Fig. 2(b) [46] as ΔT versus the mass fraction of two different molecular weight block copolymers in the same blend. While there appear to be systematic changes with copolymer molecular weight in the slopes shown in Fig. 2(b), there is insufficient data to draw any firm conclusions.

The kinetic behavior of the same system has also been studied by temperature jump time-resolved light-scattering and transmission electron microscopy (TEM) measurements [67,68]. The rate of spinodal decomposition slows as the mass fraction of the block copolymer concentration increases to 3%. Fig. 3(a) [66] shows the time dependence of the scattering

---

*According to ISO 31-8, the term "molecular weight" has been replaced by "relative molecular mass," symbol M_r. Thus, if this nomenclature and notation were to be followed in this publication, one would write $M_{r,n}$, instead of the historically conventional M_n for the number-average molecular weight, with similar changes for M_w, M_z, and M_v, and it would be called the "number average relative molecular mass." The conventional notation, rather than the ISO notation, has been employed for this publication.

†All ± values cited within the text are estimated standard uncertainties based on multiple prior measurements with the equipment utilized.

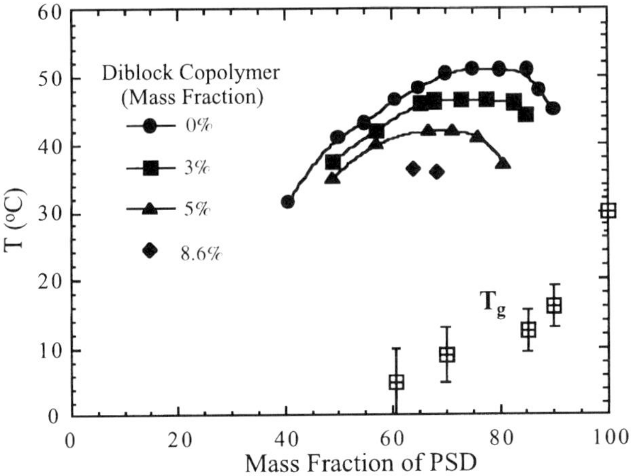

(a)

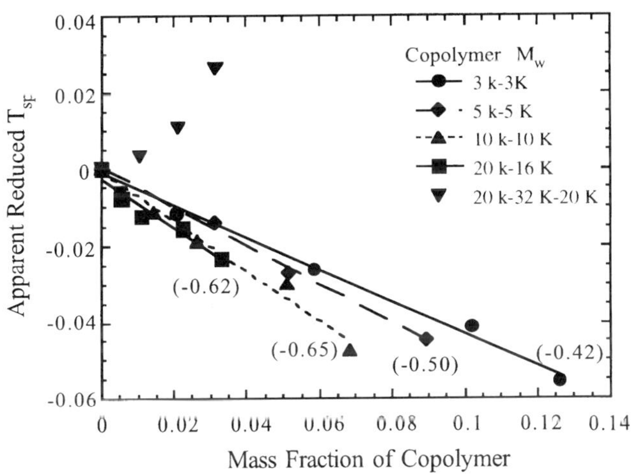

(b)

Figure 2 (a) Phase diagram of PSD ($M_w = 1 \times 10^3\,\mathrm{g\,mol^{-1}}$) and PB ($M_w = 5 \times 10^3\,\mathrm{g\,mol^{-1}}$) as a function of diblock copolymer concentration (PSD-PB, M_w of each block $= 5 \times 10^3\,\mathrm{g\,mol^{-1}}$). The glass transition temperatures (T_g) of pure blends is also shown. (b) Shift in the apparent spinodal temperature, T_{sp}, as a function of mass fraction of a styrene-butadiene block copolymer for various copolymer architectures and molecular weights.

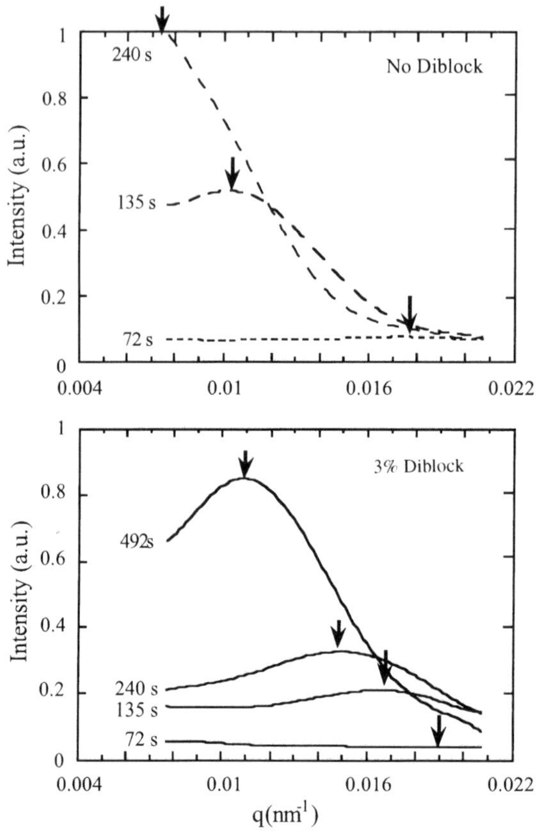

(a)

Figure 3 (a) $I(q, t)$ versus q from temperature jump light-scattering measurements for a blend of PS and PB without (top) and with diblock copolymer (bottom). Mass ratio of PS:PB is 75:25 in both samples. Both samples were quenched to an equivalent quench depth, $T_{sp} - T_f = 1.0°C$. Long-time data for the blend without block copolymer are unavailable due to the high scattering intensity and the peak position inside the beamstop. (b) Time dependence of $I(q_m)$ (bottom) and q_m (top) for various mass fractions of diblock copolymer. The mass ratio of the PS and PB homopolymers is constant (70:30) for all diblock concentrations. All samples quenched to an equivalent quench depth, $T_{sp} - T_f = 1.6°C$.

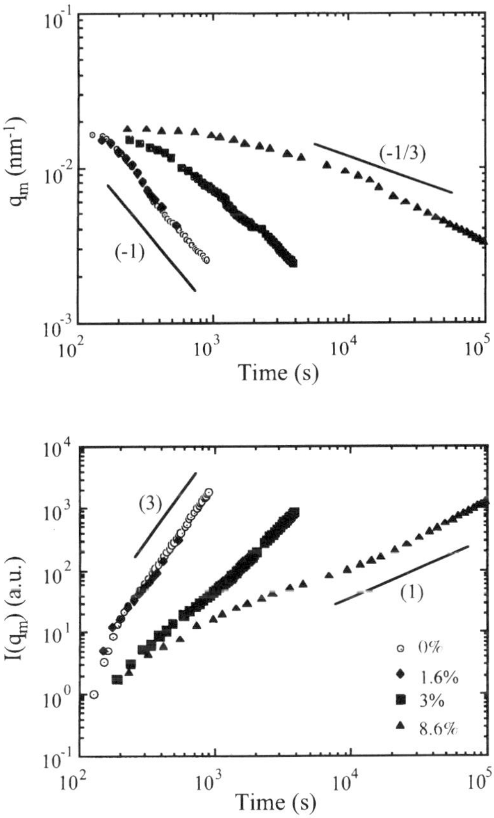

(b)

structure factor for the pure blend and a blend with 3% mass fraction of block copolymer quenched into the unstable region. In Fig. 3(b) [66], the time dependence of q_{max} and $I(q_{max})$ for different block copolymer concentrations are shown. For copolymer concentrations between 1.6 and 8.6 mass % the scaling relationships given in Eq. (16) are not observed. The spatial distribution and the partitioning of the block copolymer between the interface and the domain interior influence these results. TEM has also been used to examine samples with different block copolymer contents, quenched to the same dimension (q_m) during the late stage. It is clear from these TEM studies that phase separated domains have less well defined shapes and interfaces at higher block copolymer concentrations. These results also help to explain the nonuniversal coarsening behavior demonstrated by most binary blends in the intermediate and late stages of spinodal decomposition. Further studies, both experimental and theoretical, are needed in order to understand the distribution, and partitioning of the block copolymers and also the kinetics and structures which occur during the phase-separation process.

III. SHEAR IN THE ONE-PHASE REGION

A. Shear Rate Dependence of Fluctuation Dimensions

SANS has been used to study the concentration fluctuations of binary polymer mixtures under shear. Two different polymer systems: PSD/PVME and deuterated polystyrene/polybutadiene (PSD/PB) were studied as a function of temperature and shear rate [69]. The two samples had contrasting characteristics which provide some different tests. The PSD/PVME blend consisted of high molecular weight polymers (PSD $M_w = 2.7 \times 10^5\,\mathrm{g\,mol^{-1}}$, PVME $M_w = 1.8 \times 10^5\,\mathrm{g\,mol^{-1}}$). The phase behavior of the PSD/PVME blend exhibits lower critical solution temperature (LCST) behavior and a critical composition (80:20 mass ratio of PSD:PVME) was used in the study. The PSD/PB blend consisted of low molecular weight polymers (PSD $M_w = 1 \times 10^3\,\mathrm{g\,mol^{-1}}$, PB $M_w = 4.5 \times 10^3\,\mathrm{g\,mol^{-1}}$) and the phase diagram exhibits upper critical solution temperature (UCST) behavior. An off-critical composition of the PSD/PB blend was used (50:50 mass ratio of PSD:PB). Therefore, comparisons of the shear rate dependence of entangled versus unentangled chain molecules, LCST versus UCST behavior, and critical versus off-critical compositions were encompassed in a single study.

Due to the small wavelength of the incident neutron radiation compared with light, the shear dependence of concentration fluctuations in the one-phase region and in the strong shear limit was obtained from the q dependence of the scattering structure factor. Qualitative examination of the scattering structure factors parallel to the flow direction (x-axis) and the vorticity axis (z-axis) indicated a suppression of the scattering along the flow direction, while along the vorticity axis the structure factors are invariant to shear, for the PSD/PVME case. The PSD/PB blend shows suppression of fluctuations along the vorticity axis as well. From a detailed analysis of the scattering structure factor, $S(q)$, a crossover value of the wavenumber, q_s, was obtained as a function of temperature and shear rate. This crossover wavenumber represents the inverse size scale of the lowest fluctuation mode which is not affected by shear. Plots of log q_s against $\log(\eta\dot{\gamma}/T)$ yielded a linear plot with a slope of 1/3 for all data examined. The temperature, viscosity, and shear-rate dependence of these experimentally determined q_s values agree well with a simple rotary diffusion model and also the dynamic mode–mode coupling analysis of Kawasaki [70] and Ferrell [71]. The apparent spinodal temperature as a function of shear rate was shown to be consistent with the prediction of Onuki [24].

It is important to note that suppression of fluctuations only occurs for the PSD/PVME blend along the flow direction. Therefore, the apparent spinodal temperature along the flow direction is different from the apparent spinodal temperature along the vorticity axis. The presence of two different spinodal temperatures in the same sample depending upon the coordinate frame of measurement is physically unrealistic. This is contrasted by the behavior of the low molecular weight PSD/PB blend which demonstrates an influence by flow along both the flow and vorticity axes. This difference is attributed to the crossover from mean-field (PSD/PVME blend) to Ising-like (PSD/PB) behavior and is discussed further in the following section.

B. Crossover from Weak Shear to Strong Shear

In the Onuki–Kawasaki analysis for the shear-rate dependence of the critical dynamics [22–25], a renormalization group calculation predicts the shear-rate dependence of the critical temperature ($T_c(\dot{\gamma})$) as:

$$[T_c(\dot{\gamma}) - T_c(0)]/T_c(0) \approx r[0.0832\varepsilon + \mathbf{O}(\varepsilon^2)](\dot{\gamma}\tau_c)^{1/3\nu} \tag{17}$$

where $T_c(0)$ is the equilibrium critical temperature, $r = 1 - T_c(0)/T$ is the equilibrium reduced temperature with T being the experimental temperature, $\nu = 0.63$ is the Ising correlation length exponent, and $\tau_c = \xi^2 S(0)/\Lambda_0$ is the characteristic lifetime of the equilibrium concentration fluctuations.

This prediction has been studied with a low molecular weight PSD/PB blend [44]. Dynamic light scattering (DLS) was used to measure the equilibrium decay rate of the concentration fluctuations, $\Gamma_c \approx D_c q^2$ [72]. The influence of critical fluctuations, or "mode coupling," on the dynamics is apparent in the diffusion coefficient, $D_c = \Lambda_c/S(0)$, and the characteristic lifetime of the critical fluctuations can be calculated as:

$$\tau_c = \xi^2/D_c \tag{18}$$

for various temperatures above the equilibrium critical temperature, T_c, where the correlation length is measured by SANS and D_c by DLS. At temperatures where critical fluctuations are important, $(T \to T_c)$, $S(q, \dot{\gamma})$ will depend on $\dot{\gamma}$ through the shear rate dependence of T_c. The equilibrium critical fluctuations were measured by SANS, and the data displayed a crossover from mean-field to Ising-critical behavior roughly 10 K from the equilibrium critical temperature, $T_c(\dot{\gamma} = 0) \approx 41.2°C \pm 0.1°C$. From SANS measurements of the blend under steady shear (*in situ*), the susceptibility, $S(0, \dot{\gamma})$, and the correlation length $\xi(\dot{\gamma})$, can be extracted by fitting to the properly configured structure factor calculated from Onuki–Kawasaki theory [73]. The quantities $S(0, \dot{\gamma})$ and $\xi(\dot{\gamma})$ can be written as:

$$S(0, \dot{\gamma}) \sim [1 - T_c(\dot{\gamma})/T]^{-\gamma} \tag{19a}$$

and

$$\xi(\dot{\gamma}) \sim [1 - T_c(\dot{\gamma})/T]^{-\nu} \tag{19b}$$

where $\gamma = 2\nu$.

The break-up of large-scale critical fluctuations by flow leads to a drop in critical temperature described by Eq. (17), and the shear-rate dependence of the susceptibility and correlation length can be written as:

$$S(0, \dot{\gamma}) = S(0)[1 + \Delta r/r]^{-\gamma} \tag{20a}$$

and

$$\xi(\dot{\gamma}) = \xi(0)[1 + \Delta r/r]^{-\nu} \tag{20b}$$

with $\Delta r = [T_c(\dot{\gamma}) - T_c(0)]/T_c(0)$. The reduced susceptibility, $S_{\text{red}} = S(0, \dot{\gamma})/S(0)$, and the reduced correlation length, $\xi_{\text{red}} = \xi(\dot{\gamma})/\xi(0)$, are then given in terms of the reduced shear rate, $\sigma = \dot{\gamma}\tau_c$, by

$$S_{\text{red}} = \xi_{\text{red}}^2 \approx \{1 + [0.0832\varepsilon + \mathbf{O}(\varepsilon^2)]\sigma^{1/3\nu}\}^{-2\nu} \tag{21}$$

The quantities S_{red} and ξ_{red} are direct measures of the influence of shear on the critical fluctuations. If the flow has no effect, S_{red} and ξ_{red} are both 1, as under quiescent conditions. When the flow starts to suppress the long-wavelength critical fluctuations, S_{red} and ξ_{red} begin to decrease, reflected by the decrease in the low-q scattering intensity. In Fig. 4 [73] we have plotted S_{red} and ξ_{red}^2 as a function of σ for all of the shear rates and temperatures used in this study. The solid line is the behavior predicted by the right-hand side of Eq. (21) to first order in ε. Since S_{red}, ξ_{red}, and σ are all directly known, there are no free parameters in this fit. The agreement is excellent over six decades of reduced shear rate.

The steady shear SANS behavior of an off-critical blend solution was also examined [74]. A suppression of concentration fluctuations both parallel and perpendicular to the flow direction was also observed with the application of sufficiently large shear rates. The structure factors were

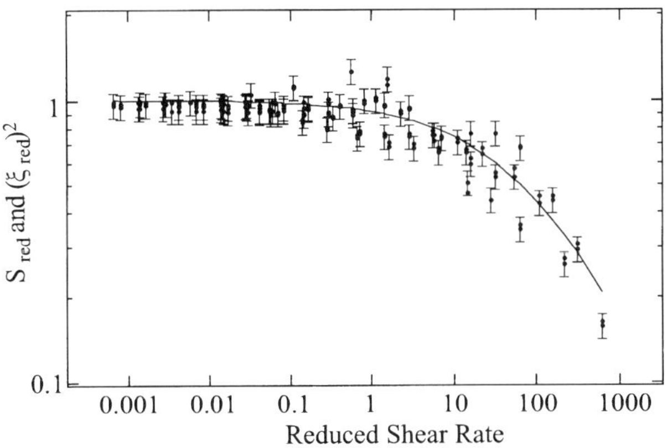

Figure 4 The reduced susceptibility and reduced correlation length squared as a function of the reduced shear rate for a PSD/PB blend. The solid line is the behavior predicted for $\mathbf{O}(\varepsilon)$ over six decades in a reduced shear rate, with no free parameters.

also fit to the Ornstein–Zernike function to extract values of the correlation length, ξ, and $S(q = 0)$ as a function of temperature and shear rate. DLS was used in the same fashion as for the analysis of the undiluted polymer blend discussed in the previous paragraph [75]. Agreement with the Onuki–Kawasaki predictions for this sample is also very good over the two decades of reduced shear rate covered in these experiments.

Figure 4 reveals the very intuitive feature that flow begins to break apart the long-wavelength critical fluctuations when the shear rate becomes comparable to the equilibrium relaxation rate of these fluctuations. This is evidenced by the fact that S_{red} and ξ^2_{red} begin to deviate from unity when the reduced shear rate $\sigma = \dot{\gamma}\tau_c$ becomes comparable to unity, or when $\dot{\gamma} \sim \tau_c^{-1}$. It is natural that the size of the critical fluctuations would be limited by the shear for $\dot{\gamma} \geq \tau_c^{-1}$, since any component with an equilibrium lifetime greater than $\dot{\gamma}^{-1}$ will essentially be suppressed by the flow. However, one important question we must ask is, "What is the shear-rate dependence of $T_c(\gamma)$ if we cross the phase boundary and start our experiment in the quiescent two-phase region?". In other words, if we start our experiment at a temperature, $\Delta T(= T_c(0) - T)$, below the quiescent critical temperature, how does the phase-separated system change in structure from two phases to one phase and also how does the critical temperature, $T_c(\gamma)$, change as a function of shear rate? We address this question in the following section.

IV. SHEAR IN THE TWO-PHASE REGION

A. Steady Shear and the Shift in the Phase Diagram

1. Light Scattering

Due to the large size of the intermediate and late stage phase-separated morphology in polymer blends, scattering studies in the two-phase region have been confined to light scattering. A number of different research groups have probed various aspects of multicomponent systems under flow conditions using this technique. Extensive research in this area has been conducted by Hashimoto and coworkers [76–82]. Many of our experiments and work by other research groups, can also be interpreted qualitatively in a fashion similar to that proposed by Hashimoto and coworkers, and we briefly summarize those results to place the remainder of this section in context.

For polystyrene/polybutadiene blends dissolved in dioctyl phthalate (DOP), in the two-phase region, Hashimoto has delineated five basic types of behavior under steady shear. It is presumed that DOP does not preferentially solvate either homopolymer, hence the sample is treated as a quasi-binary mixture. The first two regimes are defined by zero shear or very weak shearing conditions where the sample is macroscopically phase separated (Regime I). At rest, due to density differences, the sample segregates into polystyrene-rich and polybutadiene-rich layers. An incident laser beam normal to these layers produces very little scattered light intensity. In the very weak shear regime (Regime II), the layers break up into droplets, or a highly interconnected structure which displays very little anisotropy due to flow. The light-scattering patterns in this regime are also fairly isotropic. In Regime III, the droplets become elongated and anisotropy in the scattering patterns is observed. Parallel to flow, the scattering intensity is diminished, while perpendicular to flow, the scattered intensity is constant or slightly increased. In Regime IV, the scattering intensity perpendicular to flow also become diminished and the scattering anisotropy is highly pronounced. This has been interpreted as the onset of remixing, where the concentration difference between the two phases is being reduced. The highly elongated droplets observed in Regimes III and IV have been referred to as a "string phase." Finally, in Regime V, no scattered light intensity is observed except for some low intensity, critical scattering, and the sample is presumed to be homogenized on the length scales probed by light scattering.

Although many details of the behavior outlined by Hashimoto above require quantitative evaluation, this behavior has also been observed more generally in undiluted blends of PS/PB [83]. More detailed experiments and attempts to quantitatively interpret the domain elongation and break-up, and the shift of the phase diagram with shear flow in the PS/PB blend will be described in the next section.

2. Phase Contrast Optical Microscopy

A specially constructed rheometer cell with a quartz cone and plate or parallel plate fixtures was used for combined light scattering/phase contrast microscopy/fluorescence microscopy studies. Polystyrene ($M_w = 96 \times 10^3\,\mathrm{g\,mol^{-1}}$) and polybutadiene ($M_w = 22 \times 10^3\,\mathrm{g\,mol^{-1}}$) was used with DOP as a solvent. The total concentration of the polymers was a mass fraction of 8%. The critical composition of PS/PB/DOP is a 30/70 mass ratio of PS/PB. The quiescent critical temperature, $T_c(0) = 68^\circ\mathrm{C} \pm 0.2^\circ\mathrm{C}$.

In Fig. 5, phase contrast micrographs of the domain morphology as a function of shear rate are presented. At low $\dot{\gamma}$, droplets reach a steady state with minimal deformation. With increasing $\dot{\gamma}$, the drops elongate and break repeatedly. Isolated droplets with extremely high aspect ratios begin to appear at around $20\,\mathrm{s^{-1}}$. With a further increase in shear rate, string-like patterns emerge. At these higher shear rates or for smaller quench depths, the contrast between the two phases (droplet and matrix or string and matrix) decreases, indicating a decrease in the composition difference between the two phases. This structural transition can be explained [84] by combining the macroscopic Taylor model [85] and the symmetry of critical scaling from our previous results [73] and the Onuki–Kawasaki theory [22–25]. In the Taylor model, when the viscosities of the two phases are comparable, the aspect ratio $\zeta = R_{\|}/R_{\perp}$ can be expressed as:

$$\zeta = (1 + \tau_c\dot{\gamma})/(1 - \tau_c\dot{\gamma} + \ldots) \tag{22}$$

$$\tau_c = \eta R_0/\sigma \tag{23}$$

where τ_c is the characteristic stress-relaxation time of a droplet with viscosity, η, diameter, R_0, and interfacial tension, σ. At a threshold shear rate, $\gamma_c \sim \sigma/2\eta R_0$, the stress becomes comparable to the capillary pressure and the droplet breaks [85,86]. If we take an ideal situation that one droplet breaks into two identical droplets, with conservation of volume, we can obtain a set of recursion relations between R, τ_c, and $\dot{\gamma}_c$. On the other side, the surface tension in a critical fluid is $\sigma \approx k_B T/\xi^2$, where ξ is the thermal correlation length [86]. Combining the above two theories (Eqs. (22) and (23)) are assuming the symmetry is conserved below T_c, then

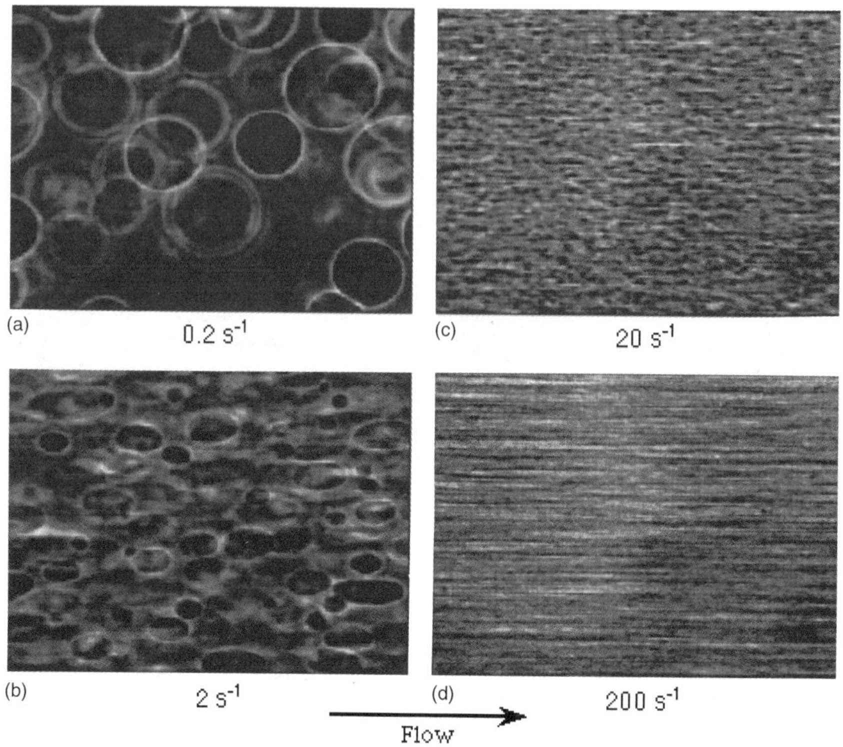

Figure 5 Digital video micrographs of a coexisting critical mixture (30:70 mass ratio of PS:PB, 8% total mass fraction of polymer in DOP at 60°C, $T_c - T_{exp} = 6\,K$) under steady shear flow. The total width of each image covers 200 μm and the velocity gradient is normal to the image. Initially, spherical phase separating droplets are stabilized by flow at low shear rates (a). They are stretched by the shear stress until they break and the process repeats itself (b and c). The string phase pattern emerges around $200\,s^{-1}$.

$$\sigma(\dot{\gamma}) \approx \sigma_0\{1 - A(\tau_\Psi\dot{\gamma})^{1/3\nu}\}^{2\nu} \tag{24}$$

σ_0 is the limiting value of the interfacial tension as $\dot{\gamma} \to 0$. In Eq. (24), A is an amplitude (~ 0.0832), ν is the critical exponent associated with ξ, and τ_Ψ is the characteristic time of the order parameter fluctuation. This softening of interfacial tension due to shear-induced mixing leads to a change of τ_c in Eq. (23). With the approximation of using the initial τ_c for τ_Ψ, the shear rate dependence of the droplet break up and the aspect ratio can be calculated as shown in Fig. 6 [84]. Fig. 6(a) shows results obtained for the blend solution ($T_{exp} = 62°C$) [84], while Fig. 6(b) shows results for a

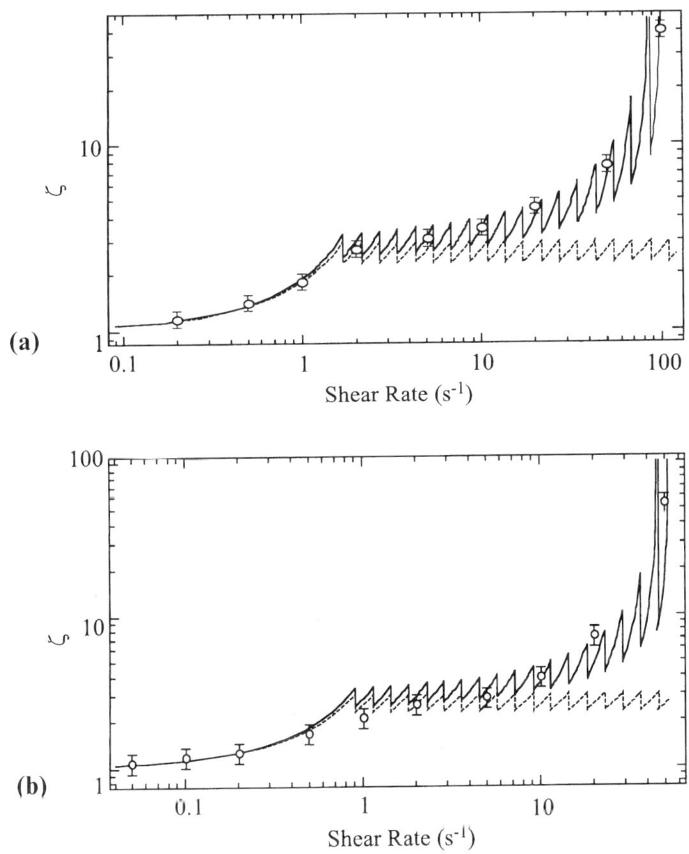

Figure 6 Double logarithmic plots of the string order parameter, ζ, as a function of shear rate for the data in Fig. 5. The lower curve represents an idealized breaking sequence for a noncritical droplet (Taylor), the upper curve represents an idealized breaking sequence for a critical dispersion (Onuki), and the markers represent average aspect ratios calculated from the digital micrographs. (a) Data for a blend solution (8 mass % total polymer) of PS ($M_w = 96 \times 10^3\ \mathrm{g\,mol^{-1}}$) and PB ($M_w = 22 \times 10^3\ \mathrm{g\,mol^{-1}}$) dissolved in DOP (mass ratio of 30:70 PS:PB). (b) Data for an undiluted blend of PS ($M_w = 3 \times 10^3\ \mathrm{g\,mol^{-1}}$) and PB ($M_w = 3 \times 10^3\ \mathrm{g\,mol^{-1}}$) (mass ratio of 60:40 PS:PB).

PS/PB blend (M_w PS $= 3 \times 10^3$ g mol^{-1}, M_w PB $= 3 \times 10^3$ g mol^{-1}, 60:40 mass ratio of PS:PB, $T_c = 134.2$°C, $T_{exp} = 132.7$°C) without solvent [83]. The similarity in the two sets of results is striking. The string-like behavior in both cases emerges at $\dot{\gamma} \sim 200\,s^{-1}$. At even higher shear rates, the contrast between the string phase and the background diminishes. This is obvious from the total light-scattering intensity (not shown here). In order to study the shear rate dependence of the critical temperature shift, fluorescence microscopy is utilized.

3. Fluorescence Microscopy

In the above study of the morphological changes (droplet elongation, break-up, and string pattern formation), a symmetry of critical scaling, above and below the critical temperature, is assumed. In other words, it is presumed that the Onuki–Kawasaki theory for the shift in the critical temperature is equally applicable with respect to T_c and the critical exponent, ν, is constant. A substantiation of these assumptions is required. Furthermore, if the shear rate dependence of T_c can be measured, then we may be able to better understand the thermodynamics of phase behavior far into the quiescent immiscible region at steady state, especially when this steady state is generated under a nonpotential shear flow field.

We have demonstrated that the shear rate dependence of the critical temperature shift, $\Delta T_c \equiv T_c(\dot{\gamma} = 0) - T_c(\dot{\gamma})$, can be obtained from fluorescence microscopy images of the two-phase blend under shear flow [87]. Using fluorophore-labeled polystyrene ($M_w = 110 \times 10^3$ g mol^{-1}) and unlabeled polybutadiene ($M_w = 22 \times 10^3$ g mol^{-1}), the change in the fluorescence intensity ratio of the two phases under steady shear was obtained and related to the change in ratio of compositions of the two phases at different shear rates. By supposing that the coexistence curve is shifted by shear without changing shape (the critical exponent, β, remains constant), the change in the coexistence composition yields the change in critical temperature with shear rate, $\Delta T_c = T_c(\dot{\gamma} = 0) - T_c(\dot{\gamma})$ [87]. The supposition that the exponent β remains constant, hence the shape of phase diagram is not changed, can also be validated. In Fig. 7, a typical fluorescence microscope image (a) is displayed together with the corresponding phase contrast microscope image (b) for a fluorophore labeled PS/PB/DOP system (50:50 mass ratio PS:PB, 8 mass % total polymer) at 22°C and a shear rate of $0.005\,s^{-1}$ ($T_c(\dot{\gamma} = 0) = 86.5°C \pm 0.5°C$). In this case, the quiescent phase diagram can be depicted according to the following relation:

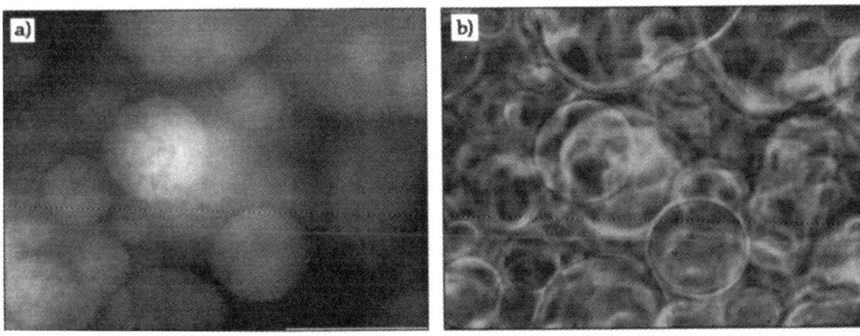

Figure 7 Microscopy images of the PS/PB/DOP system at 22°C at a shear rate of $0.005\,s^{-1}$. (a) fluorescence image; (b) phase contrast image.

$$\begin{aligned}\Delta\phi(\dot{\gamma}=0,\mathrm{T}) &= 0.19|T-T_c|^{0.35}\\ &= 0.90|1-T/T_c|^{0.35}\end{aligned} \tag{25}$$

By assuming the shear rate dependence of the critical temperature has the following form:

$$\Delta T_c(T,\dot{\gamma}) \equiv (T_c(\dot{\gamma}=0) - T_c(T,\dot{\gamma})) \sim \dot{\gamma}^{\nu} \tag{26}$$

and the shape of the phase diagram remains the same as under the quiescent case, the composition differences obtained from the fluorescence microscopy, $\Delta\phi$, can be expressed by the following equation:

$$(\Delta\phi(\dot{\gamma},T)/0.19)^{(1/0.35)} - T_c(\dot{\gamma}=0) + T(\dot{\gamma}) = -B(\gamma)^{y} \tag{27}$$

The exponent, y, was determined to be 0.42 ± 0.10, which is slightly less than but still consistent with the value of $1/3\nu = 0.53$ by using $\nu = 0.63$ in the Onuki–Kawasaki theory. In Fig. 8 [87], experimental data for temperatures of 22, 30, 40, and 50°C and the quiescent data are displayed in a double-logarithmic form according to the equation:

$$\Delta\phi(T,\dot{\gamma}) = A'|1 - T(\dot{\gamma})/T_c(\dot{\gamma})|^{0.35} \tag{28}$$

The consistency between the quiescent data and the data obtained at various temperatures and shear rates is clear. However, better accuracy of data is required in order to get a more precise value of the critical exponent. Also, results from different multicomponent samples are necessary to prove the universality of this observed behavior.

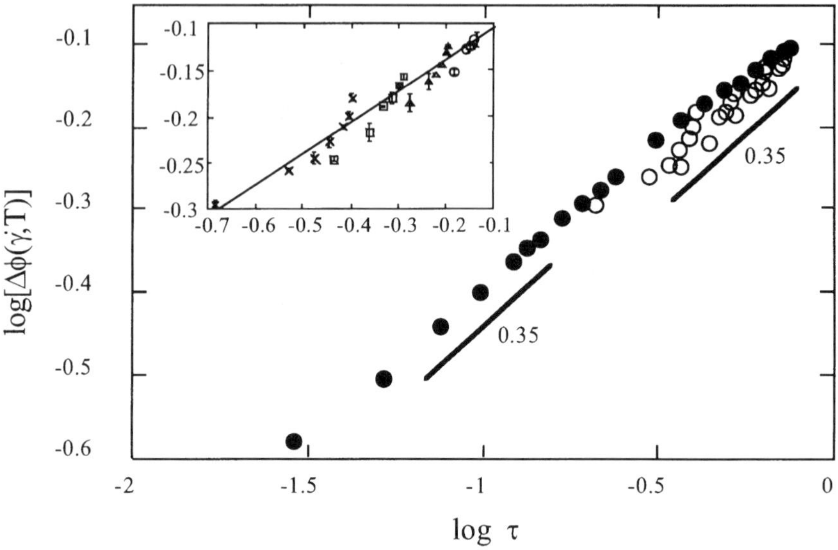

Figure 8 Phase composition data obtained from fluorescence microscopy and the deduced shear rate dependence of $T_c(T, \dot{\gamma})$ displayed as $\log [\Delta\phi(\dot{\gamma}, T)]$ versus $\log \tau$ ($\tau = |1 - T(\dot{\gamma})/T_c(T, \dot{\gamma})|$). Quiescent data (solid symbols) are shown for reference. The inset shows the temperature dependence of the shear results, where the bars on the symbols represent one standard deviation. The standard uncertainty of each point on the expanded scale is less than or equal to 0.05: ○—22°C; Δ—30°C; □—40°C; ×—50°C.

B. Behavior After Cessation of Shear

Kinetic studies of these polymer blend solutions after cessation of flow have also been conducted. The kinetics of the transition from an apparent single phase back to a two-phase mixture has also been studied extensively by Takebe et al. [78]. The phase-separation kinetics behavior was found to be isotropic with respect to the original flow direction. A "spinodal ring' was observed after cessation of shear (sometimes referred to as a "shear quench" experiment) and the peak position and intensity of the peak scale as expected for the intermediate to late stages of spinodal decomposition described above in Sec. II. A.3. We have examined the scattering behavior following cessation of shear for initial shear rates less than the critical shear rate for homogenization on a 50:50 mass ratio PS ($M_w = 89 \times 10^3\,\mathrm{g\,mol^{-1}}$) and PB ($M_w = 22 \times 10^3\,\mathrm{g\,mol^{-1}}$) blend in DOP (8 mass % total polymer). The critical temperature for this sample was

34°C ± 0.5°C and all experiments were performed at ambient laboratory temperature (21°C). The relaxation behavior, in these cases, was anisotropic for all shear rates examined. Qualitatively, three different typical relaxation scattering profiles were identified (Fig. 9). At the lowest shear rates ($< 50\,s^{-1}$), the anisotropic scattering patterns persist for very long times (~ 10 min) before becoming isotropic. Between 150 and $250\,s^{-1}$, the initial "string" pattern relaxes quickly to a "lozenge"-type scattering pattern which relaxes much more slowly (~ 5 min) to an isotropic pattern. Above $300\,s^{-1}$, a "spinodal ring" is observed. A bright streak of scattered light normal to flow superimposed on the ring in some cases, indicated that the critical shear rate for homogenization was not achieved. The

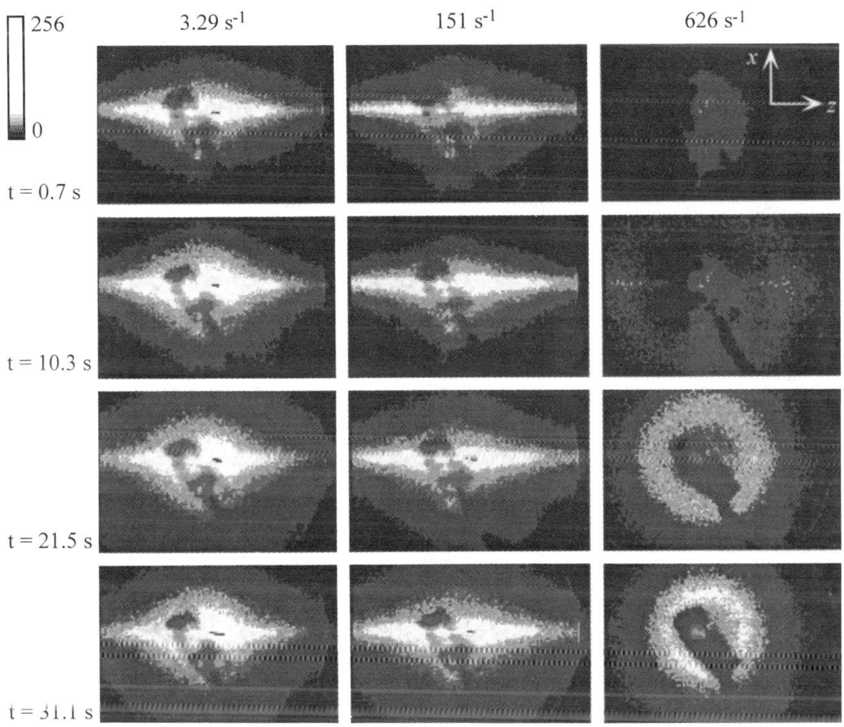

Figure 9 Light-scattering patterns after cessation of shear demonstrating three types of relaxation behavior. Patterns evolve from top to bottom in each column. Initial shear rates are given at the tops of each column and the elapsed time after cessation is given on the left

spinodal ring migrates to the minimum q value of the instrumentation in approximation 30 s. Therefore, as the value of the initial shear rate increases, the time for the scattering to become isotropic decreases. This observation conflicted with the anticipation that a system which experiences a small perturbation should return to equilibrium faster than the same system which has undergone a large perturbation. From the results described above, it appears that a greater degree of anisotropy induced at high shear rates is shorter lived than the lesser degree of anisotropy produced at lower shear rates. Hence, the production of anisotropic (oriented) two-phase materials may actually be improved by using lower shear rates. For the processing of two-phase materials, this result has important implications because two-phase materials traditionally have been oriented by applying high shear rates to a sample to induce and preserve morphological order. However, more recent results indicate that after cessation of shear rates which generate string patterns, the break-up of the strings into necklace-like patterns can be observed by PCOM. The small bead-like domains remain aligned, giving rise to a bright scattering streak in the vorticity direction, and randomize slowly. Concurrently, a spinodal decomposition-like ring also begins to appear. Whether the spinodal decomposition-like ring represents spinodal decomposition within the phase separated domains or within the matrix has yet to be conclusively demonstrated.

C. Two-Step Shear Experiments

It is possible that differences in the prior mechanical history or the presence of pre-existing structures may be the source of the slow relaxation behavior after cessation of the lower shear rates described above. To test this hypothesis, the phase-separation kinetics during a two-step, steady shearing process were examined [88]. The same sample as described in the previous section was utilized for these studies. In these experiments, the sample was allowed to reach steady state at a high shear rate ($> \dot{\gamma}_c$), then "shear-quenched" to a lower shear rate. A delay time, t_d, was employed between the two shear rates, in order to control the amount of structure development before the second shear rate, $\dot{\gamma}_2$, is applied. The samples were exposed to two types of shear histories: (1) Two-step shear, variable t_d—the delay time, t_d, was varied and $\dot{\gamma}_2$ was fixed; the data were taken only after cessation of $\dot{\gamma}_2$ so the effect of t_d could be monitored; (2) two-step shear, variable $\dot{\gamma}_2$—the values of $\dot{\gamma}_2$ were varied and the delay time was fixed at various delay times; data were taken immediately after stopping $\dot{\gamma}_1$ so that the coarsening behavior during the delay, during $\dot{\gamma}_2$, and after cessation of $\dot{\gamma}_2$ were monitored.

The results from the two-step shear experiments can be described qualitatively according to the growth patterns for the scattered light intensities as a function of scattering angle. Two basic types of growth patterns were observed (Fig. 10) [88]. In the first type of behavior, which we term "*q*-dependent growth", the scattered intensity as a function of time grows at different rates as a function of *q*. The *q*-dependent growth pattern is similar to the patterns observed during spinodal decomposition. The second type of growth pattern, we call "*q*-independent growth." For this growth pattern, the scattered light intensities increase as a function of time at approximately the same rate, regardless of the *q* value. The *q*-independent growth is

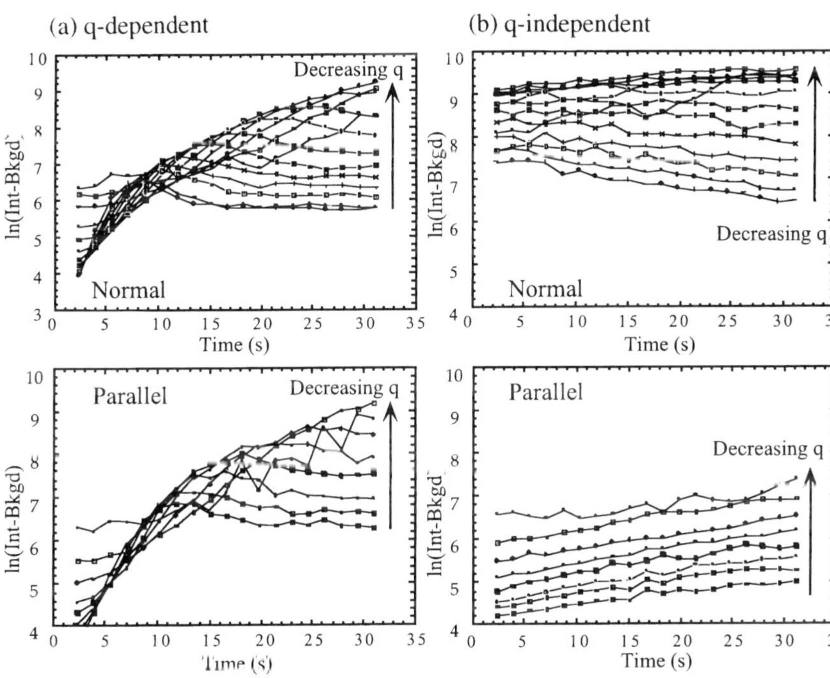

Figure 10 The time dependence of the scattered light intensities at various q values following cessation of shear for an 8% by mass 50:50 blend of PS ($M_w = 89 \times 10^3\,\mathrm{g\,mol^{-1}}$) and PB ($M_w = 22 \times 10^3\,\mathrm{g\,mol^{-1}}$) in DOP. (a) q-dependent growth patterns in the directions normal and parallel to flow. (b) q-independent growth patterns in the directions normal and parallel to flow. Minimum q values = 0.27 and 0.13 μm^{-1}; maximum q values = 0.75 and 0.97 μm^{-1} for the normal and parallel directions, respectively, in 0.07 μm^{-1} intervals.

observed after cessation of shear rates below the critical shear rate for homogenization. In the two-step experiments, with a variable delay time, only q-independent growth was observed. The same behavior was observed regardless of the value of $\dot{\gamma}_2$. Therefore, the slow relaxation of anisotropic scattering patterns produced by low shear rates to an isotropic pattern was found to be independent of the sample history prior to the deformation. The nature and extent of pre-existing structure in the sample is therefore not responsible for the slow relaxation behavior.

In the two-step experiments with a fixed delay time and variable $\dot{\gamma}_2$, values of $\dot{\gamma}_2$ which were insufficient to produce homogenization of the sample during steady shear were capable of suppressing the coarsening during $\dot{\gamma}_2$. For very low values of $\dot{\gamma}_2$, some coarsening occurred during the second shear rate, with the growth normal to flow greater than the growth parallel to flow. Where peaks in the structure factor were observed, the peak positions, q_m, and the intensity of the peaks, $I(q_m)$ could be followed as a function of time. This analysis revealed that the coarsening process is accelerated during the second shear rate, then decelerates after cessation of the second shear rate. There is an apparent coupling between the coarsening behavior in the directions parallel and normal to flow. If phase separation occurs during $\dot{\gamma}_2$ normal to flow, after cessation of $\dot{\gamma}_2$, phase separation only occurs parallel to flow in a q-independent fashion. The q-dependent growth pattern is only observed parallel to flow when the second shear rate suppresses growth in all directions.

Two possible interpretations for the apparent coupling between the parallel and normal coarsening mechanisms are possible. If we assume that once the droplets are finely dispersed to the same length scale as the polymer single-chain dimensions, then the concentration fluctuations are of the microscopic (or molecular) type, and spinodal decomposition, with the signature q-dependent growth rate, occurs after the shear quench. On the other hand, if the droplet dimensions remain macroscopic (orders of magnitude larger than the polymer chain dimensions), then q-independent droplet coarsening kinetics are observed. Therefore, the first explanation for the coupled behavior is that the sample is more homogeneous in the normal direction than in the parallel direction during shear. The seemingly related growth mechanisms between the parallel and normal directions may originate from the relative mixing conditions in the two directions, with the normal direction always being better mixed.

The second possible explanation for the coupling is also related to the relative mixing in the two directions. During lower values of $\dot{\gamma}_2$, coarsening is observed normal to flow, while no coarsening is observed along the flow direction. When the second shear rate is switched off, the driving force for phase separation parallel to flow no longer exists, since large amplitude concentration fluctuations (droplets) already exist in the normal direction. Therefore, only surface tension and droplet diffusion remain as driving forces to coalesce the droplets along the flow direction.

D. Phase Separation Kinetics During Shear with Block Copolymer

The phase separation and mixing behavior of a blend of polystyrene (PS) ($M_w = 2 \times 10^3\,\mathrm{g\,mol^{-1}}$) and polybutadiene (PB) ($M_w = 2.7 \times 10^3\,\mathrm{g\,mol^{-1}}$) with and without added diblock copolymer (PS/PB) ($M_w = 5 \times 10^3$ for each block) have been examined during shear as a function of quench depth and copolymer concentration for a fixed mass ratio of PS to PB (PS:PB = 60:40 by mass). The low molecular weights of the homopolymers were necessary to access the phase boundary for this blend, which exhibits upper critical solution temperature (UCST)-type behavior. For the pure 60:40 blend, the cloud-point temperature was $105^\circ\mathrm{C} \pm 0.5^\circ\mathrm{C}$. Steady shear experiments in the two-phase region show evidence of only partial sample homogenization in the pure blend from the partial suppression of scattered light intensity. In contrast, for samples containing copolymers at an equivalent quench depth, complete homogenization occurred at much lower shear rates.

For phase separation kinetics studies during shear, the samples were slowly cooled with the simultaneous application of shear, and the kinetics of the phase-separation behavior were monitored. These complicated thermal and mechanical histories are relevant to the effect of manufacturing processing on polymer blend morphology. The results were strongly dependent on the depth of quench. In the pure blend at a shallow quench depth of $1^\circ\mathrm{C}$, an increase in the growth rate of the scattered light intensity was observed under shear compared to the growth rate of the scattered light intensity from a quiescent sample. The highest growth rate was observed for the highest shear rate. These results indicated an enhancement of the droplet coalescence rate with shear. In contrast, for the same relative quench depth, the sample with added copolymer demonstrated a decreased rate of growth in the scattering intensity compared to the quiescent growth rate. This indicated a suppression of droplet coalescence and was consistent with atomic force microscopy results from thin films of blend samples with copolymer present, which demonstrated frustrated droplet coalescence [89]. Both results demonstrate the interplay between increased droplet coalescence due to a higher collision frequency during shear and the frustration of coalescence due to the presence of an interfacial modifying agent. Further studies are necessary to delineate the role of these two mechanisms during simultaneous mechanical deformation and phase separation.

V. CONCLUSIONS

Over the past 15 years, the phase-separation behavior of binary polymer mixtures under quiescent conditions has been studied by many laboratories, with many experimental studies confirming various aspects of theoretical predictions. Of notable significance has been the verification of the Cahn–Hilliard–Cook theory which is a linearized mean-field approach describing spinodal decomposition. Detailed theoretical descriptions for the nonlinear, intermediate, and late stages of phase separation are required beyond the scaling predictions currently available and more extensive numerical simulations are also needed.

For the case of polymer blends under shear, it has been shown that the critical temperature shift as a function of shear rate can be interpreted by the theoretical model of Onuki–Kawasaki. We have presented more recent studies of the droplet-to-string transition by extending this approach to the two-phase region. The true shift in the phase diagram as a function of shear rate has been studied by fluorescence microscopy with limited success. Although current theories can explain these results within the moderate precision of the experiments, detailed questions such as the condition for coupling of the flow field to the fourth power of the order parameter (critical fluctuation) in the equation of motion describing these phenomena remain unresolved.

Due to the wide array of variables present in these studies, a large number of topics remain unexplored. For example, a large variety of chemical additives are incorporated in commercial polymer blends and their effects on the phase behavior of these materials have only been investigated for a small number of specific systems. It is clear from these studies that the mechanical history imposed upon these mixtures has a profound influence on the resultant morphology particularly when changes in the shear rate are involved. The complex relationship between droplet break-up and string formation and shear-induced droplet coalescence as a function of shear rate, sample temperature, additive concentration, homopolymer molecular weights, and viscosity is quite evident. While some systematic studies have been initiated and reported by several research groups, there is still a substantial amount of work remaining before a coherent understanding of the shear-rate dependence in multicomponent systems can be achieved.

ACKNOWLEDGMENTS

The authors would like to acknowledge Drs L. Sung, J.-W. Yu, E. K. Hobbie, S. Kim, and C. L. Jackson, Mr. D. S. Johnsonbaugh for providing unpublished experimental results and their insight and enlightened comments during the preparation of this manuscript.

REFERENCES

1. E Stanley. Introduction to Phase Transitions and Critical Phenomena. New York: Oxford University Press, 1971.
2. D Beysens, M Gbadamassi, B Moncef-Bouanz. Phys Rev A 28:2491, 1983.
3. D Beysens, M Gbadamassi. Phys Rev A 22:2250, 1980.
4. P G de Gennes. Scaling Concepts in Polymer Physics. New York: Cornell University Press, 1979.
5. A Silberberg, W Kuhn. Nature 170:450, 1952.
6. A Silberberg, W Kuhn. J Polym Sci 13:21, 1954.
7. G VerStrate, W Philippoff. J Polym Sci Polym Lett Ed 12:267, 1974.
8. E Helfand, G H Fredrickson. Phys Rev Lett 62:2468, 1989.
9. S T Milner. Phys Rev Lett 66:1477, 1991.
10. X L Wu, D J Pine, P K Dixon. Phys Rev Lett 66:2408, 1991.
11. H Yanase, P Moldenaers, J Mewis, V Abetz, J van Egmond, G G Fuller. Rheol Acta 30:89, 1991.
12. E Moses, T Kume, T Hashimoto. Phys Rev Lett 72:2037, 1994.
13. C Rangel-Nafaile, A B Metzner, K F Wissbrun. Macromolecules 17:1187, 1984.
14. R G Larson. Rheol Acta 31:497, 1992.
15. M Tirrell. Fluid Phase Equilibria 30:367, 1986.
16. B Chu. Laser Light Scattering: Basic Principles and Practice. 2nd ed. San Diego, CA: Academic Press, Inc., 1991.
17. J S Higgins, H C Benoit. Polymers and Neutron Scattering. Oxford Series on Neutron Scattering in Condensed Matter. New York: Oxford University Press, 1994.
18. A I Nakatani. ACS Symp Ser 597:1, 1995.
19. G G Fuller. Ann Rev Fluid Mech 22:387, 1990.
20. H Janeschitz-Kriegl. Polymer Melt Rheology and Flow Birefringence: Polymers/Properties and Applications. Vol. 6. Berlin: Springer-Verlag, 1983.
21. K Søndergaard, J Lyngaae-Jørgensen, ed. Rheo-Physics of Multiphase Polymeric Systems Application of Rheo-optical Techniques in Characterization. Lancaster, PA: Technomic Publishing, 1993.
22. A Onuki, K Kawasaki. Ann Phys 121:456, 1979.
23. A Onuki, K Kawasaki. Prog Theor Phys Supp 64:436, 1978.
24. A Onuki, K Yamazaki, K Kawasaki. Ann Phys 131:217, 1981.

25. A Onuki. Physica A 140:204, 1986.
26. J W Cahn, J E Hilliard. J Chem Phys 28:258, 1958; J Chem Phys 31:688, 1959.
27. J W Cahn. J Chem Phys 42:93, 1965.
28. H E Cook. Acta Metall 18:297, 1970.
29. A Z Akcasu, I Bahar, B Erman, Y Feng, C C Han. J Chem Phys 97:5782, 1992.
30. C Herkt-Maetzky, J Schelten. Phys Rev Lett 51:896, 1983.
31. J Dudowicz, K F Freed. Macromolecules 24:5074, 1991.
32. J Dudowicz, M S Freed, K F Freed. Macromolecules 24:5096, 1991.
33. K F Freed, J Dudowicz. Macromolecules 29:625, 1996.
34. K S Schweizer, J G Curro. Adv Polym Sci 116:319, 1994.
35. E F David, K S Schweizer. J Chem Phys 100:7767, 1994.
36. I C Sanchez. Ann Rev Mater Sci 13:387, 1983.
37. B Hammouda, M Benmouna. J Polym Sci Part B Polym Phys 33:2359, 1995.
38. J K Taylor, P G Debenedetti, W W Graessley, S K Kumar. Macromolecules 29:764, 1996.
39. C C Han, B J Bauer, J C Clark, Y Muroga, Y Matsushita, M Okada, Q Tran-Cong, T Chang, I C Sanchez. Polymer 29:2002, 1988.
40. D Schwahn, K Mortensen, H Yee-Madeira. Phys Rev Lett 58:1544, 1987.
41. F S Bates, J H Rosedale, P Stepanek, T P Lodge, P Wiltzius, G H Fredrickson, R P Hjelm. Phys Rev Lett 65:1893, 1990.
42. P Stepanek, T P Lodge, C Kedrowski, F S Bates. J Chem Phys 94:8289, 1991.
43. B Chu, Q Ying, K Linliu, P Xie, T Gao, Y Li. ACS PMSE Preprints 65:209, 1991.
44. D W Hair, E K Hobbie, A I Nakatani, C C Han. J Chem Phys 96:9133, 1992.
45. B Chu, Q Ying, K Linliu, P Xie, T Gao, Y Li, T Nose, M Okada. Macromolecules 25:7382, 1992.
46. L Sung, A I Nakatani, C C Han, A Karim, J F Douglas, S K Satija, submitted to Physica B.
47. P G de Gennes. J Phys Lett (France) 38:L441, 1977.
48. R F Joanny. J Phys (France) B 11:L117, 1978.
49. K Binder. J Chem Phys 79:6387, 1983.
50. M Shibayama, H Yang, R S Stein, C C Han. Macromolecules 18:2179, 1985.
51. C C Han, A Z Akcasu. Ann Rev Phys Chem 43:61, 1992.
52. P G de Gennes. J Chem Phys 72:4756, 1980.
53. P Pincus. J Chem Phys 75:1996, 1981.
54. T Hashimoto. In: Phase Transition. London: Gordon and Breach, 1988, pp 47–119.
55. T Sato, C C Han. J Chem Phys 88:2057, 1988.
56. M Motowoka, H Jinnai, T Hashimoto, Y Qiu, C C Han. J Chem Phys 99:2095, 1993.
57. H L Snyder, P Meakin. J Chem Phys 79:5588, 1983.
58. H L Snyder, P Meakin, S Reich. J Chem Phys 79:334, 1983.
59. J S Langer, M Bar-on, H D Miller. Phys Rev A 11:1417, 1975.
60. E D Siggia. Phys Rev A 20:595, 1979.
61. J Noolandi, K M Hong. Macromolecules 17:1531, 1984.

62. Y Lyatskaya, D Gersappe, A C Balazs. Macromolecules 28:6278, 1995.
63. D Gersappe, A C Balazs. Phys Rev E 52:5061, 1995.
64. C C Lin, H S Jeon, N P Balsara, B Hammouda. J Chem Phys 103:1957, 1995.
65. R-J Roe, D Rigby. Adv Polym Sci 82:105, 1987.
66. L Sung, C C Han. J Polym Sci: Polym Phys Ed 33:2405, 1995.
67. L Sung, D B Hess, C L Jackson, C C Han. J Polym Res 3:139, 1996.
68. C L Jackson, L Sung, C C Han. Polym Sci Eng (submitted).
69. A I Nakatani, H Kim, Y Takahashi, Y Matsushita, A Takano, B J Bauer, C C Han. J Chem Phys 93:795, 1990.
70. K Kawasaki. In: C Domb, M S Green, ed., Phase Transitions and Critical Phenomena. Vol. 5a. New York: Academic Press, 1976, p 165.
71. R Ferrell. Phys Rev Lett 24:1169, 1970.
72. D W Hair, E K Hobbie, J F Douglas, C C Han. Phys Rev Lett 68:2476, 1992.
73. E K Hobbie, D W Hair, A I Nakatani, C C Han. Phys Rev Lett 69:1951, 1992.
74. E K Hobbie, A I Nakatani, H Yajima, J F Douglas, C C Han. Phys Rev E Rapid Commun 53:R4322, 1996.
75. H Yajima, D W Hair, A I Nakatani, J F Douglas, C C Han. Phys Rev B 47:12268, 1993.
76. T Takebe, T Hashimoto. Polymer 29:261, 1988.
77. T Hashimoto, T Takebe, S Suehiro. J Chem Phys 88:5874, 1988.
78. T Takebe, R Sawaoka, T Hashimoto. J Chem Phys 91:4369, 1989.
79. T Takebe, K Fujioka, R Sawaoka, T Hashimoto. J Chem Phys 93:5271, 1990.
80. T Hashimoto, T Takebe, K Fujioka. In: A Onuki, K Kawasaki, ed. Dynamics and Patterns in Complex Fluids; Springer Proceedings in Physics. Vol 52. Berlin: Springer Verlag, 1990, pp 86–99.
81. K Fukjioka, T Takebe, T Hashimoto. J Chem Phys 98:717, 1993.
82. T Hashimoto, T Takebe, K Asakawa. Physica A 194:338, 1993.
83. S Kim, E K Hobbie, C C Han. (In press), Macromolecules.
84. E K Hobbie, S Kim, C C Han. Phys Rev E 54:R5909, 1996.
85. G I Taylor. Proc Roy Soc A138:41, 1932; Proc Roy Soc A146:501, 1934.
86. M R Moldover. Phys Rev A 31:1022, 1985.
87. J-W Yu, J F Douglas, E K Hobbie, S Kim, C C Han. Phys Rev Lett 78:2664, 1997.
88. A I Nakatani, D S Johnsonbaugh, C C Han. ACS Symp Ser 597:246, 1995.
89. A Karim, J F Douglas, A P Wong, S K Satija, C C Han. Macromolecules.

8

Surface Organization and Dynamics in Multicomponent Polymer Systems

Ian Hopkinson
University of Cambridge, Cambridge, England

Randal W. Richards
University of Durham, Durham, England

I. INTRODUCTION

The surface and interfaces of polymer systems are of substantial importance for a number of reasons. The surface is where the polymer interacts with its surroundings, e.g., a surrounding atmosphere or another material, consequently the nature of the polymer surface will determine the tribological and adhesive properties as well as the resistance to chemical attack. The nature of polymer–polymer interfaces is highly relevant to the compatibility of immiscible homopolymers and the "welding" of two polymers. For a strong joint between the two polymers to be formed it is necessary that whatever the nature of the polymer at the interface, it should diffuse to a sufficient depth in both of the polymers either side of the interface and become entangled with the molecules in each of the two polymers so that sufficient mechanical strength is imparted to the composite polymer. Copolymers are frequently used in such applications and theoretical understanding of the influence of polymer architecture and composition is increasing steadily [1]. Until recent years, the nature of the surface of polymers was restricted to information from very high-resolution studies from such methods as secondary ion mass spectroscopy (SIMS) and X-ray photoelectron spectroscopy (XPS). These techniques are able to provide very specific analyses of the immediate surface layer (~ top 30 Å) but information at greater depths is difficult to obtain except by specialized methods (see below). By contrast attenuated total internal reflection Fourier transform infrared (ATR-FTIR) spectroscopy penetrates to too great a depth and does not have the spatial resolution needed to provide information on polymer surfaces and interfaces at the level needed. None of the techniques mentioned thus far is able to access *buried* interfaces easily, if at all.

Surface investigations of polymers are faced with a dilemma; information on the arrangement of the segments near to the surface or interface is needed on resolvable length scales of the order of about 20 Å, but this must be sustainable over distances which are at least the same as the radius of gyration of the polymer and frequently may need to be much larger. There are a few techniques which can provide such information and their development over the past eight to ten years has resulted in much greater insight into the organization of polymers at interfaces. The methods of choice are neuron reflectometry and ion beam analysis methods (forward recoil spectrometry, Rutherford backscattering and nuclear reaction analysis). We focus here on neutron reflectometry; the ion beam analysis methods are only briefly mentioned in passing. Furthermore we have, perhaps, rather liberally interpreted multicomponent polymers. Not only do we discuss results for copolymers but we also interpret multicomponent to include isotopic mixtures of polymers and polymers at the air–water interface. This last aspect also illustrates the versatility of neutron reflection in not being confined to clean, high-vacuum environments, a result of the weak interaction of the neutron with matter in condensed phases.

One aspect which is only briefly treated here is the surface viscoelasticity of polymers at interfaces. Capillary waves are known to exist at the surface of low molecular weight liquids such as water and increasingly we are becoming aware that polymers at this interface (either as a spread film or as a surface) can impart previously unknown properties, e.g., negative dilational viscosities. A brief mention is made of such studies at the end of this chapter. Its relevance to the main part of the discussion (apart from that section dealing specifically with polymers at the air–water interfaces), is that capillary waves will exist at the interface in polymer–polymer melts. However, there is no quantitative knowledge of these available at present, indeed the understanding of capillary waves at polymer–water interfaces is not highly developed. Consequently it was deemed advisable to include a little on this subject for those sufficiently interested to be able to enquire further.

II. NEUTRON REFLECTOMETRY

The heart of a neutron reflectivity experiment is the determination of the specular reflectivity (i.e., the ratio of the reflected beam intensity to the incident beam intensity) of a beam of neutrons impinging at a glancing angle θ to the surface of a sample as a function of the component of the scattering vector perpendicular to the sample surface, Q:

$$Q = \frac{4\pi}{\lambda}\sin\theta \tag{1}$$

where θ is the angle between the incident neutron beam and the sample surface and λ is the neutron wavelength. It is cwlear from Eq. (1) that there are two ways of probing the Q range, either θ can be varied or λ can be varied. Typically for solid interfaces a Q range of 0.005 Å^{-1} to 0.08 Å^{-1} is used, whereas for liquid interfaces a rather larger Q vector range is probed, from about 0.005 Å^{-1} to 0.6 Å^{-1}.

The reflectivity, $R(Q)$, contains information on the variation in the nuclear scattering length density normal to the sample surface, $\rho_N(z)$. The scattering length density is a nuclear property (magnetic effects are not considered here) which varies irregularly through the periodic table. The scattering length densities of ^1H and ^2H (deuterium) are very different and so a composition profile can be obtained if one of the components in a binary mixture is isotopically labeled by ^2H. The scattering length density of a material being given by:

$$\rho_N = \frac{\rho N_A}{m}\sum b_i \tag{2}$$

where ρ is the mass density, N_A is Avogadro's number, b_i are the nuclear scattering lengths of the atoms in the species of molar mass, m, for polymers the species for which ρ_N is calculated is usually the repeat unit.Values of b are tabulated for many known isotopes.

The reflection of neutrons is entirely analogous to the reflection of (s-wave) electromagnetic radiation from a surface. The principles of the analysis of neutron reflectometry will be outlined briefly, a more detailed description of the mathematics of reflection and optical matrix methods can be found in Lekner [2], Born and Wolf [3] and Heavens [4]. The major difference from the reflection of light is the replacement of the optical refractive index by the neutron refractive index, n, given by:

$$n = 1 - \frac{\lambda^2}{2\pi}\rho_N + \frac{i\lambda\rho N_A(\sigma_i + \sigma_A)}{m4\pi} \tag{3}$$

where σ_i and σ_A are the incoherent and absorption cross-sections respectively of the same species for which ρ_N applies. In general incoherent and absorption effects are not significant in polymer systems and this final, complex term can be ignored. Penfold [5] discusses these effects more fully and an example of where such effects have been usefully used to analyze organization in polymer thin films has been provided by Norton et al. [6]. The neutron refractive index is very slightly less than one for most materials, this means that total external reflection takes place below a critical incident wavevector. The critical angle below which reflection is total (i.e., reflectivity = 1) is given by:

$$\theta_c = \lambda\left(\frac{\rho_N}{\pi}\right)^{1/2} \tag{4}$$

This is obtained from Snell's law and the expansion for $\cos\theta$ at small θ. Above the critical angle the reflectivity at a single sharp interface between two media is given by:

$$\begin{aligned} r_{01} &= \frac{k_0 - k_1}{k_0 + k_1} \\ R &= r_{01} r_{01}^* \end{aligned} \tag{5}$$

r_{01} is the Fresnel reflection coefficient and r_{01}^* is its complex conjugate. $k_i (= Q/2)$ is the component of the neutron wavevector perpendicular to the sample surface in medium i. The reflectivity is a very rapidly decreasing function of Q as can be seen in Fig. 1.

The reflectivity of an arbitrary scattering length density profile can be calculated by representing the structure by an ensemble of discrete uniform layers (Fig. 1). The reflectivity of such an ensemble can be calculated recursively from a general form of Eq. (5) or, more commonly, by the use of optical matrix methods. The properties of the j-th element in the ensemble are given by the matrix:

$$M_j = \begin{pmatrix} e^{i\beta_{j-1}} & r_j e^{i\beta_{j-1}} \\ r_j e^{-i\beta_{j-1}} & e^{-i\beta_{j-1}} \end{pmatrix} \tag{6}$$

where r_j is a modified Fresnel coefficient for the interface between layers $j-1$ and j:

$$r_j = \frac{p_{j-1} - p_j}{p_{j-1} + p_j} \exp\left(-\frac{1}{2} Q_{j-1} Q_j \langle\sigma\rangle^2\right) \tag{7}$$

the exponential term is incorporated to account for the occurrence of roughness at the interface, i.e. the scattering length density varies about some

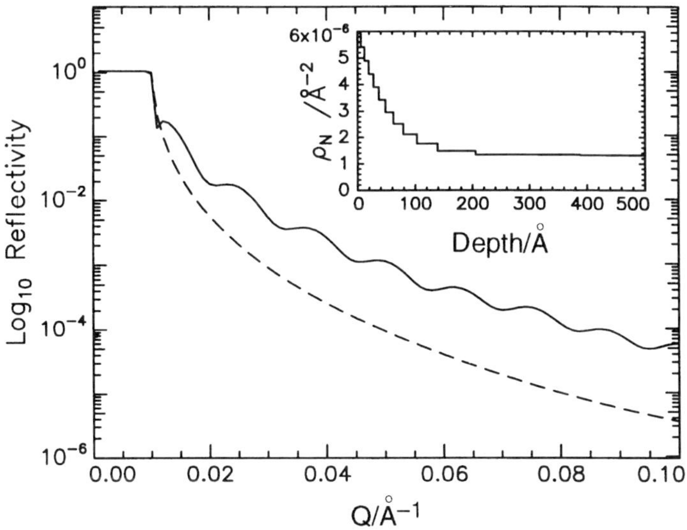

Figure 1 Reflectivity from a sharp interface (broken line) and from a typical scattering length density profile (full line). The inset shows the profile used to calculate this simulation.

mean smooth boundary in a Gaussian manner, the root mean square deviation being $\langle\sigma\rangle$, and:

$$p_j = n_j \sin\theta_j = (n^2 - n_{j-1}^2 \cos^2\theta_{j-1})^{1/2} \tag{8}$$

where θ_j has been expressed in terms of the θ and n for the previous layer, $j-1$, using Snell's law:

$$\beta_j = \left(\frac{2\pi}{\lambda}\right) n_j z_j \sin\theta \tag{9}$$

where z_j is the thickness of layer j. Each of these matrices is multiplied together to produce a characteristic optical matrix of the whole multilayer assembly. The reflectivity is then given by:

$$R = \frac{M_{21}M_{21}^*}{M_{11}M_{11}^*} \tag{10}$$

M_{11} and M_{21} are elements of the resultant matrix which is the product of the matrices for each layer.

Fig. 1 includes the calculated reflectivity for an ensemble of layers representing an exponential scattering length density profile with a decay length of 50 Å. Experimental data (i.e., the reflectivity profile, $R(Q)$) is analyzed by adopting a model of the composition variation normal to the interface, calculating the reflectivity by the procedures outlined above, and fitting the simulation to the experimental reflectivity profile by adjusting parameters of the model, e.g., layer thickness or volume fraction composition of each layer. This fitting may be done by conventional nonlinear least-squares methods, but maximum entropy [7] and simulated annealing [8] methods have also been applied. The difficulty is that reflectivity data contain no phase information and hence there may be more than one model of the scattering length density variation which will produce a fit to the reflectivity. Uniqueness of the model adopted must always be questioned. Furthermore, beyond this fundamental restriction other models of the scattering length density may be experimentally indistinguishable because of instrument resolution effects and momentum transfer limits. In practice these problems are not insurmountable when neutron reflectometry is used in conjunction with other techniques, such as ion beam analysis outlined briefly below, and using chemical and physical knowledge which restricts the number of viable scattering length density profile models.

A schematic volume fraction profile for a polymer at an interface is shown in Fig. 2 together with the characteristic surface and bulk volume

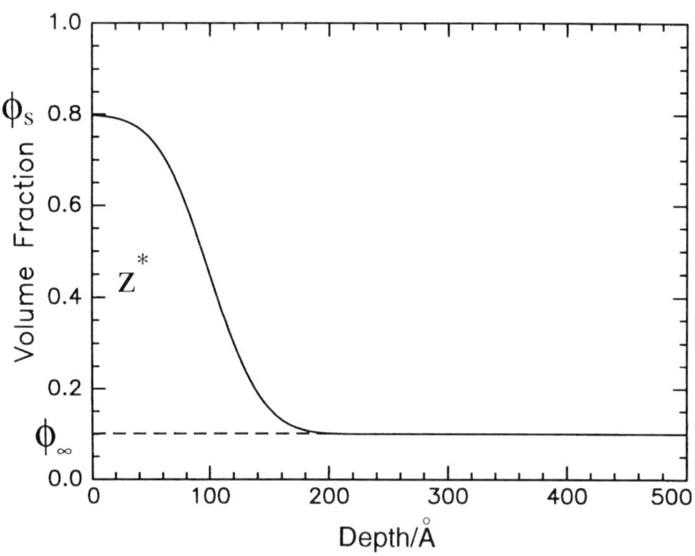

Figure 2 Schematic volume fraction profile of a polymer near an interface in excess of bulk concentration. Interfacial (ϕ_s) and bulk equilibrium (ϕ_∞) volume fractions and the surface excess z^*, are marked.

fraction parameters, ϕ_s, and ϕ_∞, respectively, and the surface excess, z^*. The surface excess, z^*, is given by

$$z^* = \int_0^\infty dz(\phi(z) - \phi_\infty)$$

and σ, the grafting density in number of polymer molecules (degree of polymerization, N) per unit area by

$$\sigma = \frac{z^* \rho N_A}{Nm}$$

L, the (ellipsometric) layer thickness is the first moment of the profile,

$$L = \frac{1}{z^*} \int_0^\infty z(\phi(z) - \phi_\infty) dz$$

Where ϕ_∞ is the equilibrium bulk volume fraction, N the degree of polymerization, ρ the polymer density and m the molecular weight of a monomer unit.

One area where the rigorous optical matrix approach may not be so informative without excessive effort at generating a model is in the study of polymers spread at the air–liquid interface where the distribution of three components is required (air, water, polymer). Ion beam and other high-resolution surface analysis techniques cannot be used in this case since they all require high vacuum. These problems can be overcome by performing experiments with isotopic labeling of the different components and using the kinematic approximation with partial structure factors [9]. The kinematic approximation, introduced by Als-Nielsen [10] and extended by Crowley [11], assumes that reflected neutrons have been scattered from a single Fermi pseudo-potential and since this is violated at small scattering vectors, near the critical edge, the kinematic approximation is restricted to the region of Q where the reflectivity is less than about 2×10^{-3}. In the context of the kinematic approximation the specular reflection is given by:

$$R(Q) = \frac{16\pi^2}{Q^2} \mid \rho(Q) \mid^2 \tag{11}$$

where $\rho(Q)$ is the Fourier transform of the scattering length density profile:

$$\rho(Q) = \int_{-\infty}^{\infty} \exp(iQz)\rho(z)\, dz \tag{12}$$

Eq. (11) can also be expressed as:

$$R(Q) - \frac{16\pi^2}{Q^4} \mid \rho'(Q) \mid^2 \tag{13}$$

where $\rho'(Q)$ is the Fourier transform of the first spatial derivative of $\rho(z)$. In the limit of high Q, $\rho'(Q)$ is proportional to $\Delta\rho$ (the magnitude of the "steps" in the $\rho(z)$ profile. By replacing $\rho(z)$ with the following:

$$\rho(z) = \sum n_i(z) b_i \tag{14}$$

where $n_i(z)$ is the *number* density of species i as a function of the depth, we obtain:

$$R(Q) = \frac{16\pi^2}{Q^2} \sum b_i b_j h_{ij}(Q) \tag{15}$$

where the $h_{ij}(Q)$ terms are the partial structure factors ($h_{ii}(Q) = | n_i(Q)) |^2$ and $h_{ij}(Q) = \mathrm{Re}\{n_i(Q) n_j(Q)\}$, where $n_i(Q)$ is the Fourier transform of the spatial variation of the number density of the species i or j. Analogously to Eqs. (11) and (13), Eq. (15) can be expressed as:

$$R(Q) = \frac{16\pi^2}{Q^4} \sum b_i b_j h'_{ij}(Q) \tag{16}$$

where $h'_{ij}(Q)$ is the gradient in the partial structure factor. If we consider the case of a polymer with components A and B, spread at an air–liquid (L) interface then the scattering length density profile is given by:

$$\rho(z) = b_{\mathrm{A}} n_{\mathrm{A}}(z) + b_{\mathrm{B}} n_{\mathrm{B}}(z) + b_{\mathrm{L}} n_{\mathrm{L}}(z) \tag{17}$$

where b_i and $n_i(z)$ are the scattering length densities and number densities of the components i. Eq. (15) now becomes:

$$R(Q) = \frac{16\pi^2}{Q^2} [b_{\mathrm{A}}^2 h_{\mathrm{AA}}(Q) + b_{\mathrm{B}}^2 h_{\mathrm{BB}}(Q) + b_{\mathrm{L}}^2 h_{\mathrm{LL}}(Q) + 2b_{\mathrm{A}} b_{\mathrm{B}} h_{\mathrm{AB}}(Q) + 2b_{\mathrm{A}} b_{\mathrm{L}} h_{\mathrm{AL}}(Q) + 2b_{\mathrm{L}} b_{\mathrm{B}} h_{\mathrm{LB}}(Q)] \tag{18}$$

If reflectivity profiles can be obtained at six different nuclear scattering length density contrasts them all of the partial structure factors can be obtained by solution of the set of simultaneous equations each of the form of Eq. (18). In principle, the partial structure factors can each be Fourier transformed to obtain Patterson autocorrelation functions, in practice the data do not extend over a wide enough Q range for this to be practicable, so again modeling is required. Some information may be obtained more directly, given some assumptions about the symmetry of the system. Components A and B are most likely to be symmetric about the centers of their distributions, i.e., they are even functions and:

$$h_{AB}(Q) = \pm(h_{AA}(Q)h_{BB}(Q))^{1/2}\cos Q\delta_{AB} \tag{19}$$

where δ_{AB} is the separation between the centers of the distributions of A and B. The liquid subphase, on the other hand, will be described by an odd function at the interface. Therefore:

$$h_{AL}(Q) = \pm(h_{AA}(Q)h_{LL}(Q))^{1/2}\sin Q\delta_{AB} \tag{20}$$

the $\pm$ sign introduces some uncertainty, but this may be resolved by physical constraints.

III. ION BEAM METHODS

As was noted earlier, neutron reflectometry experiments can be complemented by lower-resolution techniques which provide a guide to the adoption of a suitable model for fitting to the neutron reflectometry data. The principal advantage of ion beam methods is that they are real space techniques and so no "inversion" of the data from reciprocal space is required hence data analysis is much more straightforward.

(a) Secondary Ion Mass Spectrometry (SIMS) utilizes an ion beam (such as Cs^+, Ar^+, or O^-) impinging on the sample surface to eject fragments of the sample. These fragments only have a small escape depth and so the technique is very surface specific. SIMS is used in two modes: static SIMS [12], where a relatively low-energy beam is used to eject charged molecular fragments which are subsequently accelerated by an electric field and analyzed by mass spectrometry. A typical escape depth for such fragments is on the order 10–20 Å. Polymers have characteristic fragmentation patterns and so SIMS is chemically sensitive, in addition the technique is sensitive to deuteration, since this alters the charge/mass ratio of analogous fragments. Even the relatively limited information static SIMS provides on the surface composition of a polymer film can be useful in analyzing neutron reflectometry (NR) data.

The second type of SIMS experiment is the dynamic mode [13], again ions impinge on the sample surface but in this case a larger ion current is used, the effect of this is to strip away the surface of the sample over time. To obtain a depth profile, the ion beam is repeatedly rastered over an area of the sample, and the ejected ion current is measured as a function of time. Ions collected at later times correspond to molecules further below the "original" sample surface than those collected at earlier times. Typically isotopic substitution is used in these experiments and $^{2}H^{+}$ and $^{1}H^{+}$ secondary ion currents are measured. The depth resolution of the dynamics SIMS technique is of the order 200 Å, this is rather lower than the sensitivity of the static mode because the higher incident beam current causes atomic mixing in the exposed layers of the sample and care must be taken to avoid other artefacts resulting from the destructive nature of this type of experiment.

(b) Forward Recoil Elastic Scattering (FRES) [14] utilizes a high-energy beam (3.0 MeV) of $^{4}He^{++}$ impinging on the sample surface at a low incident angle. These ions undergo elastic collisions with the sample nuclei, knocking some of them out of the sample. These nuclei are detected at a forward angle using an energy-sensitive detector shielded by a Mylar film, this stops all but ^{1}H and ^{2}H nuclei reaching the detector. As a consequence of their very different masses ^{2}H and ^{1}H recoil with different energies and so signals from them are separated easily. The measured energy of the two nuclei is dependent upon the incident beam energy and the depth in the sample from which they are ejected, since the recoiling nuclei lose energy as they move through the sample to the surface. Thus depth versus composition profiles are obtained by measuring the particle flux as a function of energy; because of straggling in the Mylar film the resolution is on the order 800 Å. This can be improved to 125 Å at the surface by changing the beam energy and the sample geometry [15], a resolution of 250 Å can be obtained by dispensing with the Mylar film and using a time flight detector [16].

(c) Nuclear Reaction Analysis (NRA) [17,18] is very similar to FRES in the instrumentation used; a beam of $^{3}He^{+}$ impinging the sample at an energy of order 1.5 MeV undergoes a nuclear reaction:

$$^{3}\mathrm{He}^{+} + {}^{2}\mathrm{H} \rightarrow {}^{4}\mathrm{He} + {}^{1}\mathrm{H}^{+} \quad (Q = 18.532\ \mathrm{MeV})$$

The reaction has a maximum cross-section for a $^{3}He^{+}$ energy of 0.7 MeV. The energies of the resultant nuclei can be calculated from the conservation of momentum and energy. These energies depend on the depth at which they are produced because the incident ^{3}He loses energy as it penetrates the sample and this alters the energy distribution between the resultant nuclei. The resolution achieved in this experiment is on the order 150–300 Å, by altering the incident angle closer to the normal the depth probed can be increased at the expense of lower resolution.

IV. SURFACE ENRICHMENT IN POLYMER MIXTURES

In a binary blend the component of lower surface energy will be energetically favored at the surface, since this reduces the overall free energy of the system. Such general behavior was considered theoretically by Cahn [19] and has been studied extensively in mixtures of small molecules [20] and in ^{3}He, ^{4}He mixtures [21] as well as polymers. Polymers are of particular interest, partly because of the technological importance of polymer surfaces and partly because the characteristic length scales of polymers (generally of the order of the radius of gyration) are amenable to detailed comparisons with theory. The first theories of polymer surface enrichment were those of Nakanishi and Pincus [22] and Schmidt and Binder [23]. These were mean-field theories and as such gave a good indication of the overall "phase behavior" of surface enrichment. These models have subsequently been simplified [24,25] and extended [26], in addition there have been self-consistent field (SCF) theories [27] and Monte Carlo models [28–30]. SCF theories are rather better at describing the shape of the enrichment profile and are relatively easy to apply to a wide range of experimental systems. In principle Monte Carlo models are the most "exact" although the phase behavior is not explicitly determined and in practice the computational resources required for useful modeling are very large. Fig. 3 shows the expected surface enrichment phase behavior for a binary polymer blend, derived from mean-field theory. As we move along the coexistence curve towards critical unmixing first partial wetting behavior, where a microscopic layer of the lower surface energy component is found at the surface, and then complete wetting, where a macroscopic layer of the lower-energy component is found at the surface, are observed. The observed profile is the result of the competition between having a lower surface energy component at the surface and the maintenance of a concentration gradient in the blend, which has an energy cost. The predicted profiles are exponential in form with a decay length of the order of the radius of gyration. The shape of the profile is determined by the bulk thermodynamics of the blend (i.e., the Flory–Huggins interaction parameter χ), more compatible blends having a thinner surface-enriched layer.

Although there has been considerable work on blends of different polymers such as polystyrene (PS)/poly(vinyl methyl ether) (PVME) [31–33], poly(ethylene oxide) (PEO)/poly(methyl methacrylate) (PMMA) [34], etc. using a variety of techniques, neutron reflectometry work has been focused on isotopic blends and detailed comparisons of the expected theoretical behavior with experiment. Isotopic blends of polymers provide excellent "ideal mixtures." Typically isotopic substitution has no effect on the

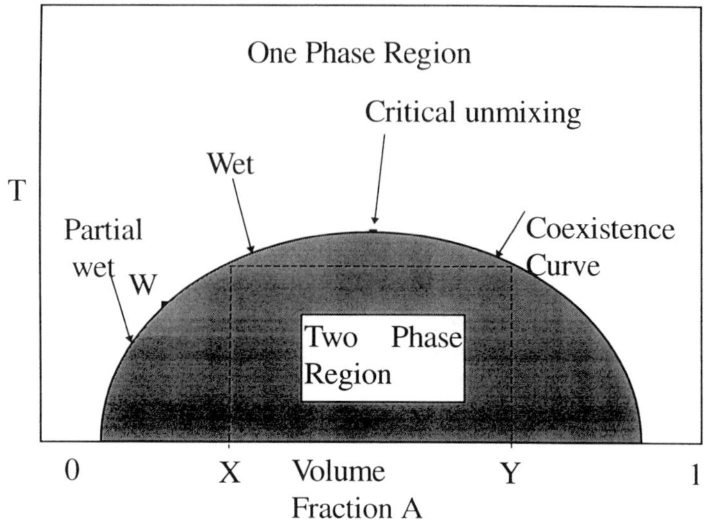

Figure 3 Idealized surface enrichment phase diagram.

phase behavior of molecular or atomic mixtures, except in the case of helium. Generally, mixing is driven by the increase in entropy of the mixed species as compared with pure components. However, in polymers the entropy of mixing is small because of the constraints applied by chain connectivity, this means that the very small unfavorable interactions introduced by isotopic substitution can force phase separation. This phenomenon was first observed in hydrogenated polystyrene (h-PS)/perdeuterated polystyrene (d-PS) by Bates and Wignall [35], who were also able to show that the size of the effect could be predicted using relatively simple arguments [36]. Subsequently this effect has been observed in a number of other systems [37–39] and precise measurements [40] of apparent χ values are a spur to produce improved theoretical models [41,42] of polymer blends.

Jones et al. [43,44] used this simple phase behavior to study surface enrichment in high molecular weight h-PS/d-PS blends. Deuteration is sufficient to give the d-PS a slightly lower surface energy than h-PS and drive it to the air–polymer interface, and initially FRES was used to establish the existence of a surface excess of d-PS. Subsequently neutron reflectometry was used to make a precise determination of the enrichment profile shape. It was found that the theory of Schmidt and Binder described the surface excess well, if a surface energy difference of 0.078 mJ m^{-2} was assumed; this is illustrated in Fig. 4. The form of the surface-enriched layer was found

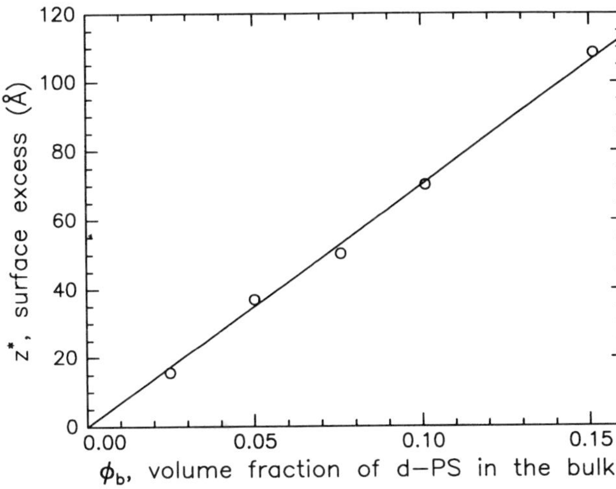

Figure 4 Measured surface excess as a function of bulk volume fraction (points) in a PS/d-PS blend; the solid line is a fit using the theory of Schmidt and Binder. Reproduced from Jones et al. [43] by permission of The American Chemical Society.

to be predicted quite well; however, there was some flattening of the profile at the surface which is not predicted by theory.

Following the work of Jones et al. the effect of chain length disparity was considered by Hariharan et al. [45], lower molecular weight chains are favored at the interface and they were able to force h-PS of low M_w to the surface. These results were supported theoretically by the self-consistent field of Hariharan et al. [46]. The effect of film thickness has also been considered both experimentally [47] and theoretically [48].

The deviations from mean-field theory, observed for the d-PS/h-PS blends have been studied more thoroughly by Norton et al. [49] for the poly(ethylene propylene) (PEP)/perdeuterated PEP (d-PEP)—in this case deviation from the mean-field theory is stronger, in part because the surface energy difference is rather larger (~0.21 mJ m^{-2}). Experimental profiles show a clear plateau region whose width is of the order of the correlation length. The comparison between the theoretical predictions of Schmidt and Binder and these profiles is shown in Fig. 5. It is suggested that the conformation of individual chains at the surface has a bearing on this observation. Individual chains in a blend are ellipsoidal in shape and recent Monte Carlo [50] models show that these ellipsoids align preferentially with their long axes parallel to the surface.

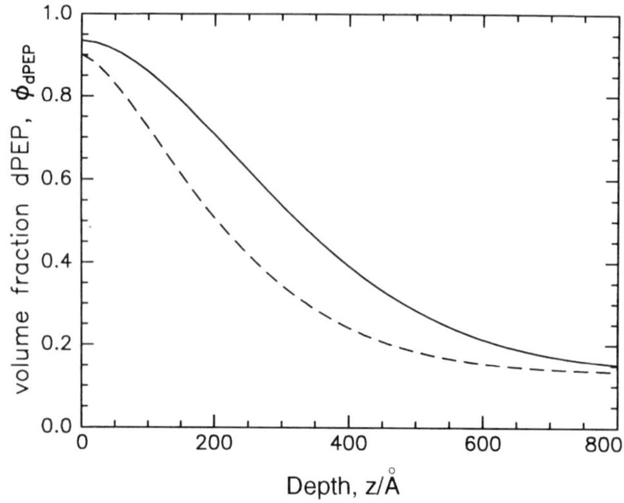

Figure 5 Near-surface concentration profiles for d-PEP/PEP blends (solid line) with theoretical predictions from the theory of Schmidt and Binder (broken line). Reproduced from Norton et al. [49] by permission of The American Chemical Society.

V. INTERDIFFUSION

Interfaces between polymers are of importance to polymer bonding (or welding) and by implication the strength of composite materials. Interdiffusion between polymer bilayers has been used to investigate the validity of the prevailing model of polymer dynamics—the reptation model. The reptation model is based on the ideas of Edwards [51] and was developed more fully by de Gennes [52]. The basis is that polymer chains are constrained to move along a tube defined by the surrounding polymers. This gives rise to a prediction that the diffusion of a polymer will vary with the timescale over which it is studied and only for the very longest times will polymer chains exhibit Fickian diffusion (i.e., mobility $\propto t^{1/2}$). At the shortest times, $t < \tau_e$, chains move within the width of the initial tube, essentially unconstrained and follow Rouse [53] dynamics. This relaxation happens rapidly and the expected exponent is 1/4. In the next regime up to the Rouse time, τ_R, the exponent is 1/8. This corresponds to the polymer chain rattling around in its initial constraint tube. For times up to the reptation time, τ_D, parts of the chain remain in the original constraint but motion beyond the ends of the tube begins—in this regime $t^{1/4}$ behavior is expected. Finally for times after which the polymer leaves its original constraint tube Fickian dynamics are observed. All of these regimes are expected for polymers above the so-called "entanglement" molecular weight, for PS and PMMA it is of the order of 300,000.

Several groups have studied the development of the interfaces between a polymer and its deuterated counterpart (h-PS/d-PS [54–56] and h-PMMA/d-PMMA [57]). The experimental procedure is to spin cast a thin polymer film (~1000 Å) and then float another similar layer onto it. This bilayer is then annealed and the width of the interface is determined using neutron reflectometry as a function of the annealing time. Fig. 6 shows the experimentally determined interfacial width obtained by Karim et al. [54] along with theoretical predictions from reptation theory. Note that the theoretical model contains no free parameters, it appears to fit the data qualitatively although the quantitative fit is not exact. Other experiments [58,59] have confirmed that Fickian diffusion is observed at longer times, neutron reflectometry is relatively insensitive to the broad interfaces that are characteristic of this final regime, and neither is it able to resolve the rapid relaxation at the shortest times; however, some supporting evidence for this regime has come from neutron spin echo experiments [60].

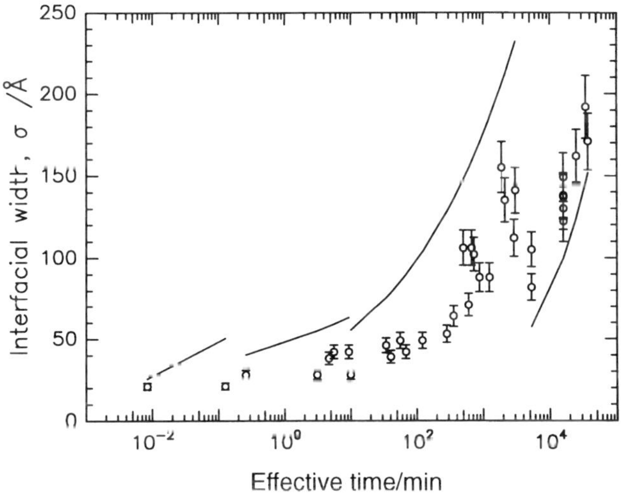

Figure 6 Interfacial widths for d PS/PS bilayers measured using neutron reflectometry, along with theoretical predictions. Adapted from Karim et al. [54] with permission from The American Chemical Society.

Clearly these experiments do give support to the reptation model. Agrawal et al. [61,62] have performed a further elegant experiment on bilayers. In this experiment isotopic triblock copolymers of polystyrene are used. For one layer a DHD triblock with the deuterated end blocks each half the molecular weight of the hydrogenous center block were used and for the other layer the "inverse" of this was used, i.e., an HDH triblock. The principle of this experiment is that if the polymer chains cross the interface with equal probability at any point along their length then the nuclear scattering length density across the interface will remain uniform. If, however, the ends of the chains cross the interface first, as would be predicted by reptation theory, then initially a discontinuous structure would develop at the interface. This structure arises because in the early stages of interdiffusion hydrogenous ends will move out of the surface region of the HDH triblock to be replaced by deuterated ends, thus increasing the nuclear scattering length density in the surface of the HDH triblock layer and conversely decreasing the nuclear scattering length density in the surface of the DHD triblock. At long times whole polymer chains will move across the interface which has no net effect on the scattering length density and the structure will be uniform once more. The structures observed by Agrawal et al. are shown in Fig. 7; the predicted discontinuous structure

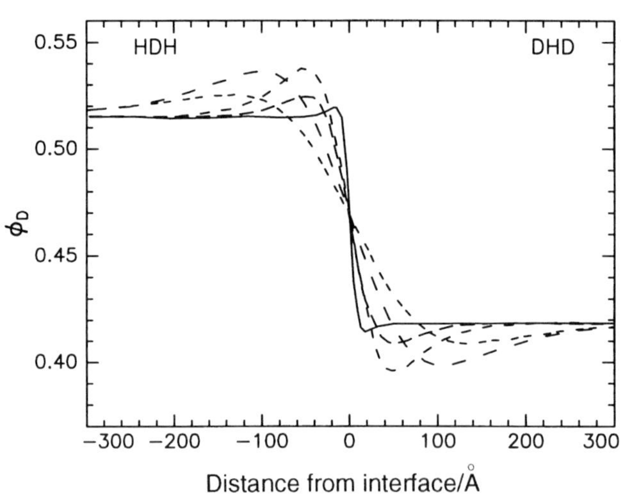

Figure 7 Interfacial scattering length density profiles for bilayers of HDH/DHD polystyrene triblock copolymers as a function of annealing. Reproduced from Agrawa et al. [62] by permission of The American Chemical Society.

develops as annealing commences and then decays away as annealing proceeds to longer times. These results give strong support to the reptation model of polymer dynamics.

VI. POLYMER BRUSHES

In this section we consider polymer brushes, i.e., polymers attached to an interface by one end with the rest of the polymer chain extending into the matrix which may be a low molecular weight solvent or another polymer. Polymer brushes are of interest because they are a route to modifying interfacial behavior and, in the polymer–solvent case, they are germane to the steric stabilization of colloids.

It is appropriate to summarize the theoretical predictions for brushes here, the first quantitative description being the mean-field description of Scheutjens and Fleer associated with a self-consistent field model [63,64]. This has been exhaustively described [65]. The scaling law approach of Alexander [66] and de Gennes [67] has the benefit of transparency but the most detailed theory is the self-consistent field approach of Milner et al. [68]. Although both scaling and self-consistent field approaches are identical in the main, they differ markedly in the shape of the segment density profile normal to the surface. Fig. 8 summarizes the scaling models. When the grafting density, σ, (number of end-attached polymer chains per unit surface area) is low the molecules do not overlap and each forms a "mushroom" on the surface. When molecular overlap is appreciable a stretching of the molecules takes place to minimize the free energy and the polymer layer becomes akin to a brush.

When the distance between neighboring molecules is large, each polymer occupies a half sphere of radius comparable to its radius of gyration in a good solvent,

$$R_g = N^{3/5}a \tag{21}$$

where N is the number of statistical steps of length a in the molecule and the volume fraction profile normal to the surface is (where σ_m is the notional maximum grafting density possible)

$$\phi(z) = \frac{\sigma}{\sigma_m}\left(\frac{z}{a}\right)^{2/3} \tag{22}$$

which is shown in Fig. 9. When the distance on the surface between grafted chains becomes less than R_g, the grafted chains can each be subdivided into the "blobs" adopted in the description of bulk semdilute solutions. Within each blob the subsection of the molecule is subjected to excluded volume

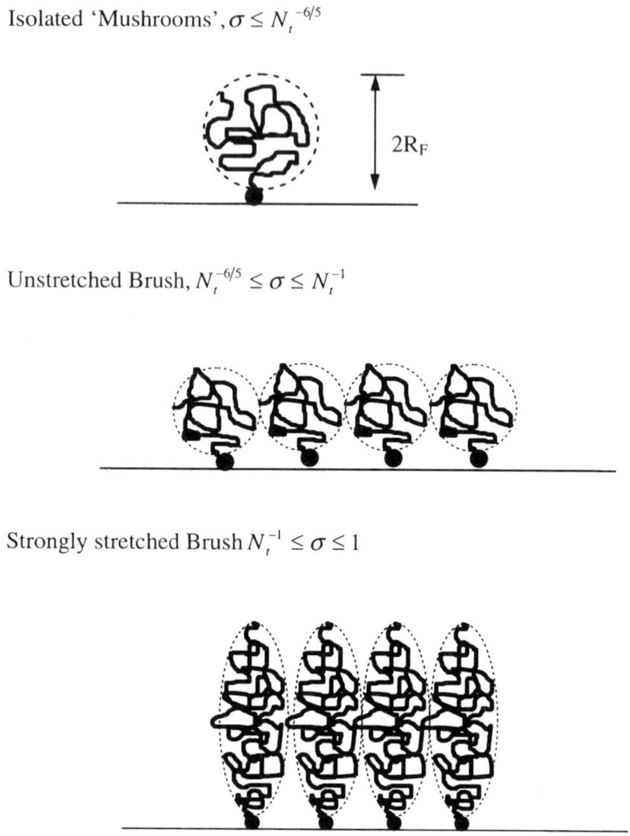

Figure 8 Sketches of polymer molecules forming brush-like layers at low, medium, and high grafting densities. N_t is the degree of polymerization of the grafted polymer.

effects and the whole chain becomes stretched. All the free chain ends are located at the same distance from the grafted wall and the layer thickness is

$$L \sim Na\sigma^{1/3} \tag{23}$$

The volume fraction profile is shown in Fig. 10. Milner et al. analytically derived the concentration profile normal to the surface. Individual configurations of grafted chains are replaced by an average path of the chains. The average path is then modeled by a particle falling to the surface in a potential force, since the grafted molecules all have the same number of units equal to the total time for the fictitious particle to reach the wall, then all particles must take the same time to reach the wall. The potential required to

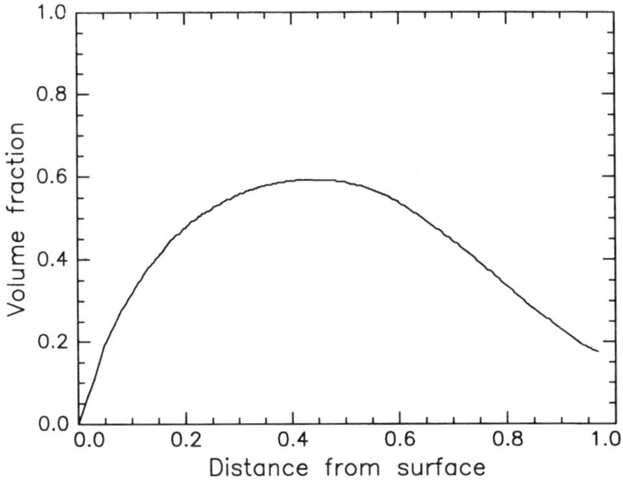

Figure 9 Volume fraction profile of segments for polymers grafted as isolated mushrooms.

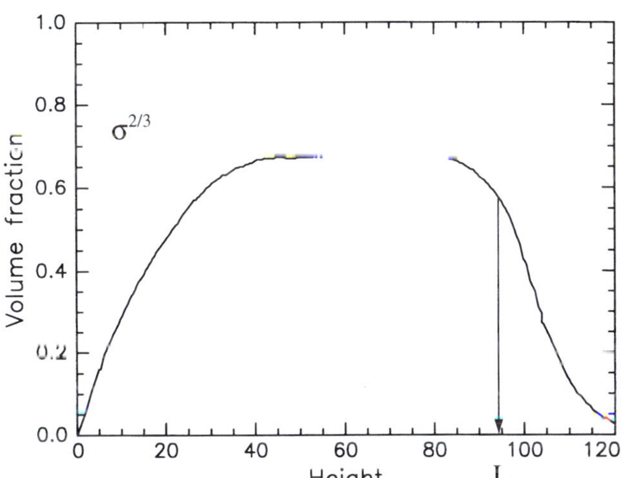

Figure 10 Volume fraction in a stretched brush layer from the Alexander–de Gennes model.

fulfill this equal time constraint is parabolic and moreover requires that the free ends of the grafted chain be distributed over the entire brush height rather than localized as in the Alexander–de Gennes models, indeed they show that configurations which do not allow for a distribution of ends are not at minima in the free energy. From this theory the equilibrium brush thickness is obtained as:

$$L = \left(\frac{12}{\pi^2}\right)^{1/3} \left(\frac{1}{2} - \chi\right)^{1/3} N\sigma^{1/3} a^{5/3} \tag{24}$$

Where χ is the polymer solvent interaction parameter and the volume fraction profile is

$$\phi(z) = \frac{\pi^2}{8N^2} \Big/ \left(\frac{1}{2} - \chi\right)[L^2 - z^2] \tag{25}$$

The form of Eq. (25) is shown in Fig. 11 together with the step profile for the same grafting density which is intrinsic to the Alexander–de Gennes approach. Other self-consistent field calculations have been developed by Shull [69] and others [70]—these are able to predict the segment density profiles over a larger range of circumstances. Additionally there have also been Monte Carlo models [71,72].

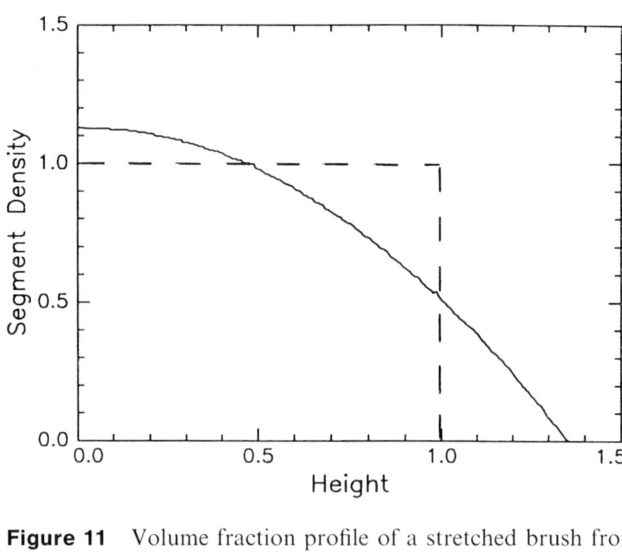

Figure 11 Volume fraction profile of a stretched brush from the analytical model of Milner et al. [68]. The broken line shows the Alexander–de Gennes model. Height is a normalized value.

We concentrate here on the use of neutron reflectometry to determine the segment density profiles of polymer–solvent and polymer–polymer systems. Although other techniques including force–distance measurements [73,74] and small-angle neutron scattering (SANS) [75,76] have been very important in the study of polymer–solvent brushes.

Karim et al. [77] have studied the effect of temperature on the behavior of polystyrene chemically grafted onto silicon in perdeuterated cyclohexane; this temperature variation corresponds to a variation of solvent quality. The density profile in perdeuterated toluene was also measured; toluene is a good solvent for polystyrene. The segment density profiles were fitted with functions of the form:

$$\phi(z) = \begin{cases} \phi_s(1 - (z/h)^2)^\alpha & z < h \\ 0 & z > h \end{cases} \tag{26}$$

where $\phi(z)$ is the polymer segment density as a function of distance from the surface, ϕ_s is the surface segment density, h, is the brush height and α is an arbitrary constant. Fig. 12 shows the fitted profiles for the d-toluene and d-cyclohexane as a function of temperature. For the good solvent, d-toluene, the brush exhibits the parabolic profile predicted by scaling theory, i.e., with $\alpha = 1$. For the d-cyclohexane the profiles become progressively flatter as the

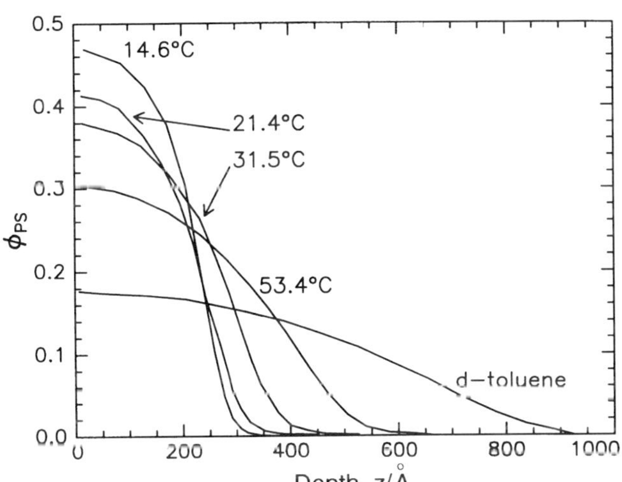

Figure 12 Segment density profiles for polystyrene brushes in d-cyclohexane as a function of temperature and for the same brush in d-toluene. Reproduced from Karim et al. [77] by permission of The American Chemical Society.

temperature and hence the solvent quality is decreased and for the θ condition an exponent of $\alpha = 0.5$ is obtained, which is as predicted by the scaling theory of Zhulina et al. [78]. Note that in all cases the brushes are strongly stretched from the equilibrium radius of gyration ($= 87$ Å), which is a necessary condition for the application of the analytical solution of the scaling theory.

Mansfield et al. [79] and Field et al. [80] have both studied the influence of molecular weight on the brush height in a solvent. The scaling theory prediction is that $h \sim N$. Both groups used physically grafted polystyrene where the graft density varies as a function of molecular weight. This variation can be incorporated into the scaling theory and the predicted scaling is that $hs^{2/3} \sim N$, where s is the average distance between graft points. Their data are summarized in Fig. 13, which clearly shows that the height of the brush scales with molecular weight as predicted by scaling theory.

If a block copolymer is spread at the air–liquid interface where the liquid subphase is a solvent for one block, then the grafting density of the solvated block may be controllable over a wide range by simply compressing the spread polymer film. This approach has been used by Lee and coworkers for a linear diblock copolymer of poly(dimethyl siloxane) (PDMS) and PS

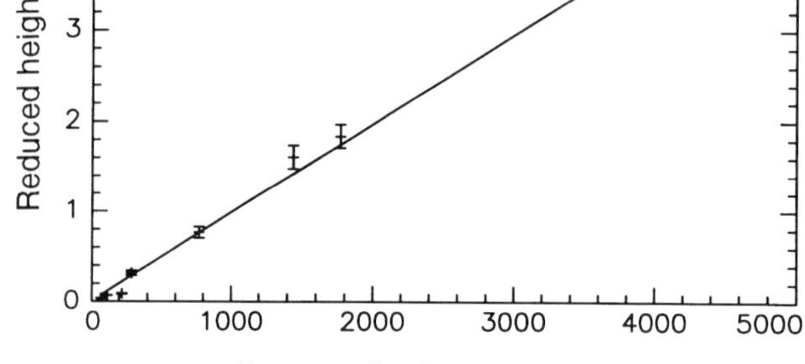

Figure 13 Scaling of brush height with molecular weight, in a solvent matrix, compiled from the data of Field et al. [80] and Mansfield et al. [79] (symbols), the straight line is a linear fit to the data.

spread on ethyl benzoate, a good solvent for PS but a nonsolvent for PDMS [81]. More recently [82], a poly(dimethyl siloxane)–deuteropolystyrene copolymer was spread on ethyl benzoate and a variety of volume fraction profiles were used to fit the data. They pointed out that good fits are obtained using a parabolic error function or stretched exponential volume fraction profiles. In all cases they were forced to include a depletion region near $z = 0$ and an exponentially decaying tail region at large z. The dependence of the brush layer thickness on both molecular weight and areal density is shown in Fig. 14 which includes the prediction of scaling theories. The solid line through the data is described by

$$h \sim \sigma^{0.22} M^{0.86}$$

The dependence of h on M and σ is rather weaker than that predicted by the Alexander–de Gennes scaling laws or the self-consistent field theory of Milner et al. ($h \sim \sigma^{0.33} M$). The self consistent field theory of Baranowski and Whitmore [83] predicts weaker dependences of layer thickness on areal density and molecular weight. The volume fraction profiles obtained by Baranowski and Whitmore are nearly identical to those obtained from the analysis of the reflectivity profiles.

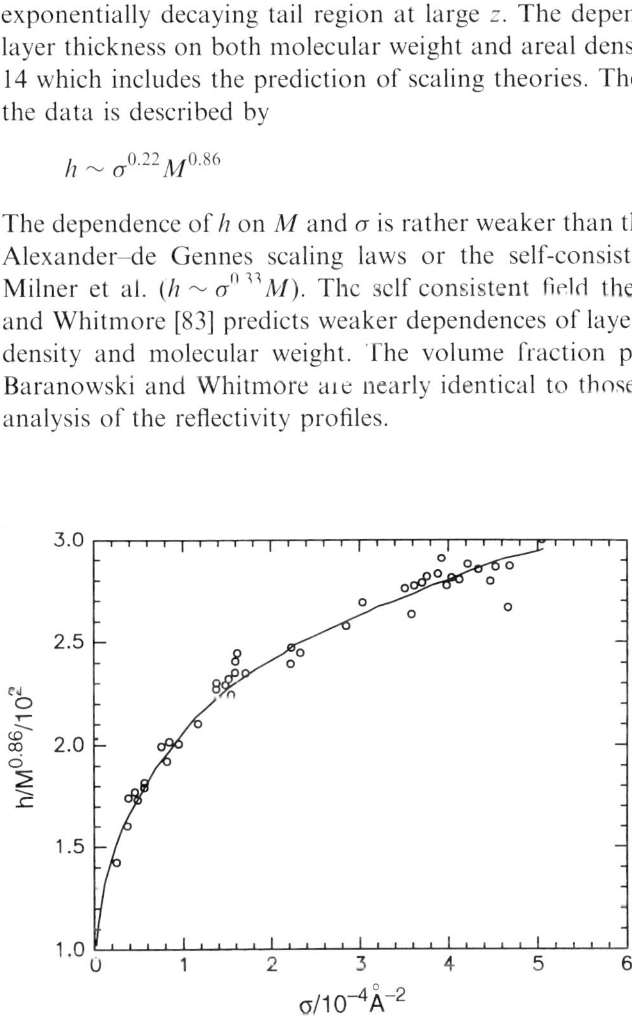

Figure 14 Dependence of brush layer thickness on molecular weight and grafting density obtained for the solvated polystyrene layer in a spread film of a diblock copolymer on ethyl benzoate. Adapted from Ref. [82].

A "phase diagram" has been proposed [66] for grafted molecules immersed in a solution of free polymer molecules of the same chemical type. This diagram of different regimes is drawn in the plane of σ and Φ where Φ is the volume fraction of free polymer (of degree of polymerization P) in solution. Fig. 15 reproduces the original diagram of de Gennes. When Φ of the free polymer is less than that in the grafted layer, the concentration profile is essentially unchanged from that in the absence of the free polymer. This corresponds to the unstretched, unmixed, and strongly stretched, unmixed regimes. When the volume fraction of the chains is greater than the threshold volume fraction of grafted chains at which stretching begins, a

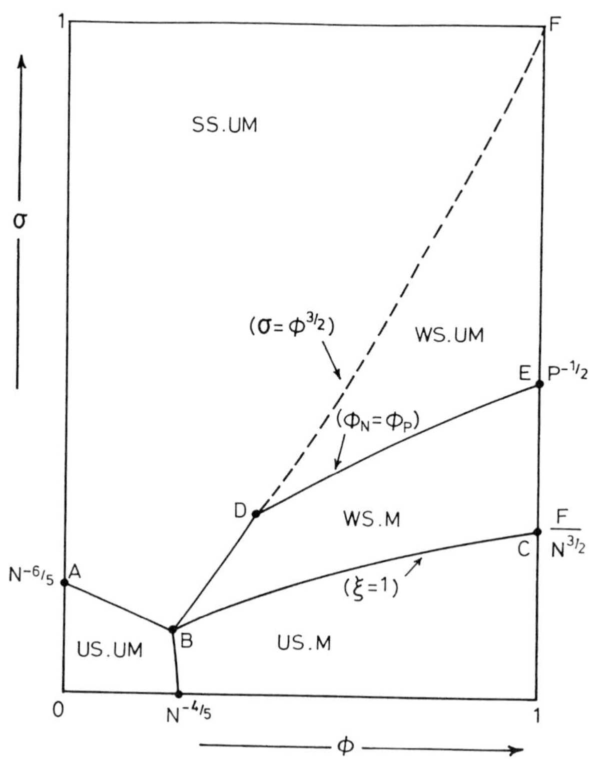

Figure 15 Regimes of behavior for grafted brushes (degree of polymerization N) in the presence of free polymer in solution. Degree of polymerization of free polymer, P, volume fraction ϕ. S—stretched; SS—strongly stretched; WS—weakly stretched; US—unstretched; M—mixed; UM—unmixed.

decrease in the layer thickness is predicted but this can be achieved in two ways. In the first, the free chain volume fraction is much greater than that of the grafted chains and the mobile chains dominate. In this weak stretching mixed regime:

$$h \sim \sigma^{1/3} N P^{-1/3} \Phi^{5/12} \tag{27}$$

When the grafted layer volume fraction exceeds that of the mobile polymer in the same spatial region there is a regime of unmixed, weakly stretched behavior where $h \sim \sigma N \Phi^{-1}$. The crossover from this region to the strongly stretched unmixed region is defined by a line defining the threshold volume fraction in the layer for strong stretching to become evident, i.e., $(\sigma/\sigma_{\mathrm{m}})^{2/3}$, however, this crossover is anticipated to be gradual. By contrast at lower values of $(\sigma/\sigma_{\mathrm{m}})$ the transition line BD defines an abrupt decrease in the value of h.

Attempts to define these regions experimentally have been made using a linear diblock copolymer of PDMS and deuterated PS spread on ethyl benzoate [84]. Reflectivity data were fitted using a model profile with an initial parabolic depletion layer, a different parabola describing the main part of the profile and an exponential tail for the distal end of the profile (Fig. 16). The presence of free polymer decreases the layer thickness (Fig. 17) and the boundary lines between some of the regimes discussed above were defined. Thus the crossover between strongly stretched unmixed and

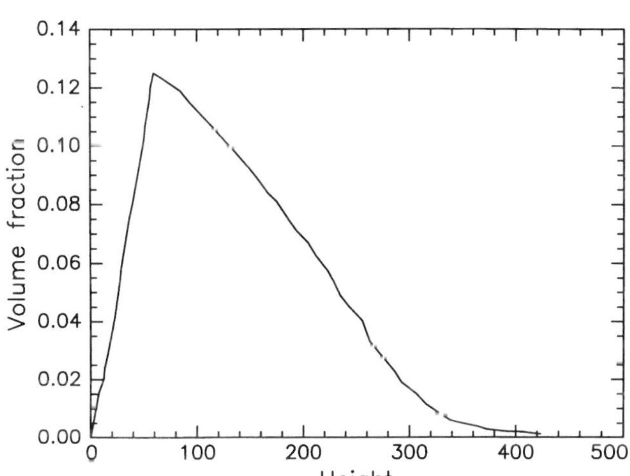

Figure 16 Volume fraction profile for a solvated polystyrene block in a poly(dimethyl siloxane)–polystyrene diblock copolymer spread on ethyl benzoate.

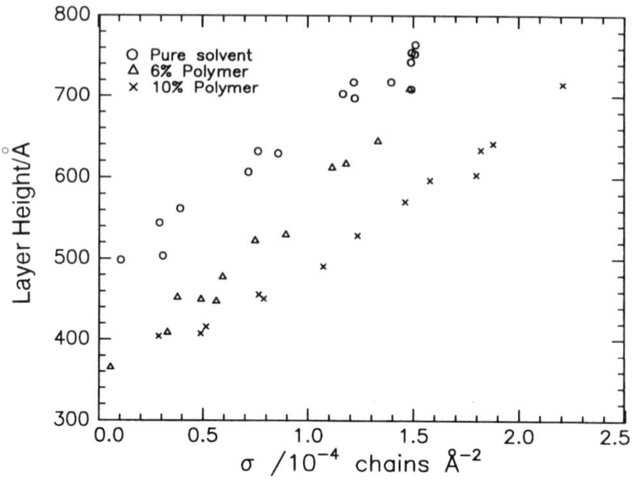

Figure 17 Brush layer thickness in the presence of a free polymer in solution. The molecular weight of the free polymer 43,000. ○—pure solvent; △—6% polymer; ×—10% polymer. Adapted from Ref. [84].

weakly stretched unmixed was taken to correspond to the volume fraction of monomer units in the grafted chain when surrounded by pure solvent and this volume fraction was determined from values of h and σ. The crossover line between the weakly stretched, unmixed (grafted chain dominant) and weakly stretched, mixed (free chain dominant) regions was defined by the volume fraction of grafted chains where free chains just begin to penetrate the layer.

The study of brushes in polymer melts has been less extensive than that of brushes in low molecular weight solvent matrices. For brushes in a polymer matrix, numerical solution of the self-consistent field has been a more appropriate theoretical method, although some important results have been obtained by scaling theory. Short matrix molecules engender a large entropic penalty when excluded from the brush and consequently penetrate the brush-like layer. This situation approximates to the brush in the presence of solvent, i.e., a "wet" brush. For matrix molecules with $P > N^{1/2}$ the matrix is less favored in the brush and for higher molecular weights the brush profile will become invariant with the matrix molecular weight, such a brush is known as "dry" although there is still an appreciable fraction of the matrix within the brush. For such brushes the height should scale as $h \sim N\sigma$. Indeed this scaling is found to hold. Fig. 18 shows a compilation of data obtained using NR by Jones et al. [85] and Hopkinson et al. [86] and

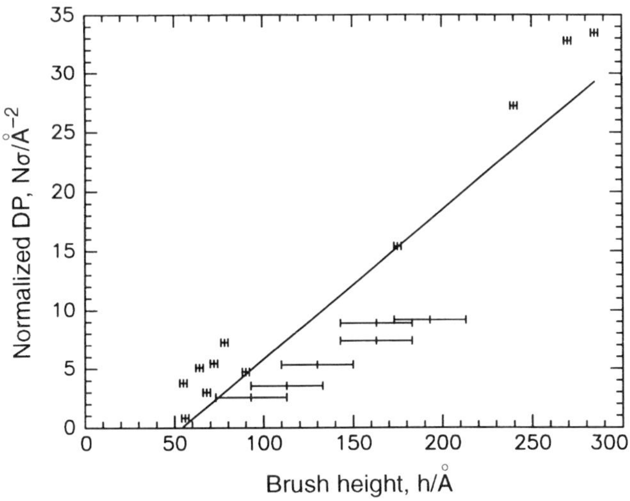

Figure 18 Scaling of brush height with molecular weight, in a polymer matrix, compiled from the data of Hopkinson et al. [86], Jones et al. [85] and Budkowski et al. [87]. The straight line is a linear fit to the data.

using NRA by Budkowski et al. [87]. All of these brushes were of polystyrene; the work of Jones et al. utilized a butadiene–silane group to attach the brush to the substrate, Hopkinson et al. used a perfluorohexane group to attach to the air–polymer interface and Budkowski et al. used a polyisoprene block to anchor a polystyrene/polyisoprene diblock copolymer to the air–polymer interface. The deviations from the linear fit for the results of Budkowski et al. probably arise from the use of the polyisoprene as an anchor block and possibly from instrumental effects from NRA. The surface activity of the perfluorohexane group was initially shown by Affrossman et al. [88] and Hunt et al. [89], Schaub et al. [90] has recently published near-surface composition profiles for lower molecular weight systems. Elman et al. [91] has investigated the distribution of the end groups through thin films; near the surface a periodic regions of enrichment and depletion are observed.

The groups of Jones and Hopkinson were both able to use the SCF theory of Shull to model the near-surface composition profiles of the brushes. The degree of agreement is good and for the perfluorinated end group a fair estimate for the sticking energy can be obtained from literature data, i.e., the relative surface energies of polystyrene and a perfluorinated group and its estimated solubility parameter.

VII. COPOLYMERS AT INTERFACES

Studies of polymer brushes and of copolymers at interfaces are in some ways very closely linked. In many situations a diblock copolymer will form a brush at an interface which is effectively identical to those discussed in the previous section. Indeed as remarked earlier, some of the data in Fig. 18 are from such a system where the polyisoprene (PI) segment of a PI–d-PS diblock copolymer forms the anchor for the brush. Similarly the end-fluorinated polystyrene polymers used by a number of groups could be described as poly(tetrafluoroethylene) (PTFE)–d-PS diblock copolymers.

Copolymers are used industrially to "compatibilize" mixtures of immiscible polymers [92]; in many instances it is desirable to mix two polymers together to form a composite material. Typically polymers are immiscible with each other and have a sufficiently high interfacial tension which makes the dispersion of one polymer in the other difficult, and not necessarily stable when achieved. However, if, for example, an AB diblock copolymer is added to the immiscible mixture of the A and B homopolymers then the copolymer will locate preferentially at interfaces between the A and B homopolymers, with the A part of the copolymer in the A homopolymer and the B part of the copolymer in the B homopolymer. The effect of this is to reduce the interfacial tension and improve the adhesion between the homopolymers since segments of the copolymer are entangled with both homopolymers. The reduction of the interfacial tension in the presence of a diblock copolymer has been observed experimentally [93–96]. For low concentrations (1–5%) of added copolymer the interfacial tension drops dramatically but then the interface saturates with the copolymer and no further reduction is observed. The improved adhesion between the homopolymer phases [97,98] has also been measured. There have been a number of theories describing the behavior of diblock copolymers at interfaces [99–101], these show that there are two competing effects in locating the copolymers at interfaces; first of all unfavorable enthalpic interactions are avoided by having the A and B segments of the copolymer in the appropriate homopolymer, secondly there is an entropic penalty in confining the junction point of the diblock copolymer to the interfacial region. Copolymers, in common with "normal" surfactants, will self-associate to form micelles above a certain critical concentration—in general this is undesirable since in a micelle they are not performing a useful function at the interface.

Lyatskaya et al. [102], Gersappe et al. [103,104] and Balazs et al. [105] have used a range of theoretical methods, to predict the optimum architecture for compatibilizing agents. Although diblock copolymers are in many ways ideal, they are expensive and may be difficult to synthesize. Comb or graft copolymers are frequently easier to synthesize, where B chains are grafted onto an A backbone. These authors concluded that combs with a few long teeth were the most effective alternative to diblock copolymers, and suggested a novel system of using AC and BC graft copolymers to compatibilize an A–B mixture of homopolymers, where the C graft is immiscible with both of the homopolymers. Such copolymers were predicted to segregate to the homopolymer interfaces with the C grafts interpenetrating in order to improve interfacial adhesion. These predictions were tested [106] on a poly(ethyl acrylate) (PEA)–PMMA mixture with added PEA–PS and PMMA–PS graft copolymers. Note that the PS is immiscible with both PEA and PMMA. An improvement in the mechanical properties of the mixture was observed and dynamic SIMS was used to show that the graft copolymers were located in the interface, as predicted.

As mentioned previously, when a copolymer is added to a system there is scope for micelles of the copolymer to form. Shull et al. [107,108] and Dai et al. [109,110] have made an extensive study of the (immiscible) polystyrene poly(vinyl pyridine) (PVP) system with added PS–PVP diblock copolymer. At higher volume fractions of PS–PVP micelles formed and were observed, using transmission electron microscopy, to segregate to both free and polymer–polymer interfaces, this is shown schematically in Fig. 19. This segregation is driven by two effects; firstly there are slightly unfavorable enthalpic interactions between the d-PS of the micelle "corona" and the h-PS matrix—these interactions are reduced if the micelle is at the surface; secondly the interface between the d-PS corona and the hompolymer is entropically unfavorable and so again a corona—interface or a corona—brush contact is preferred. Similar behavior has been observed by Bucknall et al [111] for the PS–PMMA system with added PS–PMMA diblock copolymer.

Fleischer et al. [112] have investigated a novel method for supplying the copolymer to interfaces, avoiding some of the problems that diblock copolymers present. They have studied a poly(dimethyl siloxane) (PDMS)-poly(butadiene) (PBD) system where the PBD component contained a proportion of carboxy terminated PBD and PDMS contained a proportion of amine-terminated PDMS. The intention was that the carboxy and amine end groups would react at the interface to form a copolymer *in situ*. Interfacial tension measurements showed that this complexation did occur. Qualitative modeling of the observed effect was possible using reaction equilibrium data.

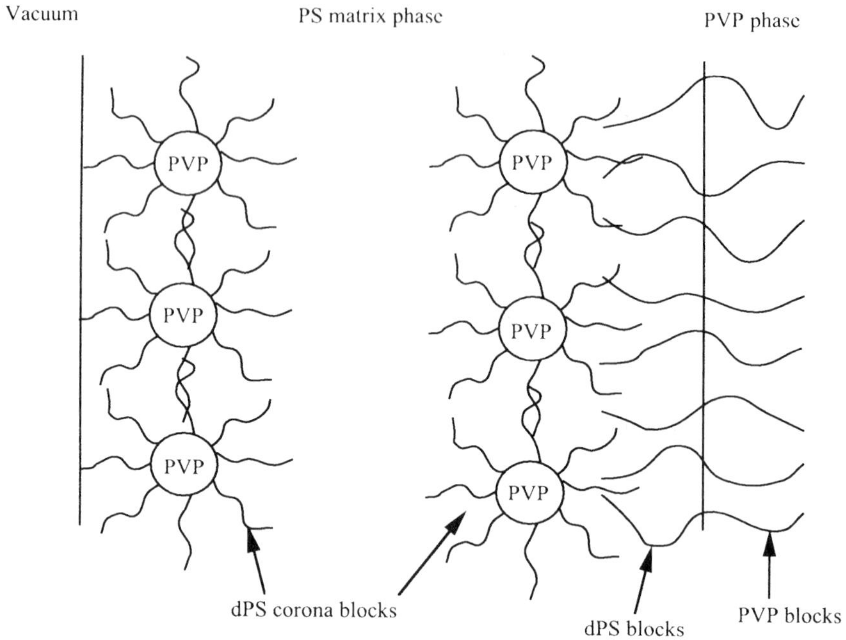

Figure 19 A schematic diagram of polymer micelles forming at interfaces.

VIII. COPOLYMERS IN THIN FILMS

Diblock copolymers are known to exhibit a wide range of morphologies in the bulk [113], which result from the competing forces of phase separation driven by unfavorable segment–segment interactions and the topological constraint of having the two parts of the diblock covalently joined. The phase behavior of a diblock copolymer is determined by two parameters, the volume fraction of the minority component, f, and the product χN. For small χN the unfavorable interactions are small and the copolymer is homogeneous, as χN increases composition fluctuations develop which distort the chains from Gaussian statistics and finally for $\chi N > 10.5$ (when $f = 0.5$) the so-called strong segregation regime is reached in which the copolymer microphase separates. Since χ typically decreases with temperature there is usually a temperature above which the copolymer is disordered and below which it is ordered and in the strong segregation limit this is known as the order–disorder transition (ODT). It is the parameter f that determines the morphology of the microphase separation, for $f \approx 0.5$ a lamellar morphology is seen, as f decreases from this value a cylindrical form is attained and for the smallest values of f spherical domains are observed; for certain values of f between those defining lamellar and cylindrical phases, a gyroid structure occurs.

In addition there have been studies of diblock copolymer morphology in thin films [114–116]. The reasons for this are twofold, firstly the influence of the interfaces on the morphology is of itself interesting, secondly, the thin-film geometry allows a more detailed study of the interfacial region between the phase-separated domains. The interfacial region will be important in determining the mechanical properties of the bulk copolymer. It has been found that the presence of the interfaces induces order in the microphase-separated domains—this has been most commonly observed in copolymers having lamellar morphology [117–121], however, there is also evidence that cylindrical domains may be orientated parallel to the surface [122] and that the interface may also have an influence on the ordering of spherical domains near surfaces.

For symmetric diblock copolymers (lamellar morphology) it is found that the domains extend continuously over a wide area and the orientational order may extend for many (of order 15) lamellae into the film. The lamellar period, L, which is the length spanning one domain of each part of the diblock perpendicular to the surface, is found to scale as $L \sim N^{0.66}$ as predicted in the strong segregation limit. The lamellae found at the interfaces of the film are of half thickness, i.e. $L/4$. For poly(styrene-*b*-methyl methacrylate) diblock copolymers the PMMA is found at the polymer–substrate interface, where the substrate is SiO_2, and polystyrene is found at the air–polymer interface, driven there by its lower surface energy. In the poly(ethyl ethylene) (PEE)–poly(ethylene propylene) (PEP) system it is the PEE block that is found at both interfaces. Even for copolymers above the ODT some near-surface organization is observed, a decaying sinusoidal distribution is found perpendicular to the surface, as predicted by Fredrickson [123].

Given the occurrence of lamellar domains aligned parallel to the surface there is a "natural" film thickness of nL, or $(n + 1/2)L$ if a different species is found at each interface, when n is an integer. The question arises as to what happens in a cast film where the original film thickness does not correspond to a natural film thickness. For PS–PMMA block copolymer films with a free surface, holes and islands will form with depth (height) L. These features form relatively slowly and do not occur for sufficiently high molecular weight systems where the energy cost of generating sufficiently large islands is too high. In systems where both interfaces are hard, hole and island formation is not possible [124,125] and the domains will stretch or compress in order to retain an integer number of domains in the film.

The formation of holes and islands is an interesting process in itself, since it can be viewed as an example of two-dimensional phase segregation [126–128], because the islands (or holes) are many times thinner than their lateral dimensions. They may exhibit either spinodal-like or nucleation and growth-like behavior depending on how far from an ideal thickness the original layer was. The edges of the holes and islands are not sharp, Carvalho and Thomas [129], have made analogies with liquid crystalline behavior to explain the complex structures observed.

IX. POLYMERS SPREAD AT AIR–LIQUID INTERFACES

The organization of polymers at the interface between two fluid media is relevant in a number of areas: liquid–liquid extraction, stabilization of emulsions, wet bath spinning, etc. By far the most extensive body of work reported has been for cases where one of the fluid media is air and the liquid phase is water, although results for copolymers at organic liquid–air interfaces have been discussed earlier. Compression of spread polymer films produces a controllable increase in surface concentration and allows the surface pressure isotherm to be established. Although surface pressure–area isotherms for spread polymer films have been reported for over 40 years, discussion is mainly qualitative—much of this work has been summarized in Ref. 130. The development of scaling theory has been the most significant recent development in the theory of polymer solutions [131] and the scaling laws can be applied to surface pressure data.

For a constant temperature and for a polymer surface concentration Γ_s, which places the polymer in the semidilute region, the scaling law surface pressure π is

$$\pi \sim \Gamma_s^{2\nu/(2\nu-1)} = \Gamma_s^y \qquad (28)$$

where ν is the scaling exponent in the relation between the bulk radius of gyration of the polymer (R_g) and its degree of polymerization, N, i.e., $R_g \sim N^\nu$. If the subphase is thermodynamically favorable for the polymer, then $\nu = 0.75$. For theta conditions at the surface a range of values have been proposed such that $0.505 \leq \nu \leq 0.59$, a generally accepted value appears to be 0.571. At film collapse, ν has a value of 0.5. A double-logarithmic plot of surface pressure as a function of surface concentration should show clear differences in the slope depending on the nature of the interaction between the polymer and subphase. For thermodynamically favorable situations $y = 3$. For the theta state y should take on a higher value with 101 and 6.6 being the limits corresponding to the range of ν values given above.

Few polymers have been subjected to a scaling analysis of their isotherms probably because few homopolymers form a stable film at the air–water interface. Vilanove and coworkers [132,133] were the first to apply scaling laws. In the semidilute state no influence of molecular weight on π was observed. Kawaguchi and coworkers [134,135] reported surface pressure studies of both poly(methyl acrylate) and PMMA. For the acrylate polymer a theta temperature of 18.5°C was suggested on the basis of the two-dimensional second virial coefficients in the equation of state for π and ν had a value of 0.51. This value of ν has been questioned by the work of Poupinet et al. [136] who obtained $\nu = 0.7$ over a temperature range of 5°C to 30°C for the same system. Very careful measurements of π for a series of different molecular weights of PMMA were also reported. For a temperature of 15.5°C a value of ν of 0.55 was obtained from both the molecular weight dependence of the two-dimensional virial coefficient, A_{22}, and the concentration dependence of π above Γ_s^*. These findings show that excluded volume exponents obtained from data in the semidilute region are as accurate as those obtained from surface pressure data when the surface concentration is less than Γ_s^*, where much greater sensitivity in the measurement of π is required. Secondly, since the A_{22} values were negative then these data support the notion that ν at the surface theta point must be greater than 0.505 and a value of 0.57 seems more acceptable.

X. ORGANIZATION OF POLYMERS AT THE AIR–WATER INTERFACE

Even when one component of the polymer film is soluble in the aqueous subphase, the high surface tension of water generally results in the formation of a "pancake" layer of the soluble portion, i.e., there is little stretching of the molecules into the subphase. There is currently no general theory of organization of polymers at the air–water interface which can be compared to experimental data. General questions that can be addressed include: queries regarding the thickness and composition of the spread films and the influence of surface concentration on these parameters. If the spread polymer is amphiphilic, the distribution in space of the hydrophobic and hydrophilic regions becomes important. Apart from the work involving PMMA and PEO, there has been no concerted, systematic examination of the organization in spread polymer films.The results for the organization of homopolymers of PMMA and PEO at the air–water interface are summarized here. Extensions to a linear diblock copolymer and a graft copolymer are discussed and the application of the kinematic approximation to the neutron reflectometry data is exemplified. This last aspect is reinforced by data for poly(lauryl methacrylate).

Surface pressure isotherms of PMMA have been known to be influenced by the stereochemistry of the polymer used [137]. However, this strong dependence does not appear to have been recognized in the scaling law analyses of the isotherms referred to earlier except by Rondelez. Analysis of spread monolayers of both isotactic and syndiotactic PMMA by neutron reflectometry showed that the thickness of the spread films was the same and of constant value at about 18 Å over a surface concentration range from 0.2 to 2.0 mg m^{-2} [138]. Surprisingly the water content of the spread films for both stereotactic isomers is approximately equal. The air content is much higher for the isotactic polymer of the spread films. Surface concentrations calculated from the layer thickness and scattering length density for the syndiotactic polymer agreed very well with the amount actually dispensed on the aqueous subphase. For the isotactic polymer, the calculated surface concentration was always less than the amount spread and above a spread concentration of about 0.5 mg m^{-2} the discrepancy was very significant. It was suggested that this was evidence of large loops and tails being formed in the subphase which were so diluted by the subphase that they made an insignificant contribution to the reflectivity.

Insufficient combinations of polymer and subphase were used for a full analysis via partial structure factors; limited use allowed determination of thickness for the spread polymer layer and the near-surface water layer [139,140]. Uniform layer models fitted the partial structure factors for both polymer and water, the thickness of each layer being approximately equal. In the absence of any information about the separation between the layers, the similarity of thicknesses was interpreted as evidence for a uniform distribution of water throughout the spread polymer film.

This reduced partial structure factor analysis has also been applied to the isotactic polymer. For surface concentrations less than 0.8 mg m^{-2} a uniform layer model fitted the partial structure factor data for the spread polymer. Above this concentration, a Gaussian partial structure factor model fitted best. There was evidence that some of the polymer penetrates the subphase and becomes so dilute that it does not contribute to the neutron reflectivity. These data also illustrate the limitations in model discrimination due to the restriction in the useful Q range because of background scattering, which is mainly incoherent.

PEO is a water-soluble polymer which also forms a stable spread film at the air–water interface. It has been suggested that the spread polymer film undergoes some form of structural change at a surface concentration of about 0.4 mg m^{-2} and this is supported by surface quasi-elastic light-scattering data and interpretation of isotherm data [141,142]. Partial structure factor analysis of neutron reflectometry data confirms that an organizational change takes place at 0.4 mg m^{-2}. Below this concentration the reflectivity data for the spread PEO film is described by a single uniform layer model. At higher surface concentrations of polymer, a Gaussian segment distribution of polymer segments is more suitable. Fig. 20 shows the distribution of polymer and water, note that the polymer is highly diluted by the subphase. Addition of magnesium sulfate to the solution produces less favorable thermodynamic conditions (0.39 M aqueous $MgSO_4$ at 315 K is a theta solvent for PEO. For a spread PEO layer with a surface concentration of 0.8 mg m^{-2}, i.e., where there is penetration of the subphase by the polymer, addition of the salt to the subphase results in a compression and densification of the surface layer.

A full application of the partial structure factor analysis of neutron reflectometry data has only been applied to two polymers to date. One of these is a block copolymer of PMMA and PEO where each block could be separately deuterated [143]. Although complete labeling enables a full description of the organization of the block copolymer spread film to be

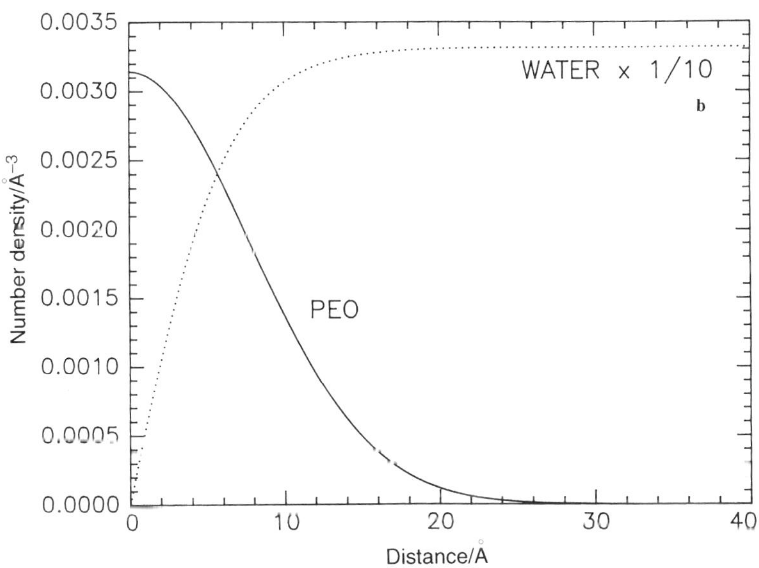

Figure 20 Distribution of ethylene oxide segments and water molecules in a spread film of PEO at the air–water interface at a surface concentration of 0.8 mg m^{-2}.

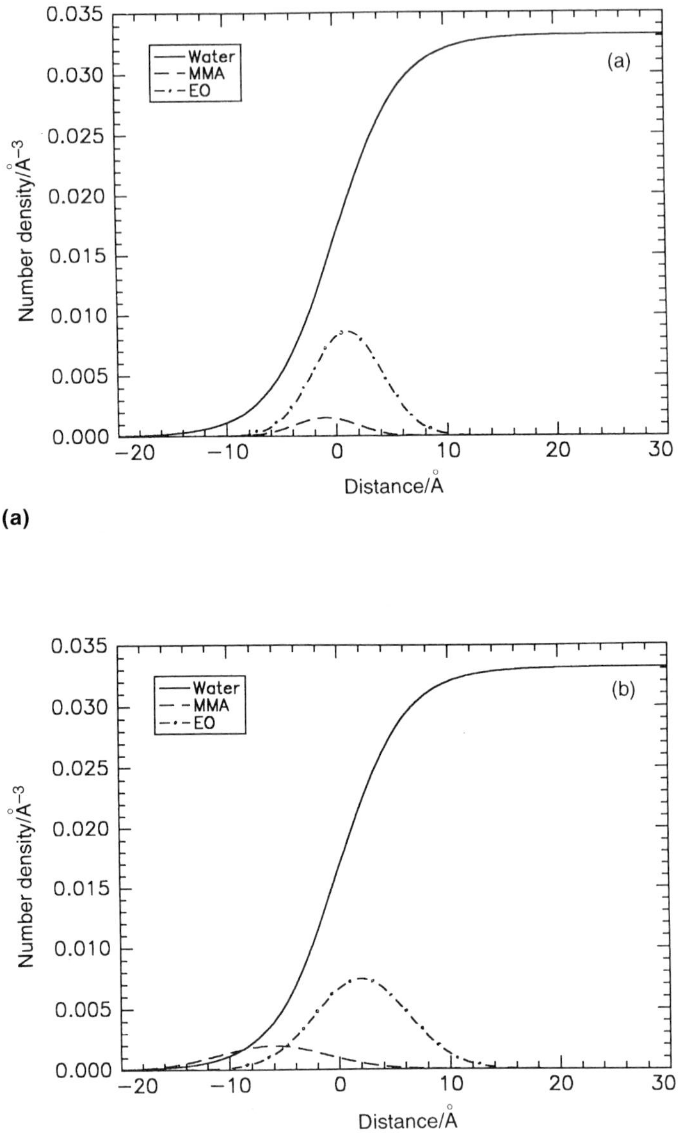

Figure 21 Segment distributions in a linear diblock polymer of PEO and PMMA (50 mol% composition) spread at the air–water interface. Surface concentrations are: (a) 0.6 mg m^{-2}; (b) 1.2 mg m^{-2}.

obtained, unambiguous conclusions can only be made if the composition of each of the labeled species is identical. Absolutely identical compositions of the variously labeled block copolymers is impossible to obtain. Some account of the variations in composition can be made by adjusting scattering lengths in Eq. (18). At reasonably high surface concentrations neutron reflectometry showed that the PMMA and PEO blocks occupied different spatial regions, the PEO being essentially submerged. At a much lower surface concentration of block copolymer, the two components are mixed, hence a demixing of the two blocks takes place above some critical surface concentration. Fig. 21 shows the distributions obtained from the partial structure factor analysis when the center-to-center separation has been obtained from the cross-partial-structures factors and the equations for these terms given by Eqs. (19) and (20).

Spread films of poly(lauryl methacrylate) (poly(dodecyl methacrylate)) have been investigated by neutron reflectometry and partial structure factor analysis because the main chain backbone and dodecyl substituent can be separately deuterated [144]. From such analysis the distributions of backbone segments, dodecyl substituent, and water molecules shown in Fig. 22 were obtained. Notable features of these distributions are: firstly the methacrylate backbone is completely immersed in the subphase—this contrasts

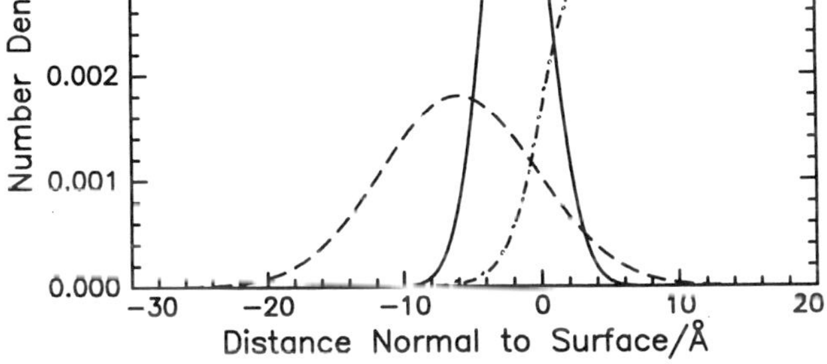

Figure 22 Backbone (–), lauryl substituent (— — —), and water (● – ● – ●) (divided by ten) for poly(lauryl methacrylate) at a surface concentration of 1 mg m^{-2} spread at the air–water interface.

with the arrangement of PMMA; secondly, the hydrocarbon substituent penetrates deeper into the subphase than the immersed backbone. At first glance this seems most unlikely, however all of this behavior can be attributed to the dodecyl substituent. Its hydrophobic nature forces the backbone into the subphase, to relieve short-range steric interactions between substituents rotations about the main chain bonds force the dodecyl units slightly deeper into the subphase before sufficient bonds in the substituent rotate to direct the dodecyl unit out of the aqueous phase. A change in direction of almost 180° requires about five bonds to be accomplished. Immersion of hydrophobic species in spread copolymers of styrene and maleic anhydride has also been observed [145]. Here again it appears that it is the varying degrees of hydrophobicity which determine which of the units on the polymer backbone are in the aqueous phase or in the air phase.

XI. POLYMER SOLUTIONS

The variation of surface tension of a polymer solution with composition is governed by the Gibbs equation in the same manner as low molecular weight molecules. If the surface tension of a polymer solution decreases with increasing concentration there will be a positive surface excess, i.e., in the region of the air–solution interface the polymer will be at a greater concentration than in the bulk. Conversely, if the surface tension increases with concentration, a depletion layer will be formed at the air–solution interface.

Two theoretical descriptions have been put forward for the organization of the surface excess layer. A mean-field description which gives

$$\phi(z) = \phi_{\rm b} \coth^2\left[\left(\frac{\phi_{\rm b}}{\phi_{\rm s}}\right)^{1/2}\left(\frac{2+D}{D}\right)\right] \tag{29}$$

where $D = a/(3\nu\phi_{\rm s})^{1/2}$ with a the monomer size, $\phi_{\rm s}$ the surface volume fraction in a solution where the bulk volume fraction is $\phi_{\rm b}$ and ν is the excluded volume parameter of the polymer in solution $(= a^3(1/2 - \chi))$. For a dilute solution or where there is strong absorption to the liquid–air interface, then a region of length scale L in the center of the surface excess region can be defined such that $L = a/(3\nu\phi_{\rm s})^{1/2}$ and in this case

$$\phi(z) = \phi_{\rm s}(D/(z+D))^2 \qquad 0 < z < L \tag{30}$$

Three length scales are defined in the scaling theory description of a surface excess layer:

1. proximal region:
   $0 \le z \le D$, where $D = a\delta^{-1}$ and $-\delta k_B T$ is the absorption energy between a monomer unit and the surface. For $z \le a$ then $\phi(z) = \phi_s = \delta$; for $a \le z \le D$ then

$$\phi(z) = \frac{a}{D}\left(\frac{a}{z}\right)^{1/3} \tag{31}$$

2. central region: $D \le z \le L$

$$\phi(z) = (a/z)^{4/3} \tag{32}$$

3. distal region: $z \ge L$

$$\phi(z) \propto \phi_b \exp(-kz) \tag{33}$$

In solutions of poly(dimethyl siloxane) in toluene, where a surface excess is formed, the scaling laws of Guiselin and coworkers [146–148] describe the volume fraction profiles normal to the surface of the solution quite well. The central region of the surface excess layer exhibits "self-similar" behavior due to repulsions betweeen monomer units of the loops and tails of the adsorbed polymer and L is approximately equal to the radius of gyration of the polymer. Reflectometry data on very dilute solutions for a range of polymer molecular weights gave an exponent of (4/3) in the central region and values of L were in very close agreement with the radii of gyration of the polymers. For these polymers the volume fraction profile could be represented by:

$$\phi(z) = 0.96 \quad 0 < z/\text{Å} \le 5$$

$$\phi(z) \simeq \left(\frac{5.2}{z + 0.1}\right)^{1.3} \quad 5 \le z/\text{Å} \le L$$

$$L \sim 0.13 M_w^{0.57}$$

Toluene solutions of polystyrene have a depletion layer because the surface tension of toluene is lower than that of polystyrene. For semidilute solutions of polystyrene in toluene, reflectometry data have been fitted by the mean-field expression for the volume fraction profile:

$$\phi(z) = \phi_b \tanh^2(z/d)$$

where d is the thickness of the depletion layer. Depletion layers have the thickness of the radius of gyration of the polymer in dilute solution. As the solution become semidilute, the depletion layer thickness approaches the dimensions of the correlation length, ξ, and since d is related to ξ, then it should display the same scaling relation. Experimentally [149], it was found that

$$\frac{d}{R_s} \sim \left(\frac{\phi_b}{\phi^*}\right)^{-0.75}$$

as predicted by the scaling relations with ϕ^* the overlap volume fraction of polystyrene in toluene.

XII. INTERFACIAL DYNAMICS OF POLYMERS AT FLUID INTERFACES

It would have been appropriate to present some discussion of the surface viscoelastic properties of polymers at fluid interfaces; however, lack of space prevents a full consideration of the subject and we present a précis here and summarize difficulties. Liquid interfaces are continually disturbed by thermal fluctuations which are called capillary waves. Parameters of the capillary waves can be obtained from surface quasi-elastic light scattering (SQELS) and subsequently analyzed to obtain the surface viscoelastic moduli. Although five surface moduli for a film-bearing fluid interface have been defined [150,151], only the transverse (capillary) mode produces the polarization change necessary to generate scattered light. Only the capillary and dilation modes of the surface waves are coupled, and therefore only these two parameters can be quantified.

Relatively few spread polymer films have been investigated by SQELS and it is difficult to draw universal conclusions from the data due to the varying set of assumptions made in calculating the viscoelastic parameters. Some of the assumptions made are quite insupportable if objective, unbiased values of the viscoelastic parameters are to be obtained. Many analyses use a Lorentzian form for the power spectrum of the surface scattered light and this has been demonstrated to be incorrect, therefore the capillary wave damping values quoted may be somewhat in error. It is clear from the work of Earnshaw and McLaughlin [152,153] that as wide a frequency range as possible should be explored; the dispersion behavior of the surface wave damping and frequency should always be compared with that predicted from the dispersion equation, especially when unexpected behavior is obtained. Unfortunately many of the data have been obtained over a restricted q range and this precludes a dispersion-based discussion. Since there is a complete absence of any molecularly based theories of either capillary wave dispersion or surface viscoelastic properties of polymer films much of the data reported in the literature is discussed in a necessarily speculative way. Little corollary has been drawn with more general treatments of vibrations and oscillations as expounded in classic texts [154–156]. The mechanical models discussed by Pippard are particularly informative. Notwithstanding these limited comments it is evident that the variation of surface moduli correlates strongly with surface organization of polymers at fluid interfaces [157–159]. Moreover the data obtained for some polymer systems has prompted a fresh consideration of the appropriate dispersion equation to be used for the analysis of SQELS data [160,161]. There remains much to be done in this field.

XIII. CONCLUDING REMARKS

It is evident that neutron reflectometry has greatly expanded the available knowledge of the organization of polymers at surface and interfaces. Although the method is a comparatively low-resolution one, it is rather better than others which are also able to penetrate polymers to depths of more than 3000 to 5000 Å. Moreover, the unique advantage of being able to highlight specific molecules, or parts of molecules, means that the most highly relevant parts of the polymer system can be probed to obtain information directly rather than infer it from effects in the surrounding matrix. A disadvantage is the inability to invert reflectometry data directly to a composition distribution, and remarks made about the need to question the uniqueness of any particular model made earlier cannot be emphasized enough. Perhaps the most unambiguous means of analyzing polymer surfaces and interfaces for their composition depth profile is to combine ion beam analysis methods with neutron reflectometry. The direct "map" of composition distribution obtained from the ion beam methods is a starting point for the analysis of the neutron reflectometry data at higher spatial resolution. We have only discussed the compositional analysis of polymers in one direction, i.e., normal to the surface or interface on which the neutron beam is incident, this results naturally from using only the specular reflection. In principle, off-specular or diffuse reflection contains information about the *in-plane* structure and organization. Only one report [162] has thus far been published where off-specular reflection results have been discussed and this was for a rather specialized low molecular weight polymer with a charged species at one end, spread at the air–water interface. Other attempts [163] at applying and analyzing off-specular reflection data are in progress. In the future the emphasis will probably turn toward more complex polymers and sample environments and the problems addressed will be more pertinent to actual applications. Notwithstanding this view, it is undoubtedly true that there are more than enough problems still to be solved on relatively simple systems to saturate the currently available neutron reflectometers around the world.

REFERENCES

1. A Balazs. J Phys Chem 100:1449, 1996.
2. J Lekner. Theory of Reflection. Dordrecht: Martinus Nijhoff Publishers, 1987.
3. M Born, E Wolf. Principles of Optics. 6th ed. Oxford: Pergamon, 1980.
4. O Heavens. Optical Properties of Thin Films. London: Butterworth, 1955.
5. J Penfold. Rutherford Appleton Laboratory Report, RAL-91-023, 1991.
6. LJ Norton, EJ Kramer, RAL Jones, FS Bates, HR Brown, GP Felcher, R Kleb. J Phys II France 28:367, 1994.
7. DS Sivia, W A Hamilton, G S Smith. Physica B 173:121, 1991.
8. K Kunz, J Reiter, A Gotzelmann, M Stamm. Macromolecules 26:4316, 1993.
9. RK Thomas. In: RW Richards, ed. Scattering Methods in Polymer Science. Chichester: Ellis Horwood, 1995, Chapter 4.
10. J Als-Nielsen. Z Phys B 61:411, 1984.
11. TL Crowley. Physica A 195:354, 1993.
12. AW Czanderna, DM Hercules, ed. Ion Spectroscopy for Surface Analysis, New York: Plenum, 1991.
13. JA McHugh. In: AW Czanderna, ed. Methods of Surface Analysis. Amsterdam: Elsevier, 1975, Chapter 6.
14. PJ Mills, PF Green, CJ Palmstøm, JW Mayer, EJ Kramer. Appl Phys Lett 45:957, 1984.
15. J Genzer, JB Rothman, RJ Composto. Nucl Instrum Methods B 86:345, 1994.
16. J Sokolov, MH Rafailovich, RAL Jones, EJ Kramer. Appl Phys Lett 54:590, 1989.
17. RS Payne, AS Clough, P. Murphy, PJ Mills. Nucl Instrum Methods B 42:130, 1989.
18. UK Chaturvedi, U Steiner, O Zak, G Krausch, G Schatz, J Klein. Appl Phys Lett 56:1228, 1990.
19. JW Cahn. J Chem Phys 66:3667, 1977.
20. D Bonn, H Kellay, J Meunier. Phys Rev Lett 73:3560, 1994.
21. B Evans, M Chan. Phys World April: 48, 1996.
22. H Nakanishi, P Pincus. J Chem Phys 79:997, 1983.
23. I Schmidt, K Binder. J Phys (France) 46:1631, 1985.
24. RAL Jones, EJ Kramer. Polymer 34:115, 1993.
25. RAL Jones, Polymer 35:2160, 1994.
26. I Carmesin, J Noolandi. Macromolecules 22:1689, 1989.
27. A Hariharan, SK Kumar, TP Russell. Macromolecules 24:4909, 1991.
28. P Cifra, FE Karasz, WJ MacKnight. Macromolecules 25:3895, 1992.
29. J-S Wang, K Binder. J Chem Phys 94:8537, 1991.

30. P Cifra, F Bruder, R Brenn. J Chem Phys 99:4121, 1993.
31. DH Pan, WM Prest. J Appl Phys 58:2861, 1985.
32. QS Bhatia, DH Pan, JT Koberstein. Macromolecules 21:2166, 1988.
33. JMG Cowie, BG Devlin, IJ McEwen. Macromolecules 26:5628, 1993.
34. P Sakellariou. Polymer 34:3408, 1993.
35. FS Bates, GD Wignall. Macromolecules 19:932, 1986.
36. FS Bates, GD Wignall. Phys Rev Lett 57:1429, 1986.
37. I Hopkinson, FT Kiff, RW Richards, SM King, H Munro. Polymer 35:1722, 1994.
38. FS Bates, SB Dierker, GD Wignall. Macromolecules 19:1938, 1986.
39. WW Graessley, R Krishnamoorti, NP Balsara, LJ Fetters, DJ Lohse, DN Sissano. Macromolecules 26:1137, 1993.
40. D Schwann, K Hahn, J Strieb, T Springer. J Chem Phys 93:8383, 1990.
41. MG Bawendi, KF Freed. J Chem Phys 88:2741, 1988.
42. SK Kumar. Macromolecules 27:260, 1994.
43. RAL Jones, EJ Kramer, MH Rafailovich, J Sokolov, SA Schwarz. Phys Rev Lett 62:280, 1989.
44. RAL Jones, LJ Norton, EJ Kramer, RJ Composto, RS Stein, TP Russell, A Mansour, A Karim, GP Felcher, MH Rafailovich, J Sokolov, X Zhao, SA Schwarz, Europhys Lett 12:41, 1990.
45. A Hariharan, SK Kumar, TP Russell. J Chem Phys 98:4163, 1993.
46. A Hariharan, SK Kumar, TP Russell. Macromolecules 24:4909, 1991.
47. A Budkowski, U Steiner, J Klein. J Chem Phys 97:5229, 1992.
48. T Flebbe, B Dünweg, K Binder. J Physique II 6:667, 1996.
49. LJ Norton, EJ Kramer, FS Bates, MD Gehlsen, RAL Jones, A Karim, GP Felcher, R Kleb. Macromolecules 28:8621, 1995.
50. Y Rouault, B Dünweg, J Baschnagel, K Binder. Polymer 37:297, 1996.
51. SF Edwards. Proc Phys Soc 92:9, 1967.
52. PG de Gennes. J Chem Phys 55:572, 1971.
53. PE Rouse. J Chem Phys 21:1272, 1953.
54. A Karim, A Mansour, GP Felcher, TP Russell. Phys Rev B 42:6846, 1990.
55. M Stamm, S Hüttenbach, G Reiter, T Springer. Europhys Lett 14:451, 1991.
56. A Karim, GP Felcher, TP Russell. Macromolecules 27:6973, 1994.
57. K Kunz, M Stamm. Macromolecules 29:2548, 1996.
58. Y Liu, G Reiter, K Kunz, M Stamm. Macromolecules 26:2134, 1993.
59. SJ Whitlow, RP Wool. Macromolecules 24:5926, 1991.
60. D Richter, R Butera, LJ Fetters, JS Huang, B Ewen. Macromolecules 25:6156, 1992.
61. TP Russell, VR Deline, WD Dozier, GP Felcher, G Agrawal, RP Wool, JW Mays. Nature 365:235, 1993.
62. A Agrawal, RP Wool, WD Dozier, GP Felcher, TP Russell, JW Mays. Macromolecules 27:4407, 1994
63. JMHM Scheutjens, GJ Fleer. J Phys Chem 83:1619, 1979.
64. JMHM Scheutjens, GJ Fleer. J Phys Chem 84:178, 1980.

65. GJ Fleer, MA Cohen-Stuart, JMHM Scheutjens, T Cosgrove, B Vincent. Polymers at Interfaces. London: Chapman and Hall, 1993.
66. S Alexander. J Physique (France) 38:977, 1977.
67. PG de Gennes. Macromolecules 13:1069–1075, 1980.
68. ST Milner, TA Witten, ME Cates. Macromolecules 21:2610–2619, 1988.
69. KR Shull. J Chem Phys 94:5723: 1991.
70. M Muthukumar, JS Ho. Macromolecules 22:965, 1989.
71. A Chakrabarti, R Toral. Macromolecules 23:2016, 1990.
72. K Binder, PY Lai, J Wittmer. Faraday Discussions 98:97, 1994, and references therein.
73. HJ Taunton, C Toprakcioglu, J Klein. Macromolecules 23:571, 1990.
74. G Hadziioannou, S Patel, S Granick, M Tirrell. J Am Chem Soc 108:2869, 1986.
75. P Auroy, Y Mir, L Auvray. Phys Rev Lett 69:93, 1992.
76. T Cosgrove, K Ryan. Langmuir 6:136, 1990.
77. A Karim, SK Satija, JF Douglas, JF Ankner, LJ Fetters. Phys Rev Lett 73:3407, 1994.
78. EB Zhulina, OV Borisov, VA Pryamitsyn, TM Birshtein. Macromolecules 24:140, 1991.
79. TL Mansfield, DR Iyengar, G Beaucage, TJ McCarthy, RS Stein, RJ Composto. Macromolecules 28:492, 1995.
80. JB Field, C Toprakcioglu, RC Ball, HB Stanley, L Dai, W Barford, J Penfold, G Smith, W Hamilton. Macromolecules 25:434, 1992.
81. MS Kent, L-T Lee, B Farnoux, F Rondelez. Macromolecules 25:6240, 1992.
82. MS Kent, L-T Lee, BJ Factor, Rondelez, GS Smith. J Chem Phys 103:2320, 1995.
83. R Baranowski, MD Whitmore, J Chem Phys 103:2343, 1995.
84. L-T Lee, BJ Factor, F Rondelez, MS Kent. Faraday Discussions 98:139, 1994.
85. RAL Jones, LJ Norton, KR Shull, EJ Kramer, GP Felcher, A Karim, LJ Fetters. Macromolecules 25:2359, 1992.
86. I Hopkinson, FT Kiff, RW Richards, DG Bucknall, AS Clough. Polymer 18:87, 1997.
87. A Budkowski, U Steiner, J Klein, LJ Fetters. Europhys Lett 20:499, 1992.
88. S Affrossman, M Hartshorne, T Kiff, RA Pethrick, RW Richards. Macromolecules 27:1588, 1994.
89. MO Hunt, AM Belu, RW Linton, JM DeSimone. Macromolecules 26:4854, 1993.
90. TF Schaub, GJ Kellog, AM Mayes, R Kulasekere, JF Ankner, H Kaiser. Macromolecules 29:3982, 1996.
91. JF Elman, BD Johs, TE Long, JT Koberstein. Macromolecules 27:5341, 1994.
92. DR Paul. In: DR Paul, S Newman, ed. Polymer Blends vol. 2. New York: Academic Press, 1978, pp 35–60.
93. SH Anastasiadis, I Gancarz, JT Koberstein. Macromolecules 22:1449, 1989.
94. P Gaillard, M Ossenbach-Sauter, G Reiss. Makromolecul Chem Rapid Commun 1:771, 1980.

95. HT Patterson, K Hu, TH Grindstaff. J Polym Sci 34:31, 1994.
96. HR Brown. Macromolecules 22:2859, 1989.
97. R Kulasekere, H Kaiser, JF Ankner, TP Russell, HR Brown, CJ Hawker, AM Mayes. Macromolecules 29:5493, 1996.
98. C Creton, EJ Kramer, G Hadziioannou. Macromolecules 24:1846, 1991.
99. J Noolandi, KM Hong. Macromolecules 17:1531, 1984.
100. L. Leibler. Makromol Chem Macromol Symp 16:1, 1988.
101. AN Semenov. Macromolecules 25:4967, 1992.
102. Y Lyatskaya, D Gersappe, AC Balazs. Macromolecules 28:6278, 1995.
103. D Gersappe, PK Harm, D Irvine, AC Balazs. Macromolcules 27:720, 1994.
104. D Gersappe, AC Balazs. Phys Rev E 52:5061, 1995.
105. AC Balazs, D. Gersappe, R Israels, M Fasolka. Macromol Theory Simul 4:585, 1995.
106. D Gersappe, D Irvine, AC Balazs, Y Liu, J Sokolov, M Rafailovich, S Schwarz, DG Peiffer. Science 265:1072, 1994.
107. KR Shull, EJ Kramer, G Hadziioannou, W Tang. Macromolecules 23:4780, 1990.
108. KR Shull, KI Winey, EL Thomas, EJ Kramer. Macromolecules 24:2748, 1991.
109. KH Dai, LJ Norton, EJ Kramer. Macromolecules 27:1949, 1994.
110. KH Dai, J Washiyama, EJ Kramer. Macromolecules 27:4544, 1994.
111. DG Bucknall, JS Higgins, J Penfold, S Rostami. Polymer 34:451, 1993.
112. CA Fleischer, AR Morales, JT Koberstein. Macromolecules 27:379, 1994.
113. FS Bates, GH Fredrickson. An. Rev Phys Chem 41:525, 1990.
114. E Vanzo. J Polym Sci A 4:1727, 1966.
115. AK Rostogi, LE St Pierre. J Colloid Interface Sci 31:168, 1969.
116. H Hasegawa, T Hashimoto. Macromolecules 18:589, 1985.
117. G Coulon, TP Russell, VR Deline, PF Green. Macromolecules 22:2581, 1989.
118. CS Henkee, EL Edwin, LJ Fetters. J Mater Sci 23:1685, 1988.
119. SH Anastasiadis, TP Russell, SK Satija, CF Majkrzak. Phys Rev Lett 62:1852, 1989.
120. MD Foster, M Sikka, N Singh, FS Bates, SK Satija, CF Majkrzak. J Chem Phys 96:8605, 1992.
121. SH Anastasiadis, TP Russell, SK Satija, CF Majkrzak. J Chem Phys 92:5677, 1990.
122. A Karim, N Singh, M Sikka, FS Bates, WD Dozier, GP Felcher. J Chem Phys 100:1620, 1994.
123. GH Fredrickson. Macromolecules 20:2535, 1987.
124. P Lambooy, TP Russell, GJ Kellog, AM Mayes, PD Gallagher, SK Satija. Phys Rev Lett 72:1994.
125. N Koneripalli, N Singh, R Levicky, FS Bates, PD Gallagher, SK Satija. Macromolecules 28:2897, 1995.
126. M Maaloum, D Ausserre, D Chatenay, Y Gallot. Phys Rev Lett 70:2577, 1993.
127. P Bassereau, D Brodbreck, TP Russell, HR Brown, KR Shull. Phys Rev Lett 71:1716, 1993.

128. G Coulon, J. Daillant, B Collin, JJ Benattar, Y Gallot. Macromolecules 26:1582, 1993.
129. BL Carvalho, EL Thomas. Phys Rev Lett 73:3321, 1994.
130. M Kawaguchi. Prog Polym Sci 18:341, 1993.
131. PG de Gennes. Scaling Concepts in Polymer Physics. Ithaca, NY: Cornell University Press, 1979.
132. R Ober, R Vilanove. Colloid Polym Sci 255:1067, 1977.
133. R Vilanove, F Rondelez. Phys Rev Lett 45:1502, 1980.
134. A Takahashi, A Yoshida, M Kawaguchi. Macromolecules 15:1196, 1982.
135. M Kawaguchi, A Yoshida, A Takahashi. Macromolecules 16:956, 1983.
136. D Poupinet, R Vilanove, F Rondelez. Macromolecules 22:2491, 1989.
137. N Beredijk, ed. Monomolecular Film Studies of Polymers. New York: Wiley-Interscience, 1963.
138. JA Henderson, RW Richards, J Penfold, C Shackleton, RK Thomas. Polymer 32:3284, 1991.
139. JA Henderson, RW Richards, J Penfold, RK Thomas. Macromolecules 26:65, 1993.
140. JA Henderson, RW Richards, J Penfold, RK Thomas. Acta Polymerica 44:184, 1993.
141. JA Henderson, RW Richards, J Penfold, RK Thomas, JR Lu. Macromolecules 26:4591, 1993.
142. RW Richards, MR Taylor. J Chem Soc Faraday Trans 92:601, 1996.
143. RW Richards, BR Rochford, JRP Webster. Faraday Discussions 98:263, 1994.
144. I Reynolds, RW Richards, JRP Webster. Macromolecules 28:7845, 1995.
145. P Hodge, CR Towns, RK Thomas, C Shackleton. Langmuir 8:584, 1992.
146. O Guiselin, L-T Lee, B Farnoux, A Lapp. J Chem Phys 95:4632, 1991.
147. O Guiselin, L-T Lee, B Farnoux. Physica B 173:113, 1991.
148. L-T Lee, O Guiselin, B Farnoux, A Lapp. Macromolecules 24:2518, 1991.
149. L-T Lee, O Guiselin, A Lapp, B Farnoux, J Penfold. Phys Rev Lett 67:2838, 1991.
150. FC Goodrich. J Phys Chem 66:1858, 1962.
151. FC Goodrich. Proc Roy Soc A 374:341, 1981.
152. JC Earnshaw, AC McLaughlin. Proc Roy Soc A 433:663, 1991.
153. JC Earnshaw, AC McLaughlin. Proc Roy Soc A 440:519, 1993.
154. AB Pippard. The Physics of Vibration Omnibus Edition. Cambridge: Cambridge University Press, 1989.
155. KU Ingard. Fundamentals of Waves and Oscillations. Cambridge: Cambridge University Press, 1988.
156. IG Main. Vibrations and Waves in Physics. 3rd ed. Cambridge: Cambridge University Press, 1993.
157. SK Peace, PhD thesis, University of Durham, 1996.
158. RW Richards, MR Taylor. J Chem Soc Faraday Trans 92:601, 1996.
159. RW Richards, BR Rochford, MR Taylor, Macromolecules 29:1980, 1996.
160. SK Peace, RW Richards. Polymer 37:4945, 1996.

161. RW Richards, MR Taylor. Macromolecules 30:3892, 1997.
162. PM Saville, IR Gentle, JW White, J Penfold, JRP Webster. J Phys Chem 98:5935, 1994.
163. RAL Jones. University of Cambridge, private communication.

9

Rheology of Multiphase Polymeric Systems

Hiroshi Watanabe
Kyoto University, Kyoto, Japan

I. INTRODUCTION

Polymeric materials are often used as composites comprising a multiphase structure. This structure strongly influences some of the properties of the composites, in particular the rheological properties at long time scales. For example, immiscible polymer blends and block copolymers often exhibit extremely slow relaxation due to the multiphase structures within. Controlled processing of such composites requires us to understand the structure–rheology relationship.

A distinguishing feature of blends and block copolymers is the wide variety of this relationship. The slow rheological responses are differently affected by various factors, the shape, size, spatial arrangements, and rigidity of the phases. The responses of the copolymers are also affected by the chain architecture. This chapter summarizes these responses and corresponding structure–rheology relationships, a subject of current research interest. First, a general framework of linear viscoelasticity is briefly summarized to specify the meaning of linearity and rheological quantities to be examined. Then, the rheological properties of the blends and copolymers are described and the factor(s) dominating those properties are explained for respective cases.

II. LINEAR VISCOELASTICITY FRAMEWORK

A. General Behavior

For measurements of a stress, we have to unavoidably apply a strain and distort a material. When the strain is small and only slightly affects the structure in the material, the linear viscoelastic responses prevail. On the other hand, nonlinearities emerge for a large strain that significantly distorts the structure.

For the simplest case of rheological measurements, we apply a step shear strain γ_0 at time $t = 0$ and detect the decay of a stress $\sigma(t)$ at $t > 0$. For sufficiently small γ_0, the linear viscoelastic responses prevail and a γ_0-independent, linear relaxation modulus $G(t)$ is defined as [1],

$$G(t) = \sigma(t)/\gamma_0 \tag{1}$$

The initial modulus, $G(0)$, represents an instantaneous increment of a (elastic) free energy due to slight distortion of the structure (or state) in the material on imposition of the small step strain. The decay of $G(t)$ reflects dissipation of this energy (recovery of the quiescent structure/state) due to molecular motion in the material. In the linear viscoelastic regime, this motion is identical to the motion at the quiescent (or equilibrium) state. Thus $G(t)$ includes information for both structure and dynamics in this state.

The linear responses are observed also for a small shear strain of arbitrary history, $\gamma(t')$ with $t' < t$. For this case, the shear stress $\sigma(t)$ at time t is given as a convolution of γ and G [1],

$$\sigma(t) = \int_{-\infty}^{t} G(t - t')\dot{\gamma}(t')\, dt' \tag{2}$$

Here, $\dot{\gamma}(t')$ is the strain rate at a past time t'; $\dot{\gamma}(t')\, dt'$ represents an infinitesimal step strain imposed at that time. Eq. (2) simply states that $\sigma(t)$ is given as a linear superposition of all strain effects in the past time, $G(t - t')\dot{\gamma}(t')\, dt'$ with $t' < t$.

In general, $G(t)$ has a distribution of relaxation modes. This distribution is conveniently represented with the relaxation spectrum $H(\tau)$ that gives the intensity of modes having logarithms of relaxation times in an interval between $\ln \tau$ and $\ln \tau + d \ln \tau$ [1],

$$G(t) = \int_{-\infty}^{\infty} H(\tau) \exp(-t/\tau)\, d \ln \tau \tag{3}$$

Thus, rheological features of the material are equivalently represented by $G(t)$ and $H(\tau)$. For a linear viscoelastic *solid*, $H(\tau)$ includes a term

$H(\infty)\delta(\tau - \tau^*)\,|_{\tau^* \to \infty}$ and $G(t)$ has a nonrelaxing, equilibrium modulus $G_e = G(\infty) = H(\infty) > 0$. A viscoelastic *liquid* has $G_e = H(\infty) = 0$.

B. Dynamic Behavior

Linear viscoelastic properties are also conveniently determined from dynamic tests. As noted from Eq. (2), $\sigma(t)$ oscillates sinusoidally with t when a sinusoidal strain with amplitude γ_0 and angular frequency ω is applied for long time,

$$\sigma(t) = \gamma_0[G'(\omega)\sin\omega t + G''(\omega)\cos\omega t] \qquad \gamma(t) = \gamma_0 \sin\omega t \tag{4}$$

From Eqs. (2)–(4), the storage and loss moduli, $G'(\omega)$ and $G''(\omega)$, are related to $H(\tau)$ as

$$G'(\omega) = \omega^2 \int_{-\infty}^{\infty} H(\tau)\frac{\tau^2}{1+\omega^2\tau^2}\, d\ln\tau \quad G''(\omega) = \omega \int_{-\infty}^{\infty} H(\tau)\frac{\tau}{1+\omega^2\tau^2}\, d\ln\tau \tag{5}$$

Thus the ω dependence of G' and G'' reflects the τ dependence of H; the linear viscoelastic features are equivalently represented by H and by G' and G'' (or the complex moduli $G^* = G' + iG''$; $i = \sqrt{-1}$).

For a liquid having the longest relaxation time τ_1, $H(\tau)$ vanishes for $\tau > \tau_1$. In the corresponding terminal relaxation regime of ω, G' and G'' exhibit the characteristic ω dependence (Eq. (6)) and an interrelationship (Eq. (7)),

$$G''(\omega) = \eta_0^2 J_e^0 \omega^2 \propto \omega^2 \quad \text{and } G''(\omega) = \eta_0\omega \propto \omega \quad \text{for } \omega \ll \tau_1^{-1} \tag{6}$$

$$\log G'(\omega) = 2\log G''(\omega) + \log J_e^0 \quad \text{for } \omega \ll \tau_1^{-1} \tag{7}$$

Here, $\eta_0\ (= [G''/\omega]_{\omega \to 0})$ and $J_e^0\ (= [G'/G''^2]_{\omega \to 0})$ are the zero-shear viscosity and steady-state recoverable compliance that are related to $H(\tau)$ as

$$\eta_0 = \int_{-\infty}^{\infty} H(\tau)\tau\, d\ln\tau \qquad J_e^0 = \left[\int_{-\infty}^{\infty} H(\tau)\tau^2\, d\ln\tau\right]\left[\int_{-\infty}^{\infty} H(\tau)\tau\, d\ln\tau\right]^{-2} \tag{8}$$

At higher $\omega\ (> \tau_1^{-1})$ where the material has corresponding relaxation modes at $\tau \cong \omega^{-1}$, G' and G'' become more weakly ω-dependent and smaller in magnitude than shown in Eq. (6), and G' becomes more weakly dependent on G'' than in Eq. (7). These features are useful when the relaxation mode distribution is examined through G' and G''. The relaxation is completed at ω where Eqs. (6) and (7) are satisfied, and vice versa. It should be also noted that J_e^0 is sensitive to the terminal relaxation mode distribution (τ dependence of $H(\tau)$ at large τ).

For specifying the end of relaxation, we can use an average relaxation time $\langle\tau\rangle_w$,

$$\langle\tau\rangle_w = J_e^0\eta_0 = \left[\int_{-\infty}^{\infty} H(\tau)\tau^2\, d\ln\tau\right]\left[\int_{-\infty}^{\infty} H(\tau)\tau\, d\ln\tau\right]^{-1} \tag{9}$$

This $\langle\tau\rangle_w$ is an average of τ weighted by a factor magnifying the slow relaxation modes, $\tau H(\tau)$. Thus, $\langle\tau\rangle_w$ is usually close to τ_1 and the relaxation of the material is completed at ω well below $\langle\tau\rangle_w^{-1}$.

Here, we note an important difference in the ω dependence of G' and G''. As noted from Eq. (5), H at large τ (slow relaxation modes) are more strongly magnified by the integration kernel for G', $\tau^2/(1+\omega^2\tau^2)$, than by the kernel for G'', $\tau/(1+\omega^2\tau^2)$. Thus weak and slow modes are much more sensitively reflected in G' than in G''. In other words, the limiting behavior, Eq. (6), is less easily attained for G'. This feature of G' becomes prominent for the immiscible blends and block copolymers that often exhibit weak but slow relaxation modes due to multiphase structures.

C. Steady Flow Behavior

Steady-state shear stress σ_s as well as the first normal stress difference N_1 are observed when a viscoelastic liquid is subjected to a shear flow at constant rate $\dot{\gamma}$ for long time. For small $\dot{\gamma} \ll \langle\tau\rangle_w^{-1}$ (or τ_1^{-1}), the flow hardly affects the structures in the liquid and the linear viscoelastic, steady flow responses (Newtonian responses) prevail. On the other hand, the structure is largely distorted to exhibit nonlinear (nonNewtonian) responses for larger $\dot{\gamma}$.

In the linear viscoelastic regime, Eq. (2) and its tensorial version extended for a second-order fluid [1] are valid. For this case, σ_s and N_1 are given as [1]

$$\sigma_s = \eta_0\dot{\gamma} \propto \dot{\gamma} \qquad N_1 = 2\eta_0^2 J_e^0\dot{\gamma}^2 \propto \dot{\gamma}^2 \tag{10}$$

Here, η_0 and J_e^0 are the viscosity and compliance defined by Eq. (8). (In the nonlinear regime, the $\dot{\gamma}$ dependence of σ_s and N_1 often becomes weaker than that given in Eq. (10).)

D. Time–Temperature Superposition

If all relaxation modes involved in the relaxation spectrum $H(\tau)$ have the *same* temperature (T) dependence in their characteristic times and intensities, $H(\tau)$ obeys the time–temperature superposition [1],

$$H(\tau)\,|_{\text{at }T} = b_T H(\tau/a_T)\,|_{\text{at }T_r} \tag{11}$$

where T_r is a reference temperature, and a_T and b_T are functions of T. For this case Eq. (5) leads to a relationship,

$$G'(\omega)\,|_{\text{at }T} = b_T G'(\omega a_T)\,|_{\text{at }T_r} \qquad G''(\omega)\,|_{\text{at }T} = b_T G''(\omega a_T)\,|_{\text{at }T_r} \tag{12}$$

Corresponding relationships are obtained for $G(t)$ as well as for σ_s and N_1 [1].

If a structure determining $H(\tau)$ does not change with T, Eq. (12) is valid and $G^*(\omega)$ curves at different T are superposed by shifts on double-logarithmic scales (cf. high-T data later shown in Fig. 4). On the other hand, changes in the structure usually lead to failure of this superposition. However, for some cases, this failure is not too severe and in a practical sense the G^* curves can be superposed in a limited range of ω.

This limited superposability can be found for heterogeneous systems that exhibit fast, intensive relaxation modes of individual chains (represented by H_{fast}) and much slower, weak modes (H_{slow}) of a large-scale structure. If the time–temperature superposition (Eq. (11)) is valid for H_{fast} but not for H_{slow}, Eq. (5) indicates that the superposition of G^* (Eq. (12)) practically works at high ω where H_{fast} dominantly contributes to G^* but it fails at low ω where the contribution of H_{slow} becomes significant. As explained earlier, the weak and slow modes (H_{fast}) are more sensitively reflected in G' than in G'' at those low ω. Thus this failure is much more clearly observed for G' than for G''.

III. RHEOLOGY OF IMMISCIBLE POLYMER BLENDS

Many studies have been carried out to investigate the rheology and structure of polymer blends, and even more extensive studies are now in progress. Main results of the earlier work (before 1990) are summarized in Refs. [2–4]. In particular, the results obtained in the period 1980–1989 are compactly reviewed in Ref. [4]. This section mainly summarizes recent results of rheological studies for immiscible polymer blends comprising multiphases in a liquid (molten) state.

The rheology of immiscible blends is strongly affected by the properties of respective phases and their interfaces as well as by the morphology (size, shape, and arrangements of the phases). In general, the morphology is readily distorted by the applied strain when all phases are in the liquid state. This leads to significantly nonlinear responses even for modest strain. The blends exhibit the linear behavior reflecting the quiescent structure only for very small strain. The remaining part of this section first examines the linear behavior and then the nonlinear behavior.

A. Linear Viscoelastic Behavior

1. General Behavior

Fig. 1 shows the ω dependence of G' and G'' at 70°C for a blend of poly(-oxyethylene)-diol (POE-DO) and poly(dimethyl siloxane) (PDMS) examined by Graebling et al. [5]. The dotted and dashed curves indicate the behavior of PEO-DO and PDMS, respectively. At the quiescent state, the blend contains discrete, spherical phases of PEO-DO (volume fraction $\phi = 0.15$) dispersed in the PDMS matrix. Both components are in the molten state at 70°C and exhibit the terminal behavior (Eq. (6)) in the entire ω-range examined. The relaxation time defined by Eq. (9) is $\langle\tau\rangle_w \cong 0.001$ s for

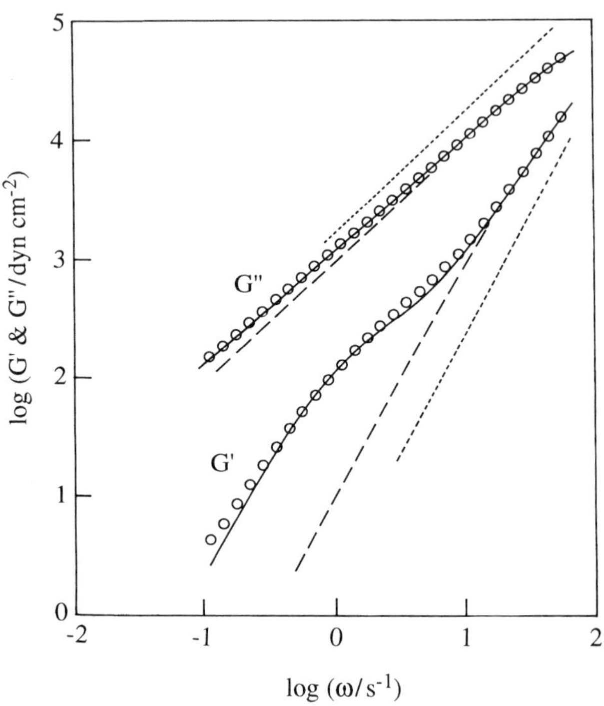

Figure 1 Frequency dependence of storage and loss moduli for a blend of POE–DO (η_0 = 1690 poise) and PDMS (η_0 = 910 poise) at 70°C. The dotted and dashed curves indicate the behavior of the component POE–DO and PDMS, respectively, and the solid curves indicate the prediction of the Palierne model. In the blend, POE–DO is a minor component (with ϕ = 0.15) and forms discrete, spherical phases of $R_{dv} = 20$ μm at the quiescent state. (Redrawn, with permission, from Ref. 5.)

POE-DO and 0.01 s for PDMS. On the other hand, for the blend, the ω dependence of G' is much weaker than ω^2 at $\omega = 1$–$10\ s^{-1}$ and the terminal behavior is attained only at lower $\omega < 1\ s^{-1}$. This result indicates that the relaxation of the blend (with $\langle\tau\rangle_w \cong 1$ s) is much slower than the component relaxation. Han and coworkers [6–8] demonstrated that such slow relaxation modes are sensitively detected through double-logarithmic plots of G' and G'' having the limiting behavior, Eq. (7). (As explained for Eq. (5), the slow modes are much more sensitively reflected in G' than in G''.)

Slow relaxation processes similar to that seen in Fig. 1 are almost commonly observed for immiscible polymer blends containing discrete phases [4–16]. Those processes have been attributed to recovery of shape of those phases deformed by the applied strain: An elastic energy stored in the deformed interface relaxes on this shape recovery. In fact, Scholz et al. [9] found that this recovery time (evaluated from microscopy observation) is close to the viscoelastic relaxation time of the blends.

2. Emulsion Model

Theories have been proposed for describing rheological properties of multiphase systems in terms of properties of respective phases and interfaces [4,17–32]. Specifically, the immiscible blends having discrete phases (droplets) are modeled as emulsions and their rheological quantities are calculated by balancing an applied stress that tends to deform the droplets and a droplet/matrix interfacial tension that tends to recover the spherical shape of the droplets. For the blends having a large volume fraction of the droplets, the hydrodynamic interaction between the droplets and the heterogeneous strain distribution in the system are the issues to be addressed theoretically. In recent studies [5,9–16], G^* data of such blends are frequently compared with models developed by Palierne [29] and by Choi and Schowalter [27].

Palierne formulated a model for concentrated emulsions composed of viscoelastic droplets and matrices [29]. For those emulsions, he considered a spherical region (conceptually similar to the Lorentz sphere [33] in dielectric calculations) that is much smaller than the system but still large enough to contain many droplets and exhibit bulk properties of the system. The linear viscoelastic, complex modulus $G^*_{\rm U}$ of the emulsion (or polymer blends) is self-consistently calculated by regarding the material out of the spherical region as a uniform continuum that interacts with the droplets in this region. When the droplets are isotropically distributed in space and the interfacial tension Γ is independent of the droplet deformation, the Palierne model predicts

$$G_b^*(\omega) = G_m^*(\omega)\left[1 + 3\sum_i \phi_i H_i(\omega)\right]\left[1 - 2\sum_i \phi_i H_i(\omega)\right]^{-1} \tag{13}$$

with

$$H_i(\omega) = \frac{[4\Gamma/R_i][2G_m^*(\omega) + 5G_d^*(\omega)] - [G_m^*(\omega) - G_d^*(\omega)][16G_m^*(\omega) + 19G_d^*(\omega)]}{[40\Gamma/R_i][G_m^*(\omega) + G_d^*(\omega)] + [3G_m^*(\omega) + 2G_d^*(\omega)][16G_m^*(\omega) + 19G_d^*(\omega)]} \tag{14}$$

Here, $G_m^*(\omega)$ and G_d^* are the complex moduli of the matrix and droplet, and ϕ_i is the volume fraction of the droplet having a radius R_i.

Although G_b^* given by Eq. (13) is a complicated function of ϕ_i, G_d^*, G_m^* and Γ/R_i, the zero-shear viscosity of the emulsion $\eta_b = [G_b''/\omega]_{\omega\to 0}$ is simply related to the viscosities of the droplet and matrix, η_d and η_m, as

$$\eta_b = \eta_m \frac{10(K+1) + 3\phi(5K+2)}{10(K+1) - 2\phi(5K+2)} \qquad K = \frac{\eta_d}{\eta_m} \qquad \phi = \sum_i \phi_i \tag{15}$$

This η_b is independent of ϕ_i, R_i, and Γ but determined only by η_m, η_d, and the total volume fraction of the droplet ϕ.

For practical use of the Palierne model, Eq. (13) can be approximated as follows. If ϕ and a volume-average droplet radius R_{dv} are fixed, G_b^* of the model hardly depends on the droplet radius distribution [5]. Thus, Eq. (13) can be approximated as $G_b^* = G_m^*[1 + 3\phi H_v]/[1 - 2\phi H_v]$ where H_v is given by Eq. (14) with R_i being replaced by R_{dv}.

We can make further simplification at low ω where both droplet and matrix have relaxed to behave as viscous, Newtonian liquids. At those ω, we can replace G_d^* and G_m^* by $i\omega\eta_d$ and $i\omega\eta_m$ ($i = \sqrt{-1}$). Then, Eq. (13) becomes

$$G_b'(\omega) = G_p \frac{\omega^2\tau_b^2}{1 + \omega^2\tau_b^2} \qquad G_b''(\omega) = \omega[\eta_b - G_p\tau_b] + G_p \frac{\omega\tau_b}{1 + \omega^2\tau_b^2} \tag{16}$$

with

$$\tau_b = \frac{\eta_m R_{dv}}{4\Gamma} \frac{(19K + 16)[2K + 3 - 2\phi(K-1)]}{10(K+1) - 2\phi(5K+2)} \qquad K = \frac{\eta_d}{\eta_m} \tag{17}$$

and

$$G_p = \frac{20\Gamma\phi}{R_{dv}} \frac{1}{[2K + 3 - 2\phi(K-1)]^2} \tag{18}$$

Thus, the Palierne model predicts that G_b^* at low ω has two contributions, one from a purely dissipative process characterized with a viscosity $(\eta_b - G_p\tau_b)$ and the other from a Debye-type single-relaxation process with the relaxation time τ_b and intensity G_p. The latter process corresponds to the recovery of the spherical shape of the droplet.

A model proposed by Choi and Schowalter [27] is also frequently used. They formulated a cell model for emulsions composed of *viscous* matrices and droplets (of radius R_d). In this model, a volume of the system distributed to each droplet is regarded as a cell, and the material out of the cell is replaced by a uniform continuum having the bulk properties of the emulsion. The strain field and G_b^* of the emulsion are calculated by considering the stress balance at the droplet/matrix interface in the cell and at the cell boundary. (Note that the Choi–Schowalter model is applicable to polymer blends at low ω where the blend components relax and behave as viscous liquids.)

When ϕ is not very large, the Choi–Schowalter model predicts that G_b^* has contributions from a dissipative process and a single-relaxation process, the latter corresponding to the droplet shape recovery [9, 27]. This G_b^* can be cast in the form of Eq. (16), with η_b, τ_b, and G_p being replaced by [9]

$$\eta_b = \eta_m \left[1 + \left(\frac{5K+2}{2K+2} \right) \phi + \left(\frac{5(5K+2)^2}{8(K+1)^2} \right) \phi^2 \right] \qquad K = \frac{\eta_d}{\eta_m} \tag{19}$$

$$\tau_b = \frac{\eta_m R_d}{4\Gamma} \left[\frac{(19K+16)(2K+3)}{10(K+1)} \right] \left[1 + \frac{5(19K+16)}{4(K+1)(2K+3)} \phi \right] \tag{20}$$

$$G_p = \frac{20\Gamma\phi}{R_{dv}(2K+3)^2} \left[1 + \left(\frac{5(19K+16)}{4(K+1)(2K+3)} \right) \phi \right]^{-2} \tag{21}$$

For relatively small ϕ, the predictions of the Choi–Schowalter model are not very different from those of the Palierne model at low ω. (For example, for $\phi = 0.15$, η_b and τ_b of the two models agree within 20% for $K > 10$ and G_p agrees within 50% for $K > 30$; cf. Eqs. (15)–(18) and Eqs. (19)–(21).) For large ϕ, differences between the two models become significant. However, for such large ϕ, actual blends often have *co-continuous* phase structures not considered in the models. For practical use of the models up to modestly large ϕ where the blends usually have discrete phases, the differences between the models (at low ω) appear to be not very significant.

3. Comparison with Experiments

For POE-DO/PDMS blends, Graebling et al. [5] compared the G^* data with G_b^* of the Palierne model (calculated from experimentally determined parameters). As noted in Fig. 1, the model (solid curves) well describes the data (symbols). Similar results were obtained also for blends of polystyrene (PS) and poly(methyl methacrylate) (PMMA) [5]. For blends containing a surfactant, the data were fitted with the model to evaluate Γ under the presence of the surfactant [5]. The resulting Γ value was considerably smaller than the value for the neat blends.

Bousmina et al. [12] found the validity of the Palierne model for PS/polyethylene (PE) blends. They also reported that a polymeric emulsifier reduces Γ but has no effect on the Γ/R_{d} ratio. Bousmina and Muller [10] found that the model is valid for PS/PMMA blends but not for concentrated blends of rubber particles in a PMMA matrix. The disagreement for the latter case was attributed to aggregates of the particles not considered in the model.

For blends of high-density PE (HDPE) in high-impact PS (HIPS), Brahimi et al. [14] attributed a disagreement between the data and the Palierne model to interactions between dispersed phases not considered in the model. For blends of polypropylene (PP) in styrene-acrylonitrile (SAN) copolymers, Maeda and Kamei [15] noted a disagreement due to changes in G^* of the components on *melt-blending*. They reported an agreement for the model prediction evaluated from G^* of the components *extracted* from the blends, demonstrating the validity of the Palierne model as well as the necessity of careful examination of the component moduli, in particular for melt-blended systems [15]. Similarly careful examination is necessary when the components are partially miscible, as shown by Nomura et al. [16] for blends of PP in ethylene-propylene rubber.

The above results demonstrate the validity of the Palierne model for immiscible polymer blends having isotropically distributed discrete phases. This in turn means that the model can be used for evaluation of Γ, as done by Graebling et al. [5] and by Bousmina et al. [12].

The Choi–Schowalter model is applicable to polymer blends at low ω where the components have relaxed. Scholz et al. [9] found the validity of this model for PP/polyamide (PA) blends at such low ω. This model was applied to PS/PMMA and emulsified SAN/PS blends in much wider ω ranges [11,13]. To extend the applicable ω range, Gramespacher and Meissner [11] and Gleinser et al. [13] assumed that G_{b}^* of the blends is given by a linear combination of contributions from the dispersed phase, matrix, and the interface. On the basis of this assumption, they compared the G_{b}^* data with

$$G_{\mathrm{b}}^*(\omega) = G_{\mathrm{p}}\left[\frac{i\omega\tau_{\mathrm{b}}}{1 + i\omega\tau_{\mathrm{b}}}\right] + \phi G_{\mathrm{d}}^*(\omega) + (1 - \phi)G_{\mathrm{m}}^*(\omega) \tag{22}$$

Here, τ_{b} and G_{p} are the relaxation time and intensity of the Choi–Schowalter model (Eqs. (20) and (21)).

If the first term in Eq. (22) (interfacial contribution) is absent, this equation is reduced to the simplest composite model (parallel model of Takayanagi; [22,23]) for co-continuous phases sustaining the same dynamic strain. Thus Eq. (22) is not generally valid for blends containing discrete phases. However, if the discrete phase and matrix have $G_d^*(\omega) \cong G_m^*(\omega)$ and the interface only slightly distorts the strain field in the blends, the dynamic strain would be almost uniformly distributed in the blends and Eq. (22) could be approximately valid. Good agreements reported by Gramespacher and Meissner [11] for their PS/PMMA blends (having $G_d^*(\omega) \cong G_m^*(\omega)$) might correspond to this case.

4. Other Relaxation Mechanisms

As explained earlier, recovery of the interfacial shape can lead to slow relaxation of immiscible polymer blends. However, this type of relaxation emerges only when the discrete phases are fairly soft and Γ is fairly small so that those phases are readily deformed: The discrete phases behave as a rigid body and the recovery of their shape does not contribute to G_b^* if Γ is very large and/or $|G_d^*| \gg |G_m^*|$ at low ω. Even for those cases the blends often exhibit relaxation processes much slower than the component relaxation [10,34–42].

For examination of such slow relaxation processes, we here focus our attention on the behavior of ABS resins extensively studied by Aoki et al. [34–42]. These resins are classified as immiscible blends of rubber particles in matrices of acrylonitrile-styrene (AS) copolymers. The rubber particles are grafted with AS chains (to enhance the affinity to the matrices). The ABS resins often exhibit relaxation much slower than the matrix relaxation. The rubber particles (discrete phases) have $|G_d^*| \gg |G_m^*|$ at low ω and the slow relaxation of the resins cannot be due to deformation of the particles. Two different mechanisms were found for such slow relaxation processes.

One mechanism is related to a network structure of aggregated rubber particles [34,35,42] that stores an elastic energy when deformed by the strain. Similar to aggregates in suspensions [43,44], the relaxation of huge networks requires a very long (practically infinite) time. Thus, the ABS resins containing such networks behave as a viscoelastic solid to exhibit quasi-equilibrium plateau of G' in experimentally accessible ω ranges [34,35,42]. The modulus for this plateau becomes larger when the matrix and graft AS chains have a large mismatch in their acrylonitrile (AN) content [36] and also when the AS grafting density for the particle deviates from an optimum value [40]. The shortage of the grafting density and the mismatch of the AN content enhance the attraction between the particles while the excess of the grafting density leads to depletion of the matrix chains from the graft layers [42], both reducing the stability of the particles to enhance aggregate formation and increase the quasi-equilibrium modulus.

The other type of slow relaxation is found for ABS resins that contain nonaggregating small rubber particles and have no quasi-equilibrium modulus [34,35,42]. The relaxation of these resins is similar to that of randomly dispersed silica particles [45] and block copolymer micellar particles [46]. As explained later in Sec. IV.D.3, the stress of these random dispersions is related to strain-induced anisotropy in the spatial distribution of particles and the slow relaxation is attributable to the particle diffusion.

B. Nonlinear Viscoelastic Behavior

1. General Behavior

Polymer blends often exhibit significantly nonlinear viscoelastic responses against large strain/rapid flow, in particular at long time scales. These responses, reflecting large changes in the blend morphology under strain/flow, are not described by the emulsion models. Such responses have been studied experimentally by Takahashi et al. [47–51] and theoretically by Onuki [30] and by Doi and Ohta [31]. As an example demonstrating the nonlinear features, Fig. 2 shows the steady-state shear stress σ_s and first

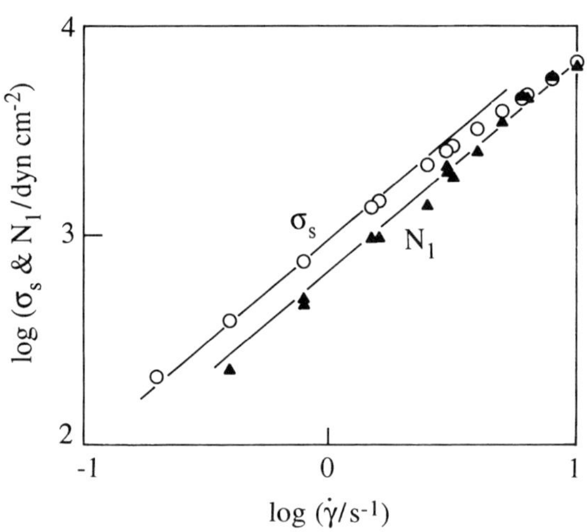

Figure 2 Steady-state shear stress and the first normal stress difference of a 1:1 blend of PDMS ($M = 50K, \eta_0 = 600$ poise) and PI ($M = 25K, \eta_0 = 500$ poise) obtained at room temperature. (Redrawn, with permission, from Ref. 49.)

normal stress difference N_1 for a 1:1 immiscible blend of PDMS and polyisoprene (PI) [49].

The component PDMS and PI of the blend examined in Fig. 2 have relaxation times << 0.1 s and thus exhibit the terminal behavior $\sigma_s \propto \dot{\gamma}$ and $N_1 \propto \dot{\gamma}^2$ (Eq. (10)) at $\dot{\gamma}$ examined [49]. Nevertheless, the blend exhibits a peculiar dependence, $N_1 \propto \dot{\gamma}$, that is never found for homogeneous liquids at low $\dot{\gamma}$. This dependence, commonly observed for strongly phase-separated, *textured* fluids irrespective of the component molecular weights [48,49], reflects the phase deformation and the increase in the interfacial area with increasing $\dot{\gamma}$. (For weakly segregated blends near a critical mixing point, Takahashi et al. [50,51] found even more complicated behavior, $N_1 \propto \dot{\gamma}$ at low $\dot{\gamma}$ and $N_1 \propto \dot{\gamma}^2$ at high $\dot{\gamma}$, the latter being due to shear-induced homogenization.)

2. Textured Fluid Model

For the simplest textured fluids, 1:1 immiscible blends of incompressible Newtonian liquids having the same zero-shear viscosity η_0 and the same density, Doi and Ohta [31] developed a model that considers strain-induced changes in the phase morphology. For representing anistropy in the shape of the interface, they defined an interfacial tensor $\mathbf{q} = \{q_{\alpha\beta}\}$ in terms of unit normal of the interface $\mathbf{n} = \{n_\alpha\}$ as

$$q_{\alpha\beta} = \frac{1}{V}\int_{\text{interface}} dS(n_\alpha n_\beta - \frac{1}{3}\delta_{\alpha\beta}) \qquad \alpha, \beta = x, y, z \tag{23}$$

Here, $\delta_{\alpha\beta}$ is the component of the unit tensor, and the integral $\int \ldots dS$ is carried out over the entire interfacial area in the blend of a volume V. The stress component $\sigma_{\alpha\beta}$ is expressed in terms of $q_{\alpha\beta}$, the interfacial tension Γ, a macroscopic velocity gradient tensor $\mathbf{K} = \{\kappa_{\alpha\beta}\}$, and a pressure P as

$$\sigma_{\alpha\beta} = \Delta\sigma_{\alpha\beta} - P\delta_{\alpha\beta} \qquad \Delta\sigma_{\alpha\beta} = -\Gamma q_{\alpha\beta} + \eta_0(\kappa_{\alpha\beta} + \kappa_{\beta\alpha}) \tag{24}$$

Here, $\Delta\sigma_{\alpha\beta}$ is the anistropic stress components due to storage of the interfacial energy $(-\Gamma q_{\alpha\beta})$ and viscous dissipation $(\eta_0(\kappa_{\alpha\beta} + \kappa_{\beta\alpha}))$.

Changes of $q_{\alpha\beta}$ with time have two contributions, one from the flow and the other from the interfacial tension. Considering these contributions to be additive, Doi and Ohta derived time-evolution equations for $q_{\alpha\beta}$ and a total interfacial area Q in the textured fluid [31]. For a change in the velocity gradient tensor from $\mathbf{K}(t)$ to $c\mathbf{K}(t)$, these equations (combined with Eq. (24)) lead to a scaling rule,

$$X(t; c\mathbf{K}(t)) = cX(ct, \mathbf{K}(t)) \quad \text{with } X = q_{\alpha\beta}, Q, \text{ and } \Delta\sigma_{\alpha\beta} \tag{25}$$

This rule, originally obtained for the 1:1 blends, can be derived for blends of various composition [48]. (Lee and Park [32] modified the time-evolution equations of the Doi–Ohta model (by adding some terms suggested from a dilute emulsion theory [18]) and calculated G' and G''. The scaling rule holds also for the Lee–Park model.)

Here, a brief comment is added for the above scaling rule. For textured fluids having no intrinsic length scale of the morphology (such as blends without emulsifiers), the flow (or strain) field determines the length scale that characterizes the morphology under flow and governs the slow rheological responses. The scaling rule is a natural consequence of this flow-determined length scale. This in turn means that the rule is not necessarily valid for fluids having intrinsic scales (such as emulsified blends having an optimum curvature of dispersed phases).

3. Comparison with Experiments

For the shear stress $\sigma_s (= \sigma_{xy})$ and the first normal stress difference $N_1 (= \sigma_{xx} - \sigma_{yy})$ during steady shear flow at rates $\dot{\gamma}$ and $c\dot{\gamma}$, Eq. (25) is simplified to

$$\sigma_s(c\dot{\gamma}) = c\sigma_s(\dot{\gamma}) \qquad N_1(c\dot{\gamma}) = cN_1(\dot{\gamma}) \tag{26}$$

This result explains the observed proportionality, $\sigma_s \propto \dot{\gamma}$ and $N_1 \propto \dot{\gamma}$ (Fig. 2). Thus, the peculiar $\dot{\gamma}$ dependence of N_1 is attributed to shear-induced changes in the interface. (Eq. (26) is also deduced from the Onuki model [30] for textured fluids near a critical mixing point, suggesting its general validity for textured fluids having no intrinsic length scale.)

Textured fluids also exhibit characteristic transient behavior after a step change of $\dot{\gamma}$ from an initial rate $\dot{\gamma}_i$ to a final rate $\dot{\gamma}_f = c\dot{\gamma}_i$ [47–49]. For the anistropic stress components after this change at $t = 0$, $\Delta\sigma_{\alpha\beta}(t; \dot{\gamma}_i, \dot{\gamma}_f)$, the Doi–Ohta model predicts a universality,

$$\frac{\Delta\sigma_{\alpha\beta}(t; \dot{\gamma}_i, \dot{\gamma}_f)}{\Delta\sigma^{o}_{\alpha\beta}(\dot{\gamma}_i)} = f(c, \gamma_f) \qquad \gamma_f = \dot{\gamma}_f \times t \tag{27}$$

where $\Delta\sigma^{o}_{\alpha\beta}(\dot{\gamma}_i)$ is the stress component during the initial, steady flow at $\dot{\gamma}_i$. Eq. (27) states that the transient stress reduced by the initial, steady stress is dependent only on the ratio of the shear rates $c = \dot{\gamma}_f/\dot{\gamma}_i$ and the total strain due to the final shear γ_f.

Takahashi et al. [47–49] experimentally examined this prediction for a variety of textured fluids. Fig. 3 shows typical results obtained for the 1:1 PDMS/PI blend subjected to the step increase of $\dot{\gamma}$ by a factor $c = 2$ [49]. The transient σ_s and N_1 reduced by the initial steady values σ_s^o and N_1^o are almost universally dependent on γ_f irrespective of the $\dot{\gamma}_f$ and $\dot{\gamma}_i$ values,

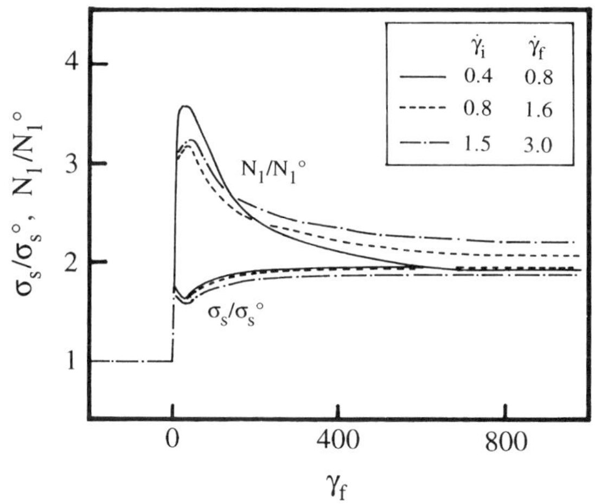

Figure 3 Transient behavior of the PDMS/PI blend examined in Fig. 2. Reduced shear stress and first normal stress difference are plotted against the total strain after a step increase of $\dot{\gamma}$. (Redrawn, with permission, from Ref. 49.)

qualitatively confirming the above prediction. However, quantitatively, σ_s and N_1 exhibit characteristic nonlinearities, undershoot and overshoot not deduced from the Doi–Ohta model. Microscopy observation revealed that the step increase of $\dot{\gamma}$ first induces elongation of the discrete phases and then rupture (into smaller droplets); these complicated changes lead to the undershoot and overshoot [49]. It is desired to incorporate these morphological changes into a refined model.

As demonstrated in Figs. 2 and 3, textured fluids containing deformable interfaces exhibit essentially nonlinear rheological responses. Related nonlinearities are observed also for strain recovery after cessation of flow. For a PS/PMMA blend, Gramespacher and Meissner [52] found that the recoverable compliance $J_r(t)$ first increases with t and then decreases and approaches its asymptote $J_r(\infty)$. This decrease, never observed for homogeneous fluids, appears to be related to differences between the strain recovery rate of the blend as a whole and the shape recovery rate of the discrete phases [52].

Here, a few comments are added for the linear viscoelastic behavior of immiscible blends. As explained earlier, the nonlinearity emerges when the length scale of the structure in the blends is determined by the shear. Small oscillatory shear requires a very long time to determine this scale by erasing the memory of the initial scale, e.g., a radius of the discrete phase at the quiescent state. Thus, this memory survives within usual time scales of dynamic tests and the resulting G^* represents linear responses of the quiescent structure. This is the case for the G^* data explained earlier.

IV. RHEOLOGY OF DIBLOCK COPOLYMERS

Block copolymers are composed of polymeric blocks that are covalently connected at their ends. This connectivity allows the constituent blocks to segregate in space only within a length scale smaller than their size (typically < 1000 Å), resulting in the well known *microdomain* structures [53]. This structure is thermodynamically determined according to the block lengths, composition, and segregation power [54–56], resulting in a variety of domain shapes (e.g., sphere, cylinder, and lamella as specified in Molau's rule [53]). The rheology of block copolymers at long time scales is strongly dependent on this domain shape as well as on other factors, such as the architecture of the copolymer chains (AB-type diblock, ABA-type triblock, etc.), viscoelastic features of each domain (e.g., glassy or rubbery), and a long-range coherent order in spatial arrangements of the domains. This section summarizes the rheological features of diblock copolymer systems, placing emphasis on the effects of these factors.

A. Quiescently Ordered Lamella at $T > T_{g,block}$

1. General Behavior

Recent studies [57–74] revealed common linear viscoelastic features for a variety of *lamella-forming* diblock copolymers at T above $T_{g,block}$ of the constituent blocks. Fig. 4 demonstrates the features for a poly(ethylene-propylene)–poly(ethylethylene) (PEP–PEE) diblock copolymer examined by Bates et al. [58]. The *order-disorder transition* (ODT) (microphase separation) temperature T_{ODT} is 124–125°C, and the PEP and PEE blocks are homogeneously mixed in a static sense at $T > T_{ODT}$. At $T < T_{ODT}$, the blocks form lamellar domains [58]. These domains were developed by cooling the copolymer quiescently (without deformation) from $T > T_{ODT}$. The domains are soft and deformable at all T examined ($> T_{g,PEP}$, $T_{g,PEE}$).

In Fig. 4, the G^* curves at various T are shifted along the ω axis to achieve the best superposition at high ω. At $T \geq 161$°C (well above T_{ODT}), G' and G'' exhibit the terminal behavior (Eq. (6)) and obey the time–temperature superposition (Eq. (12)) in the entire range of ω examined. When T is decreased toward T_{ODT}, the terminal behavior is still observed at low ω but a shoulder becomes prominent for the G' curve at intermediate ω. This shoulder reflects slow relaxation modes that emerge in the vicinity of

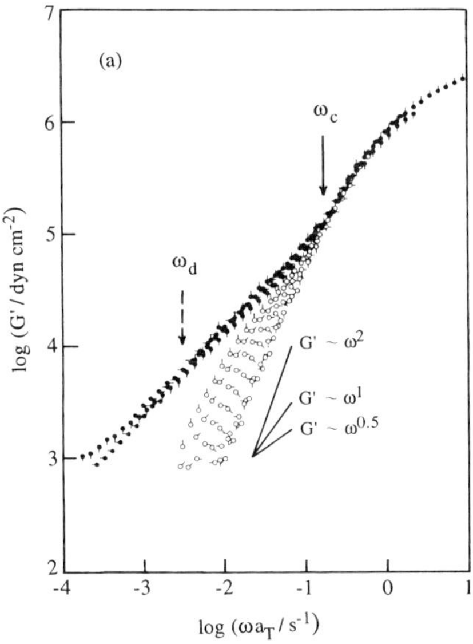

(a)

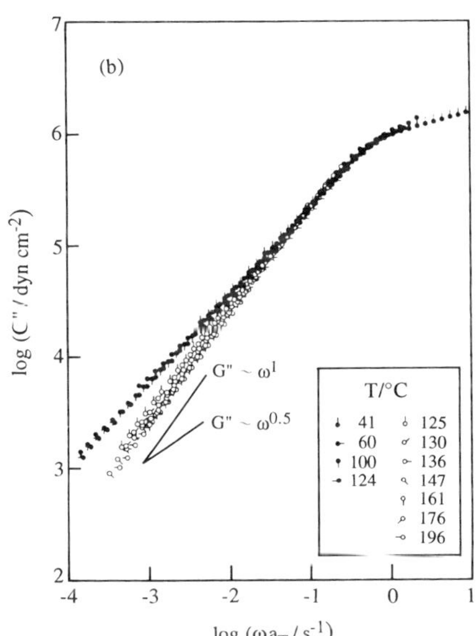

(b)

Figure 4 Frequency dependence of (a) G' and (b) G'' for a PEP–PEE diblock copolymer ($M = 54.8K$, PEP content $=$ 55%). The G' and G'' curves are shifted along the ω axis to achieve the best superposition at high ω. (Redrawn, with permission, from Ref. 58.)

T_{ODT} ($T = 125$–$147°C$). The shoulder grows with decreasing T, and the superposition of the G' curves is valid only at high ω above a characteristic ω_c (solid arrow). As explained earlier, this limited superposability suggests that a structure governing the slow relaxation modes changes with T. (Since G'' is insensitive to the slow modes, the failure of the superposition for those modes is much less clearly detected for G''.)

At $T < T_{ODT}$, the fast relaxation modes determining G^* at $\omega > \omega_c$ remain superposable. More importantly, the terminal behavior vanishes and the shoulder is followed by a rather weakly ω-dependent tail at $\omega < \omega_d$ (dashed arrow). This low-ω tail suggests that extra slow relaxation modes emerge at $T < T_{ODT}$. Such tails are often (approximately) described by a power law [57,58,63,72,73], $G'(\omega) \sim G''(\omega) \sim \omega^\alpha$ with $\alpha = 0.5$–0.7. (Note that the coincidence of the power-law indices of G' and G'' is deduced from the Kramers–Kronig relationship between G' and G'' [1]. This relationship also indicates that the value of α is related to G' and G'' as $\alpha = [2/\pi]\tan^{-1}[G''/G']$.)

As explained above, G' sensitively detects structural changes in the system. Using this feature, T_{ODT} is often determined as a temperature where a discontinuous drop is seen in plots of G' against T at sufficiently low $\omega (< \omega_d)$ [57–63,72,74]. The log G' versus log G'' plots explained for Eq. (7) also sensitively reflect the structural changes and are used for determination of T_{ODT} [57,74].

2. Structural Origins of Viscoelastic Relaxation

The viscoelastic features explained for Fig. 4 are observed for a wide variety of lamella-forming diblock copolymers at $T > T_{g,block}$ [57–74]. In the current understanding, the structure of these copolymers can be roughly classified into three regimes and the structure–rheology relationship in each regime can be summarized in the following ways.

(a) Homogeneous Regime. At T well above T_{ODT}, the copolymer chains are in the homogeneous state with negligible fluctuation of the block concentration with time [54,56,58,63]. Such chains exhibit the stress when they are anisotropically oriented by the strain and store an elastic free energy, and the orientational relaxation results in the viscoelastic relaxation. This structural origin of the stress is identical to that for homopolymer chains [1].

(b) Fluctuation Regime. At T above but close to T_{ODT}, the blocks are homogeneously mixed in a static sense (on average over long time). However, the block concentration fluctuates significantly with time and the copolymer chains are in a heterogeneous state in a dynamic sense

[54,56,58,63]. For such chains, the stress arises *via* two different mechanisms, strain-induced orientation of individual chains and distortion of the fluctuation pattern. Since the relaxation of the fluctuation pattern involves motion of many chains and is slower than the relaxation of individual chains, the former governs G^* at low ω ($< \omega_c$) (cf. Fig. 4) while the latter determines G^* at high ω ($> \omega_c$). Rosedale and Bates [57] reported that a ratio of observed increments of G' and G'' due to the fluctuation effect is fairly well described by the Fredrickson–Helfand theory [75] but the magnitudes of respective increments are underestimated.

The orientational relaxation of the chains in the fluctuation regime (b) is similar to that in the homogeneous regime (a), leading to the superposition of the G^* curves in the regime (b) at $\omega > \omega_c$ (Fig. 4). On the other hand, the amplitude and deformability of the fluctuation pattern governing G^* at low ω changes with T and the superposition fails at $\omega < \omega_c$. Concerning this fact, we note that the fluctuation-governed relaxation is not always detected experimentally. For example, Patel et al. [67] did not observe this relaxation for a lamella-forming polystyrene–polyisoprene (PS–PI) diblock copolymer ($M_{PS} = 12.5\ K$, $M_{PI} = 9.5\ K$). The difference between this PS–PI and PEP–PEE (Fig. 4) can be partly related to a difference in the T dependence of the interaction parameters χ; $\chi_{PS-PI} = -0.0937 + 60/T$ [76] and $\chi_{PEP-PEE} = 4.44 \times 10^{-4} + 4.69/T$ [57]. The PS–PI chains having the strongly T-dependent χ_{PS-PI} form lamellar domains as soon as the fluctuation becomes significant with decreasing T, thereby exhibiting the fluctuation-governed relaxation only in a narrow range of T (not detected in Ref. [67]).

(c) Ordered Regime. At $T < T_{ODT}$ the blocks are ordered and form well-developed lamellar domains [54,56,58,63]. This domain structure appears to have little effect for the viscoelastic relaxation mechanism at high ω, as suggested from dynamic birefringence (Δn^*) experiments by Kannan and Kornfield [68]. For PEP–PEE chains composed of optically similar blocks, a complex birefringence coefficient B^* ($= \Delta n^*/\gamma^*$ with γ^* being the dynamic strain) is mainly determined by the orientational anisotropy of individual chains. Kannan and Kornfield [68] found that $G^*(\omega)$ of the ordered PEP–PEE is proportional to $B^*(\omega)$ at high $\omega > \omega_{SO}$, with ω_{SO} being a characteristic frequency close to ω_c. Thus G^* at those ω is governed by the orientational relaxation of the chains, similar to the behavior in the fluctuation regime (b). This possibly leads to the time–temperature superposability for the ordered copolymer at high ω ($> \omega_c$).

Kannan and Kornfield [68] also found that the above proportionality vanishes at low $\omega < \omega_{SO}$. Thus the stress at those ω is sutained by structures larger than the chains. Koppi et al. [60] specified a characteristic ω_d (cf. Fig.

4) above and below which different types of structures sustain the stress. The relaxation at $\omega_d < \omega < \omega_c$ is attributable to motion of the lamellar domains: the lamellar interfacial energy is increased when the lamellae are bent by the strain, and the lamella motion dissipating this energy can be detected as the viscoelastic relaxation [60]. On the other hand, the extra slow relaxation seen at $\omega < \omega_d$ (cf. Fig. 4) can be related to long-range hydrodynamic interactions as well as to the defects explained below [60,64].

As schematically shown in Fig. 5, *quiescently ordered* block copolymers usually have coherent lamellar orientation only within a relatively

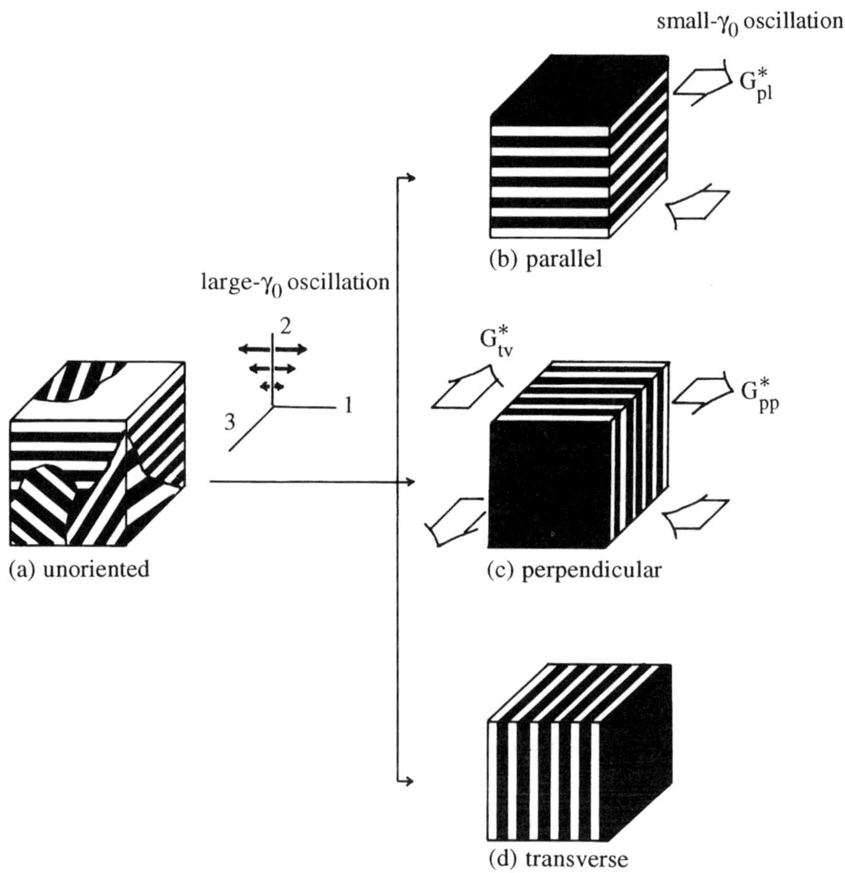

Figure 5 A schematic illustration of ordered copolymers having lamellar domains: (a) quiescently ordered structure having grains and defects; (b)–(d) shear-oriented lamellae having fewer defects.

short length scale (typically less than a few micrometers). Over larger scales, the orientation is random and isotropic. This large-scale structure is characterized with grains separated by various types of defects such as the focal conic and wall defects observed by Winey et al. [65,66]. The elastic free energy stored in distorted defects is dissipated via very slow, defect motion. This motion can be detected as the extra slow relaxation at $\omega < \omega_d$.

3. Role of Defects in the Ordered Regime

The above molecular picture of the defect-related extra slow relaxation is evidenced from results of shear-orientation experiments (explained later). The other important fact relevant to this relaxation is a viscoelastic similarity between a PS–PI copolymer ($M_{PS} = 12.5K$, $M_{PI} = 9.5K$) and 4-cyano-4′-octylbiphenyl (8CB) found by Larson et al. [64]. For the quiescently ordered PS–PI and a low-M smectic compound, 8CB, both having a layered (lamellar) structure with many defects [64], G^* are shown in Fig. 6 with unfilled symbols. Clearly, the PS–PI and 8CB exhibit very similar behavior. In the experimental window, G' and G'' appear to have the power-law type ω dependence at high ω, while G' tends to level off and G'' appears to continuously decrease at low ω. This similarity strongly suggests that the low-ω behavior of PS–PI and 8CB are commonly attributed to their layered

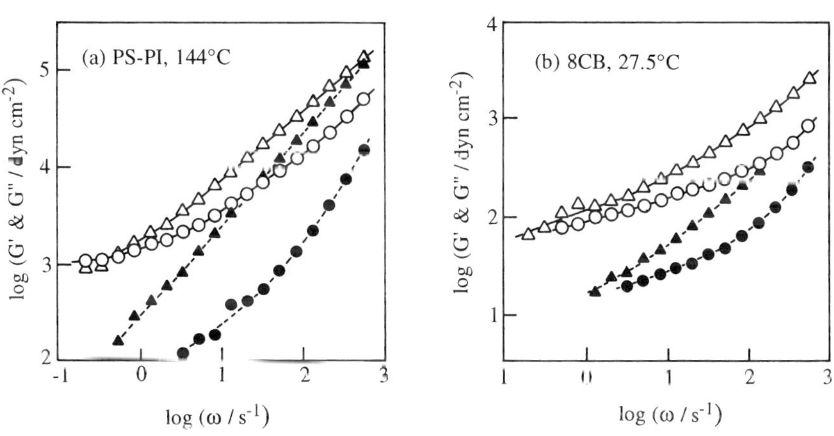

Figure 6 Frequency dependence of G' and G'' for: (a) a lamellar-forming PS–PI copolymer ($M_{PS} = 12.5K$, $M_{PI} = 9.5K$) at 144°C and (b) smectic 8CB at 27.5°C. G^* at the quiescently ordered state are indicated by the unfilled symbols, and those at a shear-oriented state, with the filled symbols: G'—circle; G''—triangle. (Redrawn, with permission, from Ref. 64.)

structure containing many defects [64]. In other words, this behavior is not related to the polymeric character of the PS–PI chains.

Theories were proposed by Kawasaki and Onuki [77] and by Rubinstein and Obukov [78] for slow relaxation behavior of the quiescently ordered block copolymer lamellae. Kawasaki and Onuki considered the relaxation via a hydrodynamic mode (overdamped second sound mode) of lamellar motion and predicted a power-law relationship,

$$G'(\omega) \cong G''(\omega) \cong [\pi/24\sqrt{2}][K_{\mathrm{lam}}\eta_{\mathrm{block}}]^{1/2}\omega^{1/2} \tag{28}$$

Here, K_{lam} and η_{block} are the bending modulus of lamellae and the viscosity of constituent blocks, respectively. Rubinstein and Obukov considered annihilation of lamellae via diffusion of the chains along the lamellar surfaces and obtained the same power-law form, $G'(\omega) \cong G''(\omega) \propto \omega^{1/2}$.

These theories (in particular Eq. (28) specifying both magnitudes and ω dependence of G^*) can explain the low-ω tails of G' and G'' exhibiting the power-law type ω dependence (cf. Fig. 4). However, we should also note that the theories consider motion of the lamellae and/or chains within a grain and are not necessarily applied at extremely low ω where many grains and defects might cooperatively determine the relaxation behavior. This low-ω limit is not necessarily covered in all experiments. However, the data of Fig. 6 and some other data (e.g., Fig. 6 in Ref. [65] and Fig. 11 in Ref. [74]) suggest a possiblity that unoriented lamellae could have equilibrium moduli $G_e = G'(0)$ and behave as a solid in that limit. At this moment, it is not clear whether this solid-like behavior is common to all unoriented lamellar systems.

4. Evaluation of the Grain Size

In viscoelastic studies of the low-ω limiting behavior of the unoriented lamellae, it is of particular importance to evaluate an average size of the grain (L_{av}) and examine whether the measured $G^*(\omega)$ corresponds to motion of chains/lamellae over the length scale $> L_{\mathrm{av}}$. Amundson et al. [79] and Balsara et al. [80] indicated that the grain structure of those lamellae exhibit depolarized light scattering. Specifically, for randomly oriented grains, Balsara et al. showed that L_{av} is given as [80]

$$L_{\mathrm{av}} = \frac{15}{4\pi^2}\frac{\lambda^2}{(\Delta n^0)^2 L}\frac{I}{I_0} \tag{29}$$

Here, λ and I_0 are the wavelength and intensity of the incident light beam, I is the intensity of the depolarized scattering that is concentrated near zero-angle of the scattering, L is the optical path length of the sample, and Δn^0 is the intrinsic birefringence of the lamellae. Δn^0 is sufficiently large to give intensive scattering for lamellae composed of optically dissimilar blocks (e.g., PS–PI and PS–PMMA). For those lamellae, we can use Eq. (29) to experimentally evaluate L_{av} from the measured I/I_0 ratio [80]. (Eq. (29) is approximately valid also for random grains of cylindrical domains. Balsara et al. [81] evaluated L_{av} for a cylinder-forming PS–PI solution and discussed the G^* data in relation to L_{av}.)

The depolarized light scattering is also used for determination of T_{ODT} for unoriented lamellae: T_{ODT} is evaluated as a temperature where I/I_0 plotted against T suddenly drops to zero [72,79]. This T_{ODT} is in good agreement with T_{ODT} determined from the G' versus T plots explained earlier [72,79].

B. Shear-Orientation of Lamella at $T > T_{g,block}$

1. General Behavior

After the pioneering work of Keller et al. [82] and Hadziioannou et al. [83], it has been recognized that block copolymer lamellae and/or cylinders are coherently oriented over large spatial scales when subjected to steady as well as large-amplitude oscillatory shear under appropriate conditions. Recently, the shear orientation and resulting changes in the viscoelastic responses of diblock copolymer lamellae have been extensively investigated in relation to the extra slow relaxation of unoriented lamellae [58,60,61,63–71,73,84–87].

Figs. 5(b)–5(d) show possible directions of ideally shear-orientated lamellae. (Some imperfection, or defect, unavoidably remains for actual lamellae.) From a simple consideration of the sliding symmetry of lamellae, one might expect that the shear induces only parallel orientation. However, as explained later, the other two types of orientation can also be achieved under appropriate shearing conditions. For these differently oriented lamellae, we can apply small oscillatory shear to measure the linear viscoelastic moduli G^*_{pl}, G^*_{pp}, and G^*_{tv} along the three directions specified in Fig. 5.

Fig. 7 shows these moduli obtained by Koppi et al. for a PEP–PEE lamella [60]. G^*_{pl} were measured for the parallel lamella (cf. Fig. 5(b)) while G^*_{pp} and G^*_{tv} were measured for the perpendicular lamella (cf. Fig. 5(c)). As seen in Fig. 7, G'_{pl}, G'_{pp}, and G'_{tv} coincide with the modulus G'_{uo} of the unoriented lamella at $\omega > \omega_c$, with ω_c being the characteristic frequency above which the unoriented lamellae exhibit time–temperature superposability (cf. Fig. 4). Similar coincidence was also found for G'' [60]. Thus the lamella alignment has little effect on the orientational relaxation of the copolymer chains that governs G^* at $\omega > \omega_c$. However, significant effects are seen at low $\omega < \omega_c$; $G'_{uo} \cong G'_{tv} > G'_{pl} > G'_{pp}$ at those ω. This inequality was also found for G'' [60].

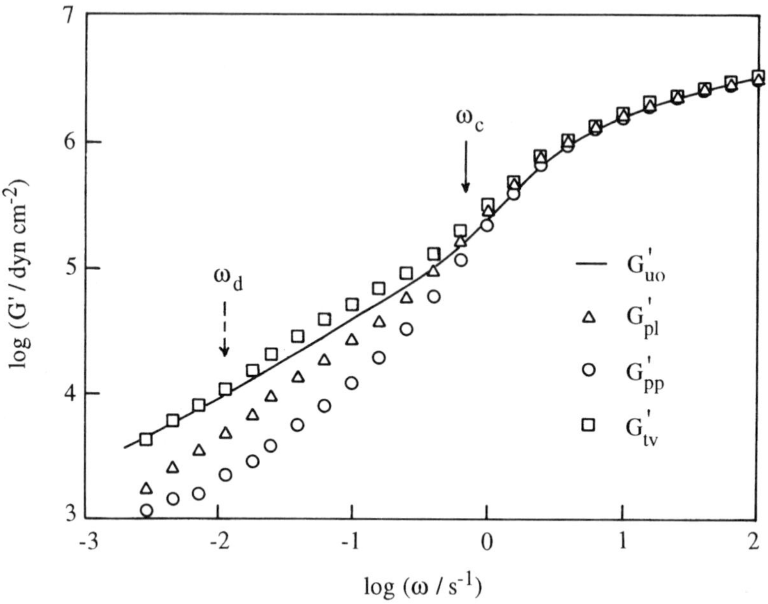

Figure 7 Comparison of G' for unoriented and oriented lamellae of a PEP–PEE copolymer ($M = 50.1K$, PEP content $= 56\%$) at 40°C. G' of the oriented lamellae are measured along the directions specified in Fig. 5. (Redrawn, with permission, from Ref. 60.)

For the PS–PI copolymer examined by Larson et al. [64], $G^*_{\rm pp}$ of the perpendicular lamellae are shown in Fig. 6(a). Clearly, $G^*_{\rm pp}$ (filled symbols) is smaller than $G^*_{\rm uo}$ of the unoriented lamella (unfilled symbols). A similar result is seen also for the oriented layers of 8CB (Fig. 6(b)), indicating that the decrease of the low-ω moduli on shear orientation is characteristic to layered structures.

The above inequality, $G^*_{\rm uo} > G^*_{\rm pl}$ and/or $G^*_{\rm pp}$ at low ω, is commonly observed for diblock copolymer lamellae at $T > T_{\rm g,block}$ [58,60,61,63–71,73]. The inequality means that the oriented lamellae having fewer defects (compared with the unoriented lamellae) store smaller elastic free energies at long time scales. This in turn suggests an important contribution of the defects to the extra slow relaxation of the unoriented lamellae. The densely populated defects in these lamellae store a large elastic energy, and dissipation of this energy due to the defect motion is observed as the significant relaxation at low ω ($< \omega_{\rm d}$).

Relevant behavior was found by Takahashi et al. [88] for a PS–PI lamellar solution under steady shear. At low $\dot{\gamma}$, the lamellae are unoriented and their first normal stress difference N_1 is proportional to $\dot{\gamma}$. This behavior is similar to that of the textured fluids (immiscible blends) described earlier, suggesting that the grains/defects give the "texture" of the lamellae and govern the slow rheological responses. (The relationship, $N_1 \propto \dot{\gamma}$, suggests that the texture size (grain size) under steady flow is determined by $\dot{\gamma}$.) At high $\dot{\gamma}$, the lamellae are orientated to exhibit a liquid-like response, $N_1 \propto \dot{\gamma}^2$ (Eq. (10)). This effect of the fast steady shear is similar to the effect of the large amplitude oscillatory shear.

2. Shear-Orientation Mechanisms

When diblock copolymer lamellae are shear-oriented at $T > T_{g,\text{block}}$ under oscillatory strains of large amplitudes (γ_0), the orientation direction changes with T and the shearing frequency ω. Some typical results are summarized below.

For PEP–PEE quiescently ordered from $T > T_{\text{ODT}}$ and then sheared at $\gamma_0 = 1$, Koppi et al. [60] found perpendicular orientation at high ω and high T (still below T_{ODT}) but parallel orientation at low ω and low T. At a given T, the parallel orientation is converted irreversibly to the perpendicular orientation with increasing ω and thus the latter is more stable than the former under the large-γ_0 shear. Koppi et al. [60] attributed the parallel orientation to shear-induced defect migration, and the perpendicular orientation, to selective melting and reformation of lamellae. The latter mechanism requires the interfaces of reformed lamellae to be undisturbed by the shear, resulting in the perpendicular lamellae having interfaces in the vorticity plane (1-2 plane shown in Fig. 5).

For a PS–PI lamella ($M_{\text{PS}} = 12.5K$, $M_{\text{PI}} = 9.5K$) quiescently ordered from $T > T_{\text{ODT}}$ and sheared at $\gamma_0 = 1$, Patel et al. [67] found the perpendicular orientation at low ω and high $T (< T_{\text{ODT}})$ and the parallel orientation at high ω and low T. At a given T, the perpendicular orientation is converted irreversibly to the parallel orientation with increasing ω. These results are in contrast to those of the PEP–PEE lamellae. Patel et al. [67] found that the orientational direction of their PS–PI lamella is strongly correlated with the magnitude of the linear moduli $|G^*|$. The parallel orientation is achieved at ω where $|G^*_{\text{pl}}| < |G^*_{\text{pp}}|, |G^*_{\text{uo}}|$ while the perpendicular orientation is preferred at ω where $|G^*_{\text{pp}}| < |G^*_{\text{pl}}|, |G^*_{\text{uo}}|$. On the basis of this result, they considered that the lamellae are oriented in the direction giving the smallest mechanical resistance. This mechanism may work for lamellar layers having a large viscoelastic contrast (like PS and PI layers having widely different T_g) but not necessarily for the layers of small contrast (like PEP and PEE).

For a PS–PI lamella ($M_{\text{PS}} \cong M_{\text{PI}} \cong 10K$) quiescently ordered from $T > T_{\text{ODT}}$, Gupta et al. [69] found $|G^*_{\text{pl}}| < |G^*_{\text{pp}}|, |G^*_{\text{uo}}|$ in the entire range of ω. Nevertheless, this PS–PI lamella exhibits perpendicular orientation at low ω (for $\gamma_0 = 1$) and parallel orientation at high ω (for $\gamma_0 = 0.5$ and 0.8), similar to the PS–PI lamella examined by Patel et al. [67]. Thus the orientation of PS–PI lamellae is not uniquely determined by the mechanical resistance. Gupta et al. [85] also found that the coherent orientation is achieved only when γ_0 is larger than a critical value.

For a PS–PI lamella ($M_{PS} = 8.6K$, $M_{PI} = 11.5K$) molded at room temperature, annealed at 85°C, and sheared (with $\gamma_0 = 1$) at 110 and 120°C, Zhang et al. [86] found the perpendicular orientation at $\omega_c > \omega > \omega_d$ and the parallel orientation at $\omega > \omega_c$ and $\omega < \omega_d$, with ω_c and ω_d being the characteristic frequencies explained for Fig. 4. The parallel orientation at low $\omega < \omega_d$ was not observed by Patel et al. [67] who covered equally low ω (down to $[10^{-3}–10^{-5}]\omega_x$ with ω_x being a characteristic ω for relaxation of individual copolymer chains). Thus the parallel orientation at very low ω does not seem to be common to all PS–PI lamellae.

Zhang and Wiesner [71] also examined the shear orientation behavior at 110 and 120°C ($\gamma_0 = 0.3$) for a high-M PS–PI lamella ($M_{PS} = 32.1K$, $M_{PI} = 26.2K$) molded at room temperature and annealed at 117°C. They found a mixture of parallel and transverse lamellae at $\omega \cong 20\omega_x$, the latter being frequently considered as the most unfavorable alignment because the shear tends to compress the transverse lamellae and increase their free energy. Similarly mixed lamellae were found by Okamoto et al. [73] for a PS–PEP ($M_{PS} = 11.8K$, $M_{PEP} = 20.1K$). These PS–PI and PS–PEP chains are in the entangled state. Using the tube concept of the Doi–Edwards theory [89], Zhang and Wiesner [71] attributed the formation of the transverse lamella to orientation of the tube trapping the entangled blocks. The tube axis is normal to the lamellar surface on average, and the orientation of the tube (and the blocks in it) along the shear direction may result in the transverse orientation of the lamella.

For a quiescently ordered lamellar solution of a PS–PI ($M_{PS} = 11K$, $M_{PI} = 17K$), Balsara et al. [87] compared the shear orientation behavior under oscillatory shear and steady shear. They found parallel lamellae for the former and less coherent, mixed parallel and perpendicular lamellae for the latter. Using the analogy of ironing a crumpled sheet, Balsara et al. [87] attributed the higher efficiency of orientation under oscillatory shear to an easier migration of defects.

As seen above, the ordered lamellae (at $T > T_{g,block}$) exhibit various types of shear orientation behavior probably due to a wide variety of the potentially available orientation mechanisms such as selective rotation of grains, selective melting and reformation of lamellae, defect migration, reduction of mechanical resistance, tube orientation (for entangled chains), and coupling of the lamella motion and the shear field. An interesting phenomenon relevant to the last mechanism is the shear-induced lamella formation at T above T_{ODT} of the quiescent system: Koppi et al. [84] observed the formation of perpendicular PEP–PEE lamellae while Balsara et al. [87] found the formation of parallel PS–PI lamellae. Theoretically, Cates and Milner [90] considered a coupling of the shear field and block concentration fluctuation. With the coupling mechanism, the fluctuation is suppressed under shear and the perpendicular lamella can be formed at $T >$ quiescent T_{ODT} [90]. In this mechanism, Fredrickson [91] incorporated a viscosity contrast of the constituent blocks and coupling of the fluctuation and mean concentration fields. He showed that the perpendicular and parallel lamellae can be formed under slow and fast shear, respectively [91].

It is very interesting to experimentally establish a unified molecular picture for the various types of shear orientation behavior of the ordered lamellae as well as for the shear-induced lamella formation. For this purpose, it would be very informative to carry out *in situ* rheo-optical measurements and detect changes in the lamella direction during a course of shear orientation. For example, Gupta et al. [69] traced changes in the orientation of PS–PI lamellae through *in situ* dynamic birefringence. They found that the orientation exhibits rapid and slow changes with time and the same, final orientation is achieved via different routes at different ω (e.g., rapid suppression of transverse and perpendicular orientations at intermediate and high ω, respectively, to achieve the final parallel orientation). This type of information would provide us with a clue to establish the unified picture.

C. Cylinders and Spheres at $T > T_{g,block}$

1. General Behavior

The rheology and structure of ordered diblock copolymers having cylindrical and/or spherical domains have also been extensively studied [46,62,81,92–107]. Similar to the lamellae, quiescently ordered cylinders have an unaligned grain structure with many defects, and coherent alignment in a hexagonal lattice is achieved on application of large shear under appropriate conditions [81,93–97]. Changes in the directions of lattice planes with the shearing frequency [94] and increases of T_{ODT} during and after the shear alignment [94,97] are also found for the cylinders.

As an example of linear viscoelastic behavior of unaligned and aligned cylinders at $T > T_{g,block}$, Fig. 8(a) shows G^* data obtained by Koppi et al. [93] for a PEP–PEE copolymer at 150°C. The system contains PEP cylinders at 150°C ($\ll T_{ODT} = 220°C$ in the quiescent state). For the aligned cylinders G^* was measured along the cylinder axis. We note that G^* is nearly the same for the unaligned and aligned cylinders at high ω (where relaxation of indi-

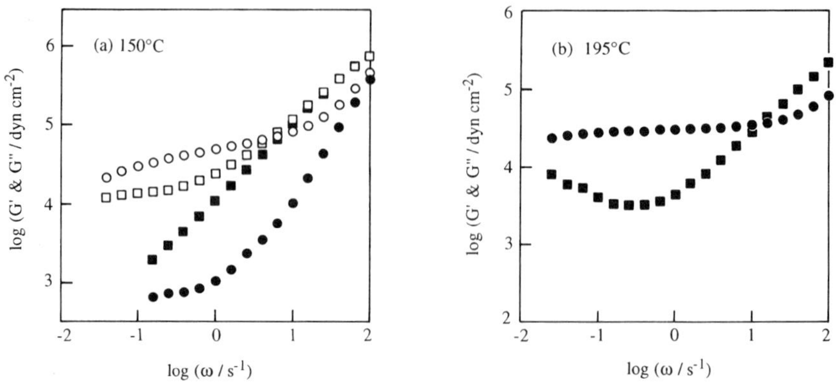

Figure 8 G^* of a PEP–PEE copolymer ($M = 98K$, PEP content = 25%) at (a) 150°C and (b) 195°C. The system contains cylindrical and spherical PEP domains at 150°C and 195°C, respectively. In (a), the responses of unoriented and oriented cylinders are shown by the unfilled and filled symbols, respectively. In (b), the responses of spheres shear-aligned at $\omega = 10\,\mathrm{s}^{-1}$ are shown. In both (a) and (b), circles and squares indicate G' and G'', respectively. (Redrawn, with permission, from Ref. 93.)

vidual chains governs G^*) but is much smaller for the latter at low ω. This result, qualitatively similar to that for the lamellae, demonstrates the importance of the defects for the slow rheological responses of the cylinders.

The aligned PEP cylinders examined in Fig. 8(a) exhibit an order–order transition to become spheres (aligned on a twinned body centered cubic (bcc) lattice) with increasing T above $T_{\mathrm{OOT}} = 175°\mathrm{C}$ [93]. Koppi et al. [93] shear-aligned those epitaxially grown spheres and then measured G^* against small strain. Fig. 8(b) shows those G^* data at 195°C. At low ω, G' of the aligned spheres is much larger in magnitude and much more weakly ω-dependent than G' of the aligned cylinders (cf. Figs. 8(a) and 8(b)). Thus, roughly speaking, solid-like responses are seen for the aligned spheres while liquid-like responses are observed for the aligned cylinders (sheared along their axis). Note, however, that the static behavior (solid or liquid) of these aligned structures is a delicate issue, as discussed in the next section.

2. Effect of Domain Shape

As explained earlier, the defect structures as well as the domain shape have a large influence on the rheological responses of ordered diblock copolymers at long time scales. The influence of the domain shape is most clearly observed for shear-oriented systems for which the influence of the defects is minimized. More or less liquid-like responses are observed for the oriented lamellae in a direction along their surface (Fig. 7) and for the aligned cylinders along their axis (Fig. 8(a)). On the other hand, roughly solid-like responses are seen for the aligned spheres (Fig. 8(b)). These results can be related to spatial symmetries of domain arrangements in the following way.

For a moment, we consider perfectly aligned domains and neglect their thermal undulation. Then, we may consider that the lamellae and cylinders form one- and two-dimensional lattices. These lattices have a sliding symmetry in directions parallel to the lamellar surface and/or cylinder axis. In other words, the lattices are not distorted by a sufficiently slow shear strain along these directions. Thus the lattices exhibit no thermodynamic restoring force against this strain and behave as a liquid in the low-ω limit. On the other hand, the spheres form a three-dimensional cubic lattice. This lattice is distorted by small shear strain of any direction and behaves as a solid in the linear viscoelastic regime.

The actual low-ω behavior of the shear-aligned copolymers is also affected by some additional factors. The first, simplest factor is a small amount of unavoidably remaining defects. Weak but extremely slow relaxation modes due to those defects may provide a solid-like character to the system. The second factor is the thermal undulation of the copolymer domains that always exists at T above $T_{g,block}$ of all constituent blocks. Coupling of the undulation and shear field may also provide a solid-like character. The third and fourth factors that become important in the vicinity of T_{ODT} are thickening of the interphase between the domains and thermal migration of the copolymer chains from one domain to the others. A large freedom in motion/location of the block junctions in the thickened interphase reduces the lattice stability and provides a liquid-like character. The chain migration, taking place at time scales of thermal equilibration of the whole system, allows local rearrangements of the domains and also provides a liquid-like character. Thus, according to the lattice symmetry the aligned lamellae and cylinders tend to behave as a liquid at low ω while the aligned spheres tend to behave as a solid, but the actual behavior is affected by the above four factors.

As noted in Figs. 6–8(a), the low-ω moduli are significantly smaller for the shear-aligned lamellae and cylinders (at $T > T_{g,block}$) than for the unoriented structures. This fact reflects the above tendency of the lamellae and cylinders to behave as a liquid. However, the terminal behavior characteristic of liquids (Eq. (6)) is not observed for those aligned structures, possibly due to the remaining defects and to the coupling between the domain undulation and shear field. (The thickening of the interphase and the chain migration appear to have no significant effect on the behavior of those structures at T well below T_{ODT}.)

In Fig. 8(b), the G^* data at high ω ($> 0.3\ s^{-1}$) demonstrate the tendency of the aligned PEP spheres to behave as a solid. However, G' gradually decreases while G'' significantly increases with decreasing ω ($< 0.3\ s^{-1}$). This behavior suggests that the PEP spheres at 195°C (in the vicinity of T_{ODT}) relax significantly at low ω (not covered in Fig. 8(b)). Thus we expect more or less liquid-like behavior of those spheres at the low-ω limit. In fact, liquid-like behavior characterized by significant decreases of G' ($< G''$) with decreasing ω and a peak in G'' is seen for a PE-PEE copolymer forming PE spheres [96]. These results may be related to the large freedom in motion/location of the block junctions in the thickened interphase and also to the copolymer chain migration.

D. Spherical Micelles at $T < T_{g,core}$

1. Elasto-Plastic Nature of Micellar Lattices

As explained earlier, slow rheological responses of the ordered diblock copolymers at $T > T_{g,block}$ [57–74,81,92–97] are governed by the symmetry of the domain arrangements (lattices) but the defects, domain undulation, freedom in motion/location of the block junctions, and chain migration may also have large influences. When one of the constituent blocks forms discrete, glassy domains, the domain undulation/deformation and the chain migration do not take place and the freedom of the junctions is frozen. This is the case in which we can most clearly observe the rheological features of the lattices.

The features of cubic lattices were studied by Watanabe et al. in the early 1980s [98–102] for quiescently ordered solutions of styrene–butadiene (PS–PB) diblock copolymers in a PB-selective solvent, n-tetradecane (C14). Those copolymers (having PS content less than 30%) formed micelles with glassy, spherical PS cores and solvated PB corona at around room temperature. The micelles were arranged on cubic lattices in moderately concentrated solutions to exhibit characteristic nonlinear responses [98]. Under slow steady shear, the micelles exhibited plastic flow [98] (with and without plug-flow behavior [108] in Couette and cone-and-plate geometries, respectively). Corresponding elasto-plastic responses against slow oscillatory strain [98], a rectilinear Lissajous pattern for small γ_0 and a lozenge-shaped pattern for large γ_0, are demonstrated in Fig. 9(a).

The rectilinear pattern indicates the linear, elastic response of the micellar lattice before yielding (for the stress amplitude σ_m smaller than the yield stress σ_y [98]). This response is characterized by ω-insensitive

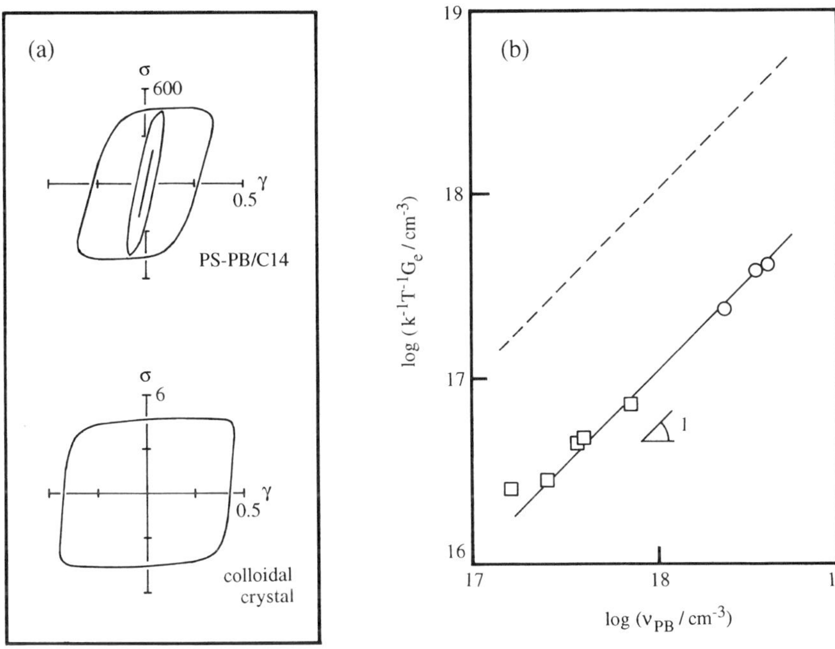

Figure 9 Dynamic behavior of PS–PB/n-tetradecane (C14) solutions containing micellar lattices. (a) Lissajous patterns (σ in unit of dyn cm^{-2}) for a 20 wt% PS–PB/C14 micellar solution ($M_{PS} = 16K$, $M_{PB} = 36K$) at 25°C and a colloidal crystal of charged PS latex (14 wt%) in an aqueous medium at 20°C. These patterns were obtained in a Couette geometry at $\omega = 0.02\,s^{-1}$. (b) Dependence of equilibrium modulus G_e of various PS–PB/C14 solutions at 25°C on the number density υ_{PB} of the corona PB blocks in the solutions. Circles indicate G_e for 20, 30, and 35 wt% solutions of PS–PB 16–36, and squares, G_e for 10 wt% solutions of PS–PB 20–46, 20–100, 32–102, 32–160, and 32–262, with sample code numbers denoting $10^{-3}M_{PS} - 10^{-3}M_{PB}$. The dashed line indicates the relationship $G_e = \upsilon_{PB}kT$ expected for perfect lattices with no defects. (Redrawn, with permission, from Refs. 98 and 99.)

storage moduli G' (shown by triangles in Fig. 10) and negligibly small G'' ($< 0.02G'$; not shown). The lozenge-shaped pattern, observed when $\sigma_m \geq \sigma_y$, corresponds to the elastic deformation (for $|\sigma(t)| < \sigma_y$) and successive plastic flow (for $|\sigma(t)| \geq \sigma_y$), each being repeated twice in the opposite directions during each cycle of oscillation. Such patterns, characterizing low-ω features of the lattices, are observed also for a colloidal crystal of charged PS latex (cf. Fig. 9(a)).

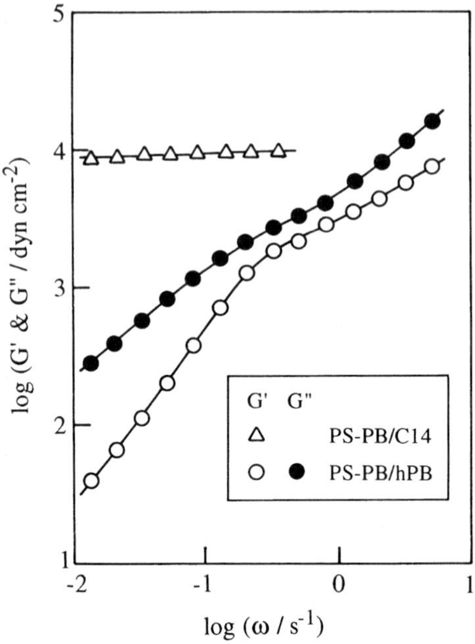

Figure 10 Comparison of linear viscoelastic moduli at 25°C for 20 wt% micellar systems of a PS–PB copolymer ($M_{PS} = 16K$, $M_{PB} = 36K$) in n-tetradecane (C14) and hPB ($M = 2K$). For the n-tetradecane system, G'' are negligibly small ($< 0.02G'$) and not shown here. (Redrawn, with permission, from Refs. 98 and 100.)

Doi et al. [109] formulated a model for lattices containing imperfections (defects) and analyzed its dynamic as well as steady-flow behavior. Their model, qualitatively reproducing the nonlinear features explained above, demonstrates that the plastic flow of the lattices is governed by a mechanical instability (lattice slippage) at the defects. Details of this slipping process can be detected from time-resolved small-angle X-ray scattering (SAXS) and/or small-angle neutron scattering (SANS) at sufficiently low ω. (The slippage is not clearly observed at high ω where individual blocks have not relaxed completely [107].)

Here, we note differences between the cubic lattices of the spherical domains examined in Figs. 8(b) and 9(a). For the soft PEP spheres in an equally soft PEE matrix, G'' significantly increases with decreasing ω; G'' is as large as $0.3G'$ at the lowest ω examined (Fig. 8(b)). Those G^* data suggest more or less liquid-like behavior in the low-ω limit (possibly due to the large freedom in the motion/location of the PEP–PEE junctions in the thickened interphase and also to the chain migration at T well above $T_{g,\text{block}}$ and close to T_{ODT}). On the other hand, this freedom is frozen and the chain migration does not take place for the PS–PB/C14 micelles having glassy PS cores. These micelles exhibit elastic behavior characterized by an ω-insensitive G' and negligibly small G'' at low ω (cf. Fig. 10). Thus the solid-like nature of the cubic lattices (reflecting the lattice symmetry explained earlier) is much more clearly observed for the PS–PB/C14 system than for the PEP–PEE system.

2. Lattice Stability

For PS–PB/C14 micellar solutions, the lattice formation is governed by the corona PB blocks tightly anchored on the glassy PS cores [98,102]. In moderately concentrated C14 matrices, the overlapping corona blocks of neighboring micelles are required to have uniform concentration distribution so as to reduce their osmotic free energy. Then, those blocks have to take mutually correlated and constrained conformations (having a large elastic free energy). This osmotic constraint for the corona block conformation is minimized when the micelles are regularly arranged in space, giving a thermodynamic driving force of lattice formation. This lattice-forming mechanism is similar to the mechanism determining the morphologies of strongly segregated block copolymers [54].

The stability of the PS–PB micellar lattices is examined through their equilibrium moduli G_e ($= G'|_{\omega \to 0}$; cf. Fig. 10). For lattice-forming micellar solutions of various concentrations c_{PS-PB} and block molecular weights M_{PS} and M_{PB} [98,99,102], Fig. 9(b) shows plots of G_e/kT (kT = thermal energy) against the number density υ_{PB} of the corona PB blocks in the solutions. Clearly, G_e is proportional to υ_{PB} irrespective of c_{PS-PB}, M_{PS}, and M_{PB}, indicating that the lattice stability (rigidity against small strain) is governed by the corona blocks. Under strong osmotic constraint, each corona block works as an entropic strand to sustain the stress $\propto kT$, giving the observed proportionality. This result is in harmony with the lattice-forming mechanism explained above. (Note, however, that the measured G_e are smaller than the equilibrium moduli expected for perfect lattices having no defects, $G_e = \upsilon_{PB}kT$ (dashed line). Thus the stability of actual lattices appears to be reduced by the defects.)

The above lattice-forming mechanism is evidenced from several experimental facts: (1) a critical concentration for the lattice formation is close to the overlapping concentration of the micelles above which the osmotic compressibility rapidly decreases to enhance the osmotic constraint for the corona block conformation [99]; (2) with increasing T, the lattice is first molten and then the micellar core and corona are homogeneously mixed [98]. This lattice instability is largely reduced when the core is chemically crosslinked [101]: the crosslink prohibits thickening of the corona/core interphase at high T and freezes the freedom in motion/location of the tethered corona ends, thereby enhancing the osmotic constraint and increasing the lattice stability; (3) the PS–PB micellar lattice is disordered (but the micelles are preserved) when the solvent, C14, is replaced by low-M homo-PB (hPB) [100]. This result reflects the osmotic nature of the driving force for the lattice formation: in the hPB matrix, the PB segment distribution is kept uniform by the hPB chains irrespective of the conformation of the corona PB blocks. Thus the osmotic constraint is screened and the driving force vanishes in the hPB matrix.

Because of the absence of the lattice structure, PS–PB/hPB systems have no plasticity and relax completely. Fig. 10 compares the linear viscoelastic moduli for such PS–PB/hPB and PS–PB/C14 systems having the same PS–PB content (20 wt%) [100]. Clearly, the low-ω moduli are much smaller for the former, despite the fact that the total polymeric content (PS–PB plus hPB) is larger for the former. This fact demonstrates that the low-ω stress of the PS–PB/C14 system is sustained by the lattice therein.

The stability of micellar lattices under fast steady shear has also been investigated through SANS experiments. Phoon et al. [105] and McConnell et al. [106], respectively, found shear-induced melting of PS–PEP and PS–PI micellar lattices (having PS cores) at high $\dot{\gamma}$. At lower $\dot{\gamma}$, these lattices exhibit enhanced long-range order [105,106]. Qualitatively similar behavior, shear-induced single crystallization, was found by Mortensen et al. [110] for PEO–PPO–PEO micelles (with PPO cores). For a PS–PI micellar system, McConnell et al. [106] also found a large hysteresis in the shear stress at intermediate $\dot{\gamma}$ that corresponds to structural changes from polycrystalline face centered cubic (fcc) to a sliding layer structure. All these structural changes, governed by the inter-micellar potential and hydrodynamic interactions, are strikingly similar to changes in colloidal crystals of charged particles [111]. Thus the micellar lattices and colloidal crystals exhibit structural and rheological similarities for many aspects (including the nonlinear dynamic responses shown in Fig. 9(a)).

3. Viscoelastic Relaxation of Micellar Dispersions

In hPB matrices, the PS–PB micelles are randomly dispersed and exhibit linear viscoelastic relaxation (unless the PS–PB content is large). As can be noted for the G^* data in Fig. 10 (circles), the mode distribution for this relaxation is much broader than that for relaxation of linear homopolymer chains. In fact, the micelles exhibit fast and slow relaxation processes of widely separated relaxation times [46,103,104]. In the early 1980s, Watanabe and Kotaka [103,104] found that both processes are related to motion of the

PB corona blocks. For the slow process, the relaxation time is much longer than the time expected for relaxation of individual corona blocks. From this result, that process was attributed to diffusion of the micelles governed by (collective) motion of many corona blocks tethered on the same PS cores [104]. However, features of the fast relaxation process of the micelles were not analyzed in the early work [103,104].

Recently, Watanabe et al. [46] reexamined the relaxation of micellar blends in much more detail. For PS–PB and PS–PI micelles having glassy PS cores and being dispersed randomly in short, nonentangling matrices of hPB and homo-PI (hPI), they found the following features [46].

(a) Fast Process. The mode distribution of the fast relaxation process is universally scaled with a factor, $[M_{cb}/c_{cb}RT]$ with c_{cb} and M_{cb} being the corona concentration and molecular weight. The relaxation time τ^* of this process increases exponentially with c_{cb} when the neighboring micelles are entangled through their corona. These features are qualitatively similar to those of star-branched homopolymers [112,113]. In addition, for large step strains, the fast process exhibits nonlinear stress damping [46] that is very similar to the damping seen for homopolymers [113,114]. All these results indicate that the fast relaxation of the micelles is a star-like, orientational relaxation of individual corona blocks having immobilized ends on the glassy PS cores. For the PS–PI micelles having dipoles parallel along the PI block contour, this assignment is further supported by the fact that motion of individual PI corona blocks is dielectrically detected at the time scales of the fast process [46].

Quantitatively, τ^* of the corona blocks are 10^3–10^4 times longer than τ_{star} of the star chains having the same entanglement density [46]. Within the framework of the tube model [89,113], tethered chains (corona blocks and/or star arms) are considered to relax via a chain retraction mechanism. The difference between τ^* and τ_{star} is attributed to effects of the micellar cores that disturb the retraction of the corona blocks and enlarge τ^* [46].

(b) Slow Process. For the PS–PI and PS–PB micelles randomly dispersed in hPB and hPI matrices, the linear viscoelastic relaxation time τ_s of the slow process was compared with the characteristic time τ_{SE} required for Stokes–Einstein diffusion over a distance close to the micelle diameter [46]. (τ_{SE} of concentrated micelles entangled through their corona was estimated from a viscosity associated to the corona relaxation, while τ_{SE} of dilute micelles was estimated from the matrix viscosity.) For the micelles in nonentangling matrices, reasonable agreements are found for τ_s and τ_{SE} [46]. For stress relaxation against large step strains, the slow process exhibits significant nonlinear damping attributable to displacement of micellar cores [46]. These results suggest that the slow

relaxation of the micelles in nonentangling matrices corresponds to their diffusion (motion of the cores), in harmony with the assignment in the early work [103,104]. (The behavior in long, entangling matrices is more complicated [46], and a further study is now in progress.)

The diffusion of the micelles leads to viscoelastic relaxation via the following mechanism. In a quiescent state, the micelles are randomly and isotropically dispersed in space to have the maximum placement entropy. When these micelles are subjected to a strain, their spatial distribution becomes anisotropic and the resulting decrease in the placement entropy raises the mechanical stress. This stress relaxes when the micelles undergo diffusion and recover their isotropic spatial distribution.

The above mechanism is quite similar to the relaxation mechanism in suspensions of solid particles having hard-core interactions (no long-range attraction/repulsion) [45]. Such suspensions exhibit nonlinear stress damping behavior [115] similar to that of the micelles in the slow relaxation regime. These similarities, together with the similarities seen between the micellar lattices and colloidal crystals, demonstrate that at long time scales rheological features of the micelles having glassy cores and suspensions of solid particles are governed by a common physics.

V. RHEOLOGY OF TRIBLOCK COPOLYMERS

A. Similarities and Differences Between Diblock and Triblock Copolymers

Rheological features of A–B–A-type triblock copolymers are similar, in some aspect, to the features of A–B-type diblock copolymers. Early work by Chung and coworkers [116], Gouinlock and Porter [117], and Pico and Williams [118] revealed a rheological transition of PS–PB–PS triblock copolymers with increasing T from nonNewtonian to Newtonian behavior. This transition results from the order–disorder transition, as is the case also for diblock copolymers (cf. Fig. 4). Solutions of PS–PB–PS in a PB-selective solvent, n-tetradecane, exhibit plastic flow at intermediate T ($< T_{ODT}$) [119], again similar to the PS–PB diblock micellar lattice systems.

However, we also note differences between diblock and triblock copolymers. The clearest difference between PS–PB and PS–PB–PS is found at T above $T_{g,PB}$ but below $T_{g,PS}$: against a modest strain, PS–PB behaves as a plastic fluid while PS–PB–PS behaves as a rubbery solid at those T if the PS blocks form discrete spherical domains. This difference is a natural consequence of a difference in the conformation of the PB blocks. In PS–PB–PS copolymers some PB blocks take a bridge conformation connecting different PS domains, thereby forming networks and exhibiting rubber-like elasticity, while PS–PB copolymers have no such bridges and exhibit plastic flow. (Note, however, that PS–PB–PS copolymers also contain loop-type PB blocks with their ends being anchored on the same PS domain. As explained later, the loop blocks also have important effects on the rheology of triblock copolymers.)

Differences between diblock and triblock copolymers are also found at $T > T_{g,block}$. Riise et al. [120] compared the shear orientation behavior of PS–PI–PS and PS–PI lamellae and found that the former exhibits perpendicular orientation at all ω and $T (> T_{g,block})$ while the latter exhibits either perpendicular (at low ω and high T) or parallel (at high ω and low T) orientations. Tepe et al. [121] examined shear orientation of a PEP–PEE–PEP lamella and found that T_{ODT} decreases with increasing shear frequency and the parallel orientation is preferred in the vicinity of T_{ODT}, in contrast to the behavior of PEP–PEE diblock lamellae. All these differences between the triblock and diblock lamellae could be attributed to the existence of the bridging middle blocks in the former.

B. Loop/Bridge Problem

It is of particular interest to investigate the flow mechanism of ordered triblock copolymers in relation to the bridge/loop conformations of the middle blocks. For bulk PS–PB–PS and PS–PI–PS forming PS cylinders, Morrison et al. [122,123] examined this mechanism at T above $T_{g,PS}$ and $T_{g,PB}$ but below T_{ODT}. They found shear-induced decreases of G^* and orientation of the cylinders. These rheological and structural changes were attributed to the combined effects of domain melting (for unfavorably oriented cylinders) and flow at grain boundaries, the latter unavoidably accompanied by pull-out of the PS blocks when the middle blocks take the bridge conformation across the boundaries.

For a PS–PB–PS copolymer forming PS cylinders, Morrison et al. [124] and Jackson et al. [125] found the other type of shear-induced structural transition in the vicinity of T_{ODT}. Above a critical shear rate, the copolymer formed new cylindrical PS domains that had a diameter much smaller than the equilibrium diameter d^* and were aligned with a lattice spacing much smaller than the equilibrium spacing D^* [125]. Grains of these thin cylinders coexisted with grains of aligned cylinders having larger diameter and spacing ($\cong d^*$ and D^*) [124,125]. Jackson et al. [125] noted a similarity between the above structural transition of PS–PB–PS and a martensitic transformation for metals under deformation. Concerning a bridge/loop population distribution for middle PB blocks, Morrison et al. [124] considered an interesting possibility: on transition, the bridges may be stretched under fast shear to become the loops (by pulling out the PS blocks from one domain and transferring them into the other domain).

Recently, Sato et al. [126] investigated rheological and dielectric properties of a PS–PI–PS copolymer in the PI-selective solvent, n-tetradecane (C14), in relation to the bridge/loop conformation of the middle PI blocks. The copolymer formed spherical PS domains at $T < T_{ODT}$. The PI blocks had dipoles parallel to their contour so that the motion of their ends was dielectrically detected.

For the PS–PI–PS/C14 system, Sato et al. found rubbery, plastic, and viscous behavior at low, intermediate, and high T [126]. The rubbery behavior (at $T < T_{g,PS}$) is attributed to a network of glassy PS domains bridged by the PI blocks. This conventional assignment is supported by the fact that no dielectric dispersion is detected at those T and thus PI block ends are fixed in space [126]. The viscous behavior (at $T > T_{ODT}$), associated with significant dielectric dispersion due to almost unconstrained motion of the PI block ends, is attributed to more or less homogeneous mixing of the PS and PI blocks.

For the PS–PI–PS/C14 system at $T < T_{ODT}$ to flow, the PS block connected to a bridge-type PI block should be pulled out from its domain and unfavorably mixed with the PI blocks (before it is transferred into the other domain). As considered in the earlier work [119], the mixing enthalpy works as a barrier for flow and provides, in principle, the plastic nature of the PS–PI–PS/C14 system. If the plasticity is actually governed by this mechanism, the PI block ends should acquire a large motional freedom and exhibit large dielectric activity during flow. However, at $T_{g,PS} < T < T_{ODT}$ where the system exhibited the plastic behavior, the flow induced no significant dielectric changes [126]. This fact suggests that the PS-PI-PS/C14 system flows heterogeneously (only at grain boundaries) and that the above PS pull-out mechanism is not the only possible mechanism for the plasticity of the system, leading to a hypothesis that loop-type PI blocks, not stretched strongly under shear and thus not pulling out the PS blocks, have an important contribution to the plasticity [126].

Watanabe et al. [127] examined this hypothesis for a PS–PI–PS copolymer having inversion of parallel dipoles at the midpoint of the PI block. This inversion enables us to dielectrically detect the midpoint motion. Since this motion is different for the bridge- and loop-type PI blocks, the loop fraction ϕ_l can be dielectrically estimated. (For a bulk PS–PI–PS lamella, Watanabe [128] obtained $\phi_l \cong 60\%$, which was in good agreement with the value predicted from the thermodynamic theory of Matsen and Schick [129].)

For a PS–PI–PS/C14 system similar to the system examined by Sato et al. [126] but having dipole inversion in the PI block, ϕ_l was estimated to be $\cong$ 60% [127]. These rather highly populated loop PI chains can provide the plasticity to the system via a mechanism similar to that for the diblock copolymer micelles: as explained earlier, the micelles form the lattices and exhibit plasticity due to the osmotic constraint for the corona conformation. Similarly, the osmotic constraint for the loop PI conformation can lead to formation of the lattice of the PS domains and provide the plastic nature of the PS–PI–PS/C14 system. This molecular picture was evidenced from an experimental fact: at relatively high T ($\cong$ 5 degrees below T_{ODT}), the yield stress of the PS–PI–PS/C14 system is largely reduced when the solvent C14 is replaced by a low-M hPI that screens the osmotic constraint [127]. (This effect of hPI is identical to the effect of hPB for the PS–PB micelles explained earlier; cf. Fig. 10.) Thus, the loops dominantly contribute to the plasticity at that T.

Here, it should be emphasized that the above experimental fact does not mean that the bridges have no contribution to the plasticity. Instead, it means that the bridge contribution is overwhelmed by the loop contribution at relatively high T where the enthalpic barrier for pull-out of the PS blocks (mixing of the PS and PI blocks) is rather small. At low T where this barrier is enhanced, the PS pull-out mechanism for the bridge-type PI chains does have a significant contribution to the plasticity of the PS–PI–PS/C14 system [127].

It should be also emphasized that the plasticity of the PS–PI–PS/C14 system due to the loop-type PI chains emerges only when the system flows, i.e., only when the coexisting bridge-type PI chains pull out the PS blocks. When this PS pull-out does not take place, the loop chains exhibit an elastic response (similar to the behavior of the micellar lattice before yielding). This is the case for the PS–PI–PS/C14 system at low T ($< T_{g,PS}$) where the rubbery behavior prevails: the bridges and loops almost equally contribute to the equilibrium modulus at those T [127].

VI. SUMMARY

We have seen a wide variety of slow rheological responses of typical multiphase polymeric systems, i.e., immiscible polymer blends and block copolymers. Sometimes the blends and copolymers exhibit similar responses but with different mechanisms: for example, the slow linear relaxation processes of the blends, unoriented copolymer lamellae, and random micellar dispersions are attributable to recovery of the domain shape, motion of the defects, and diffusion of the micelles, respectively. These differences in the mechanisms reflect the lack of the contribution from the thermal motion (for the blends containing large domains) and the lack of a lattice-forming thermodynamic force (for the micellar dispersions). The other type of example is noted for cubic lattices of spherical domains of diblock copolymers that behave as solid and liquid, respectively, at T well below and above T_g of the domains. This rheological difference can be related to a difference in the domain rigidity that determines the freedom of motion/location of the block junctions and of the chain migration.

As demonstrated in the above examples, the slow rheological responses of the blends and block copolymers are differently influenced by the structural, thermodynamic, and mechanical properties of the constituent phases to exhibit a wide variety of characteristics. Careful examination of these properties would enable us to discover new rheological features and the underlying structure–rheology relationships, the latter specifying how the structure stores and dissipates the elastic free energy.

REFERENCES

1. JD Ferry. Viscoelastic Properties of Polymers. 3rd ed. New York: Wiley, 1980.
2. DR Paul, S Newman, ed. Polymer Blends. New York: Academic Press, 1978.
3. JL White, AP Plochocki, H Tanaka. Polym Eng Rev 1:217, 1981.
4. LA Utracki. Polymer Alloys and Blends. Munich: Carl Hanser, 1989.
5. D Graebling, R Muller, JF Palierne. Macromolecules 26:320, 1993.
6. HK Chuang, CD Han. J Appl Polym Sci 29:2205, 1984.
7. CD Han, HH Yang. J Appl Polym Sci 33:1221, 1987.
8. JH Han, CC Feng, DJ Li, CD Han. Polymer 30:2451, 1995.
9. P Scholz, D Froelich, R Muller. J Rheol 33:481, 1989.
10. M Bousmina, R Muller. J Rheol 37:663, 1993.
11. H Gramespacher, J Meissner. J Rheol 36:1127, 1992.
12. M Bousmina, P Bataille, S Sapieha, HP Schreiber. J Rheol 39:499, 1995.
13. W Gleinser, H Braun, Chr Friedrich, HJ Cantow. Polymer 35:128, 1994.
14. B Brahimi, A Ait-Kadi, A Ajji, R Jérôme, R Fayt. J Rheol 35:1069, 1991.
15. S Maeda, E Kamei. J Soc Rheol Jpn 22:145, 1994.
16. T Nomura, T Nishio, S Maeda, E Kamei. J Soc Rheol Jpn 22:155, 1994.
17. A Einstein. Ann Phys 19: 289, 1906; Ann Phys 34:591, 1911.
18. GI Taylor. Proc Roy Soc London A132:41, 1932; Proc Roy Soc London A146:501, 1934.
19. H Fröhlich, R Sack. Proc Roy Soc London A185:415, 1946.
20. JG Oldroyd. Proc Roy Soc London A218:122, 1953; Proc Roy Soc London A232:567, 1955.
21. EH Kerner. Proc Roy Soc 69:808, 1956.
22. M Takayanagi, S Uemura, S Minami. J Polym Sci 8:2147, 1964.

23. S Uemura, M Takayanagi. J Appl Polym Sci 10:113, 1966.
24. H Brenner. Chem Eng Sci 19:519, 1964.
25. WR Schowalter, CE Chaffy, H Brenner. J Colloid Interface Sci 26:152, 1968.
26. RA Dickie. J Appl Polym Sci 17:45, 1973.
27. SJ Choi, WR Schowalter, Phys Fluids 18:420, 1975.
28. M Oosterbrock, J Mellema. J Colloid Interface 84:14, 1981.
29. JF Palierne. Rheol Acta 29:204, 1990.
30. A Onuki. Phys Rev A 35:5149, 1987.
31. M Doi, T Ohta. J Chem Phys 95:1242, 1991.
32. HM Lee, OO Park. J Rheol 38:1405, 1994.
33. See, for example, H Fröhlich. Theory of Dielectrics. 2nd ed. Oxford: Clarendon, 1958.
34. Y Aoki. J Soc Rheol Jpn 7:20, 1979.
35. Y Aoki. J Rheol 25:351, 1981.
36. Y Aoki, K Nakayama. J Soc Rheol Jpn 9:39, 1981.
37. Y Aoki, K Nakayama. Polym J 14:951, 1982.
38. T Masuda, A Nakajima, M Kitamura, Y Aoki, N Yamauchi, A Yoshioka. Pure Appl Chem 56:1457, 1984.
39. Y Aoki. J Non-Newtonian Fluid Mech 2:91, 1986.
40. Y Aoki. Macromolecules 20:2208, 1987.
41. T Masuda, L Lin, M Kitamura, Y Aoki. J Soc Rheol Jpn 15:158, 1987.
42. Y Aoki. J Soc Rheol Jpn 16:136, 1988.
R Hasegawa, Y Aoki, M Doi. Macromolecules 29:6656, 1996.
43. S Onogi, T Masuda, T Matsumoto. Trans Soc Rheol 17:275, 1970.
44. T Matsumoto, C Hitomi, S Onogi. Trans Soc Rheol 19:541, 1975.
45. T Shikata, DS Pearson. J Rheol 38:601, 1994 and references therein.
46. H Watanabe, T Sato, K Osaki. Macromolecules 29:104, 113, 1996.
H Watanabe, T Sato, K Osaki, M-L Yao. Macromolecules 29:3881, 3890, 1996.
47. Y Takahashi, S Kitade, N Kurashima, I Noda. Polym J 26:1206, 1994.
48. Y Takahashi, N Kurashima, I Noda, M Doi. J Rheol 38:699, 1994.
49. Y Takahashi, I Noda. In: AI Nakatani, MD Dadmun, ed. Flow Induced Structure in Polymers. ACS Symp. Ser. 597, 1995.
50. Y Takahashi, H Suzuki, Y Nakagawa, I Noda. Macromolecules 27:6476, 1994.
51. Y Takahashi, H Suzuki, Y Nakagawa, I Noda. Polym Int 34:327, 1994.
52. H Gramespacher, J Meissner. J Rheol 39:151, 1995.
53. GE Molau. In SL Aggarwal, ed. Block Polymers. New York: Plenum Press, 1970.
54. FS Bates, GH Fredrickson. Ann Rev Phys Chem 41:525, 1990.
55. L Leibler. Macromolecules 13:1602, 1980.
56. GH Fredrickson, E Helfand. J Chem Phys 87:697, 1987.
57. JH Rosedale, FS Bates. Macromolecules 23:2329, 1990.
58. FS Bates, JH Rosedale, GH Fredrickson. J Chem Phys 92:6255, 1990.
59. MG Gehlsen, K Almdal, FS Bates. Macromolecules 25:939, 1992

60. KA Koppi, M Tirrell, FS Bates, K Almdal, RH Colby. J Phys II (France) 2:1941, 1992.
61. K Almdal, KA Koppi, FS Bates, K Mortensen. Macromolecules 25:1743, 1992.
62. M Gehlsen, FS Bates. Macromolecules 26:4122, 1993.
63. JH Rosedale, FS Bates, K Almdal, K Mortensen, GD Wignall. Macromolecules 28:1429, 1995.
64. RG Larson, KI Winey, SS Patel, H Watanabe, R Bruinsma. Rheol Acta 32:245, 1993.
65. KI Winey, SS Patel, RG Larson, H Watanabe. Macromolecules 26:2542, 1993.
66. KI Winey, SS Patel, RG Larson, H Watanabe. Macromolecules 26:4373, 1993.
67. SS Patel, RG Larson, KI Winey, H Watanabe. Macromolecules 28:4313, 1995.
68. RM Kannan, JA Kornfield. Macromolecules 27:1177, 1994.
69. VK Gupta, R Krishnamoorti, JA Kornfield, SD Smith. Macromolecules 28:4464, 1995.
70. VK Gupta, R Krishnamoorti, JA Kornfield, SD Smith. Macromolecules 29:1359, 1996.
71. Y Zhang, U Wiesner. J Chem Phys 103:4784, 1995.
72. NP Balsara, D Perahia, CR Safinya, M Tirrell, TP Lodge. Macromolecules 25:3896, 1992.
73. S Okamoto, K Saijyo, T Hashimoto. Macromolecules 27:5547, 1994.
74. CD Han, DM Baek, JK Kim, T Ogawa, N Sakamoto, T Hashimoto. Macromolecules 28:5043, 1995.
75. GH Fredrickson, E Helfand. J Chem Phys 89:5890, 1988.
76. K Mori, H Hasegawa, T Hashimoto. Polym J 17:799, 1985.
77. K Kawasaki, A Onuki. Phys Rev A 42: 3664, 1990.
78. M Rubinstein, SP Obukov. Macromolecules 26:1740, 1993.
79. K Amundson, E Helfand, SS Patel, X Quan, SD Smith. Macromolecules 25:1935, 1992.
80. NP Balsara, BA Garetz, HJ Dai. Macromolecules 25: 6072, 1992.
BA Garetz, MC Newstein, HJ Dai, SV Jonnalagadda, NP Balsara. Macromolecules 26:3151, 1993.
81. NP Balsara, HJ Dai, H Watanabe, T Sato, K Osaki. Macromolecules 29:3507, 1996.
82. A Keller, E Pedemonte, FM Willmouth. Kolloid-Z Z Polym 238: 25, 1970; Nature 225:538, 1970.
83. G Hadziioannou, A Mathis, A Skoulios. Colloid Polym Sci 257:136, 1979.
G Hadziioannou, C Picot, A Skoulios, M-L Ionescu, A Mathis, R Duplessix, Y Gallot, J-P Lingelser. Macromolecules 15:263, 1982.
84. KA Koppi, M Tirrell, FS Bates. Phys Rev Lett 70:1449, 1993.
85. VK Gupta, R Krishnamoorti, Z-R Chen, JA Kornfield, SD Smith, MM Satkowski, JT Grothaus. Macromolecules 29:875, 1996.
86. Y Zhang, U Wiesner, HW Spiess. Macromolecules 28:778, 1995.

87. NP Balsara, B Hammouda, PK Kesani, SV Jonnalagadda, GC Straty. Macromolecules 27:2566, 1994.
88. Y Takahashi, M Noda, N Ochiai, I Noda. Polymer 37:5943, 1996.
89. M Doi, SF Edwards. The Theory of Polymer Dynamics. Oxford: Clarendon, 1986.
90. ME Cates, ST Milner. Phys Rev Lett 62:1856, 1989.
91. GH Fredrickson. J Rheol 38:1045, 1994.
92. K Almdal, KA Koppi, FS Bates, Macromolecules 26 4058, 1993.
93. KA Koppi, M Tirrell, FS Bates, K Almdal, K Mortensen. J Rheol 38:999, 1994.
94. FS Bates, KA Koppi, M Tirrell, K Almdal, K Mortensen. Macromolecules 27:5934, 1994.
95. T Tepe, MF Schulz, J Zhao, M Tirrell, FS Bates, K Mortensen, K Almdal. Macromolecules 28:3008, 1995.
96. J Zhao, B Majumdar, MF Schulz, FS Bates, K Almdal, K Mortensen, DA Hajduk, SM Gruner. Macromolecules 29:1204, 1996.
97. NP Balsara, HJ Dai, PK Kesani, BA Garetz, B Hammouda. Macromolecules 27:7406, 1994.
98. H Watanabe, T Kotaka, T Hashimoto, M Shibayama, H Kawai. J Rheol 26:153, 1982.
99. H Watanabe, T Kotaka. Polym J 14:739, 1982.
100. H Watanabe, T Kotaka. J Rheol 27:223, 1983.
101. H Watanabe, T Kotaka. Polym J 15:337, 1983.
102. H Watanabe, T Kotaka. Polym Eng Rev 4:73, 1984.
103. H Watanabe, T Kotaka. Macromolecules 16:769, 1983.
104. H Watanabe, T Kotaka. Macromolecules 17:342, 1984.
105. CL Phoon, JS Higgins, G Allegra, P Van Leeuwen, E Staples. Proc Roy Soc London A 442:221, 1993.
106. GA McConnell, MY Lin, AP Gast. Macromolecules 28:6754, 1995.
107. S Okamoto, K Saijyo, T Hashimoto. Macromolecules 27:3753, 1994.
108. S Öka. In: FR Eirich, ed. Rheology. vol. 3. New York: Academic Press, 1960.
109. M Doi, JL Harden, T Ohta. Macromolecules 26:4935, 1993.
110. K Mortensen, W Brown, B Nordén. Phys Rev Lett 68:2340, 1992.
111. L Chen, M Chow, B Ackerson, C Zukoski. Langmuir 10:2817, 1994 and references therein.
112. VR Raju, EV Menezes, G Marin, WW Graessley, LJ Fetters. Macromolecules 14:1668, 1981
113. DS Pearson. Rubber Chem Technol 60:439, 1987.
114. K Osaki. Rheol Acta 32:429, 1993 and references therein.
115. H Watanabe, M-L Yao, A Yamagishi, K Osaki, T Shikata, H Niwa, Y Morishima. Rheol Acta 35:433, 1996.
116. CI Chung, JC Gale, J Polym Sci Polym Phys Ed 14:1149, 1976.
CI Chung, MI Lin. J Polym Sci Polym Phys Ed 16:545, 1978.
CI Chung, HL Griesbach, L Young. J Polym Sci Polym Phys Ed 18:1237, 1980.
117. EV Gouinlock, RS Porter. Polym Eng Sci 17:535, 1977.

118. ER Pico, MC Williams. Polym Eng Sci 17:573, 1977.
119. H Watanabe, S Kuwahara, T Kotaka. J Rheol 28:393, 1984.
120. BL Riise, GH Fredrickson, RG Larson, DS Pearson. Macromolecules 28:7653, 1995.
121. T Tepe, DA Hajduk, MA Hillmyer, PA Weimann, M Tirrell, FS Bates, K Almdal, K Mortensen. J Rheol 41:1147, 1997.
122. FA Morrison, H Winter. Macromolecules 22:3533, 1989.
123. FA Morrison, H Winter, W Gronski, JD Barnes. Macromolecules 23:4200, 1990.
124. FA Morrison, JW Mays, M Muthukumar. AI Nakatani, CC Han. Macromolecules 26:5271, 1993.
125. CL Jackson, KA Barnes, FA Morrison, JW Mays, AI Nakatani, CC Han. Macromolecules 28:713, 1995.
126. T Sato, H Watanabe, K Osaki. Macromolecules 29:6231, 1996.
127. H Watanabe, T Sato, K Osaki, M-L Yao, A Yamagishi. Macromolecules 30:5877, 1997.
128. H Watanabe. Macromolecules 28:5006, 1995.
129. MW Matsen, M Schick. Macromolecules 27:187, 1994.

10
Polymers at Interfaces

Masami Kawaguchi
Mie University, Mie, Japan

I. INTRODUCTION

When a polymer is brought near an interface, a number of interesting effects can be observed. Polymer chains may be strongly adsorbed on the surface, which leads to inhomogeneous multiphase polymer systems. Polymer adsorption at interfaces is a topic of fundamental importance in a wide range of technological processes such as surface coatings, composite materials, stabilization of colloidal suspensions, and stabilization of foams. The structure of the polymer chains adsorbed at interfaces plays a crucial role in these processes. A great deal of effort has been spent to understand the polymer adsorption phenomena, including new and advanced experimental tools, analytical theory, and simulation. Textbooks and reviews have recently addressed the importance of polymer adsorption [1–11].

In this chapter we describe experimental results regarding adsorption of polymer blends of the same chemical characters, of polymer blends of different chemical natures, and of copolymers at the solid liquid interfaces. We also emphasize the necessity of understanding the structure of adsorbed chains. The most widely measured quantities are the *adsorbed amount* of polymer, Γ, the fraction of the polymer segments attached to the surface, p, and the *thickness of the adsorbed layer*, t. The respective quantities can be related to the *density profile of polymer segments* normal to the surface, $\phi(z)$ as follows,

$$\Gamma = \int_0^\infty \phi(z)\,dz \tag{1}$$

$$p = \int_0^\delta \phi(z)\,dz \Big/ \int_0^\infty \phi(z)\,dz \tag{2}$$

$$t = \int_0^\infty z\phi(z)\,dz \tag{3}$$

where z is the distance from the surface and δ is the extent of the train layer. The successful attempts to measure $\phi(z)$ have been by *small-angle neutron scattering* (SANS) and *neutron reflectometry*.

II. EXPERIMENTAL TECHNIQUES

A variety of experimental techniques such as infrared (IR), nuclear magnetic resonance (NMR), electron paramagnetic resonance (EPR) spectroscopies, optical reflection and scattering methods, SANS, neutron reflectometry, *surface force* measurements, and *atomic force microscopy* (AFM) have been developed and applied to investigate polymer adsorption phenomena at solution–solid interfaces. AFM was mostly used to study images of polymers in the dried state after adsorption on solid surfaces from the solutions. These experimental methods have led to our understanding of the detailed structure of adsorbed polymer chains involving the density profile of polymer segments from the adsorbing surface, of the interaction free energy between adsorbed polymer chains and solid surfaces, and of the forces arising from molecular interactions between adsorbed polymer chains, i.e., surface force. Table 1 lists a typical relationship between several important quantities and available techniques.

A. Spectroscopies

IR spectroscopy can be used to determine the value of p given in Eq. (2) and the fraction of the sites occupied by polymer chains, θ, from the frequency shifts of functional groups of the polymer chains and the surface sites, respectively. Furthermore, both NMR and EPR spectroscopies can distinguish the differences in mobility between polymer segments in trains and those in loops and/or tails by means of the magnetic relaxation times of proton and spin labels chemically attached to the polymer chains, respectively. The value p can thus be obtained from these quantities.

Table 1 Measured Quantities and Available Techniques

Quantities	Techniques
Adsorbed amount of polymer, Γ	Various methods for determination of polymer concentrations in supernatants
Fraction of polymer segments attached to surfaces, p	IR, NMR, ESR
Surface coverage, θ	IR
Adsorbed layer thickness, t	Optical reflection and scattering methods; hydrodynamic methods; small-angle neutron scattering; neutron reflectometry; surface force apparatus
Density profile of polymer segments normal to the surface, $\phi(z)$[a]	Small angle neutron scattering; neutron reflectometry; small-angle X-ray scattering
Surface force, $F(D)$[b]	Surface force apparatus
Surface imaging of adsorbed polymer chains	Scanning probe microscopic techniques

[a] z is the distance from the surface.
[b] D is the separation distance between mica surfaces.

B. Optical Reflection and Scattering Techniques

Optical techniques, which are useful for determining the thickness of the adsorbed layer can be divided into two major categories: reflection and scattering. The former group contains *ellipsometry*, interferometry, internal reflection interferometry, and the evanescent wave method. The *photon correlation spectroscopy* (PCS) method belongs to the latter group. In this section we briefly describe the principles of ellipsometry and the PCS method.

1. Ellipsometry

Ellipsometry is not a newly developed method; however, it provides simultaneously the refractive index and the thickness of the polymer layer adsorbed at interfaces [8,11]. The technique involves the measurement of the change in polarization of linearly polarized light upon reflection at an interface. The state of polarization of a plane coherent beam of light can be characterized by the amplitude ratio E_p/E_s and phase difference $\delta_p - \delta_s$ of the components of the electric field vector, parallel $[\mathbf{E}_p = E_p \exp\{i(\omega t + \delta_p)\}]$ and perpendicular$[\mathbf{E}_s = E_s \exp\{i(\omega t + \delta_s)\}]$ to the plane of incidence, where ω

is the angular frequency. After reflection by an interface, these components can be described by $E_p^r = R_p E_p^i$ and $E_s^r = R_s E^i s$, where R_p and R_s are the complex overall reflection coefficients for the two components. The superscripts i and r refer to incident and reflected beams, respectively. The change of the polarization after reflection is given by

$$\rho = R_p / R_s = \tan \Psi \exp i\Delta \tag{4}$$

$$\tan \Psi = (E_p^r / E_s^r)/(E_p^i / E_s^i) \tag{5}$$

and

$$\Delta = (\delta_p^r - \delta_s^r) - (\delta_p^i - \delta_s^i) \tag{6}$$

For a film-covered surface, the complex overall reflection coefficients R_p and R_s can be calculated by

$$R_j = \{r_{1,j} + r_{2,j} \exp(-i\phi)\}/\{1 + r_{1,j} r_{2,j} \exp(-i\phi)\} \tag{7}$$

where j represents either subscript p or subscript s; r_1 and r_2 are the Fresnel reflection coefficients at the bulk solution/adsorbed layer and the adsorbed layer/solid substrate interface, respectively, as shown in Fig. 1. The quantity

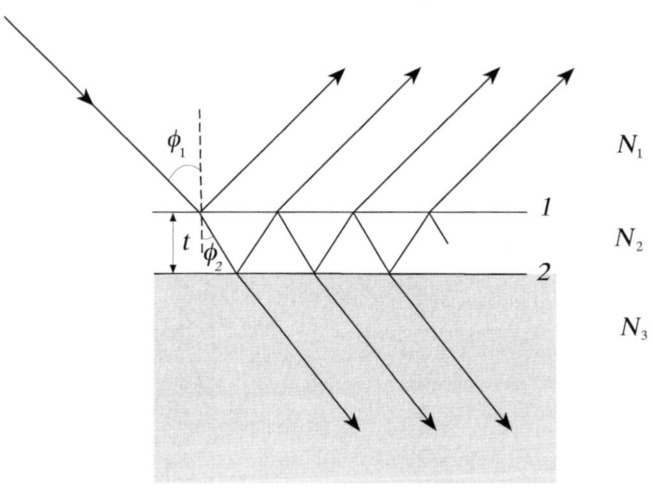

Figure 1 Schematic representation of reflected light from an adsorbed layer on a solid substrate: t—layer thickness; ϕ_1—angle of incidence; ϕ_2—angle of reflectance; N_1—refractive index of the bulk solution; N_2—refractive index of the adsorbed layer; N_3—refractive index of the solid surface; 1—an interface between the bulk solution and the adsorbed layer; 2—an interface between the adsorbed layer and the solid substrate.

ϕ represents the phase difference between two consecutive reflected rays and is given by

$$\phi = 4\pi t N_2 \cos\phi_2/\lambda \tag{8}$$

where t is the thickness of the adsorbed layer, N_2 is the refractive index of the adsorbed layer, ϕ_2 is the angle of refraction in the adsorbed layer, and λ is the angle of incidence, respectively. Furthermore, the relationship is obtained from Snell's law

$$N_1 \sin\phi_1 = N_2 \sin\phi_2 \tag{9}$$

where N_1 is the refractive index of bulk solution and ϕ_1 is the incident angle. Thus, by combining Eqs. (4), (7), (8), and (9), we can determine implicitly t and N_2 by the numerical procedure of McCrackin [12], if the other quantities are known. The resulting t and N_2 values can be regarded as effective numbers since they are based on an idealized model of a perfectly planar substrate and a homogeneous film. Using the calculated t and N_2 values, the amount of adsorbed polymer per unit surface area is given by

$$\Gamma = t(N_2 - N_0)/(dn/dc) \tag{10}$$

where N_0 is the refractive index of the solvent and (dn/dc) is the refractive index increment of the solution. Since the value of Γ is expressed as the product of t and N_2, it seems to be subjected to much less error than the parameters themselves. Thus, the adsorbed amount is often preferred over the two parameters.

There are several different techniques for the determination of the changes in polarization upon reflection. In general, the technique of null ellipsometry is widely used, as shown in Fig. 2. The polarization state

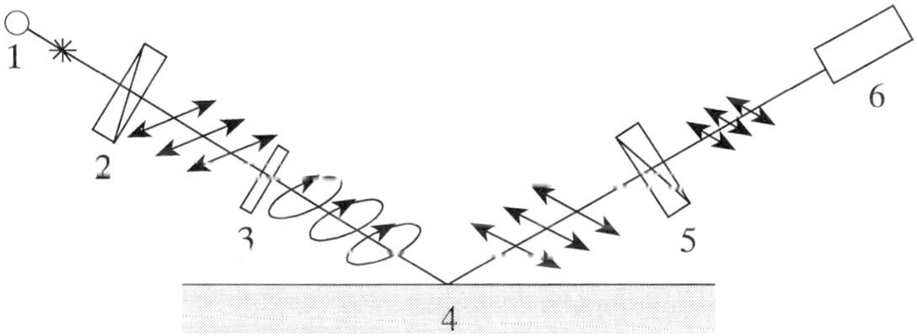

Figure 2 A schematic representation of a typical null ellipsometry: 1—light source; 2—polarizer; 3—quarter-wave plate; 4—sample; 5—analyzer; 6—photomultiplier tube.

generator consists of a collimated monochromatic light source, a polarizer, and a quarter-wave plate, in succession. The polarization state detector consists of an analyzer and a photomultiplier tube. The values of Ψ and Δ are obtained from azimuth angle measurements of the polarizer and analyzer at minimal transmission. Most ellipsometric studies of polymer adsorption have been carried out on metallic surfaces, since highly reflecting substrates are able to analyze the surface properties accurately.

2. Photon Correlation Spectroscopy

The photon correlation spectroscopy (PCS) method is a so-called quasi-electric light-scattering (QELS) or dynamic light-scattering (DLS) technique [13]. It provides the mutual diffusion coefficient, D_m, of particles, including polymer chains dissolved or suspended in bulk solution, through measuring the intensity scattered by the particle, owing to the diffusive motion as a function of the scattering angle. The measured intensity autocorrelation function can be related to the normalized autocorrelation function, g, of the scattering electric field, $\mathbf{E}$. For the homogeneous particles, the g function is expressed as a single-exponential decay

$$g(\tau) = \langle \mathbf{E}(0)\mathbf{E}(\tau) \rangle = \exp(-q^2 D_m \tau) \tag{11}$$

where q is the scattering vector, and τ is the correlation decay time which is equal to $1/q^2 D_m$.

The evaluated value of D_m is related to the hydrodynamic radius r_h of the particle by the Stokes–Einstein relation:

$$D_m = kT/6\pi\eta r_h \tag{12}$$

where k is the Boltzmann constant and η is the viscosity of the medium. The hydrodynamic thickness, t_h of the adsorbed polymer layer is defined as the difference between the hydrodynamic radius of the particle with and without the coating polymer layer. This technique has been used frequently to measure the thicknesses of polymer adsorbed on colloidal particles, such as a polymer latex and an inorganic particle.

C. Surface Force Measurements

Besides the structures in the adsorbed polymer layer, one of the major problems confronted in the study of polymer adsorption is the direct measurement of forces between surfaces bearing adsorbed polymer chains. Current measurement techniques are derived from an extension of the surface force apparatus developed by Tabor and Winterton [14], as well as by

Israerachvili and Tabor [15], to measure the surface force, $F(D)$, acting between two curved mica surfaces a separation distance, D apart [16]. The mica sheets are mounted in a crossed-cylinder configuration on cylindrical quartz lenses; their distance D is controlled by the two rigid arms using a three-stage mechanism consisting of two micrometers and a piezoelectric crystal. White-light multibeam interferometry leads to measurement of the distance of D, the refractive index $n(D)$ of the medium separating the surfaces, and their radii of curvature. Thus, the resulting changes in the distance of D between the two surfaces can be measured optically. Alternatively, the forces can be determined by suddenly reversing the voltage of the piezoelectric crystal, which expands or contracts it by a known amount. Thus, this apparatus is available for directly measuring attractive and repulsive forces [17].

D. Atomic Force Microscopy

The development of the scanning tunneling microscope has led to the invention of a host of other "scanning probe" microscopes, which rely on mechanically scanning a microfabricated tip over a sample surface [18]. AFM is one of the most successful of these new devices [19]. The concept of using a force to image a surface can be applied to magnetic and electrostatic forces as well as to the interatomic interaction between the tip and the sample.

The ability of the AFM to image nonconductors enables it to be applied to organic materials, provided that the forces involved are small enough not to destroy the sample. Imaging an individual organic molecule, however, is rather difficult because a single molecule can easily diffuse across the surface, accompanied with a scanning of an AFM tip. To prevent damage to the molecules, the interaction force must be kept below 10^{-8} N. This is generally difficult to achieve in air, due to forces arising from an adsorbed layer of water and unknown contamination. To eliminate these forces, the AFM operates under ultra-high vacuum, and in water and other fluids.

E. Techniques with Neutrons

SANS and neturon reflectometry are very useful in determining the density profile of polymer segments in a polymer layer adsorbed on the surface of a colloidal particle [20] and a flat plate [21], respectively. Details of the respective techniques are described in Chapters 4 and 8 in this book, respectively.

III. EXPERIMENTAL RESULTS

Here we will focus mainly on the experimental results for studying the structures of the polymer layers adsorbed on the solid substrates by using various spectroscopic methods, optical reflection and scattering methods, surface force apparatus, SANS, neutron reflectometry, and AFM.

A. Polymer Blends

For the adsorption of polymer blends, *competitive adsorption* and *displacement adsorption* take place due to the differences in the size of polymer chains and their chemical structures. Studies on competitive adsorption and displacement adsorption of polymer chains will lead to an understanding of the reversibility of polymer adsorption and the adsorption behavior of polydisperse polymers.

1. Polymer Blends with the Same Chemical Structures of Different Lengths

In 1939, Mark and Saito were the first to find that the low molecular weight portion of cellulose acetate dissolved in acetone was preferentially adsorbed onto charcoal [22]. Much evidence for such preferential adsorption behavior has been reported using polydisperse samples or the binary and ternary mixtures of monodisperse samples. On the other hand, some analytical models for preferential adsorption have been developed and they have been compared with the experimental results [3,7].

Several experimental data for polymer blends of monodisperse polystyrene (PS) onto various surfaces have been obtained using ellipsometry by Kawaguchi et al. [23], optical reflectometry by Dijt et al. [24], quartz crystal microbalance by Xu and Schlenoff [25], and a combination of spectroscopy and gel-permeation chromatography (GPC) by Kawaguchi and coworkers [26–28]. Ellipsometry results in trans-decalin are shown in Fig. 3. The adsorption time was covered over two days, indicating that exchange was a slow process. In the competitive adsorption process, the plateau value of t was almost the same as that of large-size PS chains for its equilibrium individual adsorption, while the value of Γ was less. Thus, in the adsorbed layer, both large and small PS chains existed, and complete preferential adsorption of long chains over short ones did not occur. In the displacement adsorption both equilibrium values of Γ and t were less than those for the competitive adsorption process, respectively.

On the other hand, the optical reflectometry technique was used to monitor the value of Γ on a silicon wafer as a function time for the shorter

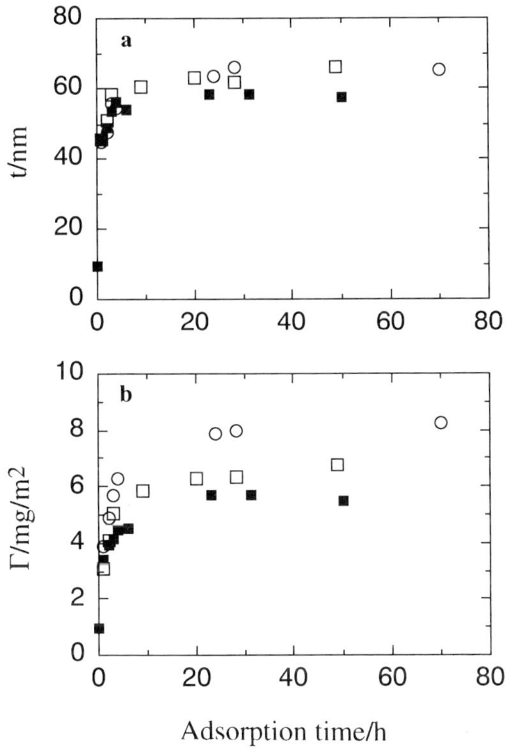

Figure 3 Kinetics of adsorption data of layer thickness, t, (a) and adsorbed amount, Γ, (b) of individual (○), competitive (□), and displacement (■) adsorption of monodisperse polystyrenes with molecular weights of 706×10^3 and 100×10^3 at a platinum plate from trans-decalin solution.

adsorption time less than 200 s for the competitive adsorption in cis/trans mixed decalin. Such a fast process might be attributed to an impinging jet flow. The exchange of smaller adsorbed chains by larger ones was observed and the total adsorbed amount increased during exchange, indicating incomplete relaxation of the polymer adsorbed layer. At the equilibrium state, the adsorbed amount coincided with that of the large size of PS for its individual adsorption.

Displacement on a gold surface in cyclohexane using the quartz crystal microbalance attained a plateau within an excess of 1 h and the time constant was shorter than for the ellipsometry results. This was attributed to the smaller interaction energy between PS and the gold surface.

UV spectroscopy and GPC were used to determine the Γ value for the competitive adsorption at porous silica surfaces under theta (in cyclohexane) and good (in carbon tetrachloride) conditions. Roughness of silica surfaces led to slower adsorption kinetics than for smooth silica surfaces, as shown in Fig. 4, where a continuous molecular weight PS sample was adsorbed on porous and nonporous silica surfaces. When a monodisperse PS chain easily penetrated into pores in the silica for a ratio of pore size in the silica to twice the radius of gyration larger in size than 4.4, the time to attain an equilibrum state was less than 10 h. Whereas, for the large PS chain with a ratio less than 4.4, forced to penetrate into pores with much deformation, an adsorption equilibrium time of 35 h was attained. For adsorption of the binary mixture the short PS was preferentially adsorbed over the large PS at the early adsorption stage. With an increase in adsorption time, the large PS was adsorbed more than the small PS at the higher initial concentration, and finally an equilibrium adsorption was attained within 35 h. Furthermore, the resulting experimental data showed a good agreement with a mathematical model describing the kinetics of competitive adsorption of polymers at a porous adsorbent surface proposed by Kislenko et al. [29].

Using optical reflectometry, competitive adsorption kinetics of polyethylene oxide (PEO) in water have been investigated by Dijt et al. [24]. During the exchange of short adsorbed PEO chains displaced by large ones, the total adsorbed amount was equal to the plateau adsorbed amount of the

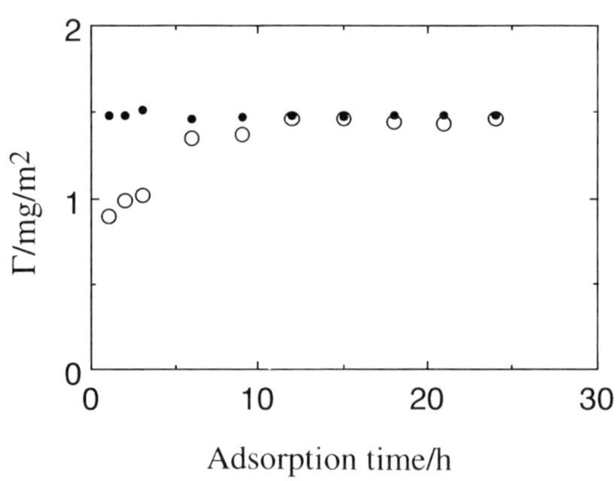

Figure 4 Kinetics of adsorption data of adsorbed amount, Γ, for a continuous molecular weight polystyrene with the molecular weight distribution index, M_w/M_n of 4.26 at smooth (•) and rough (○) silica surfaces.

short PEO. This experimental result agreed with the theoretical prediction of the exchange kinetics based on the local equilibrium concepts [30], indicating that the relaxation times of adsorbed layers of PEO are shorter than the timescale of the experiment. Thus, PEO equilibrates more rapidly than PS since it is much more flexible.

2. Competitive Adsorption and Displacement Adsorption Between Polymers of Different Chemical Nature

Since Thies found that poly(methyl methacrylate) (PMMA) was preferentially adsorbed over PS in the mixture and PS was completely displaced by PMMA at the silica surface in trichloroethylene [31], a number of experiments on the competitive adsorption and displacement adsorption between polymer chains with different chemical structures have been reported [3,7]. In general, a polymer having strong adsorbability of the monomer unit was widely believed to preferentially adsorb and to give a larger amount of polymer adsorbed on surfaces. However, no theoretical works have yet been developed that compare the preferential adsorption of polymer chains of the same chemical structures. In this section, therefore, a variety of experiments are presented to help clarify what factors govern the competitive adsorption and displacement adsorption between chemically different polymer chains.

Competitive adsorption and/or displacement adsorption studies on PS and chemically different polymers have been investigated by several research groups with various techniques. Desorption kinetics of PS by PMMA was studied by Johnson et al. [32] to determine the effect of constraints on the polymer desorption process using IR in attenuated total reflection (IRATR). The desorption of star-branched PS chains was much slower than that of linear ones, due to a stronger entanglement effect in solution.

Kawaguchi et al. [33] measured the adsorbed amount of polymers and the fraction of silanol groups occupied by polymer chains. Due to the stronger adsorbability of the ether oxygen in PEO on a silanol group than on a phenyl group in PS, as well as the greater flexibility of PEO than that of PS, PEO chains were preferentially adsorbed over PS chains, and complete desorption of the pre adsorbed PS chains by the addition of PEO occurred at very low concentrations which are far below the concentration for the plateau adsorption of the individual adsorption of PEO. Such a complete desorption took place even for the small PEO of 86×10^3 and the large PS of 3840×10^3.

The exchange kinetics between an adsorbed polymer and a chemically different, more strongly adsorbing polymer has also been reported using an optical reflectometry technique by Dijt et al. [34] Both PS and poly(n-butyl methacrylate) (PBMA) were completely displaced by poly(tetrahydrofuran) (PTHF), and their exchange rates were fast and dominantly determined by the adsorption rate of the invading species, which is controlled by mass transfer from bulk solution. Displacement of PS by PBMA, however, is different from the above: exchange of small-mass PS by PBMA was much slower and complete, while high-mass PS was hardly desorbed from the surface. Such a different exchange kinetics was attributed to the dynamic flexibility of the chains, which is higher for PTHF than for PBMA.

From the preferential absorption of PMMA over PS, it would be expected that styrene-methyl methacrylate (SM) copolymer with higher methyl methacrylate content should show stronger preferential adsorbability and higher desorption ability on a silica surface. However, the experimental results for the competitive adsorption and displacement adsorption of PS and SM copolymer at a nonporous silica surface in trichloroethylene showed just the opposite [35]; the SM copolymer with the lower methyl methacrylate content preferentially adsorbed at an equilibrium state and behaved as a better displacer, as shown in Fig. 5. The preferential adsorability and desorption ability of the SM copolymers compared with PS was mainly attributed to the larger plateau amount of the SM copolymer adsorbed for their individual adsorptions.

Adsorption–desorption kinetics of deuterio polymers and protio polymers on a silicon oxide surface by IRATR have been investigated extensively by Granik and coworkers [36–44]. For the same chemical structure, the deuterio isotope slightly favors adsorption at the surface for PS under both theta and good solvent conditions [37–39], whereas no isotope effect on adsorption was observed for PMMA [36]. The displacement time of protio-PS by deuterio-PS grew exponentially with the molecular weight of the protio-PS, and it increased with the surface residence time before replacing the solution. The displacement-time dependence was different from the desorption of PS by PMMA reported by Johnson and Granik [36], where a displacement time fitted a power law of the length of the adsorbed chains. Furthermore, the competitive adsorption of deuterio-PS and protio- and chain-end functionalized PS [40], where the respective molecular mass was almost the same, showed that the functionalized PS preferentially adsorbed over the deuterio-PS. In contrast, when the deuterio-PS had a higher molecular weight, it preferentially adsorbed over the protio-funtionalized-PS. The desorption kinetics of protio-PS by deuterior-PMMA [41] or deuterio-polyisoprene (PI) [42,43] has been widely studied as a function of temperature and it could be fitted to the stretched-exponential function, as suggested by theoretical considerations [44].

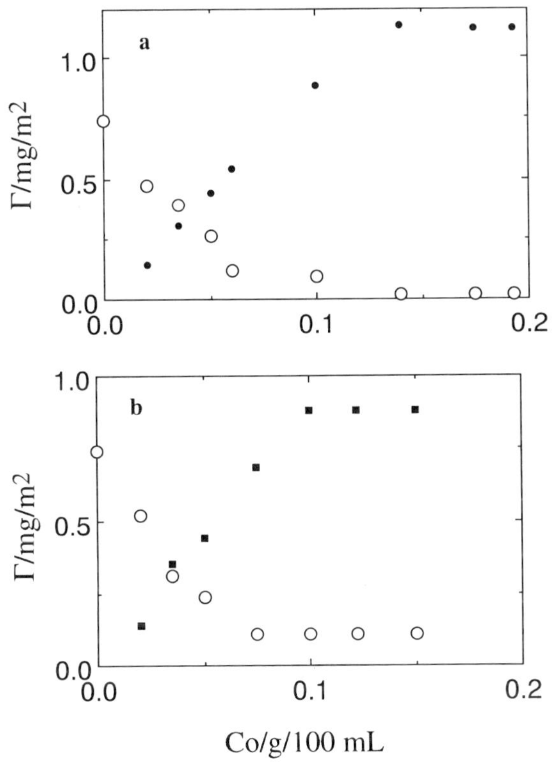

Figure 5 Displacement adsorption isotherms of polystyrene (PS) (○) and styrene-methyl methacrylate (SM) copolymers (filled symbols) as a function of dosed concentration of SM copolymer, Co: (a) SM copolymer with a styrene content of 75 mol%; (b) SM copolymer with a styrene content of 25 mol%.

B. Block Copolymers

Block copolymers in solutions have much in common with the characteristics of surfactants. In *selective solvents* where one block of the copolymer is well solvated and the other block is precipitated, the copolymer tends to self-assemble above a given concentration, such as the critical micelle concentration (CMC), to form well-defined micellar structures. For a solution with a lower polymer concentration than the CMC, when the precipitated block forms a molten layer on the surface, with the solvated block freely protruding into the solution without adsorption, block copolymer chains adsorbed on the surface should show a structural change from a *mushroom* (Fig. 6),

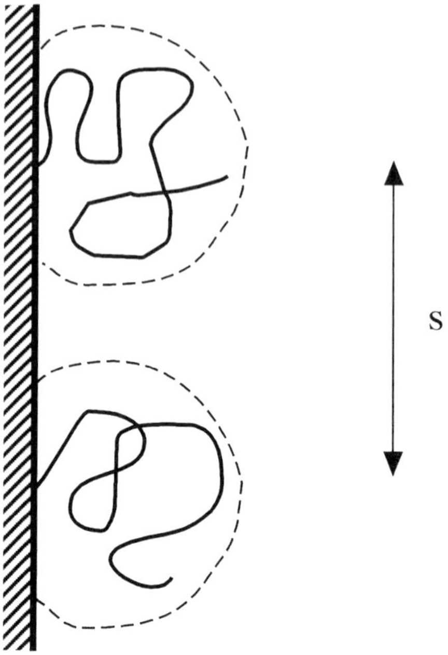

Figure 6 A mushroom model for adsorption of block copolymers when the grafting density is low. s is the interchain spacing.

which is likened to the shape of an isolated block copolymer adsorbed on the surface, to a *polymer brush* (Fig. 7) with an increase in the adsorption density [9]. Such a transition should be similar to the interfacial properties of terminally attached chains, in which one end of the polymer chain is irreversibly held at an interface (grafting chains) [45,46]. Even in nonselective solvents, similar adsorption behavior at interfaces will be absorbed when one much shorter block in the copolymer preferentially adsorbs over the other block on the surface.

The adsorption behavior of block copolymers is characterized by several variables, such as the asymmetry factor, β, the surface density, σ, and layer thickness, irrespective of the solvent. The factor β is defined by the ratio of the block sizes of the nonadsorbing block to the adsorbing block [47,48]. When block A in the AB block copolymer forms a molten layer in selective solvents, the factor β is given by

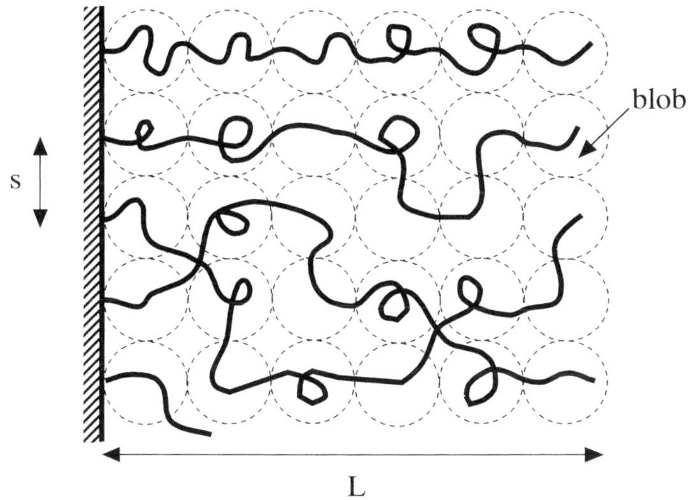

Figure 7 A polymer brush model for adsorption of block copolymers when the grafting density is high. s is the interchain spacing and L is the layer thickness.

$$\beta = N_B^{3/5}/N_A^{1/2} \tag{13}$$

where N_A and N_B are the degrees of polymerization of the A and B blocks, respectively. On the other hand, when block A preferentially adsorbs to the surface in nonselective (good) solvents, where AB block is well solvated, the factor β is given by

$$\beta = N_B^{3/5}/N_A^{3/5} \tag{14}$$

The σ value is called the grafting density, defined by the number of adsorbed polymer chains per unit area, and can be calculated from the amount of polymer adsorbed on the surface. Furthermore, the interchain spacing, s, can be related to the σ value as follows,

$$s = (1/\sigma)^{1/2} \tag{15}$$

When the value of s is assumed to be the size R_F of the B block in bulk solution, the σ value in nonselective (good) solvents is defined as

$$\sigma = (1/R_F)^2 \sim N_B^{-6/5} \tag{16}$$

The layer thickness, L, should depend on the surface density. From the Alexander–de Gennes relation [45,46] for the formation of polymer brushes, the B block chains are stretched and the B layer thickness is given by

$$L = N_B a \sigma^{1/3} \tag{17}$$

where a is the monomer size.

By contrast, in nonselective (good) solvents, the A block forms an anchor layer swollen by the solvent, and the B block forms a polymer brush for the higher surface density. The thicknesses of the anchor layer, d, and the polymer brush layer, L, are given by the mean-field theory [48]:

$$d \sim N_A^{5/6}/N_B^{1/2} \qquad L \sim aN_B/N_A^{1/3} \tag{18}$$

On the other hand, the scaling theory [48] predicts the following values of d and L:

$$d \sim N_A^{1/3}/\beta \qquad L \sim aN_B/N_A^{1/3} \tag{19}$$

The thickness of the polymer brush layer is the same, irrespective of the theory, and the thickness of the anchor layer given by the scaling theory is somewhat larger.

Typical block copolymers were widely used to investigate the adsorption behavior of a block copolymer. For the adsorption in selective solvents poly(ethylene oxide)-poly(propylene oxide)–poly(ethylene oxide) (PEO–PPO–PEO) triblock copolymers in water and block copolymers of PS and poly(2-vinyl pyridine) (P2VP) in toluene were frequently employed, whereas diblock and triblock copolymers of PEO and PS in toluene were typically available for adsorption in nonselective solvents.

1. Ellipsometry in Selective Solvents

Kinetic and equilibrium aspects of PEO–PPO–PEO triblock copolymer adsorption on silicon wafers in water have been studied by several research groups [49–51]. Since water is a poor solvent for PPO with intermediate and high molecular weights, and it is a good solvent for PEO, the PPO block forms a molten layer on the substrate. The adsorbed amounts of the block copolymers were determined as functions of the hydrophobicity gradient [49] and temperature [49–51]. A higher adsorbed amount was obtained on the hydrophobic side than on the hydrophilic side [49]. Both adsorption and desorption by rinsing with water [50] or by changes in pH [51] were fast processes. The adsorbed amounts on hydrophilic surfaces showed an abrupt increase above the critical micellization temperature, whereas on hydrophobic surfaces they remained independent of temperature [49,51]. On the other hand, there was a sharp increase in the adsorbed amounts with changes in temperature and bulk polymer concentration at hydrophobic surfaces when the PPO block is much larger than the PEO [50].

Adsorption of sodium poly(styrenesulfonate)–poly(tert-butylstyrene) (NaPSS–PtBS) block copolymers on silicon wafers in water was investigated as a function of NaCl concentration, using AFM [52]. These diblocks are highly asymmetric with a small hydrophobic PtBS block relative to the hydrophilic NaPSS block. Adsorption could not be observed in pure water, whereas it could be readily observed in the presence of 1 M NaCl. However, the NaPSS homopolymer did not adsorb on the substrate even dissolved in 1 M NaCl aqueous solution. The difference in the surface coverage between the block copolymer and the NaPSS homopolymer was confirmed by AFM on dried layers. The adsorption kinetics showed a two-stage process: an initial diffusion-limited stage, followed by a slower build-up of surface coverage in a polymer brush-limited stage.

Filippov et al. [53] determined the adsorbed amount of triblock copolymer of nonionic polyuretanes based on PEO with a structure of R-O-(DI-PEO)$_6$-DI-O-R (R is $C_{16}H_{33}$ and DI is isophorone diisocyanate) on a TiO_2 surface. The adsorption kinetics studied over a wide range of polymer concentrations was analyzed with the theory of kinetic-diffusion-controlled polymer adsorption. The existence of three distinct regimes was verified: (1) a kinetic-controlled adsorption for short times; (2) a diffusion-kinetic-controlled adsorption for intermediate times; and (3) a diffusion-controlled adsorption for long times.

2. Ellipsometry in Nonselective Solvents

Adsorption kinetics and adsorption isotherms of PEO–PS and PEO–PS–PEO block copolymers at a silicon wafer surface in toluene were studied as functions of asymmetry and chain architecture [54–56]. In toluene, the PEO block preferentially adsorbed on the substrate. The adsorption kinetics of the PEO–PS copolymers could be described by a two stage process [54]. In the beginning, adsorption was limited by the diffusion process of polymer chains on the surface. Further adsorption could only proceed by the penetration of chains through the existing monolayer combined with the conformational rearrangement to a polymer brush-like conformation. The plateau of the adsorbed amount increased with a decrease in the molecular weight of the block copolymer.

The initial adsorption process of PEO–PS–PEO triblock copolymer adsorbed on a silicon wafer in toluene also followed diffusion-limited behavior [55]. For solutions of higher polymer concentration, kinetics of the overshoot phenomena was observed, namely, the triblock copolymer passed a maximum in the adsorbed amount with increasing adsorption time [56]. Similar overshooting behavior was also seen for the adsorption of the PEO–PS block copolymer [56]. Furthermore, it was found that the adsorbed amount is in agreement with the scaling prediction of Marques and Joanny [48]: the σ value is inversely proportional to the degree of polymerization of PS for the symmetric and moderately symmetric diblock copolymers. For the triblock copolymers with comparable asymmetry, the adsorbed amount scales with $1/\beta^2$, indicating that moderately symmetric PEO–PS–PEO triblock copolymers behave like highly asymmetric diblock copolymers [56].

Adsorption of PS–PEO–PS triblock copolymers onto silicon wafers in toluene was studied as a function of the β factor by Pai-Panandiker et al. [57]. The resulting adsorbed amounts for highly asymmetric triblocks and for moderately symmetric to symmetric triblocks were consistent with the scaling predictions [48] of $\sigma \sim 1/\beta^2$ and $\sigma \sim 1/N$, respectively. The block copolymers were stretched and could form a polymer brush structure.

Stamm and coworkers [58–60] have carefully investigated the adsorption isotherm [58,59] and adsorption and desorption kinetics [60] of functionalized PS–polybutadiene (PB) on silicon wafers in toluene, and the solvent is good for both blocks but is poor for the functional groups. The size of the PB block is much smaller than that of the PS block and the functional groups attached to the PB anchor on the surface. The adsorption isotherms were similar in form for the various PS–PB block copolymers, the plateau surface density satisfied Eq. (16) but was essentially independent of the number of the functional groups [59]. From the competitive adsorption and displacement adsorption of small and large chains [60], it was found that the small chains preferentially adsorb on the surface and they completely displace the large ones in the equilibrium state. The adsorption and desorption were both controlled by a diffusion-limited process.

3. Layer Thickness by Other Optical Reflection Methods

Other optical reflection methods, such as surface plasmon [61], internal reflection interferometry [62], and scanning angle reflectometry [63], were used to study the adsorption of block copolymer on solid surfaces. These techniques give simultaneously the adsorbed amount and the layer thickness of the polymer. The kinetics of adsorption of PS–P2VP block copolymers in toluene on a silver surface has been investigated using the surface plasmon technique [61]. At higher polymer concentrations, the presence of micelles led to a high initial rate of adsorption, whereas the adsorption of a single molecule proceeded more slowly.

Gast et al. have studied the adsorption kinetics of PEO–PS block copolymers in a selective solvent of cyclopentane by internal reflection interferometry [62] and by the scanning angle reflectometer [63]. In the solvent the PEO block only adsorbed on the surfaces and the micellization of the block copolymer greatly affected the adsorption. Above the CMC, the initial adsorption rate was faster than that below the CMC, whereas the final adsorbed amounts above the CMC were lower than that below the CMC. The time evolution of the measured quantities indicated that the block copolymers adsorb like a mushroom in shape and eventually form a high-density polymer brush.

The adsorption kinetics of two PEO–PPP–PEO block copolymers on hydrophilic and hydrophobic silicon wafers has been studied to prepare an adsorbing substrate for proteins using reflectometry [64]. At the hydrophilic surface, the PEO block adsorbed in a flat pancake configuration, resulting in a layer of 1 nm thickness, whereas at the hydrophobic surface the PPO block was attached to the surface and the PEO block was extended in solution to form a polymer brush with a layer thickness of 10 nm.

4. Hydrodynamic Thickness

To measure the hydrodynamic thickness of block copolymer layers, two different techniques of DLS and capillary flow have frequently been used. Most work has been concerned with block copolymers in selective solvents. Several studies on the adsorption of PEO–PPO–PEO block copolymers at hydrophobic surfaces in water by DLS have been reported [51,65–68]. Killmann et al. [65] measured the t_h on PS latex and precipitated silica particles. They found the t_h on the latex was proportional to $M_{PEO}^{0.55}$ and there was no dependence of M_{PEO} on the t_h for the silica, where M_{PEO} is the molecular mass of PEO. The effects of polymer size [66,67] and latex size [67] on t_h were studied together with the determination of adsorption isotherms [66]. The adsorbed amount and layer thickness increased with PEO block size and polymer concentration, and both isotherms were of a high-affinity type. The plateau layer thickness was proportional to the PEO mass with a power of 0.6 [66] and its dependence on M_{PEO} decreased with an increase in particle size, starting at a power law of 0.69 on the PEO mass for the smallest latex and becoming essentially independent of M_{PEO} for the largest particle [67].

Lee et al. [68] also measured the t_h on PS latex particles to prepare a substrate for protein adsorption as a function of the molecular weight of the block copolymer. The t_h scaled with $M_{PEO}^{0.83}$ for the small size PS latex and its magnitude was larger than that for the large PS particle. The t_h on colloidal hydrophobic silica particles was as thin as 2–5 nm, irrespective of temperature [51].

Experimental evidence for a transition from a pancake, which is likened to the shape of an isolated adsorbed chain with a flat layer thickness of the order of the segment size, to a polymer brush was provided by measurements of layer thickness of triblock copolymers adsorbed on PS latex using DLS [69]. Triblock copolymers used were nonionic polyurethanes based on PEO capped on both ends with either linear alkyl groups, $C_{16}H_{33}$, $C_{12}H_{25}$ or hydrogen, whose structure is R-O-(DI-PEO)$_6$-DI-O-R (R is H, $C_{16}H_{33}$, $C_{12}H_{25}$ and DI is isophorone diisocyanate). The alkyl groups were expected to strongly adsorb on the hydrophobic surface of PS latex. The t_h value increased at low concentrations, but over a narrow concentration range it increased sharply and then reached a plateau, indicating the existence of the pancake-to-polymer brush transition. This adsorption isotherm was quite different from that of the PEO homopolymer. The transition regime varied with molecular weight and was independent of the adsorbing strength of the end group. The plateau thickness scaled with molecular weight with a power law of 0.46, irrespective of the length of the alkyl chain.

Adsorption of asymmetric poly(dimethyl amino ethyl methacrylate)–poly(butyl methacrylate) (PDMAEM–PBMA) diblock copolymers on colloidal silica particles in 2-propanol has been studied by monitoring the t_h by DLS and the adsorbed amounts [70]. A PDMAEM block only adsorbed on the surface via acid–base interactions. At a fixed degree of polymerization, both the adsorbed amount and the t_h value showed steep maxima as the PDMAEM composition was varied. The t_h scaled by a power of 0.53 with the degree of polymerization of PBMA.

Quali and Pefferkorn [71] have investigated the adsorption behavior of an asymmetric poly(styrene sulfonate)–poly(vinyl pyridine) (PSS–PVP) block copolymer adsorbed on neutral and electrically charged surfaces. The PSS–PVP sample was a polyampholyte where the short PSS block was negatively charged and the PVP block contained neutral pyridine and positively charged pyridinium groups. The adsorbed amounts of PSS–PVP were determined using polymer labeled with radioisotope. The polymer irreversibly adsorbed on a Teflon surface in pure water and on positively charged PS latex; partially reversible adsorption occurred in pure water and in the presence of salt. The t_h on the charged PS latex, determined by DLS, increased with an increase in salt concentration, a mechanism similar to that for the solubilization of polyampholytes in concentrated salt solutions [72].

The capillary flow method is based on measuring the permeability of a porous medium or a glass capillary before and after adsorption by solvent flow. When the sizes, R_h, of the pores and capillaries are known, the t_h can be calculated from the flow rates, V_0 and V, measured before and after adsorption, respectively, according to Poiseuille's law:

$$t_h = R_h[1 - (V/V_0)^{1/4}] \tag{20}$$

Anderson and coworkers [73–76] extensively measured changes in the t_h of asymmetric PS–P2VP block copolymers adsorbed on surfaces of mica pores as a function of solvent quality, which was changed by alternating the solvent itself and/or the temperature, by the capillary method after achievement of adsorption in a toluene. Other solvents used were methanol, n-heptane, and cyclohexane and the changing of the temperature was performed in toluene and cyclohexane. Methanol and n-heptane were nonsolvent for the PS block. There is no shear thinning of the t_h in toluene, indicating the formation of a dense polymer brush. The t_h value was reversibly collapsed and expanded by changing the solvent quality. A linear variation of t_h with temperature was observed in cyclohexane, whereas in toluene the t_h was independent of temperature. The expansion coefficient of the layers defined as the ratio of t_h values between a given and θ temperatures in cyclohexane was almost an order of magnitude greater than that for a free chain in bulk solution, and the temperature dependence of the expansion coefficient was in disagreement with the theoretical predictions [76].

In toluene, the t_h of the PS–P2VP block copolymers scaled with $M_{PS}^{0.815}$ and the ratio of t_h to the hydrodynamic radius of the corresponding block copolymer in bulk solution was found to be proportional to $M_{PS}^{0.255}$, indicating a greater stretching of the solvated chains than one would expect for finite-size adsorbing blocks. For symmetric poly(methacrylic acid)–poly (butyl methacrylate) (PMAA–PBMA) block copolymers adsorbed on the mica pores from methyl ethyl ketone (MEK) solutions, in which PMAA block adsorbed on the mica surface [74], the same scaling of t_h by the hydrodynamic radius was obtained.

5. Surface Force Measurements in Selective Solvents

Tirrell and coworkers [77] first tried to measure forces between mica bearing two asymmetric PS–P2VP in toluene. The resulting forces were purely repulsive and they were detectable at a distance of $D = 10\,R_F$, where R_F is the radius of gyration of a free PS chain of molecular weight equal to the PS block in toluene solution. One-half the distance of D at the first detectable onset of repulsion was taken as t_s, which can be regarded as a polymer brush layer thickness, L, in the theoretical treatments [45–48]. It was found that these PS block are stretched to 25% of their total contour lengths. Even after the replacement of the solution with cyclohexane the forces were purely attractive at 38°C and further lowering the temperature to 21°C finally showed an attraction, but its magnitude was some 5–10 times weaker than for the homopolymer adsorbed under the same conditions.

Sequentially, Tirrell et al. have investigated the forces between PS–P2VP layers [78–80], P2VP–PS–P2VP layers [78,79], PI–P2VP layers [80], asymmetric layers of PS–P2VP and PI–P2VP [80], and bimodal layers of PS–P2VP and PS–P2VP–PS [81] adsorbed on mica surfaces in toluene. For the P2VP–PS–P2VP triblock copolymers, the middle PS block is expected to behave as loops with various lengths. However, a linear variation in the range of the forces with the PS molecular weight was not observed. This behavior probably stems from a gradual transition from an overlapping P2VP of osculating disks to separated disks on the surface with increasing PS block size. Patel and Tirrell [79] have developed a simple model for forces between adsorbed block copolymers based on terminally attached chains. From this model, they obtained a pair of dimensionless variables: a dimensionless separation and a dimensionless energy. It was found that the resulting forces between various diblock [77] and triblock copolymers [78,79] adsorbed on a mica surface can be superimposed on a master curve using these two dimensionless variables. Such a reduced plot fits well with the asymmetric interactions between layers of PS–P2VP as well as those of PI–P2VP.

On the other hand, Dai et al. [82] found the first experimental evidence for a rapid conformational transition between loops and tails for the surface force–distance profile for a P2VP–PS–P2VP adsorbed layer against a bare mica surface in toluene during compression–decompression cycles when the attractive forces and outward jumps were observed under decompression. This is due to the formation of polymer bridges between different mica surfaces under compression and the subsequent stretching of the bridging chains on decompression. Its process qualitatively agreed with the scaling theory of Rabin and Alexander [83]. They also obtained similar results for the force measurements between a PEO–PS–PEO triblock copolymer layer and a clean mica surface in toluene as described below.

Ansarifar and Luckham [84] have measured the forces between PtBS–P2VP block copolymer layers adsorbed on mica surfaces in toluene. The block copolymers should have interfacial properties similar to the PS–P2VP block copolymer. The resulting forces were purely long repulsive and showed no hysteresis in the compression–decompression cycle. The t_s value is proportional to a power law of 0.7 with the molecular mass of the PtBS block. A plot of $\log\{F(D)/R\}$ as a function of $\log D$ gave an exponent of about −4 for all samples, which is in disagreement with the scaling theory of de Gennes [85].

6. Surface Force Measurements in Nonselective Solvents

In nonselective solvents, diblock and triblock copolymers of PEO and PS were widely used for the surface force measurements. The PEO blocks preferentially adsorb on a mica surface and the PS blocks serve as polymer brushes. Hair and coworkers have extensively measured the forces between asymmetric PEO–PS block copolymers adsorbed on mica surfaces in toluene [86–90] and toluene–heptane mixtures [86]. In toluene, the forces were purely repulsive, whereas in the mixtures which are worse than a θ solvent, long-range attractive forces were observed for the force–distance profiles. Addition of heptane induced a decrease in the solvent conditions and the adsorption affinity of a PS segment to a mica surface was expected to increase. PEO segments were adsorbed more strongly and they effectively displaced weaker adsorbed PS segments. The mica surface was fully covered by the block copolymers and it prohibited the occurrence of polymer bridges that are responsible for long-range attractive forces. The observed attractive forces in the toluene–heptane mixtures were essentially the result of the negative free energy of mixing between the adsorbed layers. For the experimental data in toluene, the values of t_s and surface density were compared with the theoretical predictions of Alexander [45] and de Gennes [46] as well as Marques and Joanny [48], as a function of asymmetry. With a combination of the scaling relation of $\sigma \sim 1/N$, the length of polymer brush given by Eq. (17) is expressed as

$$t_s = aM_{PS}M_{PEO}^{-1/3} \tag{21}$$

where M_{PS} is the molecular weight of the PS block. This relation was in agreement with the experimental results of the moderately asymmetric PEO–PS block copolymers when the β value defined by Eq. (14) was larger than unity and less than $M_{PEO}^{1/2}$. However, deviations from the relation were observed when the PS block was either much larger than the PEO block or approximately the same size as the PEO block. Furthermore, the measured respulsive forces [88] could be reduced to a single curve such as a plot of $(F(D)/R)M_{PS}/M_{PEO}^{-11/6}$ against the normalized $D/2t_s$, based on the theory of Marques and Joanny [48]. Furthermore, appliction of IR techniques led to measurements of the amount of the block polymer adsorbed onto mica surfaces [88,89], and then it was possible to confirm the validity of Eq. (17) and the scaling relation, $\sigma \sim 1/N$. The linear relation between t_s and $N\sigma^{1/3}$ was satisfied for the respective block copolymers and the scaling theory was well fitted to the moderately asymmetric block copolymers.

Taunton et al. [91,92] measured the forces between layers of PEO–PS blocks with a short PEO adsorb on mica surfaces in toluene. The resulting surface force–distance profile showed a purely repulsive force curve and was independent of incubation concentration and compression–decompression cycles. When a clean mica surface approached a mica surface bearing the block copolymer, only attractive forces were observed and the force curve was similar in form [91]. Furthermore, it could be reduced to a master curve of $F(D) = (kT/s^3)[(2t_s/D)^{9/4} - (D/2t_s)^{3/4}]$, which was derived from the scaling theory of de Gennes [46] based on terminally attached chains. The value of t_s was comparable to a layer thickness calculated from Eq. (17).

Dai et al. have measured the forces between a single layer of adsorbed PEO–PS diblock copolymer [93], PEO–PS–PEO [93,94], and P2VP–PS–P2VP [82] block copolymers against a bare mica surface in toluene. In addition, they also measured the forces between layers of adsorbed PEO–PS–PEO triblock copolymer in toluene as a function of incubation concentration [82]. For the diblock copolymers in toluene, no attraction was observed and strongly repulsive forces were seen at a separation of 100 nm [82], which is in good agreement with that between layers of PEO–PS block copolymers reported previously [92]. However, replacement of the solution by cyclohexane led to a small attraction at a distance of about 40 nm and much stronger repulsion at a distance much shorter than 40 nm, indicating the collapsing of the stretched polymer brushes by changing solvent quality [82]. From the surface force–distance profiles between layers of the triblock copolymers, the forces and effective layer thickness increased with an increase in incubation concentration, due to the large tail-like (a long loop in a real form) population. For the surface force measurement between an adsorbed PEO–PS–PEO triblock copolymer layer and a bare mica surface in toluene, attractive forces were detected for the first compression at a higher incubation concentration, indicating the formation of polymer bridges between the two mica surfaces. This was the first time that the formation of polymer bridges was possible for ABA triblock copolymer chains in a good solvent with both the A sticking blocks in a single polymer chain simultaneously anchoring on opposing surfaces. Successive compression–decompression cycles induced the molecular organization of triblock copolymers at the mica surfaces, leading to changes in the surface force–distance profile.

7. Density Profile of Polymer Segments Observed by SANS and Neutron Reflectometry

Cosgrove and co-workers have taken a leading role in measuring the scattering intensity of a polymer layer adsorbed on PS latex [20,95–98]. The resulting density profile of polymer segments of PEO–PPO–PEO adsorbed on deuterio PS latex have no maximum and do not extend further into the solution, thus resembling the behavior of a physically adsorbed homopolymer [98]. This is attributed to both blocks adsorbing, despite the high insolubility of the PPO block, which must be preferentially adsorbed. On the other hand, neutron reflectometry has been shown to yield unprecedented spatial resolution for the determination of the density profile of polymer segments in solid polymer films, for a polymer and surfactant spread or adsorbed at the air–water interface, and for a polymer layer at the liquid–solid interface.

Satija et al. [21] were the first to apply the neutron reflectometry to evaluate the density profile of symmetric PS–PMMA with hydrogen and deuterium labeling blocks adsorbed on a quartz surface from carbon tetrachloride solutions. Carbon tetrachloride was a selective solvent for the block copolymer and the PMMA block preferentially adsorbed on the quartz surface, indicating that deuteration of the PMMA block changed the reflectivity at higher wavevectors. On the other hand, when the PMMA block was identical, the resulting reflectivity was similarly independent of the isotope of the PS block. A plot of log (reflectivity) against wavevector, called the reflectivity curve, for the protio-PS–protio-PMMA block copolymer fitted well with an exponential decay, whereas the measured reflectivity curve of the protio PS–deuterio PMMA block copolymer clearly deviated from exponential decay.

In fitting different models for the density profile of polymer segments, the following parabolic functions were found to reproduce reasonably well the measured reflectivity curves of the protio-PS–protio-PMMA and protio-PS–deuterio-PMMA block copolymers:

$$\begin{aligned} &\phi_{\mathrm{PMMA}}(z) = \phi_{\mathrm{PMMA}}(0)(D_0^2 - z^2)/D_0^2 \qquad \text{for } z < D_0 \\ &\phi_{\mathrm{PMMA}}(z) = 0 \qquad \text{for } z > D_0 \end{aligned} \tag{22}$$

where $D_0 = 1.5 \pm 0.8$ nm and $\phi_{\mathrm{PMMA}}(0) = 0.42$. The value of D_0 is comparable to the radius of gyration of a free PMMA chain. On the other hand, when the PS block chains were sufficiently stretched and extended into the solution, the PS segment density was undetectable by neutron reflectometry.

Cosgrove et al. [99] measured the reflectivity curve of deuterio-PS–protio-P2VP adsorbed on a mica surface in toluene or cyclohexane. In cyclohexane as the solvent, both blocks were adsorbed on the mica surface and the measured reflectivity curve was fitted well with the exponential decay. Various other models such as the Gaussian [99], parabolic [99], and the Scheutjens–Fleer (S–F) model [100,101], gave essentially the same kind of fit. In the toluene solvent, the shifted Gaussian, the terminally attached S–F, and the parabolic models were equally in agreement with the measured reflectivity curve. Analysis of these fitting results showed that the volume fraction of P2VP is about 0.37 and the calculated root-mean-square thickness is 4.6 ± 0.2 nm.

The same group [102] has studied the adsorption of PS–P2VP on a quartz surface in various solvents, such as toluene, cyclohexane, and octane, together with surface force measurements of layers of the same polymer adsorbed on mica surfaces. The adsorbed amount of the block copolymer was also determined and was in good agreement with the data on mica surfaces. The density profile of polymer segments was compressed with a decrease in the solvent power. On the other hand, the surface forces in cyclohexane and in octane after replacement of toluene showed a small attraction and strong attraction, respectively.

The resulting reflectivity curves of PS–P2VP copolymers adsorbed on a quartz surface in deuterio-toluene were well described by a parabolic or error function of the density profile of polymer segments and they could not be well fitted to exponential or power law decays [103]. The layer thickness of L_0 obtained from fits to the parabolic profiles was that calculated from Eq. (17) and was also in good agreement with that measured from the surface force measurements of similar block copolymers [92], whereas the exponent for the variation of L_0 with the PS mass was 0.57 ± 0.06. The difference may be due to the fact that the surface density did not remain fixed, since the PS block is not chemically grafted on to the quartz surface, but is determined by a balance between the attraction of the PEO block and the mica surface and the repulsive osmotic interactions between the PS blocks. A combination of the determination of the adsorbed amounts of the block copolymers and Eq. (17) provided a power relation $L^{3/2}\Gamma^{-1/2} \sim M$. This power law described the data with the measured exponent of 0.97 ± 0.10.

8. Atomic Force Microscopy

Since AFM has proven to be a very successful technique for the imaging and characterization of solid surfaces from micrometer scale down to the atomic level, it has been widely used to investigate lateral surface structure for thin block copolymer films. Thin films were prepared by various methods, such as spin coating, Langmuir–Blodgett film methods, solvent casting, and epitaxial growth, and by adsorption methods.

Here we focus on AFM studies of block copolymer films formed by adsorption from their solutions. AFM observation was performed in air and liquid media and the first AFM study of the lateral surface observation of block copolymers was reported by O'Shea et al. [104]. They measured lateral images, force profiles, and the friction (lateral) forces of a PEO–PS block copolymer with a much shorter PEO block in air and liquids of xylene, n-dodecane, and 2-propanol, where xylene is a good solvent and the others are poor solvents of the block copolymer. Topographic images of the adsorbed polymer in xylene could not be obtained since the tip tends to displace the adsorbed polymers along the surface. No friction signal could be, therefore, detected within the polymer brush, whereas, in air, where polymer chains should behave as if in poor solvent conditions, some circular-shaped islands were formed by the aggregation of the polymer molecules with an increase in polymer concentration. The size of the islands was squeezed by collapsing the polymer chains upon filling with 2-propanol. The resulting force profiles showed the distinctive repulsive forces associated with polymer brushes in xylene and the shorter-range attractive forces in poor solvents.

Watanabe et al. [105] have examined the lateral structure of PS–P2VP, PB–P2VP, and PS–P2VP–PB block copolymer layers adsorbed on a mica surface from their toluene solutions. Since the PVP block preferentially adsorbed on the surfaces, it was expected that both PS and PB blocks would be at the top of the adsorbed layers. This layered model was confirmed by X-ray photoelectron spectroscopy (XPS) and contact angle measurements [105]. AFM images of the block copolymers indicated the presence of many nodules with a typical height of about 1 nm and width of about 20–50 nm. Similar nodules have been observed in the AFM image of PS–P2VP, but not in that of PB–P2VP, which suggested that the nodules of the triblock copolymers were in the domains of the PS blocks.

Lateral structures of end-grafted PS chains adsorbed on a silicon substrate in air have been studied by Zhao et al. [106]. The resulting images showed that the chains were found to clump together into islands, leading to a bicontinuous pattern with regions of high polymer density separated by almost polymer-free regions on the silicon surface with a decrease in grafting density. Such a transition is in good agreement with the recent theoretical simulations of polymer brushes [107].

Siqueira et al. [108] have studied the surface structure of PS–PB diblock copolymers with much smaller blocks of PB carried on functional groups adsorbed on silicon wafers in air as a function of surface density. Only the functionalized groups adsorbed and neither PS nor PB adsorbed on the substrates in toluene. After the adsorption of the copolymers at the plateau region in the solvent, the adsorbed layers were rinsed and dried. The images changed with the surface density σ: (1) at low values of σ, isolated islands of clusters of polymer chains were observed; (2) at intermediate σ values, a surface covered with laterally inhomogeneous structures was observed; (3) at high surface coverage, a laterally homogeneous layer was formed. These features were in qualitative agreement with the theoretical prediction of polymer brushes in a poor solvent.

The effect of solvent quality on the surface structure of a diblock copolymer has been studied by Stamouli et al. [109]. They used a fairly asymmetric PS–P2VP block copolymer, which was adsorbed on mica surfaces in toluene. The adsorbed layers were rinsed and dried prior to the observation in water by AFM. Since water is worse than a θ solvent for the block copolymer, the PS blocks fused together to form a surface octopus-like micelle.

IV. CONCLUSIONS

Microscopic structures in polymr blends and block copolymers adsorbed at solid–liquid interfaces have been explored by various techniques. Considering the competitive adsorption and displacement adsorption between polymer chains with the same chemical structures, the larger polymer chains are preferentially adsorbed over smaller ones without specific interactions between polymer chains and solid surfaces.

Adsorption behaviour due to the competition among polymer chains with different chemical structures is, in general, more complex due to the differences in the sizes of the polymer chains, the solvent quality for both polymers, and the interactions between polymer segments and solid surfaces. The adsorption preference and desorption ability have been found to correlate well with two quantities: (1) the polar character of polymer chains and (2) the adsorbed amount in the plateau region for the individual adsorption of the displacer polymer.

The characteristics of the adsorption of block copolymers strongly depend on the solvent quality. When anchoring polymer chains are short, adsorbed block copolymers can be treated as grafted polymer chains, irrespective of solvent. A number of power laws were proposed and found to be in good agreement with the experimental results. Surface force measurements and neutron reflectometry are powerful methods contributing to the understanding of the adsorption behavior of block copolymers. However, one of the most important and useful quantities, the density profile of polymer segments, cannot be uniquely determined by neutron reflectometry. Many interesting systems such as polymer chains adsorbed at gas–liquid (air–water) and liquid–liquid (oil–water) interfaces were not discussed in this chapter because of the restricted space.

REFERENCES

1. MA Cohen Stuart, T Cosgrove, B Vincent. Adv Colloid Interface Sci 24:143, 1986.
2. T Cosgrove, B Vincent. In: CA Croxton, ed. Fluid Interfacial Phenomena. New York: Wiley, 1986, pp 607–635.
3. GJ Howard. In: H-F Eicke, GD Parfitt, eds. Interfacial Phenomena in Apolar Media. New York: Marcel Dekker, 1987, pp 281–326.
4. PG de Gennes. Adv Colloid Interface Sci 27:189, 1987.
5. SS Patel, M Tirrell. Ann Rev Phys Chem 40:597, 1989.
6. WB Russel, DA Saville, WR Schowalter. Colloidal Dispersions. London: Cambridge University Press, 1989.
7. M Kawaguchi. Adv Colloid Interface Sci 32:1, 1990.
8. M Kawaguchi, A Takahashi. Adv Colloid Interface Sci 37:219, 1992.
9. A Helperin, M Tirrell, TP Lodge. Adv Polym Sci 100:31, 1992.
10. IC Sanchez, ed. Physics of Polymer Surfaces and Interfaces. Boston: Butterworth-Heinemann, 1992.
11. GJ Fleer, MA Cohen Stuart, JMHM Scheutjenes, T Cosgrove, B Vincent. Polymers at Interfaces. London: Chapman & Hall, 1993.
12. FL McCrackin, NBS Tech Note, 479, 1969.
13. KS Schmitz. Dynamic Light Scattering by Macromolecules. New York: Academic Press, 1990.
14. D Tabor, RHS Winterton. Proc Roy Soc London Ser A 312:435, 1969.
15. JN Israelachvili, D Tabor. Proc Roy Soc London Ser A 331:19, 1972.
16. JN Israelachvili. Intermolecular and Surface Forces. London: Wiley, 1985.
17. PF Luckham, BA del Costello. Adv Colloid Interface Sci 44:183, 1993.
18. G Binning, H Rohrer, C Gerber. Phys Rev Lett 49:57, 1982.
19. G Binning, CF Quate, C Gerber. Phys Rev Lett 56:930, 1986.
20. KG Barnett, T Cosgrove, B Vincent, AN Burgess, TL Crowley, T King, JD Turner, ThF Tadros. Polym Commun 22:283, 1982.
21. SK Satija, CF Majkrzak, TP Russell, SK Sinha, EB Sirota, GJ Hughes. Macromolecules 23:3860, 1990.
22. H Mark, G Saito. Monatsh Chem 68:237, 1936.
23. M Kawaguchi, S Hattori, A Takahashi. Macromolecules 20:178, 1987.
24. JC Dijt, MA Cohen Stuart, GJ Fleer. Macromolecules 27:3219, 1994.
25. H Xu, JB Schlenoff. Langmuir 10:241, 1994.
26. M Kawaguchi, S Anada, K Nishikawa, N Kurata. Macromolecules 25:1588, 1992.
27. S Anada, M Kawaguchi. Macromolecules 25:6824, 1992.
28. M Kawaguchi, Y Sakata, S Anada, T Kato, A Takahashi. Langmuir 10:538, 1994.
29. VN Kislenko, AdA Berlin, M Kawaguchi, T Kato. Langmuir 12:768, 1996.
30. JC Dijt, MA Cohen Stuart, GJ Fleer. Macromolecules 25:5416, 1992.
31. C Thies. J Phys Chem 70:3783, 1966.
32. HE Johnson, JF Douglas, S Granik. Phys Rev Lett 70:3267, 1993.

33. M Kawaguchi, A Sakai, A Takahashi. Macromolecules 19:2952, 1986.
34. JC Dijt, MA Cohen Stuart, GJ Fleer. Macromolecules 27:3229, 1994.
35. M Kawaguchi, K Itoh, S Yamagiwa, A Takahashi. Macromolecules 22:2204, 1989.
36. HE Johnson, S Granik. Macromolecules 23:3367, 1990.
37. P Frantz, S Granik. Langmuir 8:1176, 1992.
38. P Frantz, S Granik. Phys Rev Lett 66:899, 1991.
39. HM Schneider, S Granik. Macromolecules 25:5054, 1992.
40. P Frantz, DC Leonhardt, S Granik. Macromolecules 24:1868, 1991.
41. HE Johnson, S Granik. Science 255:966, 1992.
42. HM Schneider, S Granick, S Smith. Macromolecules 27:4714, 1994.
43. HM Schneider, S Granick, S Smith. Macromolecules 27:4721, 1994.
44. JF Douglas, HE Johnson, S Granick. Science 262:2010, 1993.
45. S Alexander. J Phys (France) 38:983, 1977.
46. PG de Gennes. Macromolecules 13:1069, 1980.
47. CM Marques, JF Joanny, L Leibler. Macromolecules 21:1051, 1988.
48. CM Marques, JF Joanny. Macromolecules 22:1454, 1989.
49. LMA van de Steeg, CC Golander. Colloids Surfaces 55:105, 1991.
50. F Tiberg, M Malmsten, P Linse, B Lindmann. Langmuir 7:2723, 1991.
51. M Malmsten, P Linse, T Cosgrove. Macromolecules 25:2474, 1992.
52. C Amiel, M Sikka, JW Schneider Jr, Y-H Tsao, M Tirrell, JW Mays. Macromolecules 28:3125, 1995.
53. LK Filippov, CA Silebi, MS El-Asser. Langmuir 11:872, 1995.
54. M Motschmann, M Stamm, Ch Toprakcioglu. Macromolecules 24:3681, 1991.
55. JR Dorgan, M Stamm, C Toprakcioglu. Polymer 34:1554, 1993.
56. JR Dorgan, M Stamm, C Toprakcioglu, R Jerome, LJ Fetters. Macromolecules 26:5321, 1993.
57. RS Pai-Panandiker, JR Dorgan, RT Carlin, N Uyanik. Macromolecules 28:3471, 1995.
58. DF Siqueira, U Breiner, R Stadler, M Stamm. Langmuir 11:1680, 1995.
59. DF Siqueira, K Kohler, M Stamm. Langmuir 11:3092, 1995.
60. DF Siqueira, J Reiter, U Breiner, R Stadler, M Stamm. Langmuir 12:972, 1996.
61. JF Tassin, RL Siemens, WT Tang, G Hadziioannou, JD Swalen, BA Smith. J Phys Chem 93:2106, 1989.
62. MR Munch, AP Gast. Macromolecules 23:2313, 1990.
63. FAM Leemakers, AP Gast. Macromolecules 24:718, 1991.
64. CGPH Schroen, MA Cohen Stuart, KV van der Maarschalk, A van der Padt, K van't Riet. Langmuir 11:3068, 1995.
65. E Killmann, H Maier, JA Baker. Colloids Surfaces 31:51, 1988.
66. JA Baker, JC Berg. Langmuir 4:1055, 1988.
67. LA Baker, RA Pearson, JC Berg. Langmuir 5:339, 1989.
68. J Lee, PA Martic, JS Tan. J Colloid Interface Sci 131:252, 1989.
69. HD Ou-Yang, Z Gao. J Phys II (France) 1:1375, 1991.

70. DT Wu, A Yokoyama, RL Setterquist. Polymer J 23:709, 1991.
71. L Quali, E Pefferkorn. Macromolecules 29:686, 1996.
72. EA Bekturov, SE Kudaibergenov, SR Rafikov. Rev Macromol Chem Phys C 30:233, 1990.
73. RM Webber, JL Anderson, MS Jhon. Macromolecules 23:1026, 1990.
74. PF McKenzie, RM Webber, JL Anderson. Langmuir 10:1539, 1994.
75. RM Webber, JL Anderson. Langmuir 10:3156, 1994.
76. RM Webber, CC van der Linden, JA Anderson. Langmuir 12:1040, 1996.
77. G Hadziioannou, S Patel, S Granik, M Tirrell. J Am Chem Soc 108:2869, 1986.
78. M Tirrell, S Patel, G Hadziioannou. Proc Natl Acad USA 84:4725, 1987.
79. S Patel, M Tirrell. Colloids Surfaces 31:157, 1988.
80. H Watanabe, M Tirrell. Macromolecules 26:6455, 1993.
81. S Dhoot, H Watanabe, M Tirrell. Colloids Surfaces 86:47, 1994.
82. L Dai, C Toprakcioglu, C Hadziioannou. Macromolecules 28:5512, 1995.
83. Y Rabin, S Alexander. Europhys Lett 13:49, 1990.
84. MA Ansarifar, PF Luckham. Polymer 29:329, 1988.
85. PG de Gennes. Acad Sci Paris Ser II 300:839, 1985.
86. J Marra, ML Hair. Colloids Surfaces 34:215, 1988.
87. ML Hair, D Guzonas, D Boils. Macromolecules 24:341, 1991.
88. DA Guzonas, D Boils, ML Hair. Macromolecules 24:3383, 1991.
89. DA Guzonas, D Boils, CP Tripp, ML Hair. Macromolecules 25:2434, 1992.
90. DA Guzonas, ML Hair, T Cosgrove. Macromolecules 25:2777, 1992.
91. HJ Taunton, C Toprakcioglu, J Klein. Macromolecules 21:3333, 1988.
92. HJ Taunton, C Toprakcioglu, LJ Fetters, J Klein. Macromolecules 23:571, 1990.
93. L Dai, C Toprakcioglu. Europhys Lett 16:331, 1991.
94. L Dai, C Toprakcioglu. Macromolecules 25:6000, 1992.
95. T Cosgrove, TG Heath, K Ryan, B van Lent. Polym Comm 28:64, 1987.
96. T Cosgrove, TG Heath, K Ryan, TL Crowley. Macromolecules 20:2879, 1987.
97. T Cosgrove, TL Crowley, K Ryan, JRP Webster. Colloids Surfaces 51:255, 1990.
98. T Cosgrove. J Chem Soc Faraday Trans 86:1323, 1990.
99. T Cosgrove, TG Heath, JS Phipps, RM Richardson. Macromolecules 24:94, 1991.
100. JMHM Scheutjens, G Fleer. J Phys Chem 83:1619, 1979.
101. JMHM Scheutjens, G Fleer. J Phys Chem 84:178, 1980.
102. T Cosgrove, JS Phipps, RM Richardson, ML Hair, DA Guzonas. Macromolecules 26:4363, 1993.
103. JB Field, C Toprakcioglu, RC Ball, HB Stanley, L Dai, W Barford, J Penfold, G Smith, W Hamilton. Macromolecules 25:434, 1992.
104. SJ O'Shea, ME Welland, T Rayment. Langmuir 9:1826, 1993.
105. H Watanabe, T Shimura, T Kotaka, M Tirrell. Macromolecules 26:6338, 1993.

106. W Zhao, G Krausch, MH Rafailovich, J Sokolov. Macromolecules 27:2933, 1994.
107. GS Grest, M Murat. Macromolecules 26:3108, 1993.
108. DF Siqueira, K Kohler, M Stamm. Langmuir 11:3092, 1995.
109. A Stamouli, E Pelletier, V Koutsos, E Van der Vegte, G Hadziioannou. Langmuir 12:3221, 1996.

11

Thermal Conductivity of Multiphase Polymers

Naoto Tsutsumi
Kyoto Institute of Technology, Kyoto, Japan

I. INTRODUCTION

Thermal diffusivity and thermal conductivity of materials are not only basic thermal properties, but they are also important when probing the change in microstructure of materials. Diamond has the highest thermal diffusivity of the order of $10^1\ cm^2\ s^{-1}$. For the metallic materials, thermal diffusivity ranges from the order of $10^0\ cm^2\ s^{-1}$ for metals with high thermal conductivity, such as copper aluminum and silver, to the order of 10^{-2} or $10^{-3}\ cm^2\ s^{-1}$ for amorphous metallic glassy alloys [1–16], which may be ascribed to the shorter mean free path due to the strong scattering of electrons in the glass.

In the field of the metallic alloys, thermal diffusivity and thermal conductivity have been extensively studied in relation to the changes in microstructure and in morphology [1–16]. Thus, before starting this chapter, it is important to briefly overview the studies of the thermal diffusivity and thermal conductivity for the metallic alloy systems. For copper niobium metal–metal alloys sytems [5], the dependence of thermal diffusivity on alloy composition obeys the rule of mixtures, i.e., the proportional dependence of thermal diffusivity on the alloy composition. Many alloys with quasicrystalline phase have been reported after its first disclosure [6]. The quasicrystalline phase is metastable and transforms irreversibly into the crystalline ground state upon heating [7]. Furthermore, thermodynamically stable quasicrystals were discovered [8–10]. Thermal diffusivity of quasicrystalline and related crystalline alloys is remarkably small compared to that for conventional metallic alloys [7]. Thermal diffusivity measurements show a great sensitivity to the presence of foreign phases and macroscopic defects such as pores [7]. When it is measured as a function of temperature, thermal diffusivity is able not only to trace the phase transition but also to indicate the temperature domain where precursor phenomena of the transition take place [7]. Thermal diffusivity for a uranium–zirconium alloy sharply increases in the vicinity of phase transition temperature from δ- to $U(\gamma)Zr(\beta)$-phases [11]. The dependence of thermal conductivity on the composition shows that the thermal conductivity has minimum at about 70% of Zr where the intermetallic δ-phase is formed [11]. Structural relaxation of palladium–silicon and copper–zirconium–niobium glassy alloy systems was studied through the thermal diffusivity measurement [4]. Thermal diffusivity measurement of palladium–silicon glassy alloy, when it is aging and thermally annealing, can catch the relief of the microstresses which are induced by the process of the material fabrication [4].

Thermal conductivity, thermal diffusivity, and heat capacity for single polymers have been studied in relation to their molecular and supramolecular structures. The decrease in thermal diffusivity and the increase in heat capacity with increasing temperature are typically observed at high temperatures for amorphous polymers [17]. In these temperature profiles, stepwise decrease in thermal diffusivity and stepwise increase in heat capacity were observed in the glass transition temperature region. The temperature dependence of thermal diffusivity is related to the scattering of phonons. The phonon mean free path is determined by the geometrical scattering and the scattering by collisions with other phonons [18]. The theory of the effect of anharmonic coupling on thermal resistivity predicts that phonon mean free path is proportional to the reciprocal temperature at high temperatures [18–20]. Estimations of thermal conductivity for amorphous polymers are made by means of the additive quantities of the molar sound velocity function (Rao function) [21], the molar heat capacity, and the molar volume for the chemical component of the monomeric unit of the corresponding polymers [17].

The molecular weight dependences of thermal diffusivity and thermal conductivity were studied for polyethylene [22,23], polystyrene [22,23] and poly(methyl methacrylate) (PMMA) [24,25]. Measurements of the thermal conductivity as a function of molecular weight in polymers indicate the preferential heat conduction along the molecular chain, with the conductivity depending on the square root of the molecular weight for low molecular weights and saturating at high molecular weight [22]. A free volume theory was applied to explain the molecular weight dependence of thermal diffusivity in PMMA [25]. In relating to the free volume in the polymer matrix, several studies [26–32] were focused on the pressure dependence of thermal conductivity, thermal diffusivity and heat capacity.

For crystalline polymers, several studies [33–41] have focused on the temperature dependence of thermal conductivity, thermal diffusivity, and heat capacity. The behaviors of thermal conductivity and heat capacity at low temperatures between 0.2 and 4.5 K were studied for several crystalline polymers of polytetrafluoroethylene (PTFE), polychlorotrifluoroethylene (Kel-F), nylon, and polyethylene (PE) [36]. Increase of crystallinity in semicrystalline polymers leads to the increase of thermal diffusivity, because of higher thermal diffusivity in the crystalline phase than in the amorphous phase [40,41].

The effects of anisotropy on thermal conductivity and thermal diffusivity in drawn polymer films have been studied [42–49]. Uniaxial drawing causes an increase of the thermal diffusivity and thermal conductivity along the drawn chain direction.

Multiphase polymers are promising materials which should improve on the mechanical and physical properties achieved by single polymers. Therefore many efforts have been made and a number of reports [50] have been published in those fields. However, thermal diffusivity and thermal conductivity of multiphase polymers have not been studied so extensively. For multiphase polymeric materials including composites with inorganic fillers, thermal diffusivity is typically on the order of 10^0 to 10^{-4} cm^2 s^{-1} and is significantly related to the molecular and the supramolecular structures in binary composites and polymer blends. Therefore, it is possible that the thermal diffusivity is used as a probe tool to explore the change in microstructure of the blends. In this chapter, thermal diffusivity and thermal conductivity as a probe tool in multiphase polymers are summarized from recent literature.

II. THERMAL DIFFUSIVITY AND HEAT CAPACITY MEASUREMENTS

When materials are subject to the steady-state heat flow q (cal cm^{-2} s^{-1}), temperature gradient dT/dx (K cm^{-1}) is produced in the materials along the heat flow direction, and both are combined by the equation

$$q = -K \frac{dT}{dx} \tag{1}$$

where K is a proportional constant with units of cal cm^{-1} s^{-1} K^{-1} and is referred to as the thermal conductivity of the materials [18].

From the kinetic theory in gases, thermal conductivity is expressed as follows [18]:

$$K = \tfrac{1}{3}\, Cvl = \tfrac{1}{3}\, C_{\mathrm{p}}\rho v l \tag{2}$$

where C (cal cm^{-3} K^{-1}) is specific heat per unit volume ($= C_{\mathrm{p}}\rho$, C_{p}: heat capacity (cal g^{-1} K^{-1}), ρ: density (g cm^{-3})), v is an average velocity (cm s^{-1}) of particles, such as phonons, electrons, magnons, excitons, etc., which can carry thermal energy, and l is the mean free path of particles (cm). In Eq. (2), the product of velocity and mean free path is known as thermal diffusivity D (cm^2 s^{-1}), namely,

$$D = vl \tag{3}$$

Thus thermal conductivity is related to thermal diffusivity as

$$K = \tfrac{1}{3}\, C_{\mathrm{p}}\rho D \tag{4}$$

Experimentally, thermal conductivity can be measured from the steady-state heat flow and thermal diffusivity can be determined by the transient heat transport measurement. Details of the measurement of thermal diffusivity by the transient method are shown below.

The laser flash method is a popular technique for measuring thermal diffusivity and relative heat capacity of polymeric thin films, since Parker et al. [51] first used a flash technique to determine thermal diffusivity of metallic materials. The laser flash method includes flash radiometry and laser radiometry (pulsed photothermal radiometry) which was developed by Tam and coworkers [52–54] and applied to the measurement of thermal diffusivity for polymeric materials by Choy et al. [55] and by Tsutsumi et al. [40,41,56–59]. Fig. 1 shows a typical schematic diagram of flash radiometry for measuring thermal diffusivity in polymeric materials [41,56]. Pulsed laser or xenon flash are used to produce a pulsed thermal sheet on a surface of polymeric material. The generated thermal sheet which transits through the bulk sample to reach a counter-surface, is detected by the increase in temperature of the counter-surface using a thin wire thermocouple [51,60] or infrared radiation from that surface [40,41,55–59]. The technique using infrared radiation is called laser or flash radiometry the response time of which is on the order of micro- or submicroseconds. A transient signal $S(t)$ of surface temperature is analyzed by using the one-dimensional heat diffusion model of Eq. (4) to independently determine thermal diffusivity D and the relative heat capacity HC of the polymer solid,

$$S(t) = \frac{K_{\mathrm{c}}Q}{C_{\mathrm{p}}L}\left[1 + 2\sum_{n=1}^{\infty}(-1)^n \exp\left(\frac{-n^2\pi^2 Dt}{L^2}\right)\right] \tag{5}$$

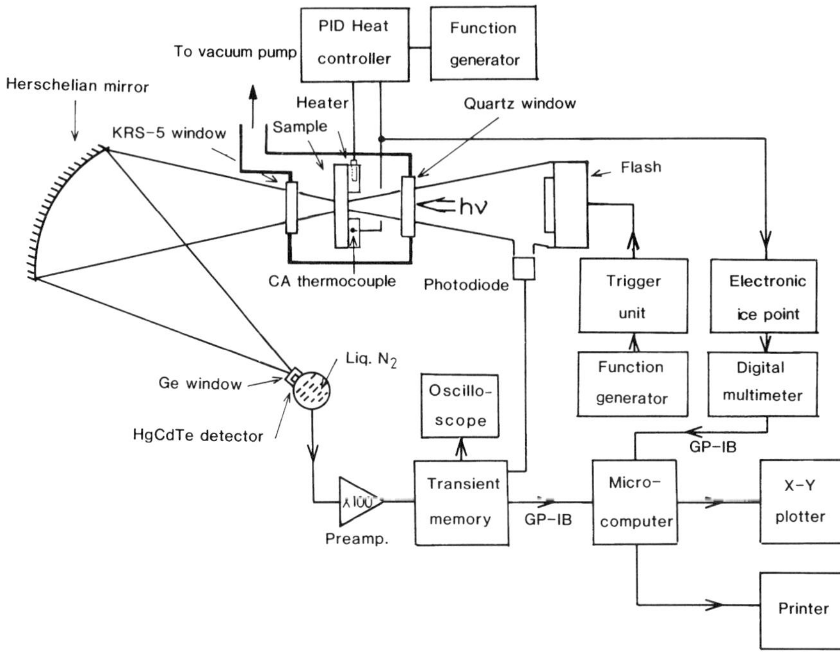

Figure 1 Schematic diagram of flash radiometry [41,56].

where $K_c(= 4G\varepsilon\sigma T_0^3)$ is a constant composed of a geometrical factor G, the emissivity ε, the Stefan–Boltzmann constant σ, and the temperature T_0 of the sample. Q is the total thermal energy absorbed by a carbon layer, C_p is the heat capacity of the sample, and L is the sample thickness. The signal fitting to Eq. (4) using a nonlinear least-squares method gives optimum values of D and $\Delta T (= K_c Q/(CL) = K' Q T_0^3/(CL)$, where $K' = 4G\varepsilon\sigma)$, and subsequently the relative capacity $HC(= C/(K'Q) = T_0^3/(\Delta TL))$ can be computed. The precise procedure for the measurement and the evaluation of D and ΔT are described in Refs. 54 and 56.

Another, simpler way for determining thermal diffusivity is the use of the characteristic time constant, $t_{1/2}$, which is defined as the time when the surface temperature increases by up to a half of its maximum value. D can be calculated using:

$$D = \frac{1.38L^2}{\pi^2 t_{1/2}} \tag{6}$$

Other techniques involving the photothermal deflection method can also be used for measuring thermal diffusivity for polymeric materials. The reader can consult the book chapter in Ref. 61 for details of these techniques, including pulse photothermal radiometry.

III. THERMAL DIFFUSIVITY IN POLYMER BLENDS

A. Immiscible Polymer Blends

1. PVDF/PC, PVDF/PS, and PVDF/Nylon11 Systems [62]

Thermal diffusivity and heat capacity for the immiscible polymer blends of poly(vinylidene fluoride) (PVDF) with polystyrene (PS), PVDF with poly (bisphenol A carbonate) (PC), and PVDF with nylon 11 are summarized in this section. PVDF (Kureha, KF-polymer, $M_w = 141{,}000$, $M_n = 64{,}000$), PS (Aldrich Chemical, $M_w = 280{,}000$), PC (General Electric, Lexan) and nylon 11 (Aldrich Chemical) were used. These blends are immiscible for the entire blend composition. D values of PVDF, PS, PC and nylon 11 are 0.96×10^{-3}, 1.23×10^{-3}, 1.48×10^{-3}, and $1.53 \times 10^{-3}\ \mathrm{cm^2\,s^{-1}}$, respectively. D almost linearly decreases with increasing PVDF composition in the blend. The plot of D versus weight fraction of PVDF can be expressed well by the equation:

$$D = D_A[\mathrm{PVDF}] + D_B(1 - [\mathrm{PVDF}]) \qquad (7)$$

where D_A is thermal diffusivity of PVDF and D_B is that of PS, PC, or Nylon 11 homopolymers, and [PVDF] is the weight fraction of PVDF in the blend.

The dependence of D on temperature is shown for (70/30), (80/20), and (90/10) PVDF/PS in Fig. 2(a). D decreases with increasing temperature. A noticeable depression in D occurs at 100°C and a distinct increase in D is observed at temperatures between 160 and 180°C. These specific points are shown by the arrows in the figure. The remarkable decrease in D at the temperature above 100°C is ascribed to the significant phonon scattering due to the large molecular motion above the glass transition of PS, and the change in D at the temperature around 175°C is attributed to the melting of the α crystallite phase of PVDF component [63]. Furthermore, the melting transition of β or γ phases of PVDF are measured as a small increase in D at the temperatures around 165°C [63,64].

The dependence of HC on temperature is shown for (40/60), (50/50) and (60/40) PVDF/PS in Fig. 2(b). The increase of HC with increasing temperature accompanies several characteristic changes upon heating, which are shown by the arrows in the figure. Stepwise increase of HC at

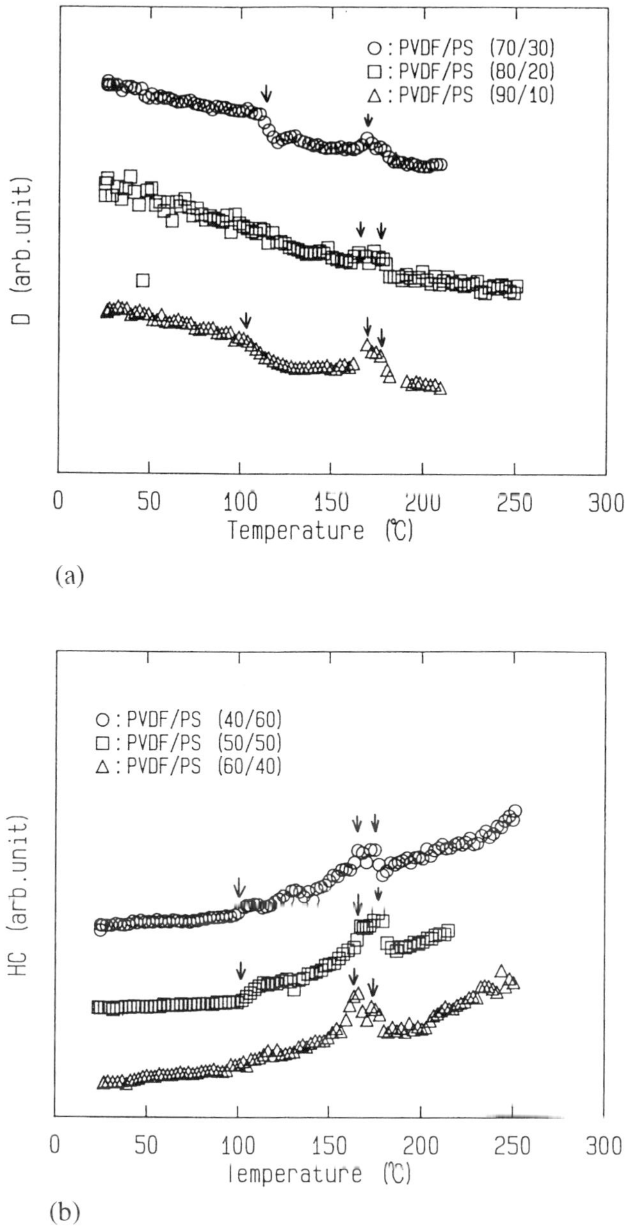

Figure 2 (a) Dependence of D on temperature for a PVDF/PS blend with various blend ratios. (b) Dependence of HC on temperature for a PVDF/PS blend with various blend ratios [62].

temperatures round 100°C is related to the glass transition of the PS component, and the increase of *HC* in the vicinity of 175°C is due to excess heat capacity on the melting of the α crystallite phase of the PVDF component. Another excess heat capacity can be observed at temperatures around 165°C, which correspond to the melting point of β or γ crystallite phases of the PVDF component. Wide-angle X-ray scattering measurement only gave the sharp diffraction patterns due to the α crystallite phase of PVDF. T_g and T_m values determined from the profiles of *D* and *HC* versus temperature and those measured by differential thermal analysis (DTA) are plotted against the weight fraction of PVDF in Fig. 3. These unchanged T_g and T_m values with the weight fraction of PVDF clearly showed the immiscibility of the two polymers in the blends. The profiles of *D* and *HC* versus temperature for the PVDF/PC blend also provide T_m due to the melting of the α crystallite phase of the PVDF component at 175°C and T_g of the PC component at 150°C which were not changed with the weight fraction of PVDF. For PVDF/Nylon 11, two T_m values due to the melting of the α crystallite phase of PVDF at 175°C and the melting of nylon 11

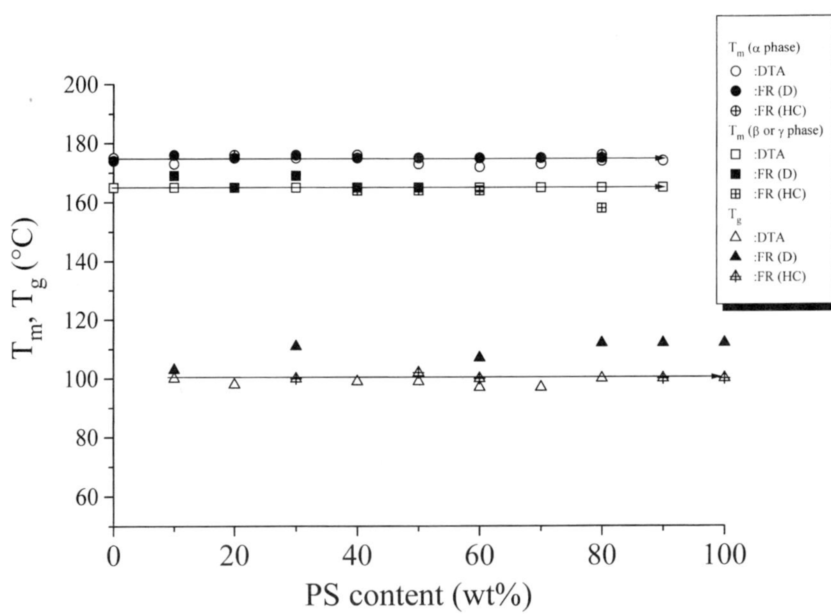

Figure 3 Plots of T_m and T_g by flash radiometry and thermal analysis when PVDF content is changed for a PVDF/PS blend system [62].

crystallite at 190°C [63] were clearly observed in the profiles of D and HC versus temperature. These T_m values were not changed with the weight fraction of PVDF.

2. PE/PS with SEBS Block Copolymer [65] and PE/SEBS and PS/SEBS Blends [66]

The effect of the compatibilizer of styrene–ethylene/butyrene–styrene (SEBS) block copolymer on thermal conductivity is studied in the polymer blends of low molecular weight PE and PS containing SEBS block copolymer [65]. PE (Mitsui Petrochemical, Highwax 100P, $M_w = 900$), PS (Sanyo Chemical Industries, Highmer ST95, $M_w = 4310$, $M_n = 1420$) and SEBS block copolymer (Shell Kagaku, Kraton G1652) with a PE/PS ratio of 71/29 were used. Phase diagrams of the blends were observed by a polarized optical microscope [65]. An immiscible blend of PE with PS has dual-phase continuity in the PE composition range from 15 to 30 wt%, as adding SEBS copolymer compatibilizer to the sample [65]. The dependence of thermal conductivity on PE weight fraction is shown in Fig. 4. At lower compatibilizer content, the plot of thermal conductivity is nearly proportional to the PE content, whereas the linearity does not hold with the PE weight fraction between 10 and 30 wt%. The compatibility between PE and PS increases with increasing the content of compatibilizer, and thus thermal conductivity deviates from the linear relationship of the rule of mixtures, i.e., the proportional relationship between thermal conductivity and the weight fraction of one polymer component in a binary polymer blend system. The details will be discussed later in the section on miscible polymer blends. The effect of the SEBS block copolymer content on the thermal conductivity of the blends was also investigated. The thermal conductivities of the blends, rich in PE (100 and 80 wt% of PE), showed a steep decrease with increasing SEBS block copolymer content, while that of other blends, with lower content of PE (50 and 20 wt%), remained approximately constant. The thermal conductivity of PS slightly increases with increasing SEBS block copolymer content. This result is explained by the thermal conductivity of SEBS, 4.57×10^{-4} cal cm^{-1} s^{-1} K^{-1}, is smaller than that of PE, 8.09×10^{-4} cal cm^{-1} s^{-1} K^{-1} and larger than that of PS, 3.83×10^{-4} cal cm^{-1} s^{-1} K^{-1} [65].

Thermal conductivities in PE/SEBS and PS/SEBS blends were performed with changing SEBS weight fraction [66]. Thermal conductivity in PE/SEBS blends monotonically decreases from 8.09×10^{-4} cal cm^{-1} s^{-1} K^{-1} for PE to 4.57×10^{-4} cal cm^{-1} s^{-1} K^{-1} for SEBS with increasing SEBS

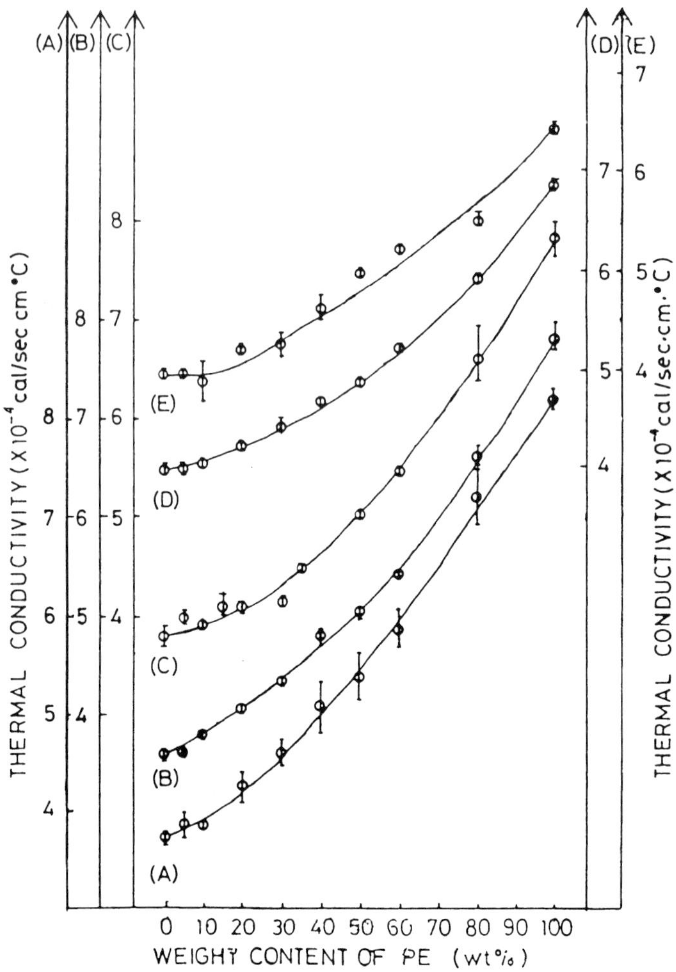

Figure 4 Thermal conductivity of a PE/PS blend containing various weight contents of SEBS block copolymer. (A) 2.0 wt%, (B) 4.8 wt%, (C) 9.1 wt%, (D) 16.7 wt%, and (E) 23.1 wt% [65].

weight fraction. Thermal conductivity in PS/SEBS blends monotonically increases from 3.83×10^{-4} cal cm^{-1} s^{-1} K^{-1} for PS to 4.57×10^{-4} cal cm^{-1} s^{-1} K^{-1} for SEBS with increasing SEBS weight fraction [66].

B. Miscible Polymer Blends

1. PVDF/PMMA System [58]

A polymer blend of PVDF (Kureha, KF-polymer, $M_w = 141{,}000$, $M_n = 64{,}000$) and PMMA (Mitsubishi Rayon, Acrypett-VHK, $M_w = 168{,}000$, $M_n = 96{,}400$) is miscible for the entire blend compositions in the molten state and in the amorphous region in the solid state. In the melt-quenched samples with PVDF concentration between 70 and 85 wt%, the β crystallite phase is preferentially formed, while samples with other PVDF compositions and slowly cooled samples form the α crystallite phase [67]. Fig. 5 shows the dependence of D on the PVDF content in samples prepared by melt-quenching and slow cooling. D was measured at room temperature. For a PVDF content below 70 wt%, D of the melt-quenched sample monotonically decreases with increasing PVDF content. On the other hand, for a PVDF content above 70 wt%, D increases with

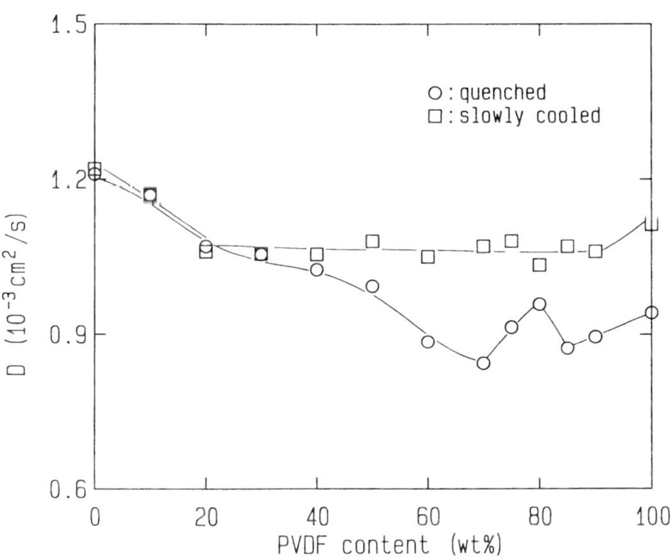

Figure 5 Plots of D versus PVDF content for a PVDF/PMMA blend provided by melt-quenching and slow cooling [58].

PVDF content due to the increase in crystallinity of the PVDF. Especially, for the blend with 70 and 85 wt% of PVDF in which the β crystallite phase is preferentially formed, D increases remarkably and subsequently decreases. This suggests that the β crystallite phase significantly affects the thermal contact between amorphous and crystallite regions.

The dependence of D on the temperature of the melt-quenched PVDF/PMMA can be classified into three groups depending on the temperature dependence of D: group A contains samples with PVDF content ranging between 100 and 70 wt%, group B samples with PVDF content of 60 and 50 wt% and group C samples with PVDF content from 40 to 0 wt%. The profile of D versus temperature of the 90/10 PVDF/PMMA blend in group A had an inflection point in the vicinity of 171°C, which corresponds to the melting of the α crystallite phase at 172°C measured by DTA. The 50/50 PVDF/PMMA in group B had three inflection points at 50, 93, and 166°C. The lowest inflection point at 50°C was due to the glass transition, corresponding to T_g at 57°C as measured by a thermomechanical analyzer (TMA). The highest inflection point at 166°C is due to the melting of the α crystallite phase, corresponding to T_m at 162°C measured by DTA. D increased in the temperature region above the middle inflection point of 93°C. The middle inflection point decreases from 93 to 76°C as the heating rate changes from 1 to 0.2°C min^{-1}, whereas no significant changes of lower and higher inflection points occur. The increase in D at the temperature above the middle inflection point is due to the onset of the crystallization.

The dependence of HC on temperature for 80/20 PVDF/PMMA prepared by melt-quenching, slow cooling, and melt-quenching and subsequent annealing at 120°C for 24 h was also investigated. Each profile of HC versus temperature has two distinct peaks and/or shoulders at temperatures of 162 and 175°C. The lower peak or shoulder at a temperature around 162°C is attributed to the melting of the β crystallite phase [68], and the higher peak or shoulder at a temperature around 175°C is ascribed to the melting of the α crystallite phase. It is noted that the profile of HC temperature gave information on the melting behavior due to the β and α crystallite phases.

2. PVC/PCL System [69]

The blend of atactic polyvinyl chloride (PVC) with polycaprolactone (PCL) is miscible in the high-content region of PVC. PVC (Shin-Etsu Chemical) and PCL (Aldrich Chemical) were used. In a lower PVC concentration region, crystallization of PCL occurs, leading to volume-filling spherulitic structures containing PVC within the PCL spherulites. Fig. 6 shows the dependence of D on the weight fraction of PCL at room temperature. As the PCL content increases up to 25–30 wt%, D decreases and then sigmoid-

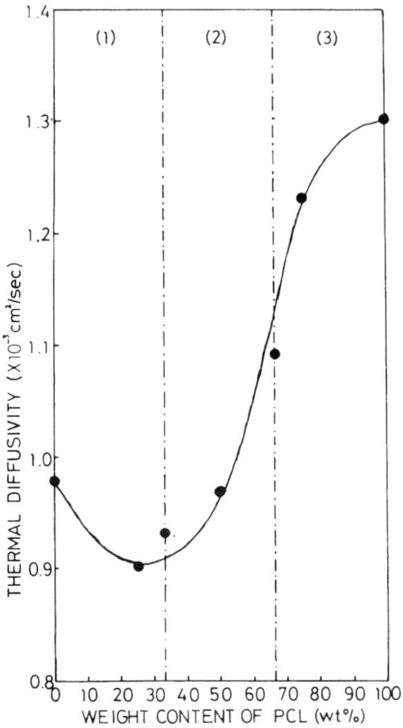

Figure 6 Dependence of D on PCL content in a PVC/PCL blend [69].

ally changes with the PCL content. The plot of D versus PCL content can be classified into three regions. D decreases with increasing PCL content in the amorphous miscible region of PVC/PCL (in region 1). In regions 2 and 3, D increases with increasing PCL content above 30 wt% which can be attributed to the increase in crystallinity of PCL component. It is noted that D decreases with increasing PCL content in region 1, where both components are miscible in the amorphous state. The decrease in D in region 1 is significantly related to the miscibility of the blend. The details will be discussed in the next section.

3. PS/PVME System [70]

The blend of PS with poly(vinyl methyl ether) (PVME) is known to be miscible, for which the phase diagram is a lower critical solution temperature (LCST). Fractionated PS with $M_w = 160{,}000$ and $M_n = 100{,}000$ and PVME (Aldrich Chemical, $M_w = 115{,}000$, $M_n = 54{,}800$) were used. Fig. 7 shows the dependence of D on the weight fraction of PVME measured at

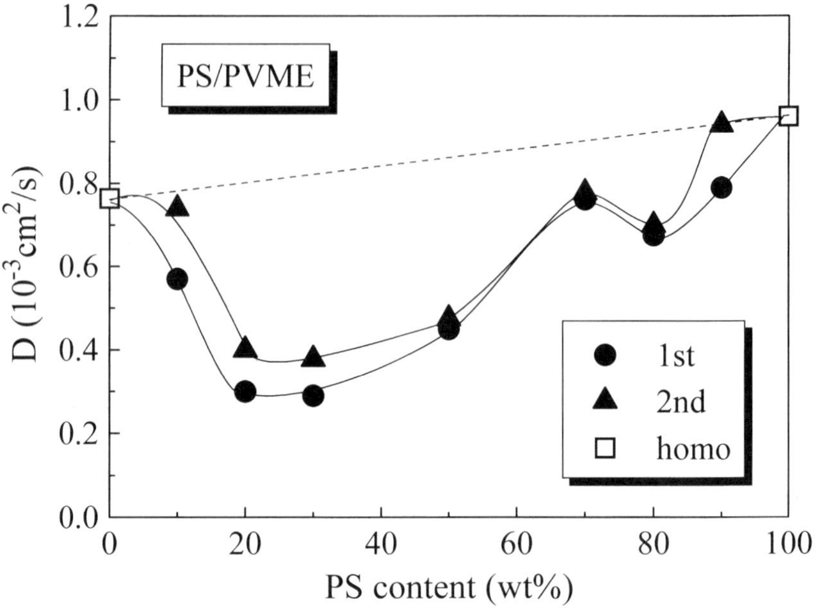

Figure 7 Dependence of D on PS content in a PS/PVME blend. 1st and 2nd mean the miscible and phase-separated states, respectively [70].

100°C. Filled circles (first run) show the relationship between D and PS content observed for the miscible blends and filled triangles (second run) show the results obtained for the blends after heating up to a temperature higher than the cloud point. The open squares are D values observed for both homopolymers. The dashed line is a D value predicted from the rule of mixtures:

$$D = D_A[PS] + D_B(1 - [PS]) \tag{8}$$

where D_A and D_B are the thermal diffusivities of PS and PVME homopolymers, respectively, and [PS] is the weight fraction of PS. It is noted that the plot of D versus [PS] largely deviates from the prediction of Eq. (8) and the D value decreases with increasing PS content and becomes minimum at a PS content of around 20 wt% from which D increases again with increasing PS content. The plot of D versus [PS] resembles the plot of cloud point versus [PS]. The PS concentration at which the large depression of D occurs coincides with critical composition in the LCST phase diagram.

As the temperature is close to the LCST, the concentration fluctuation becomes larger to form the interface between PS and PVME phases. If one assumes that the phonon is significantly scattered by the concentration fluctuation, the remarkable decrease in D would be observed when D is measured at a temperature closer to the LCST. If this assumption is adopted, the similarity between the composition dependence of D and the phase diagram would be explained. D on the second run, i.e., after phase separation by heating up to the temperature above the cloud point, is larger than those on the first run in the miscible state. This increase in D is ascribed to the decrease of the total surface area microscopically contacted to the counter-component after the phase is separated. For the immiscible blend in which a sharp and distinct interface is formed between the two phases, the total surface area microscopically contacted to the counter-component is minimized since the two components are strongly segregated. Thermal conduction preferentially occurs through the segregated continuous phases. In the case of an immiscible blend of PS and PVDF as shown above, the dependence of D on weight fraction of PS as described above is well expressed by the rule of mixtures, i.e., the sum of the thermal diffusivity of each component weighted by fraction, in Eq. (8) where D_A and D_B are the thermal diffusivities of PS and PVDF homopolymers. The phonon scattering in the interface between two components seems to contribute less to the thermal conduction.

D values of the first run for (90/10) and (10/90) PS/PVME are definitely smaller than those for PS and PVME homopolymers, respectively, whereas D values of the second run for them are on the dashed line given by Eq. (8) in Fig. 7. These results imply that thermal conduction in phase-separated PS/PVME (90/10) and (10/90) preferentially occurs through the segregated continuous PS and PVME phases. This discussion is significantly related to the similarity of the profiles of D versus temperature between the phase-separated PS/PVME (90/10) and PS homopolymer as described in the following.

The dependence of D on temperature is studied for PS/PVME (90/10), (70/30), and (50/50). D decreases with increasing temperature because the phonon scattering becomes larger at higher temperatures [18–20]. Two types of dependence of D on temperature (first and second runs) are presented. D was measured when temperature was increased up to 150°C at a heating rate of 1°C min^{-1}, which is referred to as the first run measurement, and the sample was cooled down to room temperature in a chamber cell at a cooling rate of about 1°C min^{-1}. Then the second measurement of D was carried out with increasing temperature, which is referred to as the second run measurement. The first and second profiles of D versus temperature have quite different features. The profile of D versus temperature of the third run measurements is the same as that for the second run. In the case of (90/10) PS/PVME, the first run profile has the inflection point at 44°C, corresponding to T_g of 48°C by a differential scanning calorimetry (DSC). The second profile with the inflection point at 71°C is similar to that for a PS homopolymer. The similarity of the dependence of D on temperature implies that the two-phase structure is formed in a phase-separated matrix after the first heating and the phonon is solely transported through the segregated continuous PS phase. The first profile of D versus temperature for (50/50) PS/PVME has a single inflection point in the vicinity of −1°C and the second profile gives two distinct points at temperatures around −16 and 19°C, corresponding to the respective T_g values for PVME and PS phases.

The inflection point (T_i) obtained by the first-run D measuement corresponded well to T_g obtained by DSC. Segmental motion at the transition region from glassy to rubbery states sensitively reflects the changes in D.

Fig. 8 shows the isothermal change in D for (10/90) PS/PVME at various temperatures. Isothermal measurement of D for (10/90) PS/PVME at 110°C just below the cloud point exhibits the distinct increase in D with holding time. For (20/80) PS/PVME, the increase in D was measured after 50 min isothermal annealing at 120°C just above the cloud point. Isothermal

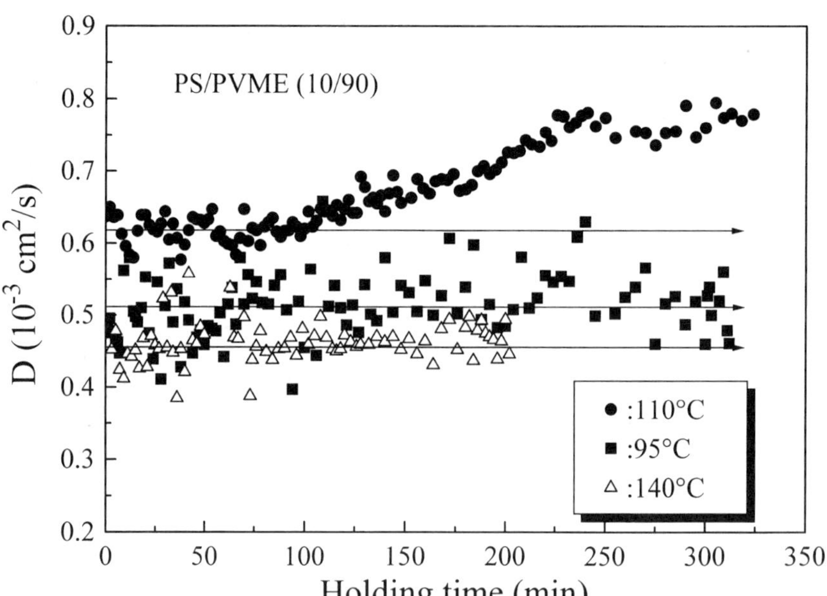

Figure 8 Plot of D versus holding time for the (10/90) PS/PVME isothermally annealed [70].

measurement of D for (30/70) PS/PVME at 115°C close to the cloud point also showed an increase in D. The samples annealed at other temperatures, 95, 106, and 140°C, did not show the specific change except for the small variance of D due to the experimental error. The increase in D under annealing is significantly related to the initial stage of the phase separation in the blends. Namely, D is sensitive to the decrease of total surface area microscopically contacted to the counter-component in the initial stage which follows the phase-separation among them.

The increase in D for (10/90) PS/PVME at 110°C can be compared with the isothermal change of the sample surface under observation by a differential interference contrast microscope. The creation of some structure can be seen under the microscope observation of (10/90) PS/PVME at 110°C at 225 min and it became distinct above 250 min. Annealing temperature, 110°C, is close to the measured LCST. For the composition of blends far from the critical composition and at a temperature in the vicinity of or above the LCST and lower than the spinodal decomposition temperature, the nucleation and growth mechanism is preferred. In this mechanism, the induced time is necessary for the nucleation and growth of the interface between two phases. Under the microscope observation, the nucleation and growth of the interface can be seen at 225 min and it became distinct above 250 min. In contrast, as shown in Fig. 8, D at 110°C increases from 100 min and levels out above 250 min. This increase in D implies that D can sensitively reflect the change in microscopic structure which follows the nucleation and growth of the interface. D measurement above 250 min, after which clear nucleation and growth can be seen in the microscope observation, does not give a significant change.

4. PMMA/PC [71]

Miscible PMMA/PC blends with LCST can be prepared by a rapid casting method using PMMA (Mitsubishi Rayon, $M_w = 44{,}000$, $M_n = 19{,}000$) and PC (Mitsubishi Gas Chemical, $M_w = 89{,}000$, $M_n = 50{,}000$). The dependence of single T_g on PC content can be observed. The single T_g profile shows that the blend is miscible for all compositions. Fig. 9 shows the dependence of D on temperature for the 50/50 PMMA/PC blend at a heating rate of 1°C min^{-1}. After gradual decreasing to about 440 K, D decreased steeply with increasing temperature in the LCST range (450–460 K), then it gradually decreased. The remarkable decrease of D in the LCST range (450–460 K) might be due to the decrease of the velocity of the phonons or phonon mean free path. This phenomenon might mean that the LCST can be determined by the measurement of thermal diffusivity.

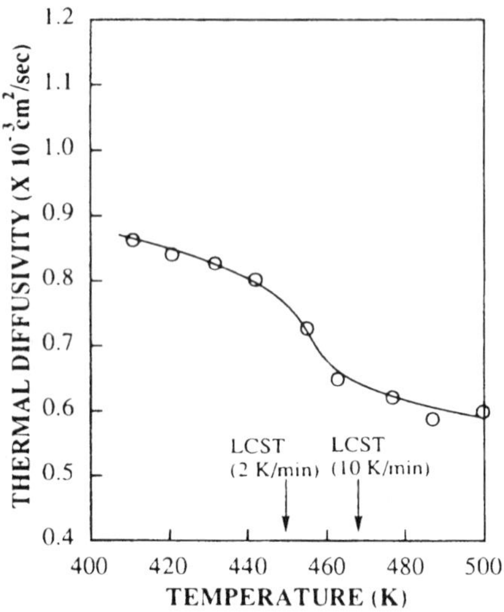

Figure 9 Dependence of D on temperature for a (50/50) PMMA/PC blend [71].

The dependence of D on PC composition is measured at $T = 291$ K and $T = 445$ K. D values of PMMA and PC homopolymers are 1.12×10^{-3} and $1.6 \times 10^{-3}\,\text{cm}^2\,\text{s}^{-1}$ at 291 K, respectively and 0.62×10^{-3} and $1.6 \times 10^{-3}\,\text{cm}^2\,\text{s}^{-1}$ at 445 K, respectively. The plot of D versus PC content at $T = 291$ K is below the plot predicted from the rule of mixtures, i.e., the proportional increase in D with PC content, whereas those at $T = 445$ K are on the plot of the rule of mixtures.

IV. THERMAL DIFFUSIVITY IN POLYMER COMPOSITES

A. Filler/Polyimide Composites [57]

Fig. 10 shows the dependence of D on the concentration of filler with high thermal conductivity, boron nitride (BN), aluminum nitride (AlN), silicon carbide (SiC), and diamond. D increases exponentially upon increasing the content of filler in the following order: D (BN composite) $\simeq D$ (AlN composite) $> D$ (SiC composite) $\simeq D$ (diamond composite). For the case of the BN composite, BN could be added up to 54 vol%, while the other filler composites could not be filled by more than circa 40 vol%. BN has a relatively small particle size with an average particle diameter of 0.33 μm.

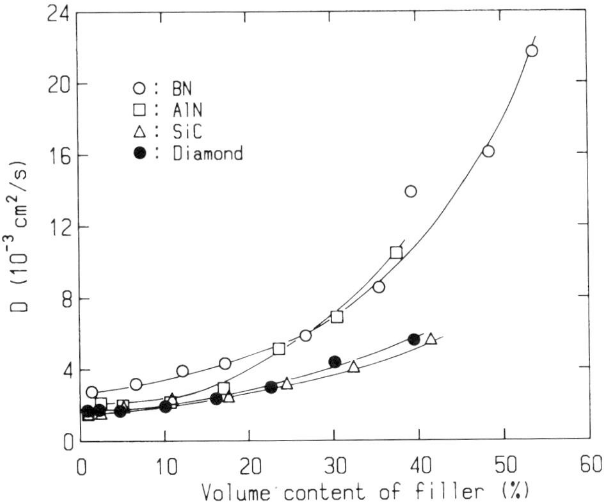

Figure 10 Dependence of D on volume content of high-conductivity filler in polyimide composites [57].

The distributions of AlN and SiC particles are almost the same and have larger particle sizes than with average particle diameters of 2.0 and 1.4 μm, respectively.

Scanning electron microscopy (SEM) pictures of the surface of BN, AlN, SiC, and diamond composites are shown in Fig. 11. Clear borders between fillers and the polyimide matrix can be seen for the composites with AlN, SiC, and diamond fillers, while BN composites do not have clear boundaries between BN filler and matrix. Thus, the large increase in D for the BN composite may be attributed to the good thermal contact between boundaries and higher thermal conduction through the aggregated BN particles.

Diamond itself has the highest D value among these fillers and thus one expects that the diamond composites would exhibit superior conductivity or thermal diffusivity among the composite systems investigated. However, the D value observed for the diamond composite is lower than those of BN and AlN composites. This could be ascribed to the significant phonon scattering at the interface between the diamond filler and polyimide matrix. The clear border at the interface as shown in Fig. 11 is significantly related to the poor thermal contact at the interface.

No significant change of sound velocity, a measure of the velocity of phonons, which was determined by a separate modulus method, was found

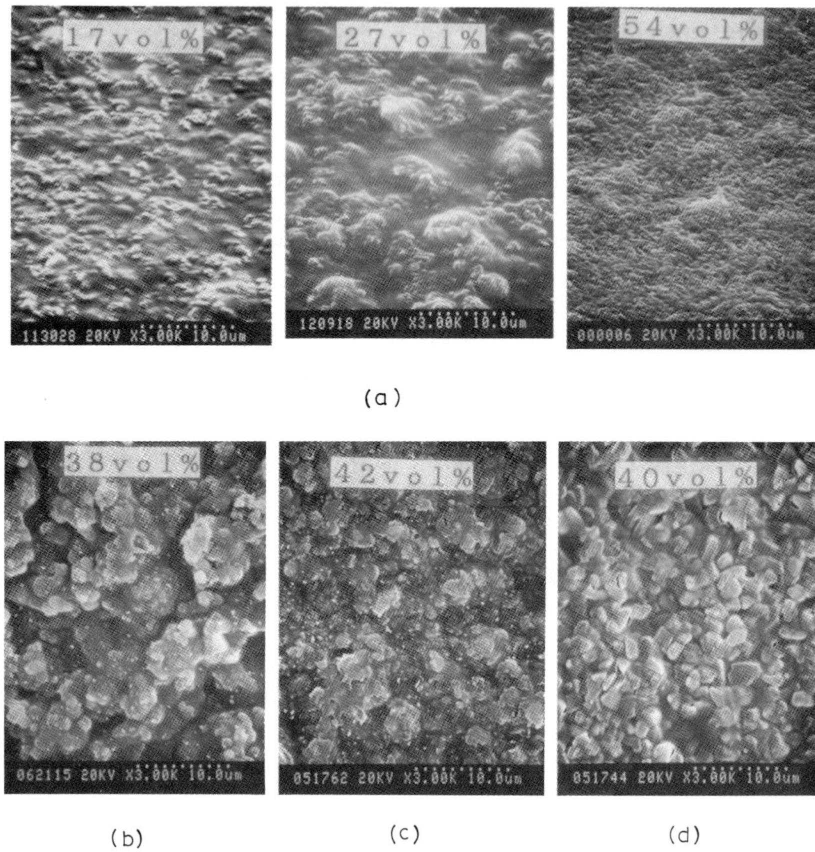

Figure 11 SEM photographs of composite surfaces: (a) BN polyimide composite; (b) AlN polyimide composite; (c) SiC polyimide composite; and (d) diamond polyimide composite [57].

irrespective of the filler content and types. Thus, the main contribution to D seems not to be the velocity of the phonons but their mean free path length.

B. PC/Liquid Crystalline Polymer [72]

When a thermotropic liquid crystalline polymer (LCP) of Vectra B950 (Hoechst Celanese Co.), copolyesteramide, is injection molded or extruded from the melt, fibrils are readily formed. Copolyesteramide consists of 60 mol% 2,6-hydroxynaphthoic acid, 20 mol% terephthalic acid and 20 mol% para-aminophenol [73]. Using this property, an *in situ* composite of a thermoplastic polymer of PC (General Electric, Lexan 151) with thermotropic LCP of Vectra B950 has been exploited. The thermal conductivity of extruded blends of LCP of Vectra B950 and PC with various volume fractions (V_f) of LCP between 0.09 and 0.80 was measured as a function of draw ratio (λ) ranging from 1.3 to 15. Fig. 12(a) shows the dependence of axis thermal conductivity (thermal conductivity along drawing direction) $K_\parallel$ on the draw ratio for a PC/LCP composite with various concentrations of LCP. $K_\parallel$ for the composites increases with increasing λ and levels off at $\lambda > 7$. This increase arises from the increase in aspect ratio of the LCP domains and the improved chain orientation within the domains. From scanning electron microscope (SEM) morphological studies of the composites, LCP domains are dispersed in a PC matrix and the aspect ratio of the domains increases with increasing λ at $V_f < 0.3$, whereas at $V_f > 0.55$ phase inversion has occurred and the LCP becomes the continuous phase. At $V_f < 0.3$, $K_\parallel$ may be understood in terms of a Halpin–Tsai equation [74] for the discontinuous fiber-reinforced composites:

$$K_\parallel = K^m \frac{1 + \zeta\eta V_f}{1 - \eta V_f} \tag{9}$$

$$\eta = \frac{K_\parallel^f / K^m - 1}{K_\parallel^f / K^m + \zeta} \tag{10}$$

$$\zeta = 2\frac{L}{d} \tag{11}$$

where K^m and $K_\parallel^f$ are the thermal conductivity of the matrix and the fiber along the fiber direction, respectively, and V_f, L, and d are the volume fraction, the length and the diameter of the fibers, respectively. The aspect ratio L/d can be estimated if the affine deformation scheme [75] is adopted. The scheme, commonly used for the flexible chain polymers [75], has been recently applied to describe the deformation of LCP domains in *in situ* composites [76]. In this scheme, the deformation of the LCP phase is identical to that of the composite rods and the volume remains unchanged. It then follows that

$$\zeta = 2\frac{L}{d} = 2\lambda^{3/2} \tag{12}$$

$K_\parallel$ values theoretically obtained from Eq. (9) with Eqs. (10) and (12) at $V_f < 0.3$ are plotted as a function of draw ratio in Fig. 12(b) (solid curve in the figures). Theoretical predictions are in agreement with the

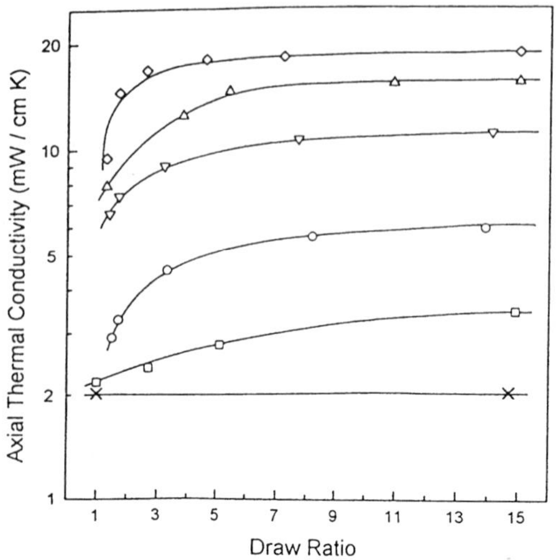

(a)

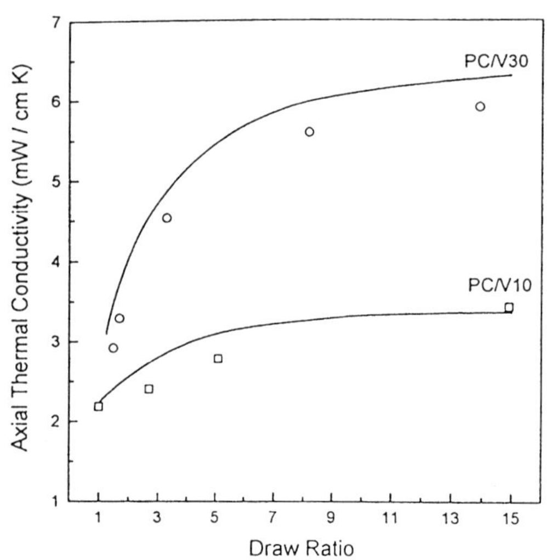

(b)

Figure 12 (a) Draw ratio dependence of the axial thermal conductivity of PC/Vectra B950 blend. ×—PC; □—(90/10) PC/Vectra; ○—(70/30) PC/Vectra; ▽—(40/60) PC/Vectra; △—(20/80) PC/Vectra; and ◇—Vectra B950. (b) Theoretical plots (solid curves) of the draw ratio dependence of axial thermal conductivity for a (90/10) and (70/30) PC/Vectra B950 blend. Open circle and square plots are experimentally obtained results [72].

experimental data, confirming that the increase in $K_{\parallel}$ results from the higher aspect ratio of the LCP fibrils and the enhanced molecular orientation within the fibrils.

SEM morphological studies also show that the LCP fibrils for the sample with a high draw ratio of $\lambda = 15$ are essentially continuous even if the volume fraction of LCP is lower than 0.3. In this case, $K_{\parallel}$ can be proportional to the volume fraction of LCP (volume fraction dependence of $K_{\parallel}$ in Fig. 12(a) at $\lambda = 15$), namely $K_{\parallel}$ can be expressed in terms of the rule of mixtures:

$$K_{\parallel} = V_f K_{\parallel}^f + (1 - V_f)K^m \tag{13}$$

Transverse thermal conductivity $K_{\perp}$ was slightly affected by drawing, whereas $K_{\parallel}$ strongly depends on the draw ratio. Therefore, the PC/LCP blends exhibit strong anisotropy on thermal conduction at high draw ratio.

C. iPP/EPR and iPP/Chalk Systems [77]

Thermal contact between components in multicomponent polymer systems is studied from thermal conductivity measurements. Two systems were chosen to study the change in thermal contacts between the components: a blend of isotactic polypropylene (iPP) and ethylene–propylene copolymer (EPR), in which the interface can be modified by partial compatibilization, and a composite of iPP filled with a chalk in which the interface can be modified by addition of liquid.

As shown in Fig. 13, at lower temperature, thermal conductivity of the blend of iPP with EPR (open triangles) is larger than that of pure iPP (filled circles), irrespective of the fact that the thermal conductivity of EPR (open circles) is lower than that of iPP. A slight increase in the crystallinity of iPP from 36.2 to 41.4 wt% was observed by adding EPR from DSC studies. The morphological influence of EPR on iPP has been the subject of several papers [78–81]. The increase of thermal conductivity of iPP by addition of EPR as compared to pure iPP may be explained in terms of the increase of crystallinity or the changes in propylene matrix morphology. Unfortunately, this morphological change in iPP/EPR has not been directly confirmed.

Thermal conductivity of the blend of iPP and ethylene–propylene rubber (open squares) is larger than that of iPP/EPR. The clear borders of EPR inclusions can be seen in SEM pictures of the fracture surface of iPP/EPR. On the other hand, in SEM pictures of the blend of iPP/ethylene–propylene rubber, the boundaries of inclusions are not significantly recognized. Thus, the higher thermal conductivity of iPP/ethylene–

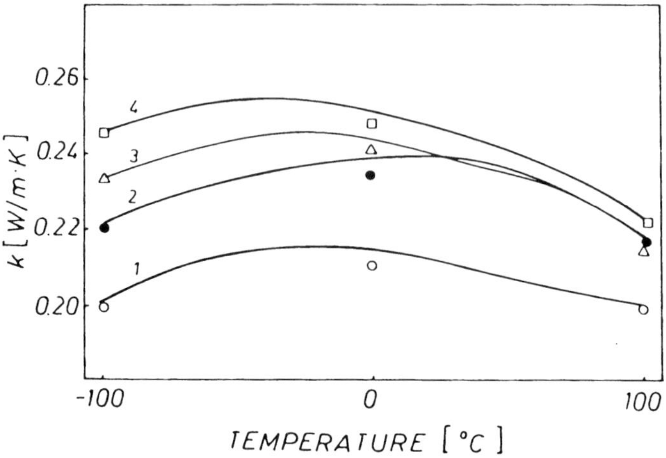

Figure 13 Temperature dependence of thermal conductivity of EPR (1), iPP (2), iPP/EPR blend (3), and iPP/ethylene–propylene rubber blend (4) [77].

propylene rubber is attributed to the improvement of thermal contact between ethylene–propylene rubber inclusions and the matrix.

Thermal conductivity of a composite of iPP and chalk with an ethylene oxide oligomer (EOO) working as a compatibilizer was larger than that of iPP/chalk without EOO. The presence of EOO on the interface improves the thermal contact between the particles and the matrix.

Since the thermal contacts between the components in heterogeneous systems are always improved by the improvement of adhesion between them, the measurement of the thermal conductivity would provide a study of the adhesion properties.

D. Polymer Carbon Fiber Composites [82]

The anisotropies of thermal diffusivity and thermal conductivity of polymer carbon fiber composites are studied using the photothermal deflection method. Polymer matrices are polyimide (PI) (Mitsui-Toatsu), poly(ether-ether-ketone) (PEEK) (ICI) and poly(ether-ketone-ketone) (PEKK) (Dupont). Three types of carbon fibers are used: T800 from Toray, AS4 continuous fibers from Hercules and AS4-LDF long discontinuous fibers from Hercules. Composites are PI–T800, PEEK–AS4, and PEKK–AS4–LDF with 62 vol% of carbon fibers. Carbon fibers are distributed in a compact manner within the material parallel to the sample surface. Thermal diffusivity values in the three directions parallel, perpendicular to

Table 1 Thermal Diffusivity Values for the Samples in Three (Parallel, Perpendicular, and Along z) Directions

Sample	Thermal diffusivity ($cm^2 s^{-1}$)		
	$D_{\parallel}$	$D_{\perp}$	D_z
PEEK-AS4	0.25	0.71	0.0027
PEKK-AS4LDF	0.36	0.82	0.0047
PI-T800	0.29	0.92	—

carbon fiber axis in the sample surface, and along the z-axis (the direction of sample thickness) are summarized in Table 1 for three samples. The results are in contrast to what one would in principle expect: thermal diffusivity perpendicular to the carbon fiber axis in the surface plane, $D_{\perp}$, is larger than that parallel to the carbon fiber axis, $D_{\parallel}$. Thermal diffusivity along the z-axis direction D_z is lower than expected, and is close to that for a polyimide matrix. Thermal conductivity can be estimated from the width of the maximum amplitude of photothermal signals. For the same frequency, the width is larger for the measurements parallel to the fiber axis than for those normal to it. This means that the thermal flux is larger in the parallel direction to the fiber axis, i.e., $K_{\perp} < K_{\parallel}$. Therefore, one has the unexpected results of $K_{\perp} < K_{\parallel}$ and $D_{\perp} > D_{\parallel}$. To explain these results, detailed studies of the effects of interfaces on the heat propagation in a composite are necessary.

For other polymer composites, readers can consult a recent review of the thermal conductivity of filled polymer composites [83].

V. CONCLUDING REMARKS AND FUTURE PERSPECTIVES

In this chapter, the effects of the microstructure in the polymer blends and composites on thermal diffusivity and thermal conductivity have been summarised from recent literature. Thermal diffusivity and thermal conductivity of materials are not only fundamental thermal properties, but are also important in studying changes in the microstructures and morphology of the matrix.

Thermal diffusivity and thermal conductivity are significantly affected by molecular motion in the glass transition and the melting of crystallite in polymer blends. The profiles of D versus blend composition for the immiscible polymer blends are remarkably different from those for the miscible polymer blends. In the immiscible polymer blend systems, the rule of mixtures, i.e., the proportional dependence of thermal diffusivity on the blend composition, was measured for the entire blend composition, which can be compared with that for the copper–niobium metal–metal alloy system [5]. For the immiscible blend, the total surface area microscopically contacted to the counter-component between two components is minimized since the two components are strongly segregated. Thermal conduction preferentially occurs through the segregated continuous phases. The phonon scattering at the interface between two components seems to contribute less to the thermal conduction. Thus, the thermal conduction in the immiscible blend is expressed by the rule of mixtures, the sum of the thermal diffusivity of each component being weighted by a fraction. In contrast , in the miscible blend systems, D is significantly depressed by blending the counter-component and the plots of D versus blend composition deviate from the rule of mixtures to lower values. In the case of a miscible blend of PS/PVME, the profile of the D versus weight fraction of PS is similar to the phase diagram. The weight fraction of PS at which a significant depression of D occurred coincides with the critical composition in the LCST phase diagram. The remarkable concentration fluctuation before the formation of the interface between two components may be responsible for the significant phonon scattering and subsequent depression in D.

The theoretical model approaches to thermal conductivity for heterogeneous polymers are summarized and applied to several filled polymer composites and polymer blends in Ref. 84. However, no model can explain the valley-type dependence of thermal diffusivity on blend composition.

Multiphase polymers are promising materials for improving the mechanical and physical properties which cannot be achieved by single polymers. Thermal diffusivity and thermal conductivity of multiphase polymers have not been studied so extensively. As described in this chapter, thermal diffusivity and thermal conductivity are sensitively affected by the change in the microstructure of materials, and these measurements are simple techniques which do not require special chemical or physical modifications of sample specimens.

Thermal diffusivity and thermal conductivity are sensitively affected by the boundary and interface conditions between two components in the binary systems, thus their measurements provide important information about the boundary or interface between binary phases. Unfortunately, at the present stage, there has been no report on measurements of thermal diffusivity and thermal conductivity directly related to the thickness of the interface in miscible and immiscible binary polymer blends. There has been no investigation of the relationship between thermal diffusivity and thermal conductivity and the spinodal decomposition. Phonon-scattering at the interface due to the change of the propagation mode of the phonon would significantly affect the thermal diffusivity and thermal conductivity. The effect of the spinodal decomposition and its morphology on thermal diffusivity and thermal conductivity will also be the subject of further investigation once appropriate conditions can be achieved.

REFERENCES

1. Q Zu, K Ichikawa. J Phys C: Solid State Phys 18:L985, 1985.
2. CL Choy, WP Leung, YK Ng. J Appl Phys 66:5335, 1989.
3. S Luby, E Majkova, E Illekova, R Sandrik, ED Anna, A Luches, A Perrone, S Enzo. Thin Solid Films 261:154, 1995.
4. L Kubičár, J Spišiak, P Duhaj. Mater Sci Eng A133:523, 1991.
5. JE Ostenson, DK Finnemore, JD Verhoeven, ED Gibson, HR Shanks. J Appl Phys 55:278, 1984.
6. D Schechtman, I Blech, D Gratias, JW Cahn. Phys. Rev Lett 53:1951, 1984.
7. JM Dubois, SS Kang, P Archambault, B Colleret. J Mater Res 8:38, 1993.
8. B Dubost, JM Lang, M Tanaka, P Sainfort, M Audier. Nature 324:48, 1986.
9. AP Tsai, A Inoue, T Matsumoto. Japan J Appl Phys 26:L1505, 1987.
10. AP Tsai, A Inoue, T Matsumoto. Phil Mag Lett 61:15, 1991.
11. Y Takahashi, M Yamawaki, K Yamamoto. J Nucl Mater 154:141, 1988.
12. JB Henderson. Therm Conduct 22:117, 1994.
13. MI Aliev, DG Arasly, RE Guseinov. Phys Status Solidi A 20:K109, 1973.
14. JB Henderson, RE Taylor, H Groot. High Temp—High Pressures 25:323, 1993.
15. RE Taylor, H Groot, JB Henderson. High Temp—High Pressures 25:569, 1993.
16. Z Li, X Jian, W Jiuru. Int J Refract Met Hard Mater 9:167, 1990.
17. DW Van Krevelen, PJ Hoftyzer, ed. Properties of Polymers: Correlations with Chemical Structure. New York: Elsevier, 1972, Chapter 14.
18. C Kittle, ed. Introduction to Solid State Physics. 7th ed., New York: Wiley, 1996, pp 131–140.
19. JM Ziman, ed. Electrons and Phonons. Oxford: Oxford University Press, 1960.
20. R Berman. Contemp Phys 14:101, 1973.
21. R Rao. Indian J Phys 14:109, 1940; J Chem Phys 9: 682, 1941.
22. D Hansen, CC Ho. J Polym Sci A 3:659, 1965.
23. GI Morgan, PD Scovell. J Polym Sci Polym Lett 15:193, 1977.
24. K Ueberreiter, J Naghizadeh. Kolloid-Z Z Polym 250:927, 1972.
25. J Naghizadeh, K Ueberreiter. Kolloid-Z Z Polym 250:932, 1972.
26. P Andersson, G Bäckström. J Appl Phys 44:705, 1973.
27. P Andersson, G Bäckström. J Appl Phys 44:2601, 1973.
28. P Andersson, B Sundqvist. J Polym Sci Polym Phys Ed 13:243, 1975.
29. RS Frost, RYS Chen, RE Baker Jr. J Appl Phys 46:4506, 1976.
30. P Andersson. J Appl Phys 47:2424, 1976.
31. RE Baker Jr, RYS Chen, RS Frost. J Polym Sci Polym Phys Ed 15:1199, 1977.

32. RS Frost, RE Baker Jr, RYS Chen. J Polym Sci Polym Phys Ed. 16:689, 1978.
33. CL Choy, K Young. Polymer 18:769, 1977.
34. CL Choy. Polymer 18:984, 1977.
35. CL Choy, EL Ong, FC Chen. J Appl Polym Sci 26:2325, 1981.
36. W Reese, JE Tucker. J Chem Phys 43:105, 1965.
37. M Giles, C Terry. Phys Rev Lett 22:882, 1969.
38. AC Anderson, W Reese, JC Wheatly. Rev Sci Instrum 34:1386, 1963.
39. CL Choy, GL Salinger, YC Chiang. J Appl Phys 41:597, 1970.
40. N Tsutsumi, T Takizawa, T Kiyotsukuri. Polym Commun 29:28, 1988.
41. N Tsutsumi, T Takizawa, T Kiyotsukuri. Polymer 31:1925, 1990.
42. STJ Peng, RF Landel. J Appl Polym Sci 19:49, 1975.
43. AG Gibson, D Greig, M Sahota, IM Ward, CL Choy. J Polym Sci Polym Lett Ed 15:183, 1977.
44. CL Choy, WH Luk, FC Chen. Polymer 19:155, 1978.
45. D Greig, M Sahota. Polymer 19:503, 1978.
46. HG Kilian, M Pietralla. Polymer 19:664, 1978.
47. CL Choy, FC Chen, WH Luk. J Polym Sci Polym Phys Ed 18:1187, 1980.
48. AG Gibson, D Greig, IM Ward. J Polym Sci Polym Phys Ed 18:1481, 1980.
49. AJ Crispin, AK Taraiya, D Greig, I M Ward. Polym Commun 28:95, 1987.
50. See other chapters in this book.
51. WJ Parker, RJ Jenkins, CP Butler, GL Abbott. J Appl Phys 32:1679, 1961.
52. AC Tam, B Sullivan. Appl Phys Lett 43:333, 1983.
53. WP Leung, AC Tam. Opt Lett 9:93, 1984.
54. WP Leung, AC Tam. J Appl Phys 56:153, 1984.
55. CL Choy, WP Leung, YK Ng. J Polym Sci Polym Phys Ed 25:1779, 1987.
56. N Tsutsumi, T Kiyotsukuri. Appl Phys Lett 52:442, 1988.
57. N Tsutsumi, N Takeuchi, T Kiyotsukuri. J Polym Sci Part B Polym Phys 29:1085, 1991.
58. N Tsutsumi, M Terao, T Kiyotsukuri. Polymer 34:90, 1993.
59. N Tsutsumi, S Ishida, T Kiyotsukuri. J Polym Sci Part B Polym Phys 32:1899, 1994.
60. FC Chen, YM Poon, CL Choy. Polymer 18:129, 1977.
61. JA Sell, ed. Photothermal Investigations of Solids and Fluids. Boston: Academic Press, 1989.
62. N Tsutsumi, H Tokunaga, T Kiyotsukuri. Unpublished data.
63. HS Nalwa, ed. Ferroelectric Polymers. New York: Marcel Dekker, 1995.
64. Y Takahashi, N Miyamoto. J Polym Sci Polym Phys Ed 23:2505, 1985.
65. Y Agari, A Ueda, S Nagai. J Appl Polym Sci 45:1957, 1992.
66. Y Agari, A Ueda, S Nagai. J Appl Polym Sci 47:331, 1993.
67. C Leonard, JL Harlary, L Monnerie, D Broussoux, B Servet, F Micheron. Polym Commun 24:110, 1983.
68. Y Hirata, T Kotaka. Polymer J 13:273, 1981.
69. Y Agari, A Ueda. J Polym Sci Part B Polym Phys 32:59, 1994.
70. N Tsutsumi, S Kizu, W Sakai, T Kiyotsukuri. J Polym Sci Part B Polym Phys 35:1869, 1997.

71. Y Agari, A Ueda, Y Omura, S Nagai. Polymer 38:801, 1997.
72. CL Choy, KW Lau, YW Wong, HM Ma. Polym Eng Sci 36:827, 1996.
73. G Creveceur, G Groeninckx. Polym Eng Sci 30:532, 1990.
74. JE Ashton, JC Halpin, PH Petit, ed. Primer on Composite Mateials: Analysis. Stamford, CT: Technomic, 1969.
75. IM Ward, ed. Mechanical Properties of Solid Polymers. 2nd ed. Chichester: Wiley, 1983.
76. Q Lin, AF Yee. Polym Compos 15:754, 1994.
77. E Piopkowska, A Galeski. J Appl Polym Sci 41:2371, 1990.
78. E Martuscelli, C Silvestre, G Abate. Polymer 23:229, 1982.
79. E Martuscelli, C Silvestre, L Bianchi. In: M Kryszewski, A Galeski, E Martuscelli, ed. Polymer Blends. New York: Plenum, 1984, vol II, p 57.
80. S Danesi, RS Porter. Polymer 19:448, 1978.
81. J Karger-Kocsis, A Kallo, A Szafner, G Bodor, Zs Seney. Polymer 20:37, 1979.
82. M Bertolotti, A Ferrari, GL Liakhou, R Li Voti, A Marras, TA Ezquerra, FL Balta-Calleja. J Appl Phys 78:5706, 1995.
83. DM Bigg. Adv Polym Sci 119:1, 1995.
84. VP Privalko, VV Novikov. Adv Polym Sci 119:31, 1995.

12

Plastic Deformation Mechanisms of Polymers and Rubber-Modified Thermoplastic Polymers: Molecular and Morphological Aspects

G. Groeninckx and D. Dompas
Catholic University of Leuven (KU Leuven), Heverlee, Belgium

I. INTRODUCTION

The mechanical behavior of polymers and polymer blends is a complicated issue. It depends on the processes of viscoelastic and plastic deformation which finally lead to the macroscopic failure of the specimen. There are two main mechanisms responsible for plastic deformation in polymers: crazing and shear yielding [1–3]. These two mechanisms can occur separately, but in some cases can also be operative simultaneously. Their occurrence depends on many extrinsic variables such as the strain rate, the temperature, the specimen geometry, the way the specimen is loaded as well as on the intrinsic material properties. Even a small change in one of these variables can alter the plastic deformation mechanism, thereby changing the mechanical performance of the material. Due to plastic deformation and necking, the strain rate will not be constant over the specimen and the specimen geometry will clearly change during deformation. Therefore, materials can only be compared for a given mechanical test and even their relative ranking with respect to the mechanical performance may differ considerably from test to test.

In all cases however, the mechanical performance is related to the type and amount of the two main deformation mechanisms: crazing and shear yielding. In this chapter, the essential features of crazing and shear yielding will be considered. Particularly, the plastic deformation mechanisms and material toughness will be predicted from molecular parameters (chain flexibility and entanglement density) in Sec. II and from the blend morphology (rubber particle size and interparticle distance) in Sec. III.

II. PLASTIC DEFORMATION MECHANISMS IN POLYMERS

A. Brittle and Pseudoductile Polymers

A craze is initiated when an applied tensile stress causes microvoids to nucleate at points of high stress concentrations in the polymer created by scratches, flaws, cracks, dust particles, molecular heterogeneities, etc. [1–7]. The microvoids develop in a plane perpendicular to the maximum principal stress but do not form a true crack since they become stabilized by fibrils of plastically deformed, oriented polymeric material spanning the craze. The resulting localized yielded region therefore consists of a system of alternating voids and polymer fibrils and is known as a craze (Fig. 1). Unlike a crack, it is capable of transmitting loads across its faces.

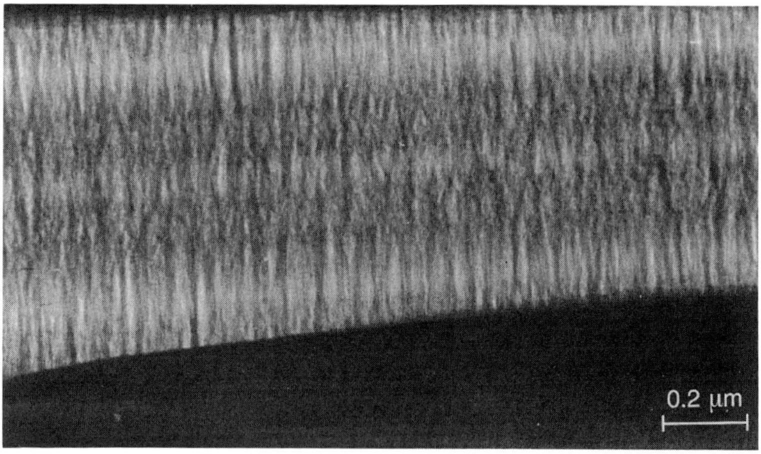

Figure 1 High-voltage electron micrograph of a craze in a 2 μm thick section of polystyrene deformed in the electron microscope. Deformation direction vertical [93].

The disadvantage of crazing is that it is frequently a precursor to brittle fracture [1–5]. This is because, although considerable plastic deformation and local energy absorption are involved in craze initiation, growth, and breakdown, this micromechanism of brittle fracture is often highly localized and confined to a very small volume of the material.

Whereas crazing involves voiding of the material and an increase of its volume, shear yielding consists of a change in shape of the specimen at constant specimen volume. On a molecular level, it is often considered as molecular slip, flow, or molecular motions of different types [8,9]. Shear yielding can be localized in "microshear" bands (Fig. 2(a)) or can be more homogeneously distributed in diffuse zones (Fig. 2(b)) [10,11].

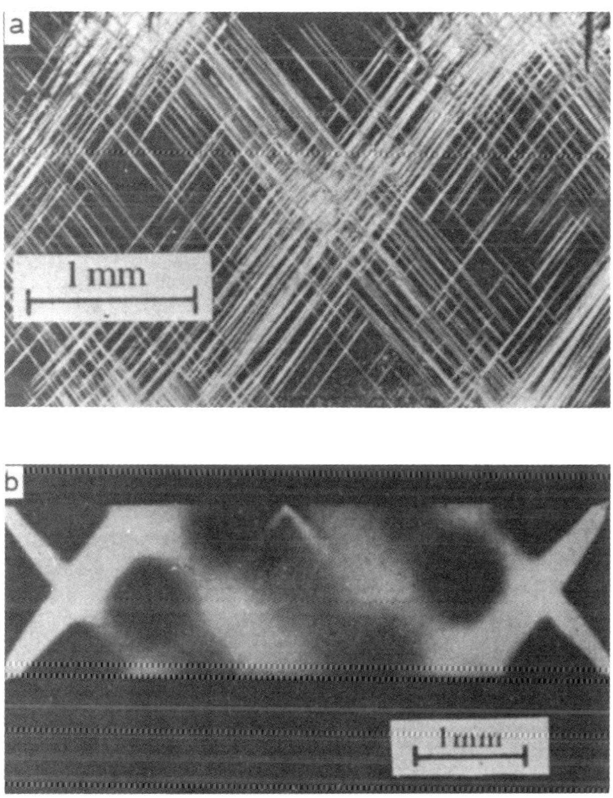

Figure 2 Sections cut from specimens that have been deformed just past the yield point at 22°C in a plane strain compression test viewed between crossed polars. (a) Polystyrene showing microshear bands, (b) PMMA showing broad diffuse zones [2].

Due to its internal void structure, a craze, in contrast to shear bands, strongly reflects light. This is observed as stress-whitening of the specimen. For the same reason, craze initiation requires the presence of a dilatational component in the stress tensor, and may be inhibited by applying hydrostatic pressure but enhanced by the presence of triaxial tensile stresses [2,3]. The latter stress state exists ahead of flaws or notches in relatively thick specimens in plane-strain conditions. Thus, the presence of sharp cracks or defects in thick specimens will favor craze initiation leading to brittle fracture, as opposed to a bulk shear yielding mechanism leading to ductile fracture.

Under a given extrinsic condition, chemically different polymers behave differently [12–14]. For instance, in a low strain rate uniaxial tensile test, polystyrene (PS) tends to craze and is brittle whereas polycarbonate (PC) tends to yield and is ductile. Accordingly, polymers are classified into two main types [12–16]: brittle (type I) and pseudoductile (type II). The prefix "pseudo" is used here to emphasize that craze/yield (brittle/ductile) behavior depends on many extrinsic variables, as mentioned before; so a pseudoductile or a brittle polymer may either craze (brittle) or yield (ductile), depending on the test conditions and the specimen geometry. Nevertheless, under ordinary conditions (a uniaxial tensile test at low strain rate and not too low temperatures), type I polymers tend to craze (brittle behavior), and type II polymers tend to yield (ductile behavior). Such a classification is convenient and useful for correlating brittle/ductile behavior with microstructural and molecular parameters.

1. Brittle polymers (type I) tend to fail by crazing, have low crack initiation energy (low unnotched toughness) and low crack propagation energy (low notched toughness). Examples are polystyrene (PS), poly(styrene-co-acrylonitrile) (SAN) and poly(methyl methacrylate) (PMMA).
2. Pseudoductile polymers (type II) tend to fail by yielding, have high crack initiation energy (high unnotched toughness) and low crack propagation energy (low notched toughness). Examples are nylons, poly(ethylene terephthalate) (PET) and polycarbonate (PC).

In the following paragraphs, it will be shown that the intrinsic brittleness or ductility of neat polymer matrices are controlled by two molecular parameters: the entanglement density (v_e) and the chain stiffness (C_∞). The entanglement density determines the craze initiation stress, while the chain stiffness parameter determines the initiation stress for shear yielding of the polymer network.

B. Crazing and Chain Entanglement Density

Crazing is initiated by chain scission [4,17–20], which leads to cavitation at the leading edge of the crazing zone. In high entanglement density polymers, the load will be distributed over different entanglements and different chains and, as a consequence, the probability of breaking a chain will be lower. It is thus expected that a high entanglement density is unfavorable for craze initiation [21], as will be demonstrated below.

The craze grows in two different ways: by craze tip advance, an expansion of the craze periphery generating more fibrils, and by craze width growth, a normal separation of the two craze interfaces behind the craze tip. The general mechanism of craze tip advance has been known for some time to be the Taylor meniscus instability process [22]. Such instabilities are commonly observed when two flat plates with a layer of liquid between them are forced apart or when adhesive tape is peeled from a solid substrate.

The craze just behind the tip is only a few nanometers wide, whereas crazes typically can attain widths of a few micrometers before fracture. Clearly, while the craze tip advance mechanism is responsible for initially generating very short lengths of fibrils, most of the fibril structure is generated by the mechanism of craze width growth. Even the stress concentrations at the craze tip can be dramatically altered by the widening of the craze well behind the tip; if the craze continues to widen without the craze tip advancing, the stress at the craze tip will rise until the craze tip advance again resumes. Hence, much of the recent work on craze growth has centered on elucidating the craze width growth mechanism and its consequences.

Kramer and Berger [23] have derived a detailed model of craze growth. The key quantity to be specified in understanding the effect of the molecular chain parameters on the craze widening stress is the surface tension of the void surface (Γ). The craze will only grow if the deformation energy associated with the applied stress is larger than the surface energy needed to create the new surface. This surface energy per unit area of the void surface (Γ) is given by [23]

$$\Gamma = \gamma + \tfrac{1}{4} d v_e U \tag{1}$$

where γ is the van der Waals surface energy, v_e is the entanglement density, d is the entanglement mesh size and U is the polymer backbone bond energy. The extra energy due to the chain scission of the entanglements crossing the interface ($\tfrac{1}{4} d v_e U$) can be quite significant. For example, the van der Waals surface energy for many polymers is about 40 mJ m^{-2} [24]. For a polymer with an entanglement density of 3×10^{25} m^{-3} (about that of entangled PS, a rather low entanglement density polymer) the second (chain scission) term approximately equals 40 mJ m^{-2} so that $\Gamma \sim 80\,\mathrm{mJ\,m^{-2}}$. Note also that $\upsilon_e d$ scales roughly as $\upsilon_e^{1/2}$; increasing the entanglement density of the network leads to a significant increase in Γ, and hence, to an increasing crazing stress.

Besides the entanglement density, the molecular weight [20] is also important. With both molecular parameters, Kramer and Berger's model [23] correctly predicts the effect of temperature on the craze/yield behavior of different types of matrices. Moreover, the effect of entanglement density on the craze microstructure [25] (craze extension ratio [17,18,26,27], mean fibril spacing, and cross-tie fibrils) can now be understood as well as phenomena such as high-temperature crazes and the transition from chain-scission-dominated crazing to chain-disentanglement-dominated crazing with increasing temperature.

Craze initiation stress and craze growth stress thus increase with increasing matrix entanglement density. A low entanglement density should be favorable for crazing, since it corresponds to a high probability of breaking a chain and a low surface tension required for craze growth. Fig. 3 plots the craze initiation stress (σ_{craze}) versus υ_e bilogarithmically for a number of polymers [12–14]. The figure confirms that the craze initiation stress scales with the square root of the entanglement density of the polymer and that,

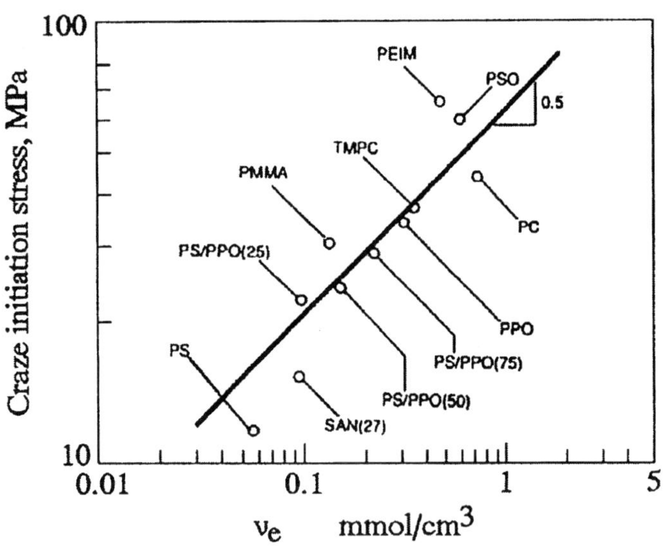

Figure 3 Craze initiation stress (σ_{craze}) versus entanglement density (υ_e) [14].

for example, the crazing stress of the ductile polycarbonate lies higher than that of the brittle polystyrene.

C. Shear Yielding and Chain Flexibility

There have been many attempts over the years to explain the process of shear yielding on a molecular level. All these models suffer, however, from the lack of sufficient experimental data on a molecular or a submicroscopic scale. These data are more readily available for the crazing mechanism explaining the better understanding of the latter on a molecular scale.

Unlike crazing, yielding is strongly temperature dependent. This has led us to think of yield as a type of viscous flow, especially since amorphous glassy polymers are basically frozen liquids that have failed to crystallize. Eyring [28] developed a theory to describe the viscous flow in liquids and it has been adapted to describe the yield behavior of glassy polymers. The theory is based on placing the segments of the polymer chain in a pseudo-lattice. Flow occurs when a segment moves from one lattice site to an adjacent site. There will be a potential barrier to overcome, because of the presence of adjacent molecular segments. The height of the barrier is given by the activation energy ΔG_0 and the frequency at which the segments move to the new position is represented by an Arrhenius-type equation. The rate at which segments jump forward over the barrier will be increased by the application of a shear stress. This has the effect of reducing the energy barrier. Bauwens [29] and Roetling [30] rearranged Eyring's equation to express the yield stress as a function of temperature and strain rate

$$\frac{\sigma_{\text{yield}}}{T} = \frac{4\sqrt{3}k}{V\eta_0} \ln\left[\frac{\sqrt{3}}{2\gamma_0 J_0}\frac{d\varepsilon}{dt} + \frac{\Delta G_0}{RT}\right] \tag{2}$$

where σ_{yield} is the yield stress, T the temperature, k the Boltzmann constant, V the activation volume, $(d\varepsilon/dt)$ the strain rate, ΔG_0 the activation energy, R the universal gas constant, γ_0 an elementary shear strain and J_0 a rate constant.

This approach successfully predicted the strain-rate and temperature dependence of the yield stress of polymers [30–35]. Eyring's equation fits the experimental data well over a broad temperature interval and over several decades of strain rate. Moreover, the activation energy (ΔG_0) in Eyring's equation is equal to the activation energy of the molecular-relaxation processes derived from mechanical and dielectrical loss spectroscopy.

Though the physical meaning of the activation volume [21] and the exact type of molecular motions [8,9] is still a point of debate, it is clear that a positive correlation is to be expected between chain mobility, yielding, and toughness. This correlation has been clearly demonstrated by Ramsteiner who determined for 12 important thermoplastics the brittle–ductile transition temperatures and the mechanical relaxation in torsion [4,36]. Wherever a pronounced low-temperature (β) relaxation peak, corresponding to a main chain motion, is observed in torsion, it coincides with the brittle–ductile transition temperature in impact bending. In the absence of such peaks (as in PMMA and PS), this transition is shifted all the way up to the beginning of the α relaxation.

It should be noted that a quantitative correlation between the brittle–ductile transition temperature and the β relaxation peak will be limited by the fact that the relaxation behavior is determined in the linear viscoelastic range of stresses and at small strain rates. These conditions are quite different from those of an impact test.

Eyring's equation for the yield stress, and Ramsteiner's correlation between main-chain β-relaxation peaks and the brittle-to-ductile transition indicate that local mobility on a molecular scale in the glassy state is essential for yielding. A measure of the rigidity of a chain is given by C_∞, the characteristic ratio of the chain or chain stiffness parameter, which is defined here as

$$C_\infty = \lim[R_0^2/n_v l^2] \tag{3}$$

where R_0^2 is the mean-square end-to-end distance of an unperturbed chain, n_v the number of statistical skeletal units, l the length of a statistical skeletal unit and $n_v l^2$ the mean-square end-to-end distance of a random-flight chain. Rigid chains such as liquid crystalline polymers (with a high end-to-end distance) will have a high C_∞, while flexible, tortuous chains (with a corresponding low chain end-to-end distance) will have a low C_∞. The initiation stress for shear yielding σ_{yield} has been shown to be strongly dependent on temperature and roughly proportional to $\Delta T = T_g - T$ and δ^2, where T is the testing temperature and δ^2 the cohesive energy density. A reduced normalized yield stress $|\sigma_{\text{yield}}|$ is thus defined as:

$$|\sigma_{\text{yield}}| = \sigma_{\text{yield}}/[\delta^2(T_g - T)] \tag{4}$$

The denominator $\delta^2(T_g - T)$ accounts for the interchain effects (friction between the chains) on the yield stress. Thus, the reduced yield stress $|\sigma_{\text{yield}}|$ should be only a function of an intrachain property, characterized by the chain stiffness parameter C_∞: the higher the chain rigidity, the lower its mobility and therefore the higher the (normalized) initiation stress for shear yielding $|\sigma_{\text{yield}}|$.

Fig. 4 plots the normalized yield stress versus C_∞ bilogarithmically for a number of polymers [12–14]. The figure confirms that the yield initiation stress increases with increasing chain stiffness of the polymer and that, for example, the yield stress for the ductile polycarbonate is lower than for the brittle polystyrene.

As mentioned before, it has been shown by Ramsteiner [4,36] that the brittle–ductile transition occurs at the secondary relaxation of the polymer if this arises from localized main-chain motions. Together with Fig. 4, the secondary relaxation temperature T_β is then logically related to the characteristic ratio of the chain by [14,37]

$$T_\beta(\text{main-chain})/T_g = (0.135 \pm 0.027) + (0.082 \pm 0.004)C_\infty \qquad (5)$$

where T_β is the β-relaxation temperature at 1 Hz frequency. Fig. 5 shows the correlation for a wide variety of polymers. Note that when $C_\infty \sim 10.5$ (PS), a secondary relaxation peak from a main-chain motion can no longer be discerned in the spectrum at 1 Hz, since then T_β coincides with T_g.

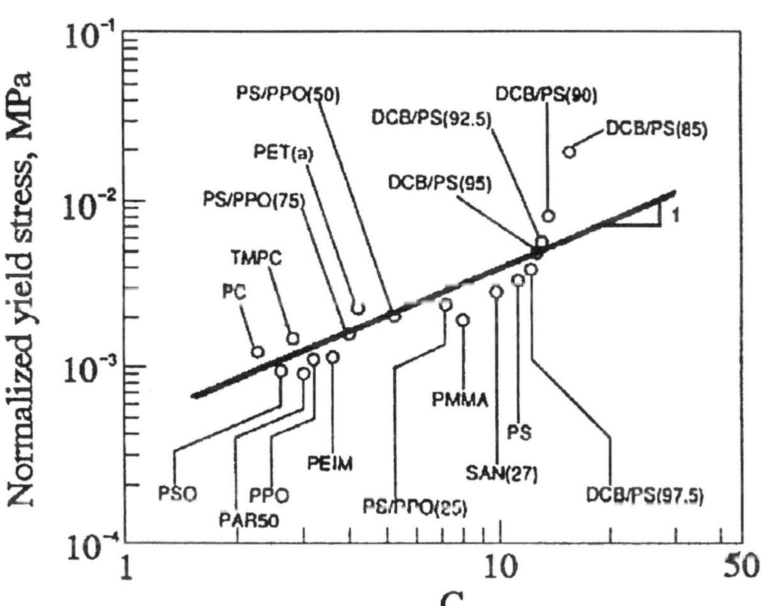

Figure 4 Normalized yield stress $[\sigma_{\text{yield}}/\delta^2(T_g - T)]$ versus characteristic chain stiffness (C_∞) [14].

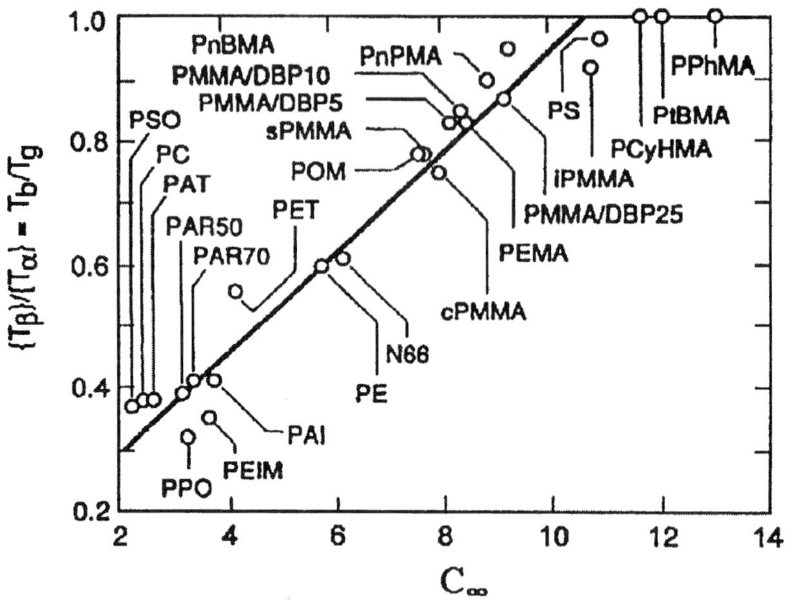

Figure 5 T_β/T_g (at 1 Hz) versus C_∞ for some polymers [14].

D. Molecular Criterion for Craze/Yield Behavior from Chain Structure Parameters

The competition between crazing and shear yielding determines which mode of fracture will predominate. Crazing tends to occur when the crazing stress is lower than the yield stress. Yielding tends to occur when the yield stress is lower than the crazing stress. Both crazing and yielding tend to occur when the crazing stress and yield stress are comparable. Combining Figs. 3 and 4 gives the molecular criterion which controls the predominant mode of fracture, i.e., [12–14]

$$\sigma_{\text{craze}}/\sigma_{\text{yield}} \alpha v_e^{1/2}/C_\infty \tag{6}$$

Since v_e and C_∞ are interrelated [12–14], this equation can be rewritten:

$$\sigma_{\text{craze}}/\sigma_{\text{yield}} \sim [3M_v/\rho_a]^{1/3} v_e \sim [\rho_a/3M_v]^{1/2} C_\infty^{-2} \tag{7}$$

where M_v is the molecular weight of a statistical skeletal unit and ρ_a the density of the (amorphous) polymer. Usually, the prefactor $3M_v/\rho_a$ is fairly constant for many polymers, except for those with unusually high ρ_a or M_v. Therefore, v_e or C_∞ is a unique molecular parameter that determines whether a polymer will tend to craze or to yield. When v_e is small (a polymer having a low entanglement density), the crazing stress will be smaller than the yield stress, and the polymer tends to craze. On the other hand, when v_e is large (a polymer having a high entanglement density), the yield stress will be smaller than the crazing stress, and the polymer tends to yield.

Table 1 lists the v_e and C_∞ values for some common polymers. The polymers are roughly divided into brittle and pseudoductile matrices. The brittle polymers (type I) have a v_e smaller than 0.15 mmol cm^{-3}. Typical examples are PS, PSAN, and PMMA, which all deform by crazing. The pseudoductile polymers (type II) deform by shear yielding and have a C_∞ smaller than 7.7. Typical examples are PVC, POM, PPO, PC, and nylon-6,6.

Both v_e and C_∞ give consistent predictions. However, it should be emphasized that the entanglement density v_e is the primary factor controlling the intrinsic crazing behavior, and the characteristic ratio C_∞ is the primary factor controlling the intrinsic yielding behavior.

The greatest intrinsic ductility limit occurs at $C_\infty = 2$, corresponding to chains of freely rotating joints with tetrahedral skeletal bonds. It is interesting to note that PC has $C_\infty = 2.4$, which is near to the greatest intrinsic ductility limit. Therefore, one expects that PC should be one of the toughest matrices, which is consistent with experiments.

The intrinsic ductility C_∞ and the entanglement density v_e can be predicted from the chemical structure of the polymers. The necesssary equations are [14,38]

Table 1 Chain Parameters for Some Polymers in the Amorphous State [14]

Polymer	ρ_a (g cm^{-3})	M_e[a] (g/mol^{-1})	C_∞	v_e (mmol cm^{-3})
Brittle polymers:				
PS	1.05	18700	10.8	0.0561
SAN25	1.08	11600	10.6	0.0931
PMMA	1.17	9200	7.9	0.127
Pseudoductile polymers:				
PVC	1.40	5560	7.6	0.252
POM	1.25	2550	7.5	0.490
N66	1.07	1990	6.1	0.537
PET	1.33	1630	4.2	0.815
PPO	1.07	3620	3.2	0.295
PC	1.20	1790	2.4	0.672

[a] M_e represents the molecular weight between entanglements.

$$C_\infty = \frac{1}{\Phi_0^{2/3}} \left[\frac{\sum K_i + Bn_\mathrm{r}}{M_\mathrm{r}} \right]^{4/3} \left[\frac{M_\mathrm{v}}{\langle l_\mathrm{v}^2 \rangle} \right] \tag{8}$$

$$\upsilon_\mathrm{e} = \rho_\mathrm{a} \left[\frac{\Phi_0^{4/3}}{3M_\mathrm{v}} \right] \left[\frac{M_\mathrm{r}}{\sum K_i + Bn_\mathrm{r}} \right]^{8/3} \left[\frac{\langle l_\mathrm{v}^2 \rangle}{M_\mathrm{v}} \right]^2 \tag{9}$$

where $\Phi_0 = 2.51 \times 10^{23}\ \mathrm{cm^3\ mol^{1/2}\ g^{-3/2}}$, B counts for the effect of tacticity and is taken to be 4.2 for conventional polymers, n_r and M_r are, respectively, the number of skeletal atoms and the molecular weight of a repeat unit, M_v and l_v^2 are respectively the average molecular weight and the mean-square length of a statistical skeletal unit, ρ_a is the amorphous density, and finally K_i is the molar stiffness constant for group i. The summation $\sum$ is taken over a repeat unit.

The characteristic chain ratio C_∞ and the entanglement density υ_e can be predicted from the chemical structure through the molar stiffness constants K_i (tabulated in Refs. 38 and 39) by group additivity with uncertainties usually less than 7% for C_∞ and 15% for υ_e, comparable with the accuracies of experimental values. In essence, it is therefore possible to predict the intrinsic ductility of a polymer from its chain microstructure even before it is synthesized.

In thermoplastic polymer–rubber blends, the dispersed rubber phase increases the toughness mainly by promoting the energy dissipation of the matrix phase. Therefore, it is expected that the preferred phase morphology and the efficiency of toughening should correlate with the chain parameters and the intrinsic brittle–ductile behavior of the matrix phase. This will be discussed in Sec. III.

III. PLASTIC DEFORMATION MECHANISMS IN RUBBER-MODIFIED THERMOPLASTIC POLYMERS

Rubber toughening is one of the key inventions in the history of the plastics industry [1]. In the late 1940s, acrylonitrile-butadiene-styrene (ABS) and high-impact polystyrene (HIPS) were invented. The success of these products has led to the development of improved grades of ABS and HIPS, and to the idea that the principle of rubber-toughening could be applied to virtually all types of thermoplastics.

The essential characteristics of rubber toughening are that the fracture resistance of the parent polymer is increased substantially, in return for a limited reduction in stiffness, yield stress, and creep resistance [1,2]. This balance of properties is achieved by dispersing the rubber as small spherical particles in the thermoplastic matrix. Optimum particle sizes vary, depending upon the chemical structure of the matrix, but are typically within the range of 0.1 to 10 μm. Rubber contents are usually between 3 and 20 vol%. Phase separation is important; a rubber that becomes dissolved in the matrix acts as a plasticizer, reducing the glass-transition temperature and the stiffness of the matrix with only a moderate increase in its toughness.

As in unmodified matrices, the fracture toughness of rubber-modified matrices is directly related to the amount and type of the occurring deformation mechanisms. Regarding the function of the rubber phase, two principally different mechanisms have been proposed.

First, some authors state that the toughness enhancement is due to the energy dissipation in the elastomeric phase. It is stated [40] that the rubber particles can bridge a crack and break after the crack front has passed them, due to the high fracture strain of the rubber. All the stored elastic energy in the highly stretched particles is then released and converted into heat. Calculations, howver, [41] reveal that the total energy associated with the deformation of the rubbery phase cannot account for more than a small fraction of the observed impact energy.

The second principal approach states that the toughness of rubber-modified blends is enhanced because the rubber particles initiate the deformation mechanism in the matrix from many places, which leads to multiple crazing or multiple shear yielding, whilst at the same time limiting the growth and breakdown of voids and crazes to prevent early crack initiation. It is now generally accepted [42] that for all thermoplastic matrices, this concept is the basic mechanism for rubber toughening. Well-known examples where this multiple-deformation mechanism occurs are multiple crazing in HIPS [43–45], a combination of multiple crazing and shear yielding in ABS [46,47] and multiple shear yielding in the more ductile polymers such as nylons [48]. In the following paragraphs, the requirements of the blend morphology for achieving the optimum mechanical performance of the rubber-modified material are considered. For matrices that deform by crazing, the rubber particle size is important; for matrices that deform by shear yielding, the average interparticle distance and the rubber particle size are important morphological parameters determining the toughness of the rubber-modified material.

A. Multiple Crazing

Early deformation studies on HIPS specimens revealed intensive stress-whitening; optical microscopy and transmission electron microscopy revealed numerous small crazes running between the rubber particles [5]. Bucknall [1] proposed that these crazes were initiated due to the stress concentration at the equator of the particles [49]. Compared to the unmodified matrix, multiple crazing occurs throughout a comparatively large volume of the rubber-modified material. This explains the tremendous increase in fracture toughness and the intensive stress-whitening in these materials.

The toughening efficiency of the rubber depends strongly on the particle size. An optimum rubber particle size of 1–5 μm has been suggested for maximum toughness [1,42]. Donald and Kramer [50] examined the craze-initiation mechanism; they found that the ease of craze initiation is greatly dependent upon the particle size and that crazes are rarely initiated by particles smaller than about 1 μm. Bucknall [1] suggested that the decrease in toughness with small rubber particles not only arises from their inability to nucleate crazes but also from their inability to be effective in craze termination.

Since energy absorption in these rubber-modified systems arises entirely from multiple crazing in the matrix and is not caused by the deformation of the elastomeric phase, the optimum particle size should depend on the intrinsic capability of the matrix to initiate crazes, i.e., on the entanglement density υ_e of the polymer. Experimentally, it has been found that

$$\log(d_{\text{optimum}}) = 1.19 - 14.1\upsilon_e \tag{10}$$

where d_{optimum} is the optimum rubber particle diameter in micrometers. This relation is shown in Fig. 6.

B. Multiple Shear Yielding

When rubber is blended with many different matrices, deforming by shear yielding, the temperature at which the fracture changes from brittle to tough (T_{BT}) shifts to lower temperatures and the fractured samples in the tough region show intensive stress-whitening. In the case of brittle samples, stress-whitening is only visible in the region just ahead of the notch. In the first studies on rubber-toughened materials, namely on HIPS and ABS, stress-whitening was proved to arise from multiple crazing in the matrix. For a long time, there has been a general belief that all rubber-modified polymers are toughened by the multiple-crazing mechanism, although a convincing proof of crazing in the deformation zone could not always be given.

During recent years, it has been demonstrated that stress-whitening in many rubber-modified systems was caused by cavitation or debonding of the rubber phase. Shear yielding is prohibited in the middle of a thick specimen because the deforming material is unable to contract laterally in order to remain at constant volume. Consequently, the material in the

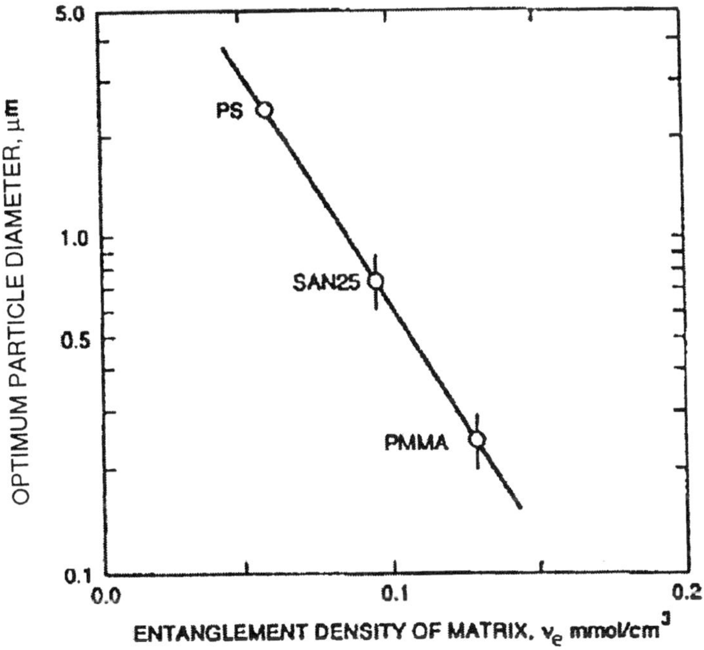

Figure 6 Optimum rubber particle size for the toughening of brittle matrices versus the entanglement density of the matrix v_e [14].

crack tip is under triaxial tension [42], inducing crazing, rubber cavitation, or debonding. If crazing occurs, then the material will break brittle. If internal cavitation or debonding occurs, then these deformation mechanisms relieve the hydrostatic tension in the material; the stress state in the thin ligaments of the matrix material between the voids is converted from a triaxial to a uniaxial tensile stress state. The latter stress state is favorable for the initiation of shear bands. In other words, the role of the rubber particles is to cavitate internally or to debond from the matrix, thereby relieving the hydrostatic tension and initiating the ductile shear yielding mechanism.

The process of rubber cavitation and associated matrix yielding is found in rubber-modified PC [51–54], PVC [55–60], PBT [61], nylon-6 [62–67], nylon-6,6 [68], and epoxy systems [69–73]. Particle debonding from the matrix on the other hand has been considered for core-shell rubber-modified PC/copolyester blends [74], PC/PE [52,75], PVC/MBS [76] and CTBN/epoxy [72]. The matrix materials in these rubber-modified systems are either crosslinked or have a high entanglement density, thus they are materials for which the crazing mechanism is suppressed [14,21,23]. The competition between voiding in or around the rubber and crazing in the matrix will therefore be determined by the enanglement density v_e of the matrix. PMMA with a moderate entanglement density is known to deform by crazing, but when rubber-toughened, it deforms by shear yielding [77,78]. SAN has an even lower entanglement density. Consequently, ABS deforms both by crazing and rubber cavitation followed by shear yielding [46,47]. The relative contribution of both mechanisms to the total deformation in ABS depends on the particle size of the rubber particles. Finally, HIPS has long been thought to deform only by multiple crazing. A recent paper [54], however, reveals that, besides crazing, rubber voiding can occur in HIPS to a very large extent.

The efficiency of the rubber again depends strongly on the particle size. For nylon-6,6–rubber blends, a sharp brittle–tough transition occurs at a critical particle size [68]. This is shown in Fig. 7, where the notched Izod impact strength is plotted versus the rubber particle size at different rubber contents.

Besides the particle size, also the rubber volume fraction influences the brittle–tough transition. Wu [68] combined the two morphological parameters and demonstrated that the condition for the onset of a sharp

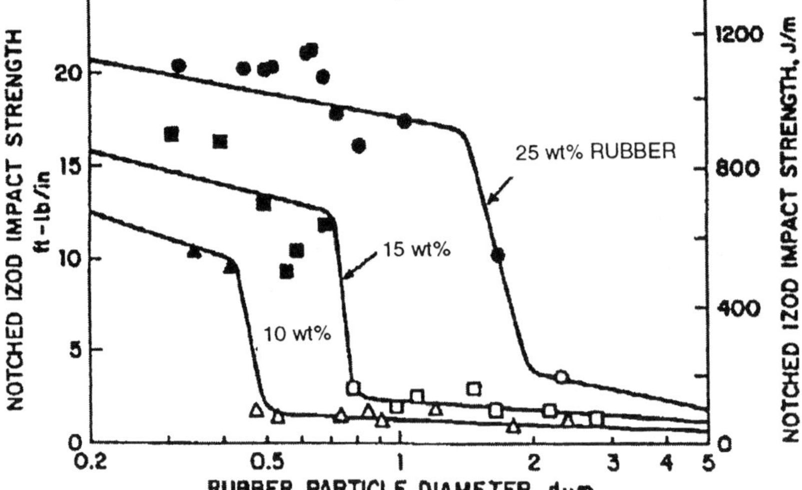

Figure 7 Impact toughness at room temperature versus rubber particle size for nylon-6,6–rubber blends [68]. 1 ft-lb in^{-1} = 53 J m^{-1} [68].

brittle–tough transition in nylon–rubber blends occurs at a critical interparticle distance (IPD_c), also called the critical matrix ligament thickness, as shown in Fig. 8.

The average value of the interparticle distance (IPD) can be calculated for a monomodal particle size distribution by assuming that the particles form a regular lattice. IPD is related to the particle size (d) and the rubber volume fraction (Φ_r) via

$$IPD = d[k(\pi/6\Phi_r)^{1/3} - 1] \tag{11}$$

where the parameter k is a measure of the packing arrangement of a particular lattice. A large value of k implies a large packing density. Usually, one takes $k = 1$, as for a simple cubic lattice. For a face centered cubic (fcc) or hexagonal closed packing (hcp) lattice, $k = \sqrt[6]{2} = 1.12$ and the average interparticle distance will be larger.

Various explanations for the existence of this critical interparticle distance parameter are described in the literature, being: (1) overlap of stress fields [68]; (2) connectivity of yielding volume elements (percolation concept) [14,79]; (3) transition from plane strain to plane stress [15,66]. However, well-dispersed nylon rubber blends with a low rubber concentration have interparticle distances which are so large compared with the particle diameter

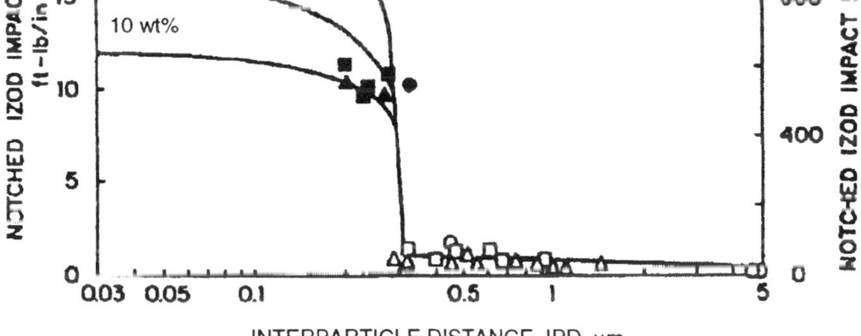

Figure 8 Impact toughness at room temperature versus interparticle distance for nylon-6,6–rubber blends [68]. 1 ft-lb in^{-1} = 53 J m^{-1} [68].

that the effect of stress field overlap can be neglected. The physical explanation in terms of only stress field overlap can be doubted. The concept is therefore explained by a combination of (2) and (3) and is shown schematically in Fig. 9. Due to rubber cavitation, the stress state will gradually change around the rubber cavities from plane strain to plane stress; this allows the thin matrix ligaments to yield locally around the rubber modifier. When the thin ligaments are interconnected to form a pervasive network (situation D in Fig. 9), the yielding can propagate and pervade over the entire deformation zone [14,79]. When this occurs, a blend will exhibit tough behavior.

The required distance between the particles will naturally be dictated by the intrinsic ductility (C_∞) of the matrix. Wu [14] verified this relation and demonstrated that

$$\log(\mathrm{IPD_c}) = 0.74 - 0.22\, C_\infty \tag{12}$$

where $\mathrm{IPD_c}$ is expressed in micrometers. This relation is shown in Fig. 10 for a number of pseudoductile matrices.

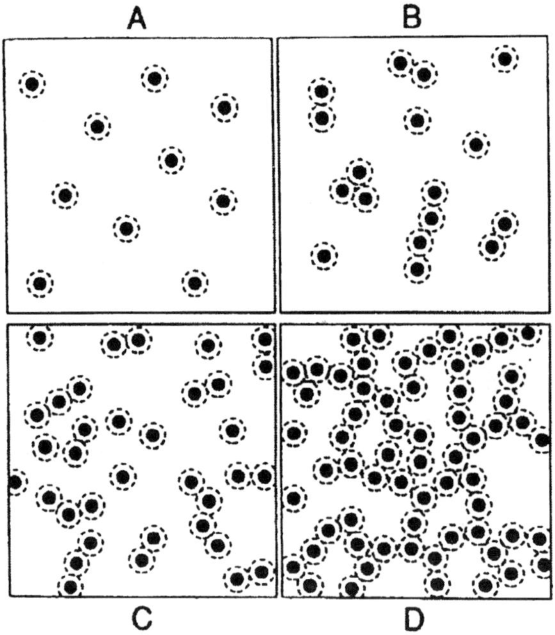

Figure 9 A schematic illustration of the continuum percolation of stress volumes [14].

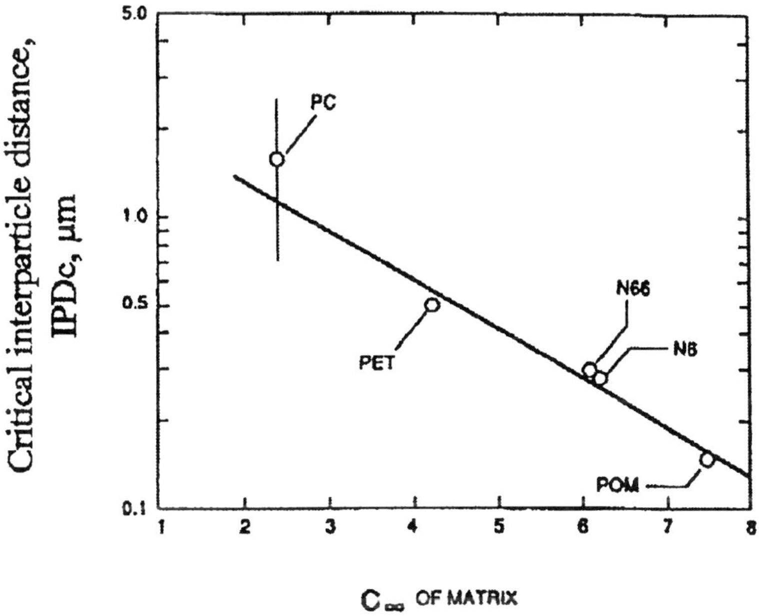

Figure 10 Critical interparticle distance (IPD_c) or critical matrix ligament thickness for the onset of the brittle–tough transition versus the characteristic ratio C_∞ of the matrix [14].

At this moment, it is not yet completely clear to what extent the impact properties are affected by the particle size *distribution*. Wu [15] stated that a wide particle size distribution is disadvantageous for the impact behavior of rubber-modified blends with pseudoductile matrices, because the interparticle distance increases with increasing size polydispersity at a given rubber volume fraction. No experimental data were presented to support this hypothesis. Other authors, however, claim that a bimodal particle size distribution [45,50] can lead to an enhancement of the toughness. At first sight, it seems that the effect from a bimodal particle size distribution is dependent on the matrix deformation mode; PS [45] and ABS [50] both deform by multiple crazing and a synergetical effect from a bimodal particle size distribution is observed. On the other hand, rubber-toughened PMMA [78] and epoxy [72] deform by multiple shear yielding that seems to be less convenient to promote a synergy effect from a bimodal distribution of particle sizes.

Secondly, Figs. 7 and 8 predict an increase of the mechanical performance with improved dispersion of the rubber in the blend, which is achieved by decreasing the particle size (Eq. (11)). However, it was recently shown [67,80] that particles below 200 nm are no longer effective in increasing the toughness in nylon-6–rubber blends. As a possible explanation, it has been suggested that very small particles do not cavitate [55,69], and therefore are not able to initiate shear yielding in the matrix. In view of the importance of rubber cavitation for the mechanical performance of rubber-modified pseudoductile matrices, a model for the internal cavitation or rubber particles has been developed recently by Dompas and Groeninckx [81] and will be presented in Sec. III.C.

The requirement of a good interfacial adhesion between the rubber phase and the matrix phase remains a point of discussion. A distinction is made between matrices deforming by multiple shear yielding and matrices deforming by multiple crazing. In the latter case, good adhesion between rubber particles and matrix is required, because the rubber particles must be effective as craze stoppers [1]. In the first case however, based on the concept that the function of the rubber particles is to create voids, a good interfacial strength is not required [66,68]. From a mechanistic point of view there is no difference between void formation by internal cavitation or by debonding.

C. A Model for the Particle Size Dependence of the Rubber Cavitation Behavior

In rubber-modified materials under triaxial tensile stresses, voiding can occur in the matrix (multiple crazing in HIPS) or can initiate inside the rubber particles (rubber cavitation). Once the rubber particles are cavitated, the hydrostatic tension in the material is relieved and the stress state in the thin ligaments of matrix material between the voids is converted from a triaxial to a more uniaxial tensile stress state. This new stress state is favorable for the initiation of shear bands. In other words, the role of the rubber particles is to cavitate internally, thereby relieving the hydrostatic tension and initiating the ductile shear yielding mechanism.

Many examples are mentioned in the literature to support this concept. The process of rubber cavitation and associated matrix shear yielding is found in rubber-modified PC [51–54], PVC [55,57,60], PBT [61], nylon-6 [62,64,65,81], nylon-6,6 [15] and epoxy systems [69,71–73]. The matrix polymers in these rubber-modified systems are either crosslinked or have a high entanglement density, thus they are polymers for which the crazing mechanism is suppressed [23]. This is not to say that rubber cavitation can only appear in high entanglement density matrices. A recent paper [54] reveals that rubber cavitation can occur in HIPS to a very large extent, but does not lead to matrix shear yielding. At first sight, rubber cavitation thus seems only necessary in matrices with high entanglement density, where it is needed for the promotion of shear yielding.

The ultimate mechanical properties of the material will depend on the blend morphology. Wu established that for blends with pseudoductile matrices the brittle–tough transition occurs at a critical interparticle distance [68].

The critical interparticle distance is thus a crucial morphological parameter controlling the toughness in rubber-modified pseudoductile matrices. The percolation concept is the physical explanation behind it [14,79]. If the rubber particles are able to cavitate internally and if the generated voids are close enough, then the thin matrix ligaments between the particles will yield locally. When all the thin matrix ligaments are interconnected, the yielding process propagates over the entire deformation zone and the blend is ductile. This happens when the ligament thickness is smaller than the critical interparticle distance. For a given volume fraction, this is achieved by decreasing the particle size and by improving the dispersion. For most systems, this concept works well and a decrease in particle size corresponds to a shift in the brittle tough transition to lower temperatures [68,82]. However, it has been found in a number of systems [46,55,62,67,80,83,84] that there exists a minimum particle size below which the brittle–tough transition no longer shifts to lower temperatures. As a possible explanation for this peculiar behavior, it has been mentioned that too small particles are not able to cavitate, and therefore do not relieve the hydrostatic tension in the material to promote ductile shear yielding.

In order to determine whether this could be true and because of the importance of rubber cavitation as a precursor for ductile behavior, the criteria for rubber cavitation have been studied by Groeninckx and cow-orkers [81,84–86] and also by Lazerri and Bucknall [87]. The model explains the increasing resistance against cavitation with decreasing rubber particle size. The consequences for the toughness of blends with pseudoductile matrices with very small particles are also discussed below. The model has been experimentally verified on rubber-modified PVC by determining the strain at which cavitation occurs [85], and by correlating this quantity to the toughening behavior of the blend [84,86].

The model outlined below is based on the following assumptions [81]. The rubber particles are spherical in shape with a diameter d_0 and the largest defects in these particles are on the order of a few nanometers. Failure will then initiate from local inhomogeneities on a molecular scale such as local differences in crosslinking density. Furthermore, it is assumed that internal rubber cavitation is an instantaneous process, with the hydrostatic tension on the particle as the driving force for cavitation. Consequently, rubber cavitation will occur when the total energy associated with the process (U_{total}) will decrease. The criterion for rubber cavitation is then given by the energy balance equation:

$$U_{total} = U_{strain} + U_{surface} < 0 \tag{13}$$

where U_{strain} is the elastic strain energy stored in the rubber particle under the applied stress and $U_{surface}$ is the surface energy associated with the cavitation. This equation means that rubber cavitation will only initiate if the energy gained by relieving the stress (U_{strain}) is greater than the energy needed to create a new surface ($U_{surface}$).

The strain energy in one rubber particle is equal to the total work done by the external tension p during the deformation assuming no cavity is created:

$$U_{strain} = -(\pi/12)K_r\Delta^2 d_0^3 \tag{14}$$

where K_r is the rubber bulk modulus and $\Delta = \Delta V/V_0$ is the volume strain.

The strain energy stored in the rubber particle is thus determined by the rubber particle size (d_0), the rubber bulk modulus (K_r) and the volume strain (Δ). This energy term is negative, indicating that it will be relieved when cavitation occurs.

However, the formation of a void in the rubber particle requires surface energy given by:

$$U_{surface} = \Gamma\pi d_i^2 \tag{15}$$

where d_i is the internal void diameter and Γ is the surface energy per unit area. d_i is related to the particle diameter (d_0) and to the volume strain (Δ) by:

$$d_i = \Delta^{1/3} d_0 \tag{16}$$

To a first approximation, the forces of interest on the void surface are the van der Waals intermolecular forces. As a consequence, the corresponding void surface tension is given by $\Gamma = \gamma_r$ where γ_r is the van der Waals surface tension of the rubber. However, this approximation is only true when the rubber particle is a liquid. Most rubbers are chemically crosslinked and the generation of a new surface requires the scission of a certain number of skeletal bonds which is largely determined by the crosslinking density (υ_x) of the rubber. The energy per unit area, associated with chain scission is then [88]:

$$\Gamma_{sc} = \tfrac{1}{4} d\upsilon_x U \tag{17}$$

where d is the mesh size of the network between the crosslinks or the root-mean-square end-to-end distance between the crosslinks in the network, υ_x

is the crosslinking density expressed as the number of crosslinks/m^3 and U is the polymer backbone bond energy. In rubbery materials one can neglect the physical entanglements; the chains will disentangle rather than break and the average force on them will be negligible when compared to the breaking force. This assumption is only valid for "true" rubbers with a low shear modulus ($G_r \approx 1$ MPa); for PE for instance, with a shear modulus of the order of 100 MPa, the average force on the chains is not negligible and the total energy required to create the surface will be higher. Neglecting this contribution in the present analysis, the surface energy per unit area is then given by:

$$\Gamma = \gamma_r + \Gamma_{sc} = \gamma_r + \tfrac{1}{4} d v_x U \tag{18}$$

It should be mentioned here that the crosslinking density in rubbery materials is often of the order of 10^{26} crosslinks/m^3 [89]. In that case, the contribution from chain scission is of the same order of magnitude as the contribution from the van der Waals surface tension ($\gamma_r \approx 0.035$ J m^{-2}).

Combining Eqs. (13), (14), (15), (16), and (18), the condition for internal rubber cavitation can be written as follows:

$$U_{total} = U_{strain} + U_{surface} = -(\pi/12) K_r \Delta^2 d_0^3 + (\gamma_r + \Gamma_{sc}) \pi \Delta^{2/3} d_0^2 < 0 \tag{19}$$

This relation predicts rubber cavitation for a given relative deformation (Δ), and from the elastic properties of the rubber (rubber bulk modulus K_r), the molecular characteristics of the rubber (γ_r, v_x, U) and the rubber particle size (d_0).

The condition for rubber particle cavitation is dependent on the volume strain. The relation between the particle size required for cavitation and the applied volume strain can be found by rearranging Eq. (19) as follows:

$$d_0 = \frac{12(\gamma_r + \Gamma_{sc})}{K_r \Delta^{4/3}} \tag{20}$$

and is depicted in Fig. 11 [81]. Eq. (20) predicts that large particles will cavitate in the early stages of the deformation process, while small particles will cavitate at a later stage. In other words, the cavitation resistance of the particle increases with decreasing particle size. The minimum particle size for cavitation therefore depends on the maximum volume strain that can be reached in the material.

The experimental verification of the criterion for rubber cavitation is shown in Fig. 11 for methyl methacrylate-butadiene-styrene (MBS) modified PVC with three different types of MBS [85].

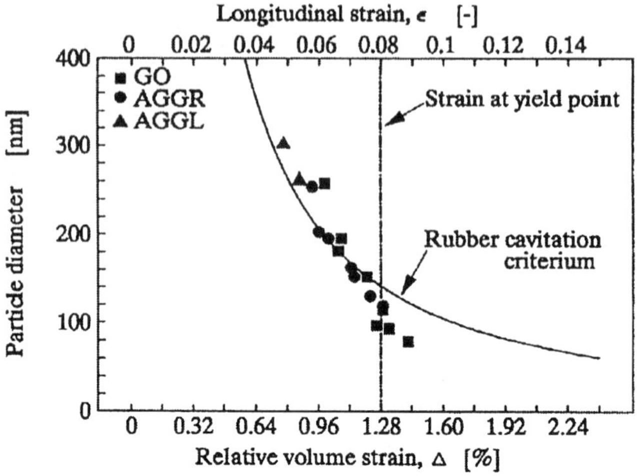

Figure 11 Relation between the minimum particle diameter required for cavitation and the longitudinal (ε) and volume strain (Δ). The theoretical line is drawn from Eq. (20) using the following parameters: $K_r = 2\,\mathrm{GPa}$, $\gamma_r = 0.035\,\mathrm{J\,m^{-2}}$, and $\Gamma_{sc} = 0.035\,\mathrm{J\,m^{-2}}$ [81]. The experimental data correspond to three different types of MBS particles in PVC [85].

D. Rubber Cavitation Behavior and Impact Mechanical Properties

One of the first studies dealing with very small rubber particles was made by Sultan and McGarry [90] on a blend of butadiene-co-acrylonitrile (CTBN) elastomer particles suspended in a crosslinked epoxy matrix. At a rubber particle size of approximately 40 nm, the plastic deformation, the fracture toughness and the mechanical properties in uniaxial and biaxial tests appeared to be very similar to the unmodified matrix material. On the other hand, in the CTBN/epoxy blends with a rubber particle size of approximately 1200 nm, yielding was reached at a lower stress level, the fracture toughness increased considerably, and the material whitened, while scanning electron microscopy revealed cavity formation at the rubber particle sites. These data confirm that for particle cavitation with associated yielding and ductile behavior, the particle size needs to have a minimum value.

The effect of rubber particle size on the cavitation behavior was also studied on blends of PVC with MBS graft copolymers by Breuer et al. [55] using transmission electron microscopy. The electron micrographs of the stress-whitened zones obtained from a uniaxial tensile test at a strain rate of 1 m s^{-1} revealed that MBS rubber particles with a size of 160 nm or larger showed internal cavitation, while only a few cavities were found in the blend with a particle size of 80 nm. The Izod impact strength of PVC/MBS blends was also investigated [82], it was found that the impact resistance increases considerably with increasing particle size, up to a particle diameter of 200 nm. In a blend of PVC with polybutadiene as dispersed elastomeric phase (particle sizes in the range of 70 to 450 nm), a maximum in the Izod impact strength was observed in the rubber particle size range of 150 to 200 nm [46].

Dompas et al. [83,86] have related the impact mechanical properties of PVC/MBS blends to the cavitation behavior of the MBS rubber particles. In Fig. 12, the notched Izod impact strength as a function of temperature is given for blends of PVC with MBS particles with average sizes in the range of 90 to 260 nm. The critical interparticle distance for ductile behavior at room temperature is estimated to be 117 nm. At a blend composition of 7.8 vol% SBR, this corresponds to a styrene-butadiene rubber (SBR) particle size of 132 nm. In contrast, it is observed that the PVC/MBS blend with a particle size of 92 nm shows brittle behavior up to 60°C. Apparently, a very small particle size does not improve the toughness of the PVC matrix, since the toughness of the blend corresponds with that of the unmodified matrix.

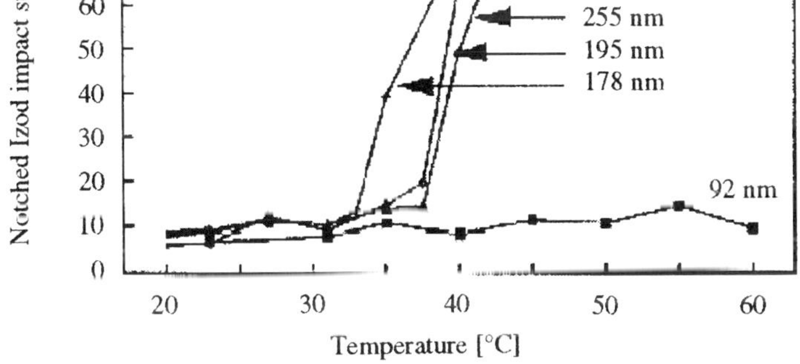

Figure 12 Izod impact strength as a function of temperature for PVC with 12 vol% MBS (7.8 vol% SBR). MBS-type = GO, average particle sizes are indicated [84].

This behavior can certainly not be explained on the basis of the critical interparticle distance concept of Wu since the brittle–tough transition temperature in this case should decrease with decreasing interparticle distance and thus with decreasing particle size.

Using the concept of the critical particle size for cavitation, the brittleness of PVC/MBS blends with an SBR particle size of 92 nm can be explained [83]. If the particle size is below the critical value ($\approx$ 150 nm), then the particles will not cavitate and the hydrostatic tension will not be relieved, which is necessary for the ductile shear yielding mechanism. On the other hand, in the ductile region for the blends with particle sizes above the critical value of 150 nm, stress whitening and lateral contraction of the Izod bar are observed, indicating that rubber cavitation and associated yielding occur in these blends.

Borggreve et al. showed that the brittle–tough transition temperature in rubber-modified nylon-6 decreased with decreasing particle size in agreement with the critical interparticle distance concept [82]. This observation was confirmed in a study on nylon-6 blended with maleic anhydride modified ethylene-co-propylene rubber (EPR-g-MA) [67]. However, in the latter study, the particle size was further decreased down to 100 nm. At very low particle sizes, below 200 nm, it was observed that the impact strength decreases sharply with decreasing particle size and approaches the value of the unmodified semicrystalline nylon-6 matrix.

The lower limit on particle size for the toughening of nylon-6 was also found in a study on nylon-6 blended with mixtures of styrene-(ethylen-co-butylene)-styrene (SEBS) and a maleic anhydride functionalized copolymer (SEBS-g-MA) [80,91]. By varying the ratio of the latter two components, the particle size of the elastomeric phase in the blend could be varied. The critical particle size for toughening was found to be approximately 250 nm.

Unmodified PMMA is a brittle material that deforms by crazing. However, rubber-toughened PMMA deforms by shear yielding. From the relatively low intrinsic brittleness of the matrix (characteristic chain ratio: $C_\infty = 7.9$ for PMMA [14]), it can be understood that the hydrostatic tension required for rubber cavitation can be lower than the hydrostatic tension required for the initiation of crazing in the matrix. Once the rubber particles have been cavitated, the hydrostatic tension in the material is relieved and shear yielding is induced. Accordingly, a minimum particle size for toughness in this system is expected. The toughening of PMMA with very small particles of a butylacrylate-co-styrene crosslinked copolymer was investigated [92]; a lower size limit of 200 to 250 nm for the toughening of this system was found.

On the basis of the criterion for rubber cavitation mentioned above, the decrease in impact strength of rubber-modified systems with increasing rubber crosslinking density can be qualitatively explained. A high crosslinking density will suppress the cavitation of the rubber particles, and as a consequence, the impact strength is expected to decrease. A recent study on nylon-6/SBS blends with different crosslinking degrees of the rubber confirms this statement [63].

The toughnesses of the different rubber-modified thermoplastic polymers (PVC, PMMA, nylon-6, PC) all seem to indicate a value for the critical particle size for rubber cavitation of about 200 nm. The critical particle size of about 200 nm is clearly expected on the basis of Eq. (19) and also by knowing that the molecular parameters of the rubber (γ_r, Γ_{sc}), the rubber bulk modulus (K_r) and the glassy matrix Poisson ratio (v_m) are fairly constant for different systems.

ACKNOWLEDGMENTS

The authors (G. Groeninckx and D. Dompas) are indebted to the Fonds voor Wetenschappelijk Onderzoek-Vlaanderen, to the KU Leuven Research Council, to Kaneka Corporation Japan, and to Kaneka Belgium for financial support given to the MSC laboratory. Very valuable and fruitful discussions related to the toughening behavior of PVC/MBS blends were held with M. Isogawa, K. Isayama, T. Hasegawa, M. Kadokura, and R. Nishimura from Kaneka Corporation, Takasago, Japan.

REFERENCES

1. CB Bucknall. Toughened Plastics. London: Applied Science Publishers, 1977.
2. AJ Kinloch, RJ Young. Fracture Behaviour of Polymers. London: Applied Science Publishers, 1983.
3. G Groeninckx, D Dompas. Advanced Polymer Science—Eupoco module 2. Leuven: Acco, 1992, vol 3.
4. HH Kausch. Polymer Fracture. Berlin: Springer-Verlag, 1987.
5. RP Kambour. J Polym Sci Macromol Rev 7:1, 1973.
6. Adv Polym Sci (Crazing in Polymers) 52/53, 1983.
7. Adv Polym Sci (Crazing in Polymers vol 2) 91/92, 1990.
8. RE Robertson. Appl Polym Symp 7:201, 1968.
9. N Brown. J Mater Sci 18:2241, 1983.
10. JCM Li, JBC Wu. J Mater Sci 11:445, 1976.
11. JCM Li. Polym Eng Sci 24(10):750, 1984.
12. S Wu. Proceedings ACS Division Polymer Materials Science and Engineering, Fall Meeting, Washington DC, 1990.
13. S Wu. Polym Eng Sci 30:753, 1990.

14. S Wu. Polym Int 29:229, 1992.
15. S Wu. J Appl Polym Sci 35:549, 1988.
16. AF Yee. PRI-International Conference on 'Toughening of Plastics', London, 1985.
17. AM Donald, EJ Kramer. Polymer 23:1183, 1982.
18. D Maes, G Groeninckx, J Ravenstijn, L Aerts. Polym Bull 16:363, 1986.
19. D Maes. PhD thesis, Catholic University of Leuven, Belgium, 1989.
20. HH Kausch. Makromol Chem Macromol Symp 41:1, 1991.
21. SM Aharoni. Macromolecules 18:2624, 1985.
22. AS Argon, M M Salama. Mater Sci Eng 23:219, 1977.
23. EJ Kramer, LL Berger. Adv Polym Sci (Crazing in Polymers vol 2) 91/92, 1990.
24. S Wu. Polymer Interface and Adhesion. New York: Marcel Dekker, 1982.
25. GH Michler. Makromol Chem Macromol Symp 41:39–54, 1991.
26. AM Donald, EJ Kramer. J Polym Sci Poly Phys Ed 20:899, 1982.
27. C Creton, E J Kramer, G Hadziioannou. Coll Polym Sci 270:399, 1992.
28. H Eyring. J Chem Phys 4:283, 1936.
29. JC Bauwens. J Polym Sci A 2 5:1145, 1967.
30. JA Roetling. Polymer 6:311, 1965.
31. C Bauwens-Crowet, JC Bauwens, G Homès. J Polym Sci A 2 7:735, 1969.
32. JC Bauwens. J Mater Sci 7:577, 1972.
33. C Bauwens-Crowet, J-M Ots, JC Bauwens. J Mater Sci Lett 9:1197, 1974.
34. S Havriliak, TJ Shortridge. J Appl Polym Sci 37:2827, 1989..
35. S Havriliak, SE Slavin, TJ Shortridge. Polym Int 25:67, 1991.
36. F Ramsteiner. Kunstoffe 73:148, 1983.
37. S Wu. J Appl Polym Sci 46:619, 1992.
38. S Wu. Polym Eng Sci 32(12):823, 1992.
39. DW Van Krevelen, P J Hoftyzer. Properties of Polymers. Amsterdam: Elsevier, 1976.
40. EH Merz, GC Claver, M Baer. J Polym Sci 22:325, 1956.
41. S Newman, S Strella. J Appl Polym Sci 9:2297, 1965.
42. CB Bucknall. Comprehensive Polymer Science. Oxford: Pergamon Press, 1989, pp 27–49.
43. CB Bucknall, P Davies, I K Partridge. J Mater Sci 21:307, 1986.
44. CB Bucknall, P Davies, I K Partridge. J Mater Sci 22:1341, 1987.
45. Y Okamoto, H Miyagi, M Kakugo, K Takahashi. Macromolecules 24:5639, 1991.
46. M Morton, M Cizmecioglu, R Lhila. Adv Chem Ser 206:221, 1984.
47. JW Boode, AEH Gaalman, AJ Pijpers, RJM Borggreve. PRI-International Conference on 'Deformation, Yield & Fracture of Polymers'. Cambridge: 1990.
48. S Wu. J Polym Sci Polym Phys Ed 21:699, 1983.
49. JN Goodier. Trans Am Chem Soc Mech Eng TASMA 55:39, 1933.
50. AM Donald, EJ Kramer. J Mater Sci 17:1765, 1982.
51. DS Parker, H-J Sue, J Huang, AF Yee. Polymer 31:2267, 1990.
52. AF Yee. J Mater Sci 12:757, 1977.
53. F-C Chang, J-S Wu, L-H Chu. J Appl Polym Sci 44:491, 1992.

54. RA Bubeck, DJ Buckley, EJ Kramer, HR Brown. J Mater Sci 26:6249, 1991.
55. H Breuer, F Haaf, J Stabenow. J Macromol Sci Phys B 14(3):387, 1977.
56. H Breuer, J Stabenow, F Haaf. PRI-International Conference on Toughening of Plastics, London, 1978.
57. F Haaf, H Breuer, A Echte, BJ Schmitt, J Stabenow. J Sci Ind Res 40:659, 1981.
58. H Breuer. In: A Maconnachie, D J Walsh, J S Higgins, ed. Polymer Blends and Mixtures. Dordrecht: Martinus Nijhoff Publishers, 1985, pp 376–392.
59. A Takaki, T Hasegawa, M Isogawa, I Narisawa. Polym Eng Sci (submitted).
60. A Tse, E Shin, A Hiltner, E Baer, R Laakso. J Mater Sci 26:2823, 1991.
61. DJ Hourston, S Lane, H X Zhang. Polymer 32(12):2215, 1991.
62. K Dijkstra. PhD thesis, University of Twente, The Netherlands, 1993.
63. Y Seo, SS Hwang, KU Kim, J Lee, SI Hong. Polymer 34(8):1667, 1993.
64. RJM Borggreve, RJ Gaymans, J Schuijer. Polymer 30:71, 1989.
65. RJM Borggreve, RJ Gaymans, HM Eichenwald. Polymer 30:78, 1989.
66. RJM Borggreve, RJ Gaymans, AR Luttmer. Makromol Chem Macromol Symp 16:195, 1988.
67. AJ Oostenbrink, K Dijkstra, S Wiegersma, A van der Waal, RJ Gaymans. PRI-International Conference on Deformation, Yield & Fracture of Polymers, Cambridge, 1990.
68. S Wu. Polymer 26:1855, 1985.
69. JN Sultan, FJ McGarry. Polym Eng Sci 13(1):29, 1973.
70. D Li, X Li, AF Yee. Proceedings ACS Division Polymer Materials: Science and Engineering, Fall Meeting, Washington DC, 1990.
71. Y Huang, AJ Kinloch. Polymer 33(24):5338, 1992.
72. RA Pearson, AF Yee. J Mater Sci 26:3828, 1991.
73. H-J Sue. J Mater Sci 27:3098, 1992.
74. EJ Moskala. J Mater Sci 27:4883, 1992.
75. H-J Sue, J Huang, AF Yee. Polymer 33(22):4869, 1992.
76. A Takaki, H Hosoi, K Saito. Plast Eng 45, 1986.
77. C Wrotecki, P Heim, P Gaillard. Polym Eng Sci 31:213, 1991.
78. C Wrotecki, P Heim, P Gaillard. Polym Eng Sci 31:218, 1991.
79. SD Sjoerdsma. Polym Commun 30:106, 1989.
80. AJ Oshinski, H Keskkula, DR Paul. Polymer (33)2:268, 1992.
81. D Dompas, G Groeninckx. Polymer 35:4743, 1994.
82. RJM Borggreve, RJ Gaymans, J Schruijer. Polymer 28:1489, 1987.
83. DL Dunkelberger, EP Dougherty. J Vinyl Technol 12:212, 1990.
84. D Dompas, G Groeninckx, M Isogawa, T Hasegawa, M Kadokura. Polymer 35:4760, 1994.
85. D Dompas, G Groeninckx, M Isogawa, T Hasegawa, M Kadokura. Polymer 35:4750, 1994.
86. D Dompas, G Groeninckx, M Isogawa, T Hasegawa, M Kadokura. Polymer 36:437, 1995.
87. A Lazzeri, CB Bucknall. J Mater Sci 28:6799, 1993.
88. EJ Kramer, LL Berger. Adv Polym Sci 91/92:1, 1990.

89. Yongsok Seo, Seung Sang Hwang, Kwang Ung Kim, Jaehwang Lee, Sung Il Hong. Polymer 34:1667, 1993.
90. JN Sultan, F J McGarry. Polym Eng Sci 13:29, 1973.
91. AJ Oshinski, H Keskkula, DR Paul. Polymer 37:4909, 1996.
92. C Wrotecki, P Heim, P Gaillard. Polym Eng Sci 31:213, 1991.
93. GH Michler. Coll Polym Sci 263:462, 1985.

13

Future Directions of Research into Multiphase Polymeric Materials

Qui Tran-Cong and Mitsuhiro Shibayama
Kyoto Institute of Technology, Kyoto, Japan

The science of multiphase polymer materials has been driven by the demands for new polymeric materials made by mixing and/or covalently linking conventional polymers. This strategy naturally comes from the recognition that the cost of searching for new combinations of polymers is less than that for developing novel synthetic methods. A great deal of research has been conducted in this direction as discussed in this book. Owing to the developments of theoretical work and characterization techniques for multiphase and/or multicomponent polymer systems, studies on polymer blends and block copolymers have flourished in the past decade. This book is devoted to the survey of this rapidly growing field of polymer science by stressing the control of the phases and morphologies in both real and reciprocal spaces.

The research of multiphase polymer materials contributes not only to developments of new materials with high performance or sophisticated functions but also to advances in the complex fluid physics and materials science of "nanostructures." In particular, a large number of analogies have been found between polymeric and nonpolymeric systems, such as metallic, glassy, and magnetic materials. It should be stressed here that owing to the unique architecture of polymers, i.e., chain-like molecules, the criterion of the mean-field assumption is more relevant than for other fluids. Therefore, polymers have been a typical model system to study critical phenomena as well.

As discussed in most chapters of this book, morphology control has played a central role in polymer materials research and has also been one of the long-standing subjects of polymer science. To vary the characteristic length scale and the regularity of the morphology, various chemical as well as physical techniques have been developed in academia as well as in industry. From the chemistry viewpoint, great efforts have been devoted to the exploitation of the "living polymerization" methods via which the architecture of polymer chains such as chain lengths or the periodicity of chemical structures can be well designed and efficiently controlled. The principal achievement is the synthesis of block copolymers having a variety of architectures such as diblocks, triblocks and starblocks as described in Chapter 4. Their physical properties vary greatly with the chain architecture as described in Chapters 8–12. Though the regularity of the morphologies can be well manipulated by these particular synthetic techniques, control of chain lengths is still limited by monomer reactivities and as a consequence, the characteristic length scales of the microdomains of block copolymers are limited to the nanometer scale. On the other hand, the practical ways which are largely utilized in industry to control the morphologies of multiphase polymeric materials are the stirring or shearing of immiscible polymer mixtures. These techniques, e.g., varying the stirring rates, provide a practical way to regulate the sizes of the phase-separated structures in the micrometer range, but unfortunately are not efficient enough to control the regularity of the morphology. Therefore, novel approaches are currently in demand to overcome these shortcomings. From today's research into and development of multiphase polymers, several future directions can be predicted.

I. NANOSTRUCTURE MANIPULATION CONTROLLED BY SPECIFIC INTERACTIONS

As discussed in Chapters 4 and 6, antagonistic interactions play a major role in the determination of the local structures on the nanometer scale, referred to as nanostructures hereafter. It has been suggested that multiphase polymer systems with controllable nanostructures can be designed by manipulating specific molecular interactions. The fact that the phase stability can be modified by an infinitesimal variation of the externally controlled parameter implies that these polymeric materials can be utilized as a molecular switch sensitive to different environments. It is also known that hydrogen-bonding interaction leads to a hysteresis effect, which can be utilized for bi- or multistable memory elements. The physics of polymer systems involving a variety of molecular interactions has not been well developed and is an unsolved problem for the next decade.

Polymers involving a wide variety of interactions are essentially "many-body" systems. Their physics has not been fully developed yet and will be one of the targets of research in the next few decades. In multiphase polymers, the local structures (nanostructures) correlate with global (mesoscopic scale) structures via cooperative phenomena such as phase separation. Due to their very slow dynamics, the structures of multiphase polymeric materials, in general, are inhomogeneous with various hierarchies. Therefore, understanding this inhomogeneity in both static and dynamic aspects remains to be exploited for useful utilization of multiphase polymer materials.

II. CREATION AND CONTROL OF COHERENT SPATIO-TEMPORAL STRUCTURES IN POLYMER SYSTEMS BY USING THERMODYNAMICALLY OPEN CONDITIONS

In general, the morphology resulting from the phase separation of multiphase polymer materials belongs to a class of pattern formation driven by equilibrium thermodynamics. A natural way of improving control techniques would be to extend the experiments to thermodynamically open conditions where the systems are operated and driven by a control parameter such as a temperature gradient, a shear field, or irradiation. An example is shown in Chapter 7 where the shear rate is used as a control parameter driving the systems far from equilibrium. Alternatively, chemical rections can also be utilized to bring the systems far from equilibrium. As described in Chapter 5, photochemical reactions play the role of a control parameter which can be manipulated independently of thermodynamic variables such as temperature, volume, and pressure. Under these conditions, by coupling various chemical reactions to phase separation, unstable modes arising in the systems can be selected by the reactions. Within the framework of thermodynamically open systems, phase separation induced by temperature gradients can also provide unique critical phenomena for multiphase polymeric materials in solutions and in the bulk state. Fundamental theories and experimental data obtained from directional solidification experiments in metallic alloys carried out in the past might also provide some helpful hints for new directions of study on multiphase polymeric materials. The spatio-temporal structures emerging from multiphase polymers under open conditions have a rich dynamics and are not governed by the free energy minimum of the system. Most of the work carried out in this direction has been initiated in the field of metallurgy since the late 1960s. Here, a large number of experiments and theories as well as computer simulations were performed in order to understand the self-organization phenomena observed in metals or metallic alloys driven far from thermodynamic equilibrium by a temperature gradient (the phenomenon of directional solidification) or under irradiation with high-energy particles (the so-called irradiation effects).

III. SUMMARY

From the viewpoint of materials science, the synthesis of materials possessing "intelligent" properties such as self-generation or self-regulation that would be adaptable to different environments should be one of the goals for polymeric material research in the next century. The rich dynamic behavior of polymer mixtures submitted to these open conditions together with the matured knowledge acquired from nonlinear dynamics should provide a promising approach to polymer materials in the future.

Index